Advances in Studies of Aging and Health

Volume 4

This book series is a theme-driven compilation, publishing the latest international research findings by key topics upon roles of social, behavioral, psychological, and contextual factors in associating or determining health and wellbeing of older people, with biomedical components taken into consideration. It applies an internationally comparative perspective and a biopsychosocial framework to understand healthy, active, productive, and successful aging, in order to build up a comprehensive and practical source of knowledge for researchers and practitioners in the field of aging and health.

This book series publishes original studies and tracks new research progresses by the major topics in the field of aging and health; systematically reviews recent research advancements of new concepts, frameworks, and methodology; and frameworks; explores the aging issues through an interdisciplinary perspective, and provides fresh empirical findings and theoretical insights in an internationally comparative manner. The goal is to effectively disseminate new policies and intervention practices on aging, in order for the improvement of quality of life of older persons.

The major themes of this book series include but are not limited to disability, cognitive function, chronic diseases and conditions, happiness and psychological wellbeing, self-rated health, frailty, mortality, living arrangements and intergenerational transfers, retirement and pension reforms, financial health and poverty, elderly abuse and human rights, ageism, long-term care, end-of-life care, aging in place, environmental health, age-friendly cities/communities, social participation, centenarian studies, big data in aging research, resilience and vulnerability, health life expectancies, active aging, successful aging, and healthy longevity.

Giacinto Libertini • Graziamaria Corbi •
Olga Shubernetskaya • Klara Komici •
Nicola Ferrara

The Essential Revolution of Evolutionary Medicine

Interpreting Diseases and Population Aging

Springer

Giacinto Libertini
ASL NA2 Nord, Italian Health National System, Frattamaggiore
Naples, Italy

Olga Shubernetskaya
M.M. Shemyakin–Yu.A. Ovchinnikov Institute of Bioorganic Chemistry
Moscow, Russia

Nicola Ferrara
Director Saint Joseph Nursing Home
Naples, Italy

Graziamaria Corbi
Department of Translational Medical Sciences
Federico II University of Naples
Naples, Italy

Klara Komici
Department of Medicine and Health Sciences
University of Molise
Campobasso, Italy

ISSN 2522-5146 ISSN 2522-5154 (electronic)
Advances in Studies of Aging and Health
ISBN 978-3-031-84876-6 ISBN 978-3-031-84877-3 (eBook)
https://doi.org/10.1007/978-3-031-84877-3

This Springer imprint is published by the registered company Springer Nature Switzerland AG
The registered company address is: Gewerbestrasse 11, 6330 Cham, Switzerland

To Stanley Boyd Eaton, Marjorie Shostak, and Melvin Joel Konner for their vital contribution to the birth of Evolutionary Medicine

Foreword

The renowned geneticist Theodosius Dobzhansky published in 1973 an assay titled "Nothing in Biology makes sense except in the light of Evolution," pointing to the central position of the Theory of Evolution in understanding Life on our planet. Although this statement has been largely cited and accepted by biologists, it is somehow surprising how the Darwin's Theory of Evolution is struggling to make its way in many fields of Biology, including Medicine. The Evolutionary Medicine, the topic of this book, is a relatively new discipline, despite Evolutionary Biology is a well-established science since the re-discover of Mendel's inheritance laws in year 1900 and the formalization of what is called "The modern synthesis" in the early twentieth century.

There are many reasons why the Theory of Evolution is hard to accept, but the main issue is that it proposes that the living beings are basically self-building systems, by a process of billions of attempts and errors, often selected by the environment, a process that ended up in the exploding biodiversity we know today. The process is so deeply affected by chance that there is no reason to think that the entire Evolution on Earth will follow the same pathway if we hypothetically go back in time and start again the history of our planet.

The concept of self-building systems is the trickiest one to accept, as it threatens one of the most intuitive approaches we have when interpreting the world around us. We are a species evolved to build tools, from pebbles to spacecrafts, and our brain is very keen in interpreting what it observes as machines built by a constructor for a given function. Life is not like that. Life is unnecessarily complicated, far from being perfect, as it proceeds by attempts and errors, so what is present today is not the best possible achievement, but a compromise of history, success, failures, and good luck. Understanding life and evolution is accepting that living beings are not machines, at least are not perfect machines built by a skilled constructor. This is like accepting that "a tornado sweeping through a junkyard might assemble a Boeing 747 from the materials therein." This argument, first presented against evolution (The junkyard tornado, also known as Hoyle's fallacy), surprisingly tells a true story, besides that the "tornados" lasted for billion years, so a single unlikely event (the assembly of the Boeing) becomes possible over countless attempts. I used to say to my students: to win a lottery is very unlikely event, but if you buy billions of tickets you win for sure sooner or later. Hard to imagine such big numbers, another "bug" of our brain. Accepting the Theory of Evolution as our way to interpret life, including ourselves, may be somehow disturbing, as it is counterintuitive. I think the struggle that this Theory had to be accepted since its proposal resides in this "unpleasant" feeling it produces. Some react by rejecting it (as is the case for some ultraconservative religious groups), others resolve the issue by avoiding applying the ideas and approaches of Evolutionary Biology in their day-by-day scientific activity. I really think this is what happened in Medicine. Besides countless examples, the COVID Pandemic has once more pointed out that an evolutionary approach to diseases is the best way to reduce their effects in our species. We can use evolutionary approaches to study the origin of COVID, it's capacity to spread, the chances of arising new variants, etc., just to name some. For this reason, this book has an important role, and it is absolutely timely, and the Authors have accepted a very important endeavor, which I hope will contribute to a change of paradigm in modern Medicine.

Department of Biological, Geological and Environmental Sciences
University of Bologna
Bologna, Italy

Marco Passamonti

Introduction

Evolutionism as a Revolution in Biology

When Darwin's idea of evolution by natural selection, the subsequent related advances and studies in depth, and the birth of genetics enriched previous biological knowledge, this was not just an addition.

Biology was completely changed by the ideas of evolutionism. For example, in previous times, among the various species one simply saw the affinities or even identities of characteristics and classified the species on the basis of these affinities. With the evolutionism, it became clear that the affinities traced back to a common origin, that is, that there were common ancestors both between similar species and between species that were very different from each other. Furthermore, the affinities were not limited to external aspects but extended to countless previously unknown biochemical characteristics and to every other feature of the living world. More and more, the morphological, physiological, behavioral characteristics, and every other aspect of the species revealed themselves as the fruit of the inheritance of the ancestors and the action of natural selection. Biology became more and more imbued with the ideas of evolutionism so much so that today it is unthinkable to imagine any study of living beings that does not take into account the concepts of evolutionism.

In short, the knowledge of evolutionism has not simply been an enrichment of biology but a profound revolution from the roots of the biological world that in various ways concerns every branch of knowledge and research in that field.

While biology has been completely revolutionized by Darwin's ideas and the subsequent multiple and profound developments, medicine, which is a particular branch of biology with applicative purposes, appears to be only partially and superficially touched by evolutionism. If you tell a doctor today that evolutionism is essential for medical studies and for all that follows as a practical application, he will surely consider this statement as erroneous or at least completely exaggerated.

He will tell you that certainly in some particular cases knowledge of evolutionism is useful, but that in much of medicine such knowledge is entirely secondary or irrelevant. For example, he may tell you that various types of genetically based anemia are the consequence of genetic adaptations to get at least a partial resistance to malaria. He may also tell you that certain types of malformation (for example those resulting from defective development of the branchial arches) or certain anatomical characteristics (for example the anatomical conformation of the retina) are the consequence of the phylogenetic history of our species. But beyond these particular cases, in general what theoretical and practical importance could the study and application of evolutionary concepts have for medicine?

Yet, as we will see, the importance of evolutionism is not limited to a few particular phenomena of limited importance but affects the very roots of medicine.

It is worth pointing out that medicine was born long before evolutionism and that the objective of this discipline is mainly the practical treatment of suffering and not the mere theoretical knowledge of diseases.

The pre-Darwinian historical origin of medicine is immediately revealed by examining how diseases are classified. In fact, the various official classifications that have followed one another over the years and that are codified worldwide divide diseases on the basis of organs and physiological systems that are affected and on some criteria of opportunity and affinity but without there being a scientific criterion that justifies this subdivision. This is similar to what was the rule for the classification of species in the pre-Darwinian era when empirical and practical criteria allowed the various groups of living species to be distinguished without there being a precise scientific criterion for this subdivision. On the contrary, with evolutionism, species are divided according to their phylogenetic origin and this objective criterion is the rule universally followed to classify species.

Furthermore, in the study of diseases, modern medicine neglects or underestimates the importance of considering the primary origin of a disease in the context of evolution. This has important consequences for both the understanding and prevention of a disease.

Let's look at the case of a very common disease, or rather syndrome, namely essential arterial hypertension, or hypertension for brevity.

The description found on Wikipedia (consulted on 20/10/2024) is:

> Essential hypertension (also called primary hypertension, or idiopathic hypertension) is a form of hypertension without an identifiable physiologic cause. It is the most common type affecting 85% of those with high blood pressure. The remaining 15% is accounted for by various causes of secondary hypertension. Essential hypertension tends to be familial and is likely to be the consequence of an interaction between environmental and genetic factors. Hypertension can increase the risk of cerebral, cardiac, and renal events.

The information provided by Wikipedia has no recognized scientific value but is continuously subject to corrections by anyone who is an expert in a subject, and therefore the aforementioned description represents quite faithfully the interpretation of hypertension according to current medicine. However, according to the criteria of evolutionary medicine, this description is seriously deficient with regard to the origin of the disease and misleading for the general interpretation of this syndrome. The questions that evolutionary medicine poses for this disease, and for many others, are the following:

1. If it is true that natural selection favors individuals best suited to survive, how is it possible that it does not seem to be able to defend us from hypertension, a disease so frequent and with fearful consequences?
2. What is the incidence of this disorder in populations that live in conditions similar to those to which our species is adapted?
3. If in such living conditions hypertension is very rare or of irrelevant frequency, why is this not at the center of attention, thus attributing the origin of the disease to altered living conditions and not to different heterogeneous factors?
4. Why is medical attention focused on the treatment of the syndrome and not on the variations in modern lifestyle compared to ancestral conditions, variations that are the likely primary cause of the syndrome and that can be corrected with appropriate measures?

Indeed, for the second question, the answer has been known for at least 30 years and is clear and unequivocal.

As widely documented, in modern populations, systolic and diastolic blood pressure increase with the passage of time (Whelton et al., 1994). On the contrary, as also widely documented (Poulter & Sever 1994), in various isolated populations that follow lifestyles closer to the ancestral ones (e.g., !Kung of Botswana and ten other populations reported in the review), there is no increase in blood pressure in relation to age. Furthermore, the same study (Poulter & Sever, 1994) reports the increase in blood pressure in the transition from primitive living conditions to those of a modern urban condition.

As for the other three questions, they are ignored by modern medicine, or appear not to be considered worthy of attention, or in any case do not determine changes in current practice.

As can be seen from this single example, the attention of modern medicine, following a logic preceding the revolution of evolutionism, seeks the causes of a disease without considering the mechanisms of evolution that instead offer an immediate explanation of the primary origin of the disease.

The False Triumph of Modern Medicine

The results achieved by modern medicine are extraordinary and would once have been inconceivable. Today we have exceptional diagnostic techniques that can investigate any biochemical, morphological, and genetic alteration. Diagnostic imaging techniques allow us to see inside any part of the human body without damaging it in any way. Surgical techniques are increasingly advanced and interventions that were once impossible or devastating are performed even with simple small cuts through which surgical probes are introduced and act with the help of robotic means. We have powerful drugs that can cure even very serious diseases. A detailed list of the conquests of medicine would be very long and astonishing. In short, we can define modern medicine as an enormous triumph of modern science and technology.

Cases that are not yet well treated or solved by modern medicine by virtue of further medical advances in the not too distant future will be adequately addressed.

However, this triumphal and enthusiastic description of modern medicine neglects and underestimates what epidemiological data coldly tell us.

As medicine became more and more effective and capable of previously impossible feats, the incidence of many diseases increased enormously and many previously undescribed diseases appeared and spread. The list of such diseases is considerable. Diseases such as diabetes, hypertension, cardiovascular diseases, allergic diseases, autoimmune diseases, neoplasms, refractive defects, hearing problems, psychiatric diseases, hemorrhoids, sacrococcygeal fistulas, constipation, caries, malocclusion (misaligned teeth), alveolar pyorrhea, which were or are rare or never described for people who lived or are living in primitive conditions have become common and with increasing frequency.

These diseases practically did not exist or do not exist in societies that lived or live in conditions closer to the ancestral ones to which our species is adapted. Today it is statistically normal to suffer from one or more of these diseases. Modern medicine fights these diseases with a success that varies depending on the type of disease and individual cases. If we limit ourselves to considering the improvements obtained by modern medicine in individual cases, the aforementioned triumphal evaluation of the results achieved appears justified. But if we also evaluate the epidemiological data on the number of sick people, even considering the people who have recovered and those with the disease partially controlled or in a quiescent phase, we obtain completely discouraging results. It is like a war in which we boast of countless battles won while in reality the front continues to retreat.

Sometimes this harsh reality is masked or diminished by the realistic prospect of further medical advances and even the development of highly personalized treatments or treatments for specific diseases. For example: (1) the identification of precise individual genetic alterations that can be counteracted by individual genetic treatments; and (2) the development of increasingly specific monoclonal antibodies for particular autoimmune diseases and types of cancer.

These developments are generally very or extremely expensive for each treatment of a single case and are not feasible in non-rich countries or even for individuals who are not covered by sufficient insurance policies.

Furthermore, these treatments are a kind of fog that hides the main problem: why are cases and types of disease increasing despite the ever-increasing effectiveness of medical treatments?

From these considerations arises the need for a different approach that preliminarily provides an answer to these questions.

Evolutionary Medicine as Rational Solution for the Problems of Current Medicine

Evolutionary medicine is a possible rational solution to the failure of current medicine.

The main purpose of this book is to show that evolutionary medicine is not simply an interesting curiosity of only theoretical value but a practical way to overcome the tragic limits of current medicine.

With the simple consideration of evolutionary mechanisms, it is possible to analyze and define the primary cause of the various types of disease, which should be classified precisely according to the type of primary cause.

From the knowledge of the primary causes it is possible to devise a strategy to combat these causes and consequently to prevent many types of disease. This strategy, if managed effectively, reduces the number of new cases of disease and the number of cases of sick people to be treated.

It is good to say that not all types of disease are preventable and that even for preventable diseases a part is caused by genetic alterations that are much less preventable.

The key element is the portion of cases that are preventable with appropriate preventive measures.

The most evident type of cases among the preventable ones are those caused by alterations in lifestyle. A well-known case in current medicine is that of diseases caused by the habit of smoking cigarettes. In smoke there are a myriad of substances for which the body is not adapted and which are therefore harmful. The diseases caused by smoking (chronic bronchitis, emphysema, cardiovascular diseases, cancers of the lung and other organs, etc.) are easily preventable by avoiding cigarette smoking.

In modern medicine, this important prescription arises from the empirical observation of the countless cases of diseases related to the habit of smoking and therefore the prescription to avoid smoking arises from empirical data. For evolutionary medicine, the habit of smoking is not something to which we are evolutionarily adapted. Consequently, *a priori*, it is to be considered—until proven otherwise—something potentially harmful to be avoided. Then, in the comparison between modern populations and others that live in conditions closer to those to which we are adapted, the empirical observation of an increase in various pathologies and the association between the incidence of these pathologies and the habit of smoking indicates that the potential harm of smoking is real harm. Consequently, to combat a series of diseases related to smoking, the priority action is to inform as much as possible about the dangers related to the habit of smoking and to discourage this habit. As a secondary remedy, and as a partial remedy to a lack of primary prevention, it is necessary to proceed with secondary prevention measures and with the appropriate treatment when a disease caused by smoking occurs.

This example can be extended to other types of diseases that are spreading because: (1) the primary cause of a disease is not known or highlighted; (2) primary prevention measures are not applied, that is, actions on the primary cause of a disease are not taken into account; and (3) attention and efforts are concentrated on the treatment of cases of disease, each of which generally indicates the absence or failure of actions for the primary prevention of the disease.

These simple synthetic considerations indicate that evolutionary medicine is not limited to a simple enrichment of current medical knowledge, but implies a radical rethinking, or rather a profound revolution, of the logic according to which medicine operates.

Structure of This Book

Let us now see how the exposition of evolutionary medicine is organized in this book.

In Chap. 1, there is the definition of evolutionary medicine and the discussion of the fundamental concepts of evolutionary medicine. Then, there is the comparison between the current classification of diseases and similar conditions (a classification empirical and of historical origin but not based on an objective scientific criterion), and a suggested classification of such conditions based on the objective criterion of their primary causes. The chapter concludes with a brief history of the origins and development of evolutionary medicine.

Chapter 2 discusses the subject of diseases caused by alterations of the genotype, the frequency to be expected for a single disease of this type, and the large overall frequency to be expected for this category of diseases as a whole.

In Chap. 3, there is the first part of the fundamental category of diseases that originate from alterations of our ecological niche. Given that the term holobiont indicates an organism plus the multitude of other living species that live within or on the same, the chapter, which does not include the alterations of the holobiont discussed later, exposes this type of disease with its various subcategories. This category of diseases includes pathologies of great frequency and economic and social impact, all dependent on alterations of the ancestral lifestyle or the environment to which we are adapted. These diseases are largely preventable with the appropriate corrections or at least compensations of the aforementioned alterations.

In Chap. 4, there is the second part of the category of diseases that originate from alterations of our ecological niche, that is, those that originate from alterations of our holobiont. Here too we have a vast category of diseases of great frequency and economic and social impact due to alterations of the ancestral lifestyle or of the environment to which we are adapted, which have consequences on our holobiont and therefore on our health. These diseases are also largely preventable with the appropriate corrections or at least compensations of the alterations mentioned above.

Chapter 5 debates diseases resulting from relationships with other species, such as infections. Diseases resulting from conditions beyond the scope of our adaptations are also discussed.

Chapter 6 addresses physiological phenomena that can cause suffering that are not diseases but adaptations to specific conditions. It is discussed whether and to what extent these phenomena should be counteracted, without causing harm in an attempt to "cure" phenomena that are not pathological.

Chapter 7 introduces the concept of phenoptosis, which includes aging and its complex manifestations, and then expounds some phenoptotic phenomena other than aging.

Chapter 8 discusses aging as a phenoptotic phenomenon. In this regard, the "why" of aging is debated, that is, the mechanisms of natural selection that under certain conditions favor aging. Furthermore, it is explained the unsustainability of the ancient interpretation of aging as an exclusively harmful phenomenon not sufficiently counteracted by natural selection. The "how" of aging is then debated, that is, the specific mechanism of aging at the cellular level and consequently at the level of tissues and organs and of the entire organism. Finally, the relationship between aging and its progressive age-related manifestations and the frequent lack of distinction between physiological aging and the acceleration of aging caused by alterations in lifestyle are discussed. This important topic has been discussed in more detail in the previous book in this series (Libertini et al., 2021), while a substantially complete but more synthetic exposition is reported here.

Chapter 9 presents a varied set of topics that are also the subject of particular studies in terms of evolutionary medicine, including: (a) anatomical constraints and curiosities determined by our phylogenetic history; (b) physiological characteristics that become problems, even very serious ones, in a modified ecological niche; (c) diseases or disorders resulting from conflicting demands of our evolution; (d) enzymatic deficiencies that are not defects but merely the result of our previous evolutionary history; and (e) problems related to the period of birth and the first months of life and caused by altered living conditions.

Chapter 10 addresses the crucial issue of how evolutionary medicine can be implemented in practical reality, not as an addition of knowledge to current medicine, but as a radical transformation of it. In fact, today's medicine is largely oriented toward the treatment of diseases and not on prevention and it is unlikely that a doctor can privilege prevention and avoid treating patients. Treatment and prevention must walk on two parallel tracks in which mechanisms and responsibilities are distinct. The common fundamental principle is that treatment is always due when there is a need, that is, when someone suffers, but that too often the treatment of a disease represents the failure of its prevention.

References

Libertini, G., Corbi, G., Conti, V., Shubernetskaya, O, & Ferrara, N. (2021). *Evolutionary gerontology and geriatrics. Why and how we age*. Springer.

Poulter, N. R., & Sever, P. S. (1994). Populations with low blood pressure and the impact of rural-urban migration. In Swales, J. D. (ed.) *Textbook of hypertension*. Blackwell Scientific Publications

Whelton, P. K., Jiang He, & Klag, M. J. (1994). Blood pressure in western countries. In Swales. J. D. (ed.) *Textbook of hypertension*. Blackwell Scientific Publications

Contents

1 Basic Concepts and a Brief History of Evolutionary Medicine 1
1.1 Definition of Evolutionary Medicine 1
1.2 Some Basic Concepts 5
1.2.1 Effects of a Random Modification in a Complex System 5
1.2.2 The Ecological Niche of a Species 6
1.2.3 Normality in Evolutionary Terms 7
1.2.4 The Concept of "Mismatch" 10
1.2.5 The Concept of "Holobiont" 11
1.2.6 Other Basic Concepts 14
1.3 Current Non-evolutionary Classification of Diseases and Similar Conditions 15
1.4 Possible Evolutionary Classification of Diseases and Similar Conditions 19
1.5 Origins and Developments of Evolutionary Medicine 20
References 32

2 Diseases Deriving from Alterations of the Genotype 39
2.1 Mutations and Other Alterations of the Genotype 39
2.2 Calculation of the Frequencies Expected for Diseases Caused by Alterations of the Genotype 41
2.3 Prevention and Care of the Diseases Deriving from Alterations of Genotype 50
References 53

3 Diseases Deriving from Alterations of the Ecological Niche—Part I: Not Regarding the Holobiont 57
3.1 General Concepts About the Diseases Deriving from Alterations of the Ecological Niche 57
3.2 Possible Alterations of the Ecological Niche Not Regarding the Holobiont 61

3.2.1 Diet Alterations 62
3.2.2 Diet Alterations, Together with Other Factors 72
3.2.3 Exposure to Harmful Substances 78
3.2.4 Excessive Noise Levels 79
3.2.5 Altered Exposure to Light in the First Years of Life 80
3.2.6 Multiple Carcinogenic Factors 86
3.2.7 Altered Social Interactions 90
3.2.8 Infectious Epidemics 93
References 103

4 Diseases Deriving from Alterations of the Ecological Niche—Part II: Regarding the Holobiont 123
4.1 Possible Alterations of the Ecological Niche Regarding the Holobiont 123
4.1.1 Alterations of the Holobiont—A: Reduced Exposure to Allergens 123
4.1.2 Alterations of the Holobiont—B: Elimination of Intestinal Parasites 129
4.1.3 Alterations of the Holobiont—C: Alterations of the Microbiome at Birth 144
4.1.4 Alterations of the Holobiont—D: Alterations of Skin and Mucosal Microbiomes 149
4.1.5 Alterations of the Holobiont—F: Alterations of the Gut Microbiome and Obesity 153
4.1.6 Alterations of the Holobiont—F: The Case of *Helicobacter Pylori* 154
References 157

5 Diseases Deriving from Relations with Other Living Beings and from Conditions Beyond the Adaptation Range 169
5.1 Infections, Infestations, and Other Damages Caused by Other Species 169
5.2 Diseases Deriving from Conditions Beyond the Adaptation Range 173
References 176

6 Physiological Phenomena that Cause Troubles and Suffering or Even Death: Part I: Non-phenoptotic Phenomena 179
6.1 Defenses against Infections, Toxic Substances, and Other Noxious Factors 179
6.2 Troubles Related to Pregnancy 186
6.3 Mental and Behavioral Disorders as an Adaptation to Particular Situations 191
References 195

7 Physiological Phenomena that Cause Troubles and Suffering or Even Death: Part II: Phenoptotic Phenomena Different from Slow Phenoptosis 201
7.1 Phenoptotic Phenomena 201
7.1.1 Selective Mechanisms Favoring Phenoptotic Phenomena 202
7.1.2 Classification of Phenoptotic Phenomena 203
7.2 Phenoptotic Phenomena in our Species Different from Slow Phenoptosis 210
References 214

8 Physiological Phenomena that Cause Troubles and Suffering or Even Death: Part III: Slow Phenoptosis or Aging 219
8.1 Slow Phenoptosis or Aging: Why we Age 219
8.1.1 Two Opposite Interpretations of Aging 219
8.1.2 Untenable Objections against Aging as an Adaptive Phenomenon 220
8.1.3 Arguments against Non-adaptive Aging Theories 223
8.1.4 Adaptive Aging Theories 227
8.1.5 Kin Selection Aging Theory 228
8.1.6 Inverse Relation Between Extrinsic Mortality and Proportion of Deaths Due to Intrinsic Mortality 237
8.2 Slow Phenoptosis or Aging: How we Age at the Cellular Level 239
8.2.1 Telomere Theory of Aging 240
8.2.2 Subtelomere-Telomere Theory of Aging 243
8.2.3 Subtelomere-Telomere Theory of Aging Integrated with TERRA Sequences 252
8.2.4 Some Characteristics of Aging Mechanisms that Are Incompatible with the Non-programmed Paradigm of Aging 258
8.3 Effects of Cellular Aging on Tissues and Organs 263
8.3.1 Cell Turnover 263
8.3.2 Replicative Senescence and Decline of Cell Turnover 265
8.3.3 The "Atrophic Syndrome" 266
8.3.4 "Direct" and "Indirect" Aging 268
8.3.5 Why some Types of Cells Are Not Subject to Turnover 270
8.3.6 Effects on some Cell Types with Turnover 272
8.3.7 Effects on some Cell Types without Turnover 275
8.3.8 Endothelial Cell Turnover: A Useful Indicator for Aging Process 279
8.4 Effects of Cellular Aging on the Whole Organism 280
8.4.1 Age-Related Fitness Reduction under Evolutionarily "Normal" Conditions 282
8.4.2 Estimation of Fitness Variation under Modern Conditions 284

8.4.3 Variation of Survival Capacity under Modern Life Conditions 285
8.4.4 Importance of Aging as a Cause of Death 286
References 288

9 Other Topics Studied by Evolutionary Medicine 311
9.1 Constraints Determined by Our Phylogenetic History 311
9.1.1 Crossing Between Respiratory and Digestive Tracts 311
9.1.2 The Route of the Inferior or Recurrent Laryngeal Nerve 314
9.1.3 The Route of Vas Deferens from the Epididymis to the Prostate 314
9.1.4 Inverted Arrangement of the Retina 316
9.1.5 Transition to Bipedalism and Birth Canal 320
9.2 Normal Physiological Characteristics that Become Problems in a Modified Ecological Niche 322
9.2.1 Necessity of Essential Nutrients 322
9.2.2 End Arteries 324
9.2.3 Impacted Wisdom Teeth and Malocclusion 326
9.2.4 Cecal Appendix 329
9.2.5 Fluid Retention in the Case of Heart Failure 330
9.3 Diseases Caused by Conflicting Evolutionary Exigencies 331
9.3.1 Modified Hemoglobins or Erythrocytes Protecting Against Malaria 331
9.3.2 Cystic Fibrosis 334
9.3.3 Morning Sickness 335
9.3.4 Restrictions in the Biological Availability of Iron 335
9.4 Enzymatic Deficiencies That Are Not Defects 336
9.4.1 Lactose Intolerance 336
9.4.2 Altered Metabolization of Alcohol 337
9.5 Childbirth and Newborn's First Months of Life 339
9.5.1 Before Birth 339
9.5.2 During Childbirth 342
9.5.3 First Days and Months of Life of the Newborn 347
References 351

10 Implementation of Evolutionary Medicine 363
10.1 Evolutionary Medicine, a New Medical Discipline, or a New Medicine? 364
10.2 Identification of the Primary Causes of a Disease 366
10.3 Difficulties Against the Implementation of Evolutionary Medicine 369
10.4 Necessity of Evolutionary Medicine 371
10.5 Effective Implementation of Evolutionary Medicine 373
10.6 Conclusion 378
References 383

Chapter 1
Basic Concepts and a Brief History of Evolutionary Medicine

In a recently published book (Libertini et al., 2021a), Gerontology and Geriatrics are discussed in evolutionary terms. It was only possible to address these issues by exposing the theme of evolutionary medicine, of which the subject of the book was certainly a fundamental part.

This book has an opposite starting point, evolutionary medicine, and in its exposition, gerontology and geriatrics considered in evolutionary terms surely cannot be disregarded.

As the themes of the two books partially coincide, it is normal and not accidental that some of the arguments, facts, and bibliographic references are the same.

This book focuses on the topics that are certainly not in common.

1.1 Definition of Evolutionary Medicine

When Alfred Wallace (Fig. 1.1) and Charles Darwin (Fig. 1.2)—independently arrived at similar results (Darwin & Wallace, 1858), deepened in the following year by Darwin (1859)—proposed that the characteristics and origins of the species of living beings were determined by natural selection, this suggestion was undoubtedly a great revolution for biology.

A long line of distinguished scholars (some illustrated in Figs. 1.3, 1.4, 1.5 and 1.6) expanded and consolidated this revolution.

A new scientific era was born in which every biological phenomenon progressively was reformulated considering natural selection. This was happily expressed synthetically by one of the most famous scholars: "Nothing in biology makes sense, except in the light of evolution" (Dobzhansky, 1973).

However, if it is true that the whole of biology requires the knowledge and consideration of the mechanisms of biological evolution, it follows that also the

G. Libertini et al., *The Essential Revolution of Evolutionary Medicine*, Advances in Studies of Aging and Health 4, https://doi.org/10.1007/978-3-031-84877-3_1

Fig. 1.1 Alfred Russell Wallace (1823–1913)

Fig. 1.2 Charles Robert Darwin (1809–1882)

biological phenomena at the origin of the diseases and the rationale of the means to combat them, i.e., the object of study and activity of medicine, must somehow consider the mechanisms of evolution. This argument has been brilliantly summed up in another well-known sentence by Ajit Varki: "Nothing in medicine makes sense, except in the light of evolution" (Varki, 2012).

The definition of evolutionary medicine formulated by Nesse agrees with these concepts: "Evolutionary Medicine is the enterprise of using evolutionary biology to address the problems of medicine" (Nesse, 2008, p. 417).

More extensively, the considerations above allow us to define evolutionary medicine as the use of the knowledge of evolutionary mechanisms: (i) to understand the origin of the primary causes of diseases and to rationally define the best means, (ii) to prevent diseases, (iii) to slow down or block the progression of pathogenic processes, and (iv) to treat the manifestations and effects of diseases.

Fig. 1.3 Friedrich Leopold August Weismann (1834–1914)

Fig. 1.4 Gregor Johann Mendel (1822–1884)

It is essential to highlight that Evolutionary Medicine should not be understood as another alternative Medicine (such as iridology, ayurvedic medicine, naturopathy, homeopathy, traditional Chinese Medicine, Energy Medicine, etc.) nor considered as a new specialization that completes current Medicine for a limited group of diseases.

On the contrary, Evolutionary Medicine should be interpreted as the enrichment and necessary completion of current medicine with the concepts of evolutionism that becomes an integral and essential part of overall medical knowledge for disease prevention and treatment.

Indeed, when medicine consistently integrates the concepts of evolutionism into its body of knowledge, it becomes a discipline more completely scientific, just as, for example, medicine has enhanced its scientific strength with the knowledge of chemistry. In fact, a medicine that supposed to ignore the principles of chemistry

Fig. 1.5 Theodosius Grigorevich Dobzhansky (1900–1975)

Fig. 1.6 John Burdon Sanderson Haldane (1892–1964)

would be impaired both in scientific knowledge and practical application and would be inconceivable today.

A possible objection is that, limiting to the example given, while chemistry is surely indispensable for medicine, evolutionism would have marginal importance for medicine, showing its usefulness only for some topics of medicine but a scarce utility for medicine in general and the prevention and treatment of most diseases.

A critic with a superficial knowledge of evolutionism could say: what importance could it ever have for a doctor in front of a suffering patient the knowledge of the evolution of living beings and the very origin of man from the ancestor of hominids and monkeys?

As we will see in the exposition of this work, the importance of the knowledge of evolutionary mechanisms concerns the mechanisms at the origin of diseases, i.e., the primary causes of the diseases, the prosecution and worsening of pathological processes, and the measures that can be adopted to prevent and treat diseases.

1.2 Some Basic Concepts

It is now essential to define and discuss, or at least mention, some fundamental concepts of evolutionary medicine.

1.2.1 Effects of a Random Modification in a Complex System

Entropy can be defined as the measure of the disorder of a system. Based on simple statistical considerations, it is known that entropy always tends to increase in a closed system with no action (i.e., energy input) from the outside (Dugdale, 1996).

So, if there is a random modification of a complex system, the most probable consequence of this modification, disregarding the cases in which it does not affect the degree of disorder of the system, will be an increase in the disorder of the system, here described as "effects of a random modification in a complex system".

For example, in a racing car highly engineered for competitive performance, if we make a random modification in any of its parts, disregarding the cases in which the modification has no effect, the most probable consequence will be a deterioration in performance and only in some rare cases the random modification could lead to an improvement in performance (Fig. 1.7).

The concept that a random modification in a complex system, if not neutral, is usually harmful to the system's functioning, also applies to biological phenomena.

If we consider the complex system made up of the genome of a species and its phenotypic expression, in the case of a random modification of the genome (e.g., a mutation), the most probable consequence, neglecting the cases in which the

Fig. 1.7 A random modification of a race car, which is surely a complex machine, is a likely cause of a malfunction

mutation has no effect and can be defined as neutral, will be a functional alteration (see *Chap. 2—Diseases deriving from alterations of the genotype*).

In the discussions about evolutionary medicine, in various cases, it will be helpful and necessary to consider complex systems in which a possible modification constitutes a probable alteration.

As a matter of fact, in addition to the genome mentioned above, we also have complex systems made up of interactions:

(i) between the genome and the conditions in which the phenotype is expressed, i.e., what we can define as the ecological niche of the species (see *Chap. 3—Diseases deriving from alterations of the ecological niche—Part I: Not regarding the holobiont* and *Chap. 4—Diseases deriving from alterations of the ecological niche—Part II: Regarding the holobiont*);
(ii) between an organism and the myriad of species that coexist with the organism constituting a holobiont (see Sect. 1.2.5 *The concept of "holobiont"*); and.
(iii) between the species (or rather, the holobionts) that live in an ecosystem;

where the more probable outcomes of random modifications are alterations of the systems.

Indeed, as we will see in later parts of this work, the simple concept of the "effects of a random modification in a complex system" has important implications for evolution in general and evolutionary medicine in particular.

1.2.2 The Ecological Niche of a Species

The individuals of a species live in an established range of conditions and relationships, which can be defined as the "ecological niche" of a species.

The concept of ecological niche certainly includes the physical conditions of the species' environment in which the species lives but is by no means limited to this.

The ecological niche encloses the foods eaten by the individuals of the species, how they compete with other individuals of the same species and other species, how they run, attack, escape, rest, breathe, struggle for survival and reproduction, etc.

The ecological niche also includes the intricate relationships with other species that live on the external and internal surfaces of an individual's body or that in any way have a relationship with it.

The study of the wide variety of ecological niches transcends the scope of this work.

However, it is important to underline some basic concepts:

(1) an ecological niche can be very restricted for definite characteristics (for example, a species that parasitizes only one other species), or it can have less restricted limits (for example, a species that preys on many other species). In general, an ecological niche is defined by a particular range of conditions, more or less wide depending on each condition;

(2) a species tends to adapt as much as possible to its ecological niche;
(3) an ecological niche of a species and the peculiar adaptations of the species to it constitute a complex system.

1.2.3 Normality in Evolutionary Terms

In statistics, the concept of normality is simple and straightforward.

In general, something is defined as normal if it is habitual or more frequent.

In quantitative terms, if a parameter may have only integer values (for example, the number of legs or eyes), normality is the most frequent integer value of the parameter. Consequently, for example, it is easy to observe for our species that it is normal to have two legs and two eyes. The few subjects who, due to some accident, have lost one or two legs or eyes do not modify this observation.

Likewise, normality can be defined in various precise ways for a parameter, such as height, which does not show only integer values. A simple way is to report the mean value (μ) of the parameter measured in a set of values. Plotting the probability of the single values on a curve is a more complex but better way. Figure 1.8 shows a symmetric distribution of values defined as the Gaussian or normal distribution for a non-integer-valued parameter. In this distribution, indicating the standard deviation of a sample with σ,[1] 68% of the sample falls within the range $\mu \pm \sigma$, and 95% of the sample falls within the range $\mu \pm 2\,\sigma$ (Armitage et al., 2002). There are other types of distribution of values, for example, asymmetrical curves, but even in these cases, it is possible, based on arbitrarily established criteria, to define values that can be considered statistically normal.

The mean value of a parameter and the distribution curve depend on the set of values considered, and a particular set of values leads to particular cases of "normality". E.g., considering a cemetery, the normality is the condition of corpses or mortal remains, and considering the patients of a mental hospital, the normality is to have a certain degree of impairment of mental functions.

To overcome these "strange" results (which are entirely rational in statistical terms), it might seem sufficient to consider a large set of values without restricting them to categories with particular characteristics. However, this is by no means sufficient in terms of evolutionary medicine.

---

[1] The standard deviation (σ), or root mean square, of a set of values is defined by the following formula:

$$\sigma = \sqrt{\sum_{i=1}^{N} (X_i - \mu)^2 / N}$$

where: N = number of values; x_i = value i (with i from 1 to N); μ = mean of the values.

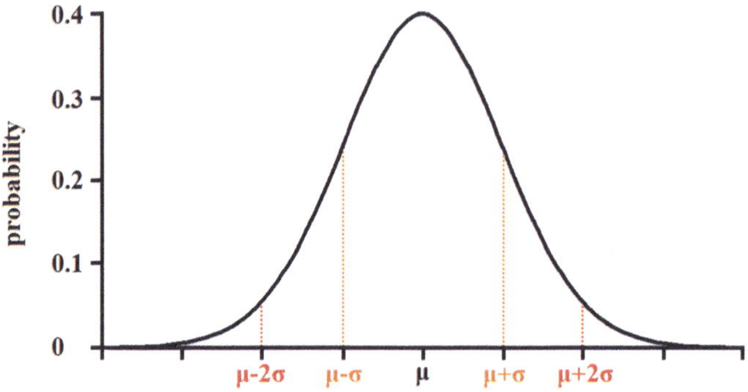

Fig. 1.8 A Gaussian or normal distribution (Armitage et al., 2002)

An example from a scientific work (Truswell et al., 1972) will be very useful to highlight the insufficiency of this approach. In the study, systolic and diastolic blood pressure values were measured in the contemporary population of London and in a small hunter-gatherer population living in Paleolithic-like conditions (the !Kung of Botswana, also known as Bushmen). In the modern London population, systolic blood pressure (P) rose progressively with age, following the popular empirical formula P = 100 + age, and diastolic blood pressure also rose with age. Conversely, among the Bushmen, blood pressure remained roughly constant around values of 120 mmHg at all ages, and diastolic blood pressure also remained roughly constant around 65–75 mmHg (Fig. 1.9).

It was evident that for the modern population, both systolic and diastolic blood pressure values increased with age, raising cardiovascular risk (Adhikary et al., 2022), while for the primitive population, blood pressure and cardiovascular risk did not increase with age.

By adopting the simple statistical criterion, the normal values were those of the London population, which is much more numerous than that of the small tribes of the Bushmen. However, it was clear medical nonsense to define as normal those values that increased the risk of cardiovascular diseases.

Similar results can be obtained by comparing the vital parameters and health conditions (or the health risks) of modern populations and populations living in primitive conditions. An example of these studies is the one conducted on the Ache population of Paraguay, another group of people who lived as hunter-gatherers in conditions similar to those of the Paleolithic (Hill & Hurtado, 1996). For such populations, diseases such as diabetes mellitus, hypertension, cancer, etc., were practically unknown (Hill & Hurtado, 1996).

So, there is a great contradiction: regarding health, the normal values of vast populations living under modern conditions are different and worse than those observed in small populations living under primitive conditions. If we accept the purely statistical criterion, we must consider the parameters and values concerning the conditions of modern populations as normal, while those shown by primitive populations should be considered "abnormal".

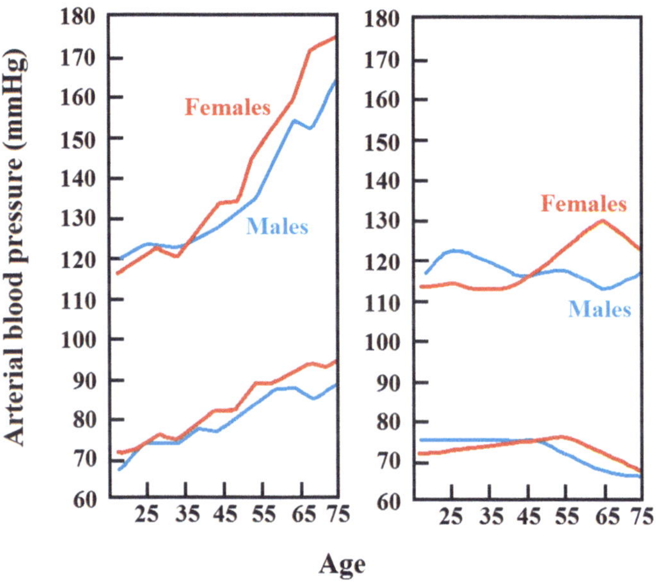

Fig. 1.9 On the left: systolic and diastolic arterial blood pressures in London citizens; on the right: the same for a!Kung population of Botswana; data from (Truswell et al., 1972)

To overcome this contradiction, instead of considering the statistical values of modern populations as normal, empirical solutions that want to solve the problem without providing a rational solution have been devised.

Indeed, the concept of health range (Jaminet & Jaminet, 2012; Klimis-Zacas & Wolinsky, 2005; McGrady & Moss, 2013) or optimal health range (Chaker et al., 2017) has been proposed. It defines as normal not the ranges of values observed in the whole population but those observed in "the 2.5th and 97.5th percentile interval of the distribution in an apparently healthy population" (Chaker et al., 2017) and so likely associated with optimal health and a minimum of disease frequency.

However, the "health range" concept is based on practical evaluations, and not on a scientific theory. So, the definition of a parameter's "normal" value, for example, blood pressure, cholesterol, or weight, becomes somewhat arbitrary and can generate controversy.

Evolutionary medicine offers without forcing the necessary rational definition of "normality" as a concept based on a scientific theory such as evolutionism.

As a consequence of natural selection, a species adapts as best as possible to the conditions in which it lives, i.e., to its ecological niche. The best possible adaptation means the best possible health conditions for individuals of the species.

Correspondingly, the concept of normality must consider the statistical values relating to individuals living in conditions as close as possible to those of the ecological niche to which the species is adapted. Therefore, the original ecological niche is the rational reference to the condition of "normality", and the values observed in this condition, or the conditions closest to it, must be considered normal in evolutionary terms.

This definition solves the contradiction of the results between those observed in primitive populations and those under modern conditions, certainly more or less distant from those of the original ecological niche to which the species has adapted. It also overcomes the need to arbitrarily define "health ranges" without having a scientific rationale to justify them.

1.2.4 The Concept of "Mismatch"

The ecological niche of a species and the adaptation of the species to it constitute a complex system.

This adaptation requires a complex and intricate modeling of the characteristics of the species, i.e., of its genome, to the innumerable and often contradictory requirements of the ecological niche.

If, for any reason, there are changes in the conditions of the ecological niche, the most probable consequences, excluding the cases in which the changes are neutral, is that there will be an alteration for the adaptation of the species and so diseases, which were defined, in 1983, as "Diseases deriving from alterations of the ecological niche" (Libertini, 1983).

A few years later, this altered correspondence between the requirements of the ecological niche and the adaptation of the species was defined as "discordance": "When conditions of life for any animal population deviate from those to which it has genetically adapted, biological maladjustment – discordance – is inevitable. The human species is no exception. For us, discordance between our current life-style and the one in which we evolved has promoted the chronic and deadly 'diseases of civilization': the heart attacks, strokes, cancer, diabetes, emphysema, hypertension, cirrhosis, and like illnesses that cause 75 percent of all mortality in the United States and other industrialized nations. Most of these illnesses simply did not exist, or were very rare, in the society of our Paleolithic ancestors. They, like us, had the genetic potential for these disorders, but it took *our* life-style to bring those genetic susceptibilities to the fore" (Eaton et al., 1988a).

The concept of "discordance", with a more fortunate word, was also defined as "mismatch" by the same authors: "Although our genes have hardly changed, our culture has been transformed almost beyond recognition during the past 10,000 years, especially since the Industrial Revolution. There is increasing evidence that the resulting mismatch fosters 'diseases of civilization' that together cause 75 percent of all deaths in Western nations, but that are rare among persons whose lifeways reflect those of our preagricultural ancestors." (Eaton et al., 1988b).

The concept of "mismatch" (see Fig. 1.10) is an immediate consequence of the three previously exposed concepts: (i) effects of a random modification in a complex system, (ii) normality in evolutionary terms, and (iii) ecological niche. It is a central concept of evolutionary medicine. The concept of mismatch and the ideas

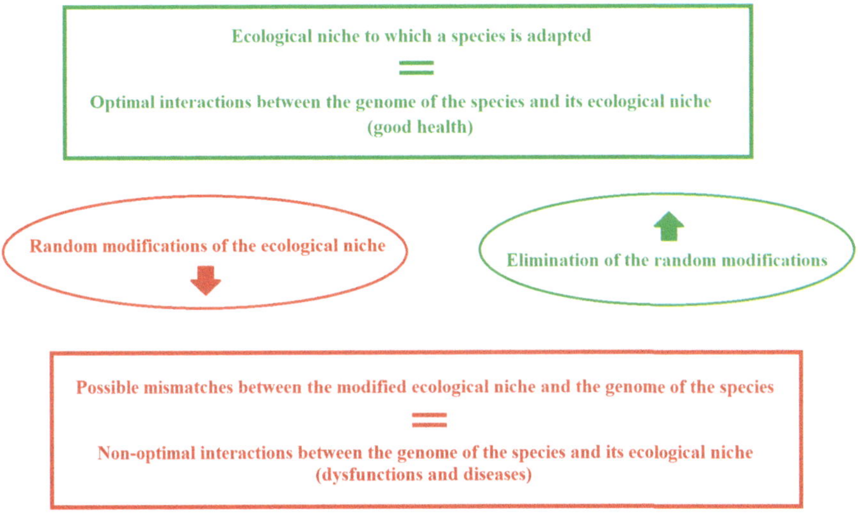

Fig. 1.10 Random modifications of the ecological niche are possible causes of mismatches between the adaptation of the species and its modified ecological niche

from which it derives allow a different and more rational reading of many types of experimentally documented phenomena.

1.2.5 The Concept of "Holobiont"

The holobiont concept was proposed for the first time by the theoretical biologist Adolf Meyer Abich in 1943 (Baedke et al., 2020), and subsequently independently reproposed by Lynn Margulis in 1991 (Margulis & Fester, 1991). The definition of holobiont has changed over time (Theis et al., 2016) and the concept has also been defined as supraorganism/superorganism (Glendinning & Free, 2014).

However, in general it is understood that a holobiont includes a eukaryote host and a large number of species of bacteria, viruses, fungi, and other species that live on or inside it (Sánchez-Cañizares et al., 2017) and that can also be multicellular (Fiuza et al., 2021). Furthermore, there is growing evidence of the existence of alterations of the holobiont that cause host disease (Castillo-Álvarez & Marzo-Sola, 2019).

For simplicity, limiting the discussion to our species, according to the traditional conception, a human being is an autonomous organism opposed by countless micro-organisms (bacteria, archaea, viruses, fungi, mycoplasma, protozoa, etc.) and macro-organisms (intestinal worms, mites, etc.), which try to use its resources by often being harmful. In the best cases, as for various bacterial species, they are non-pathological commensal species (although being potentially harmful), and in some rare cases, they may have some utility for the host, e.g., vitamins synthesized by

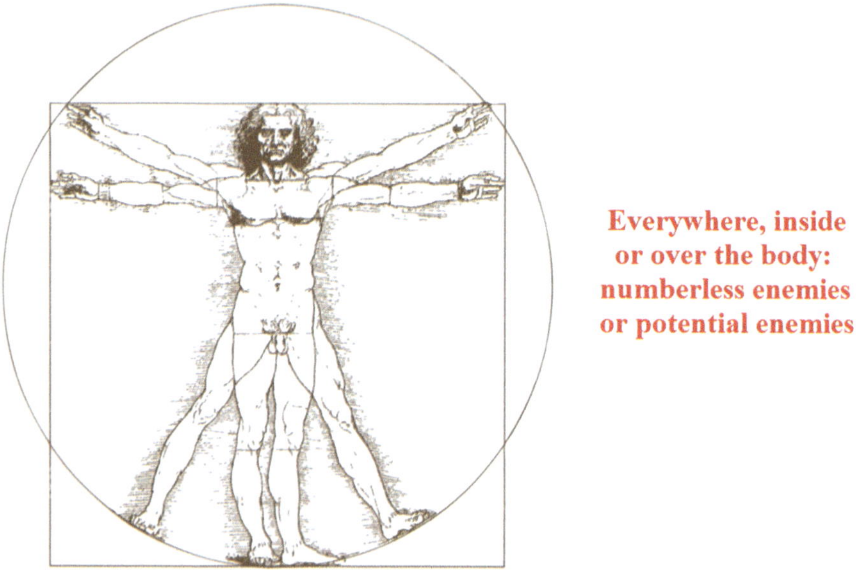

Fig. 1.11 The traditional idea of the human being, an autonomous entity besieged and threatened by countless hostile parasitic species

particular gut bacteria (LeBlanc et al., 2013; Zhan et al., 2022). According to this conception, we have a host, best shaped by natural selection, surrounded, besieged, and threatened in many ways by countless other species (Fig. 1.11). Under this popular conception, the "ideal" condition would be to live alone without the dangers and harms of these innumerable other organisms.

The reality is different and much more complicated. Each subject houses countless bacteria, archaea, viruses, fungi, intestinal parasites, mycoplasma, protozoa, etc., on the internal and external surfaces of the body. Even the interstitial fluids or some human body cells accommodate other living beings on a provisional or long-term basis.

Bacteria have been estimated to be around 3.8×10^{13}, a number equal to or greater than the estimated number of cells in the body, around 3×10^{13} (Sender et al., 2016). The number of viruses present in or over each human has been estimated at around 10^{13} (Liang & Bushman, 2021), i.e., a number similar to that of bacteria.

Viruses are present in any part of the human body (Abeles & Pride, 2014; Pride et al., 2012), even in those parts commonly considered to be sterile, such as cerebrospinal fluid (Ghose et al., 2019), bloodstream (Liang & Bushman, 2021; Moustafa et al., 2017) and breast milk (Pannaraj et al., 2018). It has been estimated that about 8% of human DNA is of viral origin (Wildschutte et al., 2016).

Each internal and external body surface of our organism houses specific ecosystems composed of numerous species of microorganisms. For example, the oral cavity, which we try to sterilize with mouthwashes and toothpaste, gives shelter to "fungi, viruses and protozoa" and "over 700 species of bacteria" (Deo & Deshmukh, 2019).

About the gut ecosystem: "The best-defined contribution of the microbiota of the GI [gastrointestinal] tract is a metabolic one: these microorganisms have a combined metabolic capacity equivalent to that of the liver, justifying their description as an additional human organ..." (Glendinning & Free, 2014).

The ecosystems of gut, respiratory, and skin microbiota have a similar complexity, and it has been estimated that the species composing all the human microbiota have genetic information that is over 150 times that of the human genome (Grice & Segre, 2012). These microbial communities are in symbiosis with the host, and their balance is important for the health of the host organism (Hou et al., 2022).

In addition to microbial species, in populations not living according to modern lifestyles (likely in conditions closer to the evolutionarily "normal" ones), helminth infestations are widespread (Dunn et al., 2016; Veesenmeyer, 2022). In general, these infestations are the object of attention for the pathologies caused by them, but it is also known that these infestations are indispensable for immune regulation (Gazzinelli-Guimaraes & Nutman, 2018; Mishra et al., 2014).

Even in modern times, as a rule, at least two species of mites (*Demodex folliculorum* and *D. brevis*) are present in the hair bulbs. These mites are small arthropods belonging to the subclass *Acari* and are highly adapted to our species, which hosts 1000-2000 mites for each individual (Litwin et al., 2017). Their possible role in some pathological conditions is being investigated (Chen & Plewig, 2015). However, the symbiotic relationship between these mites and the human organism, and between analogous species of mites and their specific hosts, needs to be clarified.

The whole constituted by our species (defined as "host") and all the other organisms with which there are close relationships (defined as "bionts", including the host as the main biont) is incredibly complex in terms of both the number of species and the balanced tangle of countless interactions.

"At least as many of the cells, and the vast majority of unique genes, in the human body are microbial... As such, we can view ourselves as holobionts..." (Charbonneau et al., 2016).

The following example is indicative of the complexity and ambiguity of the relationships between the bionts and the host.

A mother's milk contains many different types of special oligosaccharides (Human Milk Oligosaccharides, HMOs). In 2011, there were 70 identified HMOs (Wu et al., 2011), and in 2018 this number increased to around 200 (Urashima et al., 2018). They serve "as prebiotics to stimulate the growth of beneficial intestinal bacteria, as receptor analogues to inhibit binding of pathogens, and as substances that promote postnatal brain development. There is further evidence that HMOs participate in modulating the human immune system." (Wu et al., 2010).

As HMOs in babies are probiotics for specific bacteria like *bifidobacteria* and *lactobacilli* (Wu et al., 2010), i.e., the mother "feeds" these bacteria, implicitly regulating the formation of healthy microbiota, the aforesaid bacteria somehow must be useful to the newborn baby. Otherwise, the characteristics of HMOs would not be justifiable in evolutionary terms.

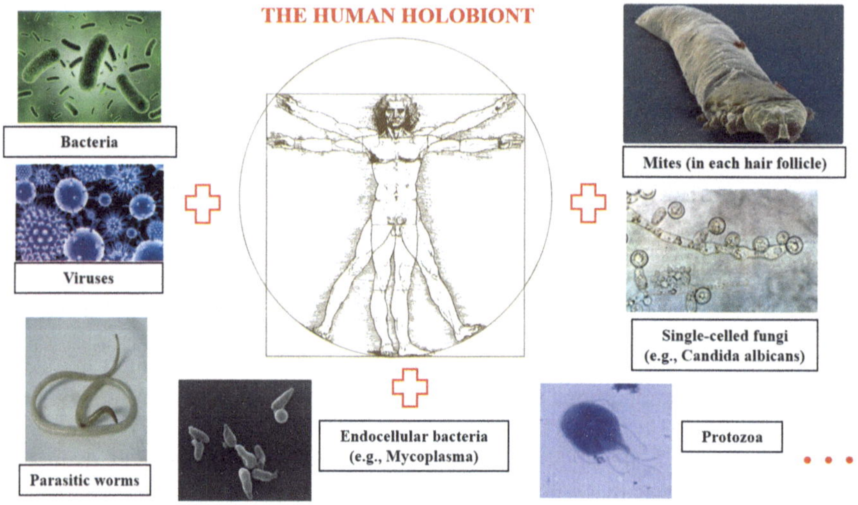

Fig. 1.12 A representation of the human holobiont

Humans have coevolved with the other bionts of our holobiont since an immemorial time before the birth of our species.

In studying a holobiont, the individual-level selection is insufficient for understanding its origin, characteristics, and evolution. Moreover, supra-individual forms of selection that have the limits of considering only the host are also insufficient for this purpose.

In the present work, the concept of holobiont is essential, because numerous and critical diseases derive from alterations of the holobiont (see *Chap. 4—Diseases deriving from alterations of the ecological niche—Part II: Regarding the holobiont* that discusses diseases caused by alterations of the ecological niche). For example, the eradication of intestinal worms and alterations of the gut microbiome may cause autoimmune diseases (Vieira et al., 2014), among which type 1 diabetes (Boerner & Sarvetnick, 2011).

A representation of our holobiont, the human holobiont, is depicted in Fig. 1.12.

1.2.6 Other Basic Concepts

There are at least two other concepts that are fundamental to evolutionary medicine. They will only be mentioned here, while a more detailed discussion will be in the appropriate chapters or sections:

(1) Phenoptosis

The first concept represents the extension of the idea of natural selection as limited to individual selection only. The concept of phenoptosis (Libertini, 2012; Skulachev,

1997, 1999) includes all the genetically determined characteristics of a living being that are harmful or even fatal to the living being itself or to closely related individuals. The most important phenomenon that falls under the definition of phenoptosis is aging.

Phenoptotic phenomena, by definition harmful to the individual, cannot be explained in terms of individual selection but can be well justifiable by supra-individual selection (e.g., kin selection and particular forms of group selection).

The phenoptotic phenomena that cause troubles and suffering, or even death, are discussed in *Chap. 7—Physiological phenomena that cause troubles and suffering or even death—Part II: Phenoptotic phenomena different from slow phenoptosis* and *Chap. 8—Physiological phenomena that cause troubles and suffering or even death—Part III: Slow phenoptosis or aging*, to which reference should be made to know more than this brief description.

(2) Conditions beyond the adaptation range

Each species necessarily has a limited range of conditions to which it can be adapted. When an individual encounters conditions beyond this range (for example, falls from an excessive height, burns from fire, ingested poisons, etc.), the organism cannot fully face them, so disease or even death occurs. The topic will be discussed in Sect. 5.2 *Diseases deriving from conditions beyond the adaptation range*.

1.3 Current Non-evolutionary Classification of Diseases and Similar Conditions

Today, it is customary and universal to officially classify diseases and conditions that cause suffering based on criteria that are empirical or follow the historical development of medicine. Let us consider the current International Classification of Diseases (ICD-11) (WHO, 2018) and also the previous classifications, among which it is sufficient to mention ICD-10 (WHO, 2016). They classify the diseases and disorders causing functional disorders or even death based on some criteria, which are surely not determined by a scientific theory and, on the contrary, could be described as empirical, utilitarian, and influenced by tradition.

Indeed, in ICD-11, the classification of diseases and similar conditions appears to follow, in order of priority, the following three criteria:

(i) Division according to the affected functional systems (e.g., respiratory, circulatory, digestive, etc.; main codes 3-16);
(ii) Division according to the branches of the medical specialties (codes 1, 2, 17-20);
(iii) Use of additional codes for all conditions that are not included in the other codes (codes 21-24) or for special needs (codes 25, 26, V, X).

Table 1.1 shows the general classification of diseases in ICD-11.

Table 1.1 ICD-11 (International Classification of Diseases, from December 2018) (WHO, 2018)

01—Certain infectious or parasitic diseases
02—Neoplasms
03—Diseases of the blood or blood-forming organs
04—Diseases of the immune system
05—Endocrine, nutritional, or metabolic diseases
06—Mental, behavioral, or neurodevelopmental disorders
07—Sleep-wake disorders
08—Diseases of the nervous system
09—Diseases of the visual system
10—Diseases of the ear or mastoid process
11—Diseases of the circulatory system
12—Diseases of the respiratory system
13—Diseases of the digestive system
14—Diseases of the skin
15—Diseases of the musculoskeletal system or connective tissue
16—Diseases of the genitourinary system
17—Conditions related to sexual health
18—Pregnancy, childbirth, or the puerperium
19—Certain conditions originating in the perinatal period
20—Developmental anomalies
21—Symptoms, signs, or clinical findings not elsewhere classified
22—Injury, poisoning, or certain other consequences of external causes
23—External causes of morbidity or mortality
24—Factors influencing health status or contact with health services
25—Codes for special purposes
26—Supplementary Chapter Traditional Medicine Conditions -Module I
V—Supplementary section for functioning assessment
X—Extension codes

The subdivision of each disease group follows a similar logic. Table 1.2 shows that the subdivision of the group of diseases defined by the main code 13 is based on anatomical or functional distinctions and is subordinate to other arbitrary criteria.

About the previous ICD-10 classification, the division of diseases followed the same logic as the ICD-11 and is, in general, practically the same.

An important fact to underline is that there is no code or subcode for aging in ICD-11, as well as in ICD-10 and previous classifications.

In fact:

- in ICD-11, there are (i) various "age-related conditions"; (ii) among the General symptoms, MG2A Ageing associated decline in intrinsic capacity; and (iii) among the Extension Codes, – Aetiology—Occupational relevance: XT9T Ageing-related;
- in ICD-10, there is, among the General symptoms and signs: R54 Senility;

while a specific code for aging or senescence is inexistent.

Table 1.2 Subdivision of code 13 Diseases of the digestive system

– Diseases or disorders of orofacial complex
– Diseases of esophagus
– Diseases of the stomach or the duodenum
– Diseases of small intestine
– Diseases of appendix
– Diseases of large intestine
– Diseases of anal canal
– Diseases of liver
– Diseases of gallbladder or biliary tract
– Diseases of pancreas
– Diseases of peritoneum
– Diverticular disease of intestine
– Ischemic vascular disorders of intestine
– Hernias
– Inflammatory bowel diseases
– Functional gastrointestinal disorders
– Postprocedural disorders of digestive system
– DE2Y Other specified diseases of the digestive system
– DE2Z Diseases of the digestive system, unspecified

Aging is excluded as a distinct process, and, in fact, the term, by the current prevailing ideas in the scientific world, is considered only as a useful abbreviation to indicate the cumulative effects of various degenerative processes. This absence is in full agreement with the traditional hypotheses regarding the factors that determine aging. For a review, see Chap. 4 in (Libertini et al., 2021a). A paradoxical consequence of this view is that, according to official statistics, no one dies with aging as the cause of death (World Cause of Death, 2020).

In fact, in 2020, the official world statistics obtained from WHO data report 84 causes of death that do not include "aging" (Table 1.3).

According to this conception, the countless elderly people who die each year are classified among the deaths due to the failure of an organ or a function or to some specific disease but never as a consequence of aging.

The arguments and empirical data that lead to rejecting this conception are discussed, as well as in other works, in a book from 3 years ago (Libertini et al., 2021a), in a recent work (Libertini et al., 2022), and in *Chap. 8—Physiological phenomena that cause troubles and suffering or even death—Part III: Slow phenoptosis or aging* of this book. Here, it is only highlighted that the rejection or acceptance of the idea of aging as a distinct adaptive and programmed phenomenon implies a fundamental distinction between two views of evolutionary medicine.

Table 1.3 World cause of death in 2020 (World Cause of Death, 2020)

1	Coronary Heart Disease	8,884,887	43	Malnutrition	200,392
2	Stroke	6,193,978	44	Maternal conditions	196,008
3	Lung disease	3,227,873	45	Drug use	181,758
4	Influenza and pneumonia	2,593,098	46	Ovary cancer	178,457
5	Lung cancers	1,784,109	47	Measles	165,756
6	Alzheimer & dementia	1,639,085	48	Alcohol	156,546
7	Diarrheal diseases	1,519,229	49	Other neoplasms	138,472
8	Diabetes mellitus	1,496,094	50	Skin cancers	125,894
9	Kidney disease	1,334,324	51	Epilepsy	115,890
10	Liver disease	1,315,359	52	Fires	114,144
11	Road traffic accidents	1,282,150	53	Pertussis	111,317
12	Tuberculosis	1,208,044	54	Skin disease	96,581
13	Hypertension	1,148,939	55	Uterine cancer	89,528
14	Colon-rectum cancers	916,166	56	Poisonings	84,278
15	Low birth weight	910,396	57	Encephalitis	77,757
16	Stomach cancer	830,682	58	War	72,457
17	Suicide	703,220	59	Tetanus	47,437
18	Falls	684,277	60	Rheumatoid arthritis	43,775
19	HIV/AIDS	674,662	61	Syphilis	43,427
20	Birth trauma	652,489	62	Hepatitis B	36,036
21	Breast cancer	640,112	63	Appendicitis	31,368
22	Other injuries	631,108	64	Dengue	30,020
23	Liver cancer	577,430	65	Multiple sclerosis	24,384
24	Congenital anomalies	532,854	66	Hepatitis C	22,147
25	Violence	474,691	67	Anemia	21,636
26	Esophagus cancer	462,995	68	Schistosomiasis	11,792
27	Asthma	455,153	69	Upper respiratory	10,054
28	Pancreas cancer	447,208	70	Chagas disease	7558
29	Malaria	410,762	71	Prostatic hypertrophy	6671
30	Prostate cancer	389,433	72	Schizophrenia	6058
31	Oral cancer	388,996	73	Leishmaniasis	5780
32	Lymphomas	383,063	74	Osteoarthritis	5100
33	Inflammatory/heart	349,908	75	Diphtheria	4964
34	Parkinson's disease	328,645	76	Ascariasis	2150
35	Endocrine disorders	316,274	77	Trypanosomiasis	1616
36	Leukemia	307,596	78	Oral conditions	1236
37	Cervical cancer	304,946	79	Chlamydia	793
38	Rheumatic heart disease	288,348	80	Otitis media	761
39	Drownings	235,642	81	Leprosy	253
40	Meningitis	233,303	82	Iodine deficiency	36
41	Peptic ulcer disease	220,263	83	Vitamin A deficiency	5
42	Bladder cancer	201,999	84	Hookworm disease	1

1.4 Possible Evolutionary Classification of Diseases and Similar Conditions

Before the birth of evolutionism, living species were classified according to empirical criteria of similarity. After the definition and achievement of evolutionism, the classification of living beings is based on the concept of evolution by natural selection with the derived idea of common ancestors for the various species and groups of species, which is both a basic idea of evolutionism and a solid scientific criterion. In fact, by studying the affinities among DNA and proteins of the various species, it is possible to reconstruct ever more detailed and reliable phylogenetic trees (Burki et al., 2020; Kapli et al., 2020; Patané et al., 2018).

As far as the classification of diseases is concerned, even the most modern classifications, although elaborated and perfected by the best experts and medical centers in the world, are rather arbitrary constructions that can be considered as the fruit of a conception preceding the birth of evolutionism.

If the classification of diseases must be based on a rational basis, which is necessarily scientific, following Varki's aphorism, it seems logical to organize or attempt to organize, the diseases in terms consistent with evolutionary mechanisms and in a way that is, in principle, useful for disease definition, prevention, and treatment.

A rational classification of diseases and similar phenomena that cause suffering and even death, appears to be based on their primary causes defined and analyzed through the mechanisms of evolutionism.

The first two categories of diseases have as primary causes those deriving from alterations of the genotype (Code I) and those deriving from alterations of the ecological niche (Code II).

The third category derives from relations with other living beings (Code III).

The fourth category is determined by conditions beyond the adaptation range (Code IV).

The fifth and sixth categories consist of Physiological phenomena that cause troubles and suffering or even death and, therefore, do not fall within the definition of diseases in its strict sense. The first of these two categories is limited to non-phenoptotic phenomena (Code V), while the other includes phenoptotic phenomena, of which the most important is aging (Code VI).

Table 1.4 reports a general division. Chapters 2–8 describe the distinct categories of diseases.

The proposed classification is based on the primary causes of the diseases interpreted in the light of evolutionism. The classification is rational, based on scientific criteria according to a theory largely accepted as scientific. The main advantage of this classification is that a rational interpretation of the primary causes of the diseases gives a strategy for their prevention, when possible and necessary, and a rational base for the treatment of the pathological manifestations.

On the contrary, the current classification is entirely focused on the treatment of the disease by a specialist (or specialists) who is (or are) most pertinent to the affected organs/systems.

Table 1.4 Evolutionary classification of diseases and similar phenomena

I—Diseases deriving from alterations of the genotype
(1) caused by: DNA/chromosomal/mitochondrial abnormalities; harmful genes that are beneficial in certain conditions
II—Diseases deriving from alterations of the ecological niche
(1) caused by alterations of the diet; reduced and/or abnormal physical activity; abuse of particular substances; altered social relations and other problems causing mental diseases; excessive crowding; excessive noise; other alterations of the lifestyle; conditions that cause infective epidemics, etc.
(2) caused by alteration of the holobiont
III—Diseases deriving from relations with other living beings
(1) caused by bacteria; viruses; fungi; protozoa; worms; insects, spiders, ectoparasites and other living beings
IV—Diseases deriving from conditions beyond the adaptation range
(1) caused by traumas; burns; downing and asphyxia; etc.
V—Physiological phenomena that cause troubles and suffering or even death—Non-phenoptotic phenomena
(1) defenses against trauma, infections, toxic substances, etc. (pain, fever, cough, sneezes, nausea, vomiting, diarrhea, etc.)
(2) Mental and behavioral manifestations (anxiety, fear, depression, jealousy, psychological pain, sadness, etc.) as adaptations to particular situations (Nesse, 2019)
(3) pregnancy, childbirth, and the puerperium
VI—Physiological phenomena that cause troubles and suffering or even death—Phenoptotic phenomena
(1) Phenoptotic phenomena, slow phenoptosis or aging excluded
(2) Slow phenoptosis or aging

1.5 Origins and Developments of Evolutionary Medicine

Now, it is appropriate to illustrate some of the origins and developments of evolutionary medicine.

In a 2008 book, a section of the Introduction is devoted to a "Brief History of Evolution in Medicine" (Trevathan et al., 2008a, pp. 4–9), and a chapter provides more information on the same topic (Nesse, 2008). These two contributions, however, explore only a part of the history of the origins and development of evolutionary medicine and must be integrated with some other information.

It is necessary to start with a Darwin, not the famous Charles Robert Darwin, but his grandfather Erasmus Robert Darwin (1731–1802) (Fig. 1.13). This polymath and illustrious scholar, physician, natural philosopher, physiologist, poet, inventor, thinker of the Midlands Enlightenment, member of the Lunar Society, and a fellow from 1761 of the Royal Society (Hassler, 1973), indeed little known and mentioned compared to his nephew, is to be considered an anticipator of some ideas of both evolution and evolutionary medicine.

He wrote a book, *Zoonomia, or the Laws of Organic Life* (Darwin, 1794–1796), where the preface "contains a remarkably prescient description of evolution, as well

Fig. 1.13 Erasmus Robert Darwin (1731–1802)

as its potential as a foundation for medicine (Darwin, 1794–1796, pp. vii-viii)…" (Nesse, 2008). E.g.: "the strongest and most active animal should propagate the species, which should thence become improved" (Darwin, 1794–1796), anticipating the ideas of the survival of the fittest and evolution.

Somehow, in the same book, he also forestalled some ideas of evolutionary medicine. Part 2 is entitled: "A Catalogue of Diseases Distributed into Natural Classes According to Their Proximate Causes, With Their Subsequent Orders, Genera and Species, and With Their Methods of Cure." and, in the Preface of Part 1, he states, in a passage also reported in (Nesse, 2008, p. 420): "The want of a theory, deduced from such strict analogy, to conduct the practice of medicine is lamented by its professors; for, as a great number of unconnected facts are difficult to be acquired, and to be reasoned from, the art of medicine is in many instances less efficacious under the direction of its wisest practitioners.... A theory founded upon nature, that should bind together the scattered facts of medical knowledge, and converge into one point of view the laws of organic life, would thus on many accounts contribute to the interest of society… it would enable every one of literary acquirements to distinguish the genuine disciples of medicine from those of boastful effrontery, or of wily address; and would teach mankind in some important situations the knowledge of themselves." (Darwin, 1794–1796).

Although Medicine has enormously improved over two centuries, even today, the need, already seen by Erasmus Darwin, for a rational organization of pathology based on a scientific theory and not on empirical criteria remains. Nesse observed that: "Unfortunately, little has changed. Many physicians still see the body as a designed machine instead of an evolved organism, and myriads of medical facts remain unconnected, awaiting a scientific framework." (Nesse, 2008, p. 420).

Dudley Joe Morton (1884–1960) (Fig. 1.14), a professor of anatomy and an orthopaedist, was one of the first to indicate the importance of the study of evolution in medicine and anticipated some ideas of evolutionary medicine (Morton, 1926): "He observed that medicine was primarily concerned with the recognition and

Fig. 1.14 Dudley Joe Morton (1884–1960)

treatment of that which lies beyond the range of normal human variation. He recognized that understanding what was a true departure from normal, and hence appropriate for clinical intervention, depended directly on knowledge of what constitutes the normal range of human variation and knowledge of what factors maintain normal conditions. He recognized that to maintain *normalcy*, medicine had to develop practices that reinforced the natural safeguards the body possessed against forces that would disrupt normalcy." (Trevathan et al., 2008a, p. 6). However, he was referring mainly to the normality of anatomical (or, in general, physiological) conditions and not to the normality of living conditions.

As highlighted by Trevathan et al. (2008a), before the concept and term of evolutionary or Darwinian medicine were proposed, some research can be considered pertinent to medicine, but for the subject and method of the study, they can be considered just as well as research of evolutionary medicine.

In 1954, Allison demonstrated that individuals with the sickle-cell anemia gene (characterized by sickle-shaped erythrocytes more vulnerable to lysis) were more likely to survive, i.e., they were favored by selection, in areas where malaria had a high incidence (Allison, 1954a, b). A few years later, Livingstone showed that the spread of agriculture led to the destruction of the forest with the formation of puddles of water where it was possible the reproduction of mosquitoes of the species *Anopheles gambiae*, insect vector of *Plasmodium falciparum*, the protozoan that causes malaria. Individuals with the sickle-cell anemia gene in heterozygous form had a mild form of anemia. However, they were also more resistant to malaria as the erythrocytes easily lysed when infected by the protozoan. On the contrary, subjects with the gene in homozygous form developed a fatal form of anemia. Consequently, the sickle-cell anemia gene was maintained at a certain frequency by opposing selection pressures (Livingstone, 1958).

About the idea of features of human design that may be considered maladaptive, Nesse and Williams highlighted that "A 1941 book by George Hoben Estabrooks, *Man, The Mechanical Misfit* (Estabrooks, 1941), describes many of the structural defects and compromises in human anatomy, especially those that result from turning a horizontal four-footed animal into an upright two-footed one. The weight of the top part of the body greatly compresses the vertebrae in the lower spine, and standing upright requires more muscular effort than a horizontal posture would …" (Nesse & Williams, 1994, p. 131). However, in the note pertinent to the same book, it is observed: "Our use of the book of George Estabrooks … is at variance with its spirit. While it does describe many design flaws of the human body, its main message is the misfit between that design and the uses to which it is put in modern times." (Nesse & Williams, 1994, p. 263, note about p. 131).

The interpretation of Nesse and Williams of Estabrooks' suggestions is reported by Lewis "as prefiguring the notion of maladaptation to same extent" (Lewis, 2008, p. 400). Nevertheless, the note above seems to indicate that according to the same authors, for Estabrooks, it could be more accurate to speak of alterations of the ecological niche instead of speaking of design flaws.

Lewis (2008, p. 400) indicated other works that anticipated some issues of evolutionary medicine:

(1) Haldane pointed out in 1949 that "the struggle against diseases, and especially infectious diseases, has been a very important evolutionary agent and that some of its results have been rather unlike those of the struggle for life in its common meaning." (Haldane, 1949). The issue of the relationships between various species (in the specific case between a host and its parasites) is of enormous interest in the field of evolutionary medicine;
(2) Two years later, Krogman explored the theme of man's imperfect adaptation to using two limbs instead of four to walk (Krogman, 1951).

An important contribution to the origins of evolutionary medicine has been given by a group of three authors, Stanley Boyd Eaton, Melvin Joel Konner, and Marjorie Shostak, who are repeatedly cited in Trevathan et al.'s book (2008a). Due to the importance and clarity of their contributions before the official birth of evolutionary medicine, these authors should be considered as founders and not forerunners of evolutionary medicine. Among other things, based on precise arguments and ample empirical evidence, as already reported in Sect. 1.2.4 *The concept of "mismatch"*, they argued that: "When conditions of life for any animal population deviate from those to which it has genetically adapted, biological maladjustment - discordance - is inevitable.…" (Eaton et al., 1988a, p. 5). Due to this discrepancy between the lifestyle [perhaps it is better to say the ecological niche] to which the species is adapted and the new living conditions [the modified ecological niche], they proposed the well-chosen and lucky definition of "mismatch" (Eaton et al., 1988b), a central concept, perhaps the most important concept, of evolutionary medicine. It should be noted that the concept of "mismatch" necessarily implies the concept of "normality in evolutionary terms", even if this concept is not explicitly defined in the works mentioned earlier by Eaton et al.

A 1980 article by Paul Ewald (1980) underlined the importance of studying the manifestations of infectious diseases (e.g., "fever, rhinorrhea, and diarrhea") "from an evolutionary perspective", as often they are "adaptations of the host to counteract

Fig. 1.15 Weston Andrew Valleau Price (1870–1948)

harmful aspects of the disease" or "adaptions of the pathogen to manipulate the host". This article aroused the attention of Williams and Nesse to write their 1991 paper (Williams & Nesse, 1991), which is considered the official birth of evolutionary medicine. This is attested by the same authors that report: "Ewald's works suggested that evolutionary ideas might well have significance for many medical problems, not just those that arise from infection" (Nesse & Williams, 1994, Preface p. ix).

The information provided so far regarding the origins of evolutionary medicine before its official birth is very important but has some central gaps.

Two other works anticipate evolutionary medicine and even offer something more.

The first is a famous book, *Nutrition and Physical Degeneration* of Weston Price, published in the USA in 1939 (Price, 1939), then published and reprinted several times by the Price-Pottenger Nutrition Foundation (www.price-pottenger.org).

The author, Dr. Weston Andrew Valeau Price (1870-1948) (Fig. 1.15), was a Canadian dentist who studied and then worked as a successful and respected dental surgeon, mainly in Cleveland, USA.

He founded the research institute National Dental Association, of which he was the chairman from 1914 to 1928, and that became the research section of the American Dental Association (New York Times, 1948).

Price, in the search for the causes of dental diseases, basing his studies on the fact that these illnesses were widespread in the USA and other modern populations while appearing much less frequent in primitive populations, focused his research on possible alterations of the diet related to civilization. For this purpose, after his retirement and using the earnings from his profession, he began a series of journeys to many different places in the world, studying and documenting the conditions of the teeth in various primitive populations still existing in his time (e.g., among Native Americans, Pygmies, Polynesians, Australian aborigines, etc.), but also in some modern populations (e.g., the inhabitants of the Lötschental valley in

Switzerland) and for some ancient populations whose skulls were preserved in museums (e.g., pre-Columbian populations in America; Australian aborigines before contact with Europeans). In addition to the teeth, he also studied other health conditions or diseases related to dietary variations.

After collecting a considerable collection of materials (including about 15,000 photographs and 4000 slides now kept by the Price-Pottenger Nutrition Foundation), he published his book in 1939, showing what appeared evident to him based on the collected documentation, namely the link between primitive peoples and their good state of teeth and other body functions as opposed to the diet and lifestyles of "civilized" peoples and their poor health conditions.

Price extensively documented that dental caries and various forms of malocclusion (overlapping teeth, pathological positions of the so-called wisdom teeth, etc.), but also other diseases such as rickets, impaired development of the facial bones (crucial for the proper development of teeth), tuberculosis and vulnerability to debilitating illnesses, although in the 1920s and 1930s afflicted a large part of the population in "civilized" countries, were rare or absent in primitive populations. Furthermore, he documented that when these populations switched to a Western lifestyle, particularly by using modern foods and abandoning the previous diet, the aforementioned diseases also spread among them. The phenomenon, therefore, did not depend on genetic factors, i.e., on some peculiar genetic characteristics of these populations. Price concluded that Western lifestyles and diets (including food preparation that deprived them of essential components) were the cause of many diseases that afflicted the civilized world. On the contrary, many primitive populations, still unaffected by rapid and unbalanced changes imposed by local overpopulation and mass production, appeared resistant to such diseases.

It should be noted that Price did not have such biological knowledge to be defined as an evolutionary biologist. Furthermore, his book is dated in various respects, and some ideas regarding race and populations living in ancestral conditions are tainted by some conceptions of the time, which today would be judged as unacceptable.

It is also easy to criticize Price's work and results by saying that often do not offer precise numbers and statistical analyses, so he should be regarded as an amateur.

In a paper (Jarvis, 1981), Price was accused of cultivating "The Myth of the Healthy Savage" and that, therefore, his conclusions were the consequence of preconceived ideas, including that teeth alterations were due to the modern diet.

However, Price's data strongly contradict the claim that his ideas are preconceived and not based on evidence. For example, regarding the conditions of the teeth in pre-Columbian populations and among the Australian Aborigines:

(1) "... there are some excellent collections of skulls in museums in Peru, with the skulls in position where they can be readily studied for the shape of the dental arches. When we have in mind that from 25 to 75 per cent of individuals in various communities in the United States have a distinct irregularity in the development of the dental arches and facial form, the cause and significance of which constitutes one of the important problems of this study, the striking contrast

found in these Peruvian skulls will be seen to constitute a challenge for our modern civilizations. In a study of 1,276 skulls of these ancient Peruvians, I did not find a single skull with significant deformity of the dental arches." (Price, 1939, Chap. 13—Ancient civilization of Peru).

(2) "Another important source of information regarding the Aborigines of Australia was provided by a study of the skeletal material and skulls in the museums at Sydney and Canberra, particularly the former. I do not know the number of skulls that are available there for study, but it is very large. I examined many and found them remarkably uniform in design and quality. The dental arches were splendidly formed. The teeth were in excellent condition with exceedingly little dental caries." (Price, 1939, Chap. 10—Isolated and modernized Australian Aborigines).

Indeed, the criticisms and limits before highlighted affect details and marginal aspects of his research but do not invalidate the reliability of the data reported and the main conclusions that he draws from them.

His research is extraordinary and largely unrepeatable as the living conditions of the primitive populations studied have mostly passed to modern living conditions, with health conditions closer to those of populations "civilized" or even worse.

In short, without sophisticated survey techniques or statistical analysis, with his impressive documentation, Price demonstrates that we have left the condition of peoples who live according to their ancestral customs and are well adapted to their ecological niche and passed to a state of strong imbalance between adaptation of our organism and lifestyles, including above all nutrition.

This condition, which will later be defined as a "mismatch", causes numberless widespread diseases and should be at the center of attention of those who want to fight disease, as well highlighted by some authoritative commentators on Price's results:

(A) "Since we have known for a long time that savages have excellent teeth and that civilized men have terrible teeth, it seems to me that we have been extraordinarily stupid in concentrating all of our attention upon the task of finding out why our teeth are so poor, without ever bothering to learn why savage teeth are good." (From the Foreword of Earnest A. Hooton, Harvard University, November 21, 1938, to Price's book) (Price, 1939, p. XXI)

(B) "His travels proved his theory: if you eat your native ethnic diet in an unprocessed form, you will have good mental, physical, and dental health. If you eat highly processed foods, which adds questionable agents and removes essential nutrients, your health deteriorates." (From the comment of Patrick Quillin, former Vice President of Nutrition for Cancer Treatments Centers of America, to the 2011 edition of Price's book) (Price, 1939, p. XXXVII).

In 1950, in a review, it was said that "Dr. Price might well be called *The Charles Darwin of Nutrition*" (Jones et al., 1950).

Another book, presented in 1983, among other topics, comprises many ideas and arguments that surely are in the range of evolutionary medicine. It was written in

Italian, was largely ignored by the academic world, and only in 2011 was it translated into English (Libertini, 1983). In this book, *Chap. V—Evolutionism and Pathology* is dedicated to evolutionary or Darwinian medicine, although this term is not used, as it will be proposed a few years later (Nesse & Williams, 1994; Williams & Nesse, 1991).

In the definition of the subject of the chapter, it is said:

> "This work is based, among other things, on the concept - already stressed by others - that evolution is the most general theory of all biology …
>
> With such an approach, the question arises of whether the pathology can be formulated in its outline in evolutionary terms. In other words, if the disease is an anomaly, an exception or rather a phenomenon that is an integral part of the evolutionary process. …
>
> From an evolutionary point of view, diseases are not something that breaks out of the mold but are, rather, a whole series of categories of phenomena which are evolutionarily 'predictable' in their general essence. … The evolutionary approach to the concept of disease is the most rational and general one possible. Any other more limited approach, even one which is more useful as regards a single pathological problem, simply because it is more limited and selectively oriented, should not be conceived in terms that run contrary to the theory of evolution. …
>
> The evolutionary approach to the concept of disease spontaneously raises suggestions and basic questions about the prevention and cure of the various categories of diseases." (Libertini, 1983, p. 83 in the English edition)

In the subsequent discussion, it was precisely indicated that the "starting point" is the ecological niche to which the species is adapted (a condition above better defined as "normality in evolutionary terms"). The discussion continued by underlining that, for a species, both the genotype and the set of relationships of genotype with the ecological niche to which the species is adapted constitute "a highly ordered structure" and that "- see Appendix 5 (The natural trend toward disorder increase) - in an ordered system the entropy always tends to increase" (Libertini, 1983, p. 84).

From this, it follows that: random modifications of (i) the genotype or (ii) the ecological niche cause, as the most probable outcomes, "Diseases deriving from alterations of the genotype" and "Diseases deriving from alterations of the ecological niche", respectively (Libertini, 1983, pp. 84–89).

About the diseases deriving from alterations of the ecological niche, it was precisely said that:

> "However, considering that the totality of the individual-ecological niche relations is a highly ordered structure, reminding again of that which is expounded in Appendix 5, an easy prediction is that a random modification of the ecological niche, given that it reduces the order of the system, alters, in most cases, the equilibrium between species and ecological niche - read: adaptation -, i.e., it entails a lesser aptitude for persistence in the individuals of the species. Lesser fitness means, by definition, damage or the possibility of damage for the individuals of the species. …
>
> The human species … provides significant examples of great modifications in the ecological niche that have led to, either by themselves or in concomitance with other factors, the outbreak of real epidemics. I could mention:
>
> a) Smoking and lung cancer and chronic bronchitis;
>
> b) The high calorie diet and atherosclerotic disease and diabetes mellitus type II;
>
> c) Diets low in vegetable waste and constipation, haemorrhoids, rhagades, anal fistulas, diverticulosis of the colon and, possibly, cancer of the rectum;

> d) The gathering of the population in large urban areas and the tremendous infectious epidemics of the pre-industrial era (and the less dramatic ones of modern times);
> e) The stress of urban and 'civilized' life and mental and psychosomatic diseases;
> f) The intake of and contact with drugs, industrial chemical substances, etc. and related pathologies." (Libertini, 1983, p. 87),

This is a description of the concept of "mismatch", although the term is not used, and then a short list of alterations of the ecological niche that cause diseases.

Consequently, on the rational basis of evolutionism, also considering other arguments about the relationships between the host and its parasites, an evolutionary classification of the disease was proposed:

> "From this formulation, indeed, as spontaneous, natural, and empirically confirmable facts, certain categories of events will arise, each with its own distinct definition, but which can be covered by a single, overall definition under the term 'disease'…
> Diseases deriving from alterations of the genotype...
> Diseases deriving from alterations of the ecological niche…
> Diseases deriving from the relations with other living beings…
> Diseases deriving from 'excesses of the ecological niche'…" (Libertini, 1983, pp. 84–89)

The book, therefore, clearly prefigured ideas and arguments that would later be integral parts of evolutionary medicine.

Moreover, in addition to this, the work addressed a theme, the interpretation of aging as an adaptive and programmed phenomenon, which still today is not part of the topics considered in the range of evolutionary medicine. Aging will be discussed more fully in *Chap. 8—Physiological phenomena that cause troubles and suffering or even death—Part III: Slow phenoptosis or aging*, but in this section, with regard to the origins of evolutionary medicine, a mention is necessary.

Summarizing what is detailed in a recent book already mentioned (Libertini et al., 2021a), we have that:

(1) Aging in its nineteenth-century interpretations, and in those of the first part of the twentieth century, and also in many subsequent interpretations, is explained as a phenomenon in which one or more factors determine cumulative damage. For this large group of hypotheses, definable as "damage accumulation hypotheses" [see (Libertini et al., 2021a, Chap. 1—Introduction) for the references]: (i) any possible influence of evolution is disregarded; (ii) in any theory, the factor hypothesized as the cause of aging is considered as inevitable; and (iii) the idea that aging could be somehow beneficial and so the result of evolution is neglected, as implicitly inconceivable;
(2) After Medawar's proposals (Medawar, 1952), the possible role of natural selection in aging began to be evaluated. In the years up to 1983, three hypotheses were proposed (and they have also been the subject of subsequent investigations): mutation accumulation hypothesis (Edney & Gill, 1968; Hamilton, 1966; Medawar, 1952), antagonistic pleiotropy hypothesis (Williams, 1957), and disposable soma hypothesis (Kirkwood, 1977; Kirkwood & Holliday, 1979). For these theories, aging is caused by the cumulative effects of detrimental genes that are insufficiently eliminated because natural selection becomes

> weaker at older ages (as the number of surviving individuals decreases in these ages) or due to conflicting needs (e.g., genes that are beneficial at younger ages and disadvantageous at older ages). In some way, these hypotheses tried to account for natural selection but shared with the previous group of damage accumulation hypotheses the idea that aging was unquestionably harmful. So, natural selection was considered only as acting to oppose aging, and the possibility that natural selection could favor aging was implicitly excluded. This speculation meant to exclude that aging could be a physiological function of the organism, i.e., a phenomenon that was adaptive, programmed, and regulated by distinct functions of specific genes. Consequently, aging was seen as the cumulative manifestation of a plurality of functional disorders determined by genes insufficiently eliminated by natural selection.

The book proposed an idea opposite to these previous theories, i.e., that aging in particular conditions was advantageous in terms of kin selection, a type of well-known supra-individual selection. So, aging should be an adaptive and programmed function determined and regulated by specific genes under opposed selective pressures.

This thesis was deepened in a subsequent article published in 1988 (Libertini, 1988).

However, we are now close to the years in which evolutionary medicine is traditionally said to have been born.

The article proposed by Williams and Nesse in 1991 (Williams & Nesse, 1991) and the book by the same authors in 1994 (Nesse & Williams, 1994) are considered the official birth of what the authors define as Darwinian medicine.

The article highlighted that the mechanism of evolution by natural selection "has recently gained new power to explain many aspects of disease". So, it should be widely used to better understand and explain diseases. Then, some issues that could be deepened with the help of Darwinian medicine were discussed. In the abstract:

> 1.... Signs and symptoms of the host-parasite contest can be categorized according to whether they represent adaptations or costs for host or parasite....
>
> 2.... Plant-herbivore contests may often underlie chemical injury from the defensive secondary compounds of plant tissues. Nausea in pregnancy, and allergy, may be adaptations against such toxins.
>
> 3.... Common genetic diseases often result from genes maintained by other beneficial effects in historically normal environments. The diseases of aging are especially likely to be associated with early benefits.
>
> 4. Abnormal environments. Human biology is designed for Stone Age conditions. Modern environments may cause many diseases - for example, deficiency syndromes such as scurvy and rickets, the effects of excess consumption of normally scarce nutrients such as fat and salt, developmental diseases such as myopia, and psychological reactions to novel environments." (Williams & Nesse, 1991)

The conclusion was that: "The substantial benefits of evolutionary studies of disease will be realized only if they become central to medical curricula, an advance that may at first require the establishment of one or more research centers dedicated to the further development of Darwinian medicine." (Williams & Nesse, 1991).

In the book, many other objects of study where this approach could be useful or interesting were highlighted. For example: "maladaptive human design features" (p. 131) already highlighted in the book by George Estabrooks (1941), e.g., problems deriving from the crossing between respiratory and digestive tracts, which is a consequence of the evolutionary history of these systems (pp. 123–127); problems deriving from the fact that, in the eye, blood vessels and nerves are placed in front of the retina instead of the back of it as in the squid and other mollusks ("inverted retina"), also determined by the evolutionary history of the eye (pp. 127–129).

In the following years, there has been a proliferation of papers and books that followed the path indicated by Williams and Nesse's work above. Among the works written by several authors, it is useful and necessary to mention the remarkable books edited by Stearns (1999), Stearns and Koella (2008), Trevathan et al. (1999) and (2008b).

In these years, the definition of "Darwinian medicine" was largely replaced by "Evolutionary medicine". E.g., the two books above with Trevathan as the first editor, have the titles "Evolutionary Medicine" and "Evolutionary Medicine and Health: New Perspectives", respectively.

In general, the number of the results of a Pubmed search for articles containing the strings: "Darwinian medicine", "evolutionary medicine", "Darwin" AND "medicine", "evolution" AND "medicine", is shown in Fig. 1.16 and Table 1.5.

These results show that:

(i) the definition "evolutionary medicine" is largely prevalent compared to "Darwinian medicine";
(ii) already before 1991 and 1994, there were articles that could somehow be dedicated both to the theme of evolution and to that of medicine;
(iii) the number of articles dedicated in some way to the aforementioned topics exploded from 1994 onwards.

About the topics discussed in 1983 Libertini's book, in the period after 1994:

(1) the concept of phenoptosis, also including aging, was proposed (Skulachev, 1997), and defined as the "programmed death of an organism" (Skulachev, 1999). In particular, aging was defined as a form of "slow phenoptosis" (Skulachev, 2002);
(2) the concept of phenoptosis was then extended by the following definition: "Phenoptosis is the death of an individual caused by its own actions or by actions of close relatives (siblicide; in particular, the parent-caused death of an offspring or filial infanticide) and not caused primarily by accidents or diseases or external factors, which is determined, regulated or influenced by genes favoured by natural selection." (Libertini, 2012) and a classification of phenoptotic phenomena was proposed in the same paper;
(3) works were published that investigated: (i) the "why" of aging as a programmed phenomenon in evolutionary terms (Libertini, 2006, 2008, 2015; Libertini et al., 2017), (ii) the "how" of aging, i.e., the specific mechanisms, genetically determined and regulated, by which the organism ages (Libertini, 2009a, 2014,

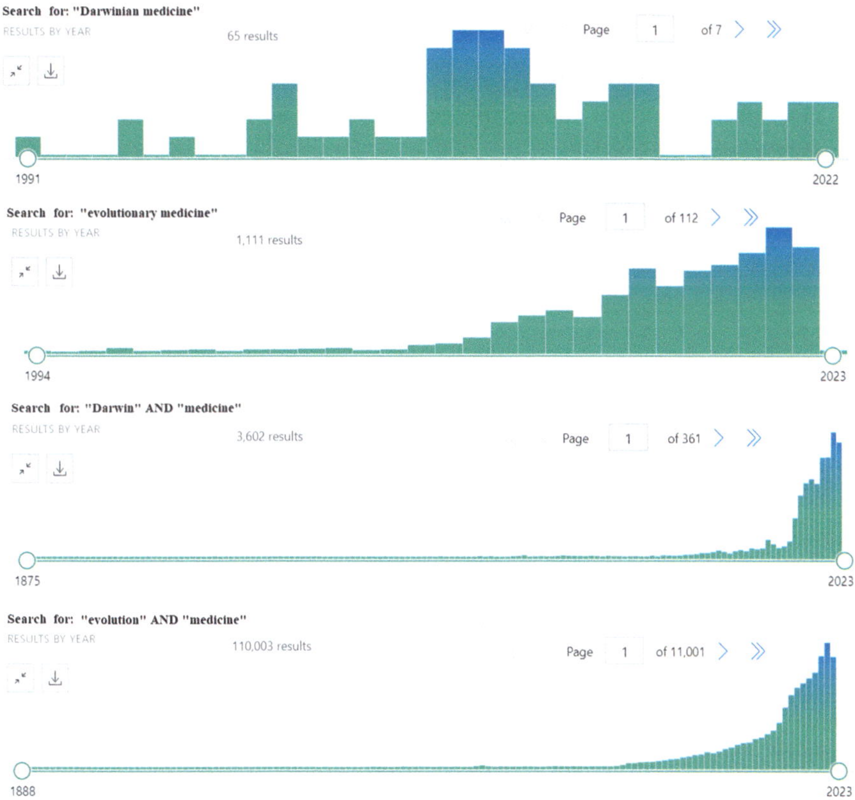

Fig. 1.16 PubMed search results (accessed on December 22, 2022) for various strings and combination of strings

Table 1.5 PubMed search results

Search strings	Period 1875–1990	Period 1991–1993	Period 1994–	Total
"Darwinian medicine"	–	1	64	65
"Evolutionary medicine"	–	–	1.111	1.111
"Darwin" AND "medicine"	77	14	3.511	3.602
"Evolution" AND "medicine"	3.521	1.540	104.945	110.006

2018; Libertini & Ferrara, 2016a; Libertini et al., 2020; Libertini, Shubernetskaya, et al., 2021b; Libertini et al., 2022), and also (iii) the relationships of aging with various widespread diseases (Libertini et al., 2019);

(4) in a 2009 work (Libertini, 2009b), a revision of the evolutionary classification of diseases published in 1983 was proposed, enriching it with concepts taken from the works mentioned above and from others, but mainly incorporating aging in a rational way among the phenomena that afterward will be classified under the concept of phenoptosis;

(5) These concepts were deepened in two recent works (Libertini et al., 2022; Libertini et al., 2021a). In the former, the 2009 evolutionary classification of diseases was updated by integrating phenoptotic phenomena, i.e., aging and other phenomena that cause suffering and death and fall within the unifying concept of phenoptosis;

(6) Moreover, in the same book, it was stated that: "... the transformation of the entire medicine in evolutionary medicine is not an unfounded claim based on a one-sided approach but rather a rational need, given huge foreseeable benefits." (Libertini et al., 2021a, p. 113). This objective of a radical redefinition of the whole of medicine appears different from that of having a new discipline, evolutionary medicine, at the center of the curricula of medical (Williams & Nesse, 1991), but in fact in an ancillary position and limited to particular topics.

In the last few years, we have two books that try to frame medical branches in terms of evolutionary medicine. The first concerns psychiatry (Nesse, 2019), while the second considers gerontology (Libertini et al., 2021a).

References

Abeles, S. R., & Pride, D. T. (2014). Molecular bases and role of viruses in the human microbiome. *Journal of Molecular Biology, 426*(23), 3892–3906. https://doi.org/10.1016/j.jmb.2014.07.002

Adhikary, D., Barman, S., Ranjan, R., & Stone, H. (2022). A systematic review of major cardiovascular risk factors: A growing Global Health concern. *Cureus, 14*(10), e30119. https://doi.org/10.7759/cureus.30119

Allison, A. C. (1954a). The distribution of the sickle-cell trait in East Africa and elsewhere, and its apparent relationship to the incidence of subtertian malaria. *Transactions of the Royal Society of Tropical Medicine and Hygiene, 48*(4), 312–318. https://doi.org/10.1016/0035-9203(54)90101-7

Allison, A. C. (1954b). Protection afforded by sickle-cell trait against subtertian malareal infection. *British Medical Journal, 1*(4857), 290–294. https://doi.org/10.1136/bmj.1.4857.290

Armitage, P., Berry, G., & Matthews, J. B. S. (2002). *Statistical methods in medical research* (4th ed.). Blackwell Science. ISBN 978-8126564811.

Baedke, J., Fábregas-Tejeda, A., & Nieves Delgado, A. (2020). The holobiont concept before Margulis. *Journal of Experimental Zoology—Part B (Molecular and Developmental Evolution), 334*(3), 149–155. https://doi.org/10.1002/jez.b.22931

Boerner, B. P., & Sarvetnick, N. E. (2011). Type 1 diabetes: Role of intestinal microbiome in humans and mice. *Annals of the New York Academy of Sciences, 1243*, 103–118. https://doi.org/10.1111/j.1749-6632.2011.06340.x

Burki, F., Roger, A. J., Brown, M. W., & Simpson, A. G. B. (2020). The new tree of eukaryotes. *Trends in Ecology & Evolution, 35*(1), 43–55. https://doi.org/10.1016/j.tree.2019.08.008

Castillo-Álvarez, F., & Marzo-Sola, M. E. (2019). Disease of the holobiont, the example of multiple sclerosis. *Medicina Clínica (Barcelona), 152*(4), 147–153. [English, Spanish].. https://doi.org/10.1016/j.medcli.2018.08.019

Chaker, L., Korevaar, T. I. M., Rizopoulos, D., Collet, T.-H., Völzke, H., Hofman, A., et al. (2017). Defining optimal health range for thyroid function based on the risk of cardiovascular disease. *The Journal of Clinical Endocrinology and Metabolism, 102*(8), 2853–2861. https://doi.org/10.1210/jc.2017-00410

Charbonneau, M. R., Blanton, L. V., DiGiulio, D. B., Relman, D. A., Lebrilla, C. B., Mills, D. A., & Gordon, J. I. (2016). A microbial perspective of human developmental biology. *Nature, 535*(7610), 48–55. https://doi.org/10.1038/nature18845

Chen, W., & Plewig, G. (2015). Are Demodex mites principal, conspirator, accomplice, witness or bystander in the cause of rosacea? *American Journal of Clinical Dermatology, 16*(2), 67–72. https://doi.org/10.1007/s40257-015-0115-y

Darwin, E. C. (1794–1796). *Zoonomia, or the laws of organic life.* Printed for J. Johnson, 2nd corrected ed., London. Available from Project Gutenberg, https://www.gutenberg.org/files/15707/15707-h/15707-h.htm and https://www.gutenberg.org/files/27600/27600-h/27600-h.htm

Darwin, C. R. (1859). *On the origin of species by means of natural selection, or the preservation of the favoured races in the struggle for life*. John Murray.

Darwin, C. R., & Wallace, A. R. (1858). On the tendency of species to form varieties; and on the perpetuation of varieties and species by natural means of selection. *Journal of the Proceedings of the Linnean Society, Zoology, 3*, 46–63.

Deo, P. N., & Deshmukh, R. (2019). Oral microbiome: Unveiling the fundamentals. *Journal of Oral and Maxillofacial Pathology, 23*, 122–128. https://doi.org/10.4103/jomfp.JOMFP_77_18

Dobzhansky, T. (1973). Nothing in biology makes sense except in the light of evolution. *The American Biology Teacher, 35*(3), 125–129.

Dugdale, J. S. (1996). *Entropy and its physical meaning* (2nd ed.). Taylor and Francis. ISBN 978-0748405682.

Dunn, J. C., Turner, H. C., Tun, A., & Anderson, R. M. (2016). Epidemiological surveys of, and research on, soil-transmitted helminths in Southeast Asia: A systematic review. *Parasites & Vectors, 9*, 31. https://doi.org/10.1186/s13071-016-1310-2

Eaton, S. B., Shostak, M., & Konner, M. (1988a). *The paleolithic prescription: A program of diet and exercise and a design for living*. Harper and Row. ISBN 978-0060916354.

Eaton, S. B., Konner, M., & Shostak, M. (1988b). Stone agers in the fast lane: Chronic degenerative diseases in evolutionary perspective. *The American Journal of Medicine, 84*(4), 739–749. https://doi.org/10.1016/0002-9343(88)90113-1

Edney, E. B., & Gill, R. W. (1968). Evolution of senescence and specific longevity. *Nature, 220*(5164), 281–282. https://doi.org/10.1038/220281a0

Estabrooks, G. (1941). *Man, the mechanical misfit*. Macmillan.

Ewald, P. W. (1980). Evolutionary biology and the treatment of signs and symptoms of infectious disease. *Journal of Theoretical Biology, 86*(1), 169–176. https://doi.org/10.1016/0022-5193(80)90073-9

Fiuza, B. S. D., Fonseca, H. F., Meirelles, P. M., Marques, C. R., da Silva, T. M., & Figueiredo, C. A. (2021). Understanding asthma and allergies by the lens of biodiversity and epigenetic changes. *Frontiers in Immunology, 12*, 623737. https://doi.org/10.3389/fimmu.2021.623737

Gazzinelli-Guimaraes, P. H., & Nutman, T. B. (2018). Helminth parasites and immune regulation. *F1000Res, 7*.:F1000 Faculty Rev-1685. https://doi.org/10.12688/f1000research.15596.1

Ghose, C., Ly, M., Schwanemann, L. K., Shin, J. H., Atab, K., Barr, J. J., et al. (2019). The Virome of cerebrospinal fluid: Viruses where we once thought there were none. *Frontiers in Microbiology, 10*, 2061. https://doi.org/10.3389/fmicb.2019.02061

Glendinning, L., & Free, A. (2014). Supra-organismal interactions in the human intestine. *Frontiers in Cellular and Infection Microbiology, 4*, 47. https://doi.org/10.3389/fcimb.2014.00047

Grice, E. A., & Segre, J. A. (2012). The human microbiome: Our second genome. *Annual Review of Genomics and Human Genetics, 13*, 151–170. https://doi.org/10.1146/annurev-genom-090711-163814

Haldane, J. B. S. (1949). Disease and evolution. *La Ricerca Scientifica, 19*, S68–S76.

Hamilton, W. D. (1966). The moulding of senescence by natural selection. *Journal of Theoretical Biology, 12*(1), 12–45. https://doi.org/10.1016/0022-5193(66)90184-6

Hassler, D. M. (1973). *Erasmus Darwin*. Twayne Publishers, Inc.

Hill, K., & Hurtado, A. M. (1996). *Ache life history*. Aldine De Gruyter. ISBN 978-0202020363.

Hou, K., Wu, Z. X., Chen, X. Y., Wang, J. Q., Zhang, D., Xiao, C., et al. (2022). Microbiota in health and diseases. *Signal Transduction and Targeted Therapy, 7*(1), 135. https://doi.org/10.1038/s41392-022-00974-4

Jaminet, P., & Jaminet, S. C. (2012). *Perfect health diet*. Simon and Schuster. ISBN 978-1451699142.

Jarvis, W. T. (1981). The myth of the healthy savage. *Nutrition Today, 16*(2), 14–15., 18, 21-22. https://doi.org/10.1097/00017285-198103000-00004

Jones, I. H., Muckleston, H. S., Lewis, E. R., & Owen, G. R. (1950). Nutrition and the eye, ear, nose and throat (with excerpts from the literature). *The Laryngoscope, 60*(12), 1210–1226.

Kapli, P., Yang, Z., & Telford, M. J. (2020). Phylogenetic tree building in the genomic age. *Nature Reviews. Genetics, 21*(7), 428–444. https://doi.org/10.1038/s41576-020-0233-0

Kirkwood, T. B. (1977). Evolution of ageing. *Nature, 270*(5635), 301–304. https://doi.org/10.1038/270301a0

Kirkwood, T. B., & Holliday, R. (1979). The evolution of ageing and longevity. *Proceedings of the Royal Society of London. Series B, Biological sciences, 205*(1161), 531–546. https://doi.org/10.1098/rspb.1979.0083

Klimis-Zacas, D., & Wolinsky, I. (2005). *Nutritional concerns of women*. CRC Press. ISBN 978-0849313376.

Krogman, W. M. (1951). The scars of human evolution. *Scientific American, 185*(6), 54–57. https://www.jstor.org/stable/24950554

LeBlanc, J. G., Milani, C., de Giori, G. S., Sesma, F., van Sinderen, D., & Ventura, M. (2013). Bacteria as vitamin suppliers to their host: A gut microbiota perspective. *Current Opinion in Biotechnology, 24*(2), 160–168. https://doi.org/10.1016/j.copbio.2012.08.005

Lewis, S. (2008). Evolution at the intersection of biology and medicine. In W. R. Trevathan, E. O. Smith, & J. J. McKenna (Eds.), *Evolutionary medicine and health: New perspectives* (pp. 399–415). Oxford University Press. ISBN 978-0195307061.

Liang, G., & Bushman, F. D. (2021). The human virome: Assembly, composition and host interactions. *Nature Reviews. Microbiology, 19*(8), 514–527. https://doi.org/10.1038/s41579-021-00536-5

Libertini, G. (1983). *Ragionamenti evoluzionistici [book in Italian]*. Società Editrice Napoletana. English edition (2011): *Evolutionary arguments on aging, disease, and other topics*. Crownsville (MD): Azinet Press. ISBN 978-0978870928.

Libertini, G. (1988). An adaptive theory of the increasing mortality with increasing chronological age in populations in the wild. *Journal of Theoretical Biology, 132*(2), 145–162. https://doi.org/10.1016/s0022-5193(88)80153-x

Libertini, G. (2006). Evolutionary explanations of the "actuarial senescence in the wild" and of the "state of senility". *TheScientificWorldJOURNAL, 6*, 1086–1108. https://doi.org/10.1100/tsw.2006.209

Libertini, G. (2008). Empirical evidence for various evolutionary hypotheses on species demonstrating increasing mortality with increasing chronological age in the wild. *TheScientificWorldJOURNAL, 8*, 183–193. https://doi.org/10.1100/tsw.2008.36

Libertini, G. (2009a). The role of telomere-telomerase system in age-related fitness decline, a tameable process. In L. Mancini (Ed.), *Telomeres: Function, shortening and lengthening* (pp. 77–132). Nova Science Publ. ISBN 978-1606923504.

Libertini, G. (2009b). Prospects of a longer life span beyond the beneficial effects of a healthy lifestyle. In J. V. Bentely & M. A. Keller (Eds.), *Handbook on longevity: Genetics, diet and disease* (pp. 35–95). Nova Science Publ. ISBN 978-1607410751.

Libertini, G. (2012). Classification of Phenoptotic phenomena. *Biochemistry (Moscow), 77*(7), 707–715. https://doi.org/10.1134/S0006297912070024

Libertini, G. (2014). Programmed aging paradigm: How we get old. *Biochemistry (Moscow), 79*(10), 1004–1016. https://doi.org/10.1134/S0006297914100034

Libertini, G. (2015). Non-programmed versus programmed aging paradigm. *Current Aging Science, 8*(1), 56–68. https://doi.org/10.2174/1874609808666150422111623

Libertini, G. (2018). Programmed aging paradigm and aging of perennial neurons. In S. I. Ahmad (Ed.), *Aging. Exploring a complex phenomenon* (pp. 91–116). CRC Press - Taylor & Francis Group. ISBN 978-1138196971.

Libertini, G., & Ferrara, N. (2016). Aging of perennial cells and organ parts according to the programmed aging paradigm. *Age (Dordrecht, Netherlands), 38*(2), 35. https://doi.org/10.1007/s11357-016-9895-0

Libertini, G., Rengo, G., & Ferrara, N. (2017). Aging and aging theories. *Journal of Gerontology and Geriatrics, 65*(1), 59–77. http://www.jgerontology-geriatrics.com/article/view/158/188

Libertini, G., Corbi, G., Cellurale, M., & Ferrara, N. (2019). Age-related dysfunctions: Evidence and relationship with some risk factors and protective drugs. *Biochemistry (Moscow), 84*(12), 1442–1450. https://doi.org/10.1134/S0006297919120034

Libertini, G., Corbi, G., & Ferrara, N. (2020). Importance and meaning of TERRA sequences for aging mechanisms. *Biochemistry (Moscow), 85*, 1505–1517. https://doi.org/10.1134/S0006297920120044

Libertini, G., Corbi, G., Conti, V., Shubernetskaya, O., & Ferrara, N. (2021a). Evolutionary gerontology and geriatrics: Why and how we age. In *Advances in studies of aging and health* (Vol. 2). Cham. ISBN 978-3030737740.

Libertini, G., Shubernetskaya, O., Corbi, G., & Ferrara, N. (2021b). Is evidence supporting the subtelomere-telomere theory of aging? *Biochemistry (Moscow), 86*(12), 1526–1539. https://doi.org/10.1134/S0006297921120026

Libertini, G., Corbi, G., Shubernetskaya, O., & Ferrara, N. (2022). Is human aging a form of Phenoptosis? *Biochemistry (Moscow), 87*(12-13), 1446–1464. https://doi.org/10.1134/S0006297922120033

Litwin, D., Chen, W., Dzika, E., & Korycińska, J. (2017). Human permanent Ectoparasites; recent advances on biology and clinical significance of Demodex mites: Narrative review article. *Iranian Journal of Parasitology, 2*(1), 12–21. PMCID: PMC5522688.

Livingstone, F. B. (1958). Anthropological implications of sickle cell gene distribution in West Africa. *American Anthropologist, 60*(3), 533–562. https://doi.org/10.1525/aa.1958.60.3.02a00110

Margulis, L., & Fester, R. (1991). *Symbiosis as a source of evolutionary innovation*. MIT Press. ISBN 9780262132695.

McGrady, A., & Moss, D. (2013). *Pathways to illness, pathways to health*. Springer. ISBN 978-1441913784.

Medawar, P. B. (1952). *An unsolved problem in biology. London: H. K. Lewis. Reprinted in: Medawar, P. B. (1957). The uniqueness of the individual*. Methuen. ISBN 978-0367281236 (for the 2021 edition).

Mishra, P. K., Palma, M., Bleich, D., Loke, P., & Gause, W. C. (2014). Systemic impact of intestinal helminth infections. *Mucosal Immunology, 7*(4), 753–762. https://doi.org/10.1038/mi.2014.23

Morton, D. J. (1926). The relation of evolution to medicine. *Science, 64*(1660), 394–396. https://doi.org/10.1126/science.64.1660.394

Moustafa, A., Xie, C., Kirkness, E., Biggs, W., Wong, E., Turpaz, Y., et al. (2017). The blood DNA virome in 8,000 humans. *PLoS Pathogens, 13*, e1006292. https://doi.org/10.1371/journal.ppat.1006292

Nesse, R. M. (2008). The importance of evolution for medicine. In W. R. Trevathan, E. O. Smith, & J. J. McKenna (Eds.), *Evolutionary medicine and health: New perspectives* (pp. 416–431). Oxford University Press. ISBN 978-0195307061.

Nesse, R. M. (2019). *Good reasons for bad feelings: Insights from the frontier of evolutionary psychiatry*. Penguin. ISBN 978-1101985663.

Nesse, R. M., & Williams, G. C. (1994). *Why we get sick: The new science of Darwinian medicine*. Times Books. ISBN 978-0812922247.

New York Times. (1948, Jan 24). "*Weston A. Price*".

Pannaraj, P. S., Ly, M., Cerini, C., Saavedra, M., Aldrovandi, G. M., Saboory, A. A., et al. (2018). Shared and distinct features of human milk and infant stool viromes. *Frontiers in Microbiology, 9*, 1162. https://doi.org/10.3389/fmicb.2018.01162

Patané, J. S. L., Martins, J., Jr., & Setubal, J. C. (2018). Phylogenomics. *Methods in Molecular Biology, 1704*, 103–187. https://doi.org/10.1007/978-1-4939-7463-4_5

Price, W. A. (1939). *Nutrition and physical degeneration*. Paul B. Hoeber. [Other editions in 1945, 1970, 1989, 1997, 2004, 2006, 2008, 2011; the last edition – ISBN 978-0916764203 - has been used].

Pride, D. T., Salzman, J., Haynes, M., Rohwer, F., Davis-Long, C., & White, R. A. (2012). Evidence of a robust resident bacteriophage population revealed through analysis of the human salivary virome. *The ISME Journal, 6*, 915–926. https://doi.org/10.1038/ismej.2011.169

Sánchez-Cañizares, C., Jorrín, B., Poole, P. S., & Tkacz, A. (2017). Understanding the holobiont: The interdependence of plants and their microbiome. *Current Opinion in Microbiology, 38*, 188–196. https://doi.org/10.1016/j.mib.2017.07.001

Sender, R., Fuchs, S., & Milo, R. (2016). Revised estimates for the number of human and bacteria cells in the body. *PLoS Biology, 14*(8), e1002533. https://doi.org/10.1371/journal.pbio.1002533

Skulachev, V. P. (1997). Aging is a specific biological function rather than the result of a disorder in complex living systems: Biochemical evidence in support of Weismann's hypothesis. *Biochemistry (Moscow), 62*(11), 1191–1195. protein.bio.msu.ru/biokhimiya/contents/v62/full/62111394.html.

Skulachev, V. P. (1999). Phenoptosis: Programmed death of an organism. *Biochemistry (Moscow), 64*(12), 1418–1426. protein.bio.msu.ru/biokhimiya/contents/v64/full/64121679.html.

Skulachev, V. P. (2002). Programmed death phenomena: From organelle to organism. *Annals of the New York Academy of Sciences, 959*, 214–237. https://doi.org/10.1111/j.1749-6632.2002.tb02095.x

Stearns, S. C. (Ed.). (1999). *Evolution in health and disease* (1st ed.). Oxford University Press. ISBN 978-0198501102.

Stearns, S. C., & Koella, J. C. (Eds.). (2008). *Evolution in health and disease* (2nd ed.). Oxford University Press. ISBN 978-0199207459.

Theis, K. R., Dheilly, N. M., Klassen, J. L., Brucker, R. M., Baines, J. F., Bosch, T. C., & Cryan, J. F. (2016). Getting the Hologenome Concept Right: an Eco-Evolutionary Framework for Hosts and Their Microbiomes. *mSystems, 1*, e00028–e00016. https://doi.org/10.1128/mSystems.00028-16

Trevathan, W. R., McKenna, J. J., & Smith, E. O. (Eds.). (1999). *Evolutionary medicine*. Oxford University Press.

Trevathan, W. R., Smith, E. O., & McKenna, J. J. (2008a). Introduction and overview of evolutionary medicine. In W. R. Trevathan, E. O. Smith, & J. J. McKenna (Eds.), *Evolutionary medicine and health: New perspectives* (pp. 1–54). Oxford University Press. ISBN 978-0195307061.

Trevathan, W. R., Smith, E. O., & McKenna, J. J. (Eds.). (2008b). *Evolutionary medicine and health: New perspectives*. Oxford University Press. ISBN 978-0195307061.

Truswell, A. S., Kennelly, B. M., Hansen, J. D. L., & Lee, R. B. (1972). Blood pressure of !Kung bushmen in northern Botswana. *American Heart Journal, 84*(1), 5–12. https://doi.org/10.1016/0002-8703(72)90299-2

Urashima, T., Hirabayashi, J., Sato, S., & Kobata, A. (2018). Human Milk oligosaccharides as essential tools for basic and application studies on Galectins. *Trends in Glycoscience and Glycotechnology, 30*, SJ11–SJ24. https://doi.org/10.4052/tigg.1734.1SJ

Varki, A. (2012). Nothing in medicine makes sense, except in the light of evolution. *Journal of Molecular Medicine (Berlin, Germany), 90*(5), 481–494. https://doi.org/10.1007/s00109-012-0900-5

Veesenmeyer, A. F. (2022). Important nematodes in children. *Pediatric Clinics of North America, 69*(1), 129–139. https://doi.org/10.1016/j.pcl.2021.08.005

Vieira, S. M., Pagovich, O. E., & Kriegel, M. A. (2014). Diet, microbiota and autoimmune diseases. *Lupus, 23*(6), 518–526. https://doi.org/10.1177/0961203313501401

WHO. (2016). *ICD-10*, https://icd.who.int/browse10/2019/en#/. Accessed 15.12.22.

WHO. (2018). *ICD-11*, https://icd.who.int/browse11/l-m/en. Accessed 15.12.22.

Wildschutte, J. H., Williams, Z. H., Meagan Montesion, M., Subramian, R. P., Kidd, J. M., & Coffin, J. M. (2016). Discovery of unfixed endogenous retrovirus insertions in diverse human populations. *Proceedings of the National Academy of Sciences (PNAS), 113*(16), E2326–E2334. https://doi.org/10.1073/pnas.1602336113

Williams, G. C. (1957). Pleiotropy, natural selection and the evolution of senescence. *Evolution, 11*, 398–411. https://doi.org/10.2307/2406060

Williams, G. C., & Nesse, R. M. (1991). The dawn of Darwinian medicine. *The Quarterly Review of Biology, 66*(1), 1–22. https://doi.org/10.1086/417048

World Cause of Death. (2020). *Data from WHO*, www.worldlifeexpectancy.com/world-rankings-total-deaths. Accessed 15.12.22.

Wu, S., Tao, N., German, J. B., Grimm, R., & Lebrilla, C. B. (2010). Development of an annotated library of neutral human milk oligosaccharides. *Journal of Proteome Research, 9*(8), 4138–4151. https://doi.org/10.1021/pr100362f

Wu, S., Grimm, R., German, J. B., & Lebrilla, C. B. (2011). Annotation and structural analysis of sialylated human milk oligosaccharides. *Journal of Proteome Research, 10*(2), 856–868. https://doi.org/10.1021/pr101006u

Zhan, Q., Wang, R., Thakur, K., Feng, J. Y., Zhu, Y. Y., Zhang, J. G., & Wei, Z. J. (2022). Unveiling of dietary and gut-microbiota derived B vitamins: Metabolism patterns and their synergistic functions in gut-brain homeostasis. *Critical Reviews in Food Science and Nutrition, 22*, 1–13. https://doi.org/10.1080/10408398.2022.2138263

Chapter 2
Diseases Deriving from Alterations of the Genotype

Alterations of an organism's genotype (or genome) are caused by point modifications of the DNA, i.e., mutations, or even by massive alterations such as duplication or loss of a chromosome, duplication or loss or inversion of a segment of DNA, etc. For simplicity of discussion, we will discuss mainly about mutations, although in logical terms, the discussion will often refer to the other alterations as well.

2.1 Mutations and Other Alterations of the Genotype

For biological evolution to be possible, some fundamental facts are necessary and well-known:

1) The characteristics of a living being must somehow be determined by specific information, which is well known to be in DNA nucleotide sequence and often divided into several chromosomes;
2) This information must be preserved from alterations in the passage to successive generations, in cell duplication, and when the cell is not in division. So, there are specific mechanisms for the protection of information integrity (Carusillo & Mussolino, 2020; García-Rodríguez et al., 2018; Lanz et al., 2019);
3) These mechanisms are not perfect, and random or, to some degree, controlled alterations of genetic information occur (mutations and other alterations of the genotype);
4) As the genotype of a living being is a complex structure, a random modification of it, apart from the cases in which it has no effect (neutral mutations), in most cases will have a harmful effect and only in a minority of cases an advantageous effect (see Sect. 1.2.1 *Consequence of a random modification in a complex system*);
5) Mutations and their effects are the substrates that allow natural selection and, therefore, evolution.

There are two different ways of considering the mutations.

G. Libertini et al., *The Essential Revolution of Evolutionary Medicine*, Advances in Studies of Aging and Health 4, https://doi.org/10.1007/978-3-031-84877-3_2

- The first is that mutations, being generally a random phenomenon, are inevitable and intrinsic to the nature of the information that determines the characteristics of a living being. This situation does not contradict the fact that mutations are the indispensable substrate of natural selection. However, the concept of the biological inevitability of mutations is an essential starting point for the hypothesis that aging is unavoidable due to the accumulation of the effects of mutations acting at older ages and so insufficiently eliminated by natural selection ("mutation accumulation theory" (Medawar, 1952; Mueller, 1987).
- The second is that there are mostly harmful mutations and their onset is opposed by natural selection. At the same time, as mutations are necessary for natural selection and evolution, a higher frequency of mutations offers more possibilities for natural selection and evolution (s. Fig. 2.1). Therefore, being mutations both advantageous and disadvantageous, the mechanisms protecting the integrity of information and thus restraining the number of mutations are subjected to opposing selective pressures.

Following the second view, the thesis that the frequency of mutations overcoming the mechanisms that protect information integrity is regulated by natural selection might seem like a hypothesis without any evidence. However, a recent authoritative work (Cagan et al., 2022) demonstrated that, when comparing different species, there is an inverse correlation between lifespan and detectable mutation rates per year in somatic cells. In particular: "Despite widely different life histories among the species we examined - including a variation of around 30-fold in lifespan and around 40,000-fold in body mass - the somatic mutation burden at the end of

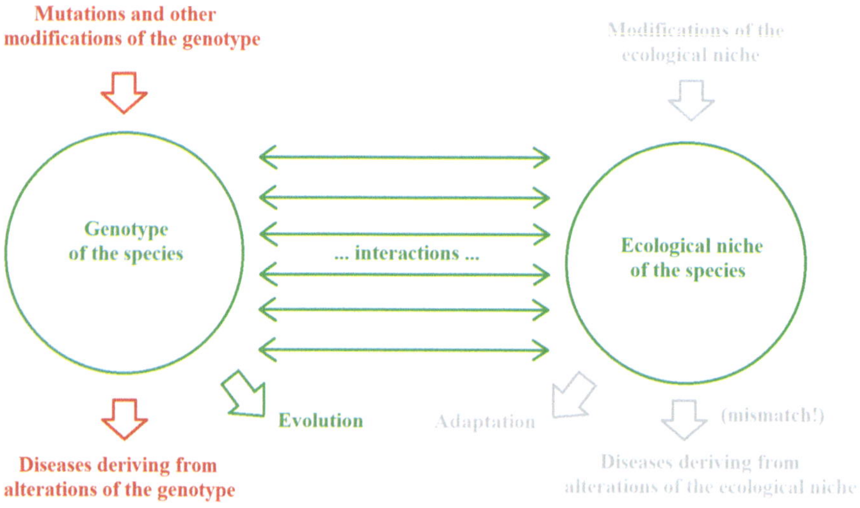

Fig. 2.1 In most cases, a random mutation (or other genotype modification) alters the genotype and causes a dysfunction (i.e., a disease). In rare cases, a mutation improves a function and/or increases fitness, providing the selection material for evolution. Evolution would be impossible without mutations

lifespan varied only by a factor of around 3. These data unveil common mutational processes across mammals, and suggest that somatic mutation rates are evolutionarily constrained..." (Cagan et al., 2022).

The greater restriction of somatic mutation rates in animals with greater longevity (and presumably also in those with greater body mass) would make it possible to avoid having excess cells with a critical level of cancer-causing mutations (Gorelick & Naxerova, 2022).

This consideration demonstrates that natural selection can perfectly regulate the number of mutations—"the somatic mutation burden" (Cagan et al., 2022)—through mechanisms that allow for lesser or greater preservation of the nucleotide sequence.

Let us now consider the entirely hypothetical case of a species in which the preservation mechanisms of the nucleotide sequence were perfect. For this species, there would be no individuals with dysfunctions due to genotype alterations. At the same time, the genetic variety on which natural selection could operate would be increasingly limited, and the species would face higher risks of becoming extinct eventually.

Indeed, the evolutionary ability to regulate the frequency of mutations not corrected by appropriate mechanisms is a case in which, in a species, selective mechanisms prove to be able to regulate the capability and speed of evolution, i.e., evolvability (Brookfield, 2009; Payne & Wagner, 2019; Pigliucci, 2008; Riederer et al., 2022).

As a parallel consideration, the fact that the mutation rate is influenced by natural selection is an element in contrast with the thesis of mutation accumulation as a cause of aging. In this regard it should be remembered that individuals with genetic defects causing higher mutation rates in somatic cells do not show accelerated aging (Robinson et al., 2021), and this "argues against a straightforward connection between the accumulation of mutations and aging" (Gorelick & Naxerova, 2022).

These preliminary considerations show that, within the process of natural selection and evolution, a part of the individuals of any species will suffer from alterations caused by mutations, although the number of mutations is restrained by natural selection, and that this load of individuals with functional alterations due to mutations is an indispensable price to allow evolution.

2.2 Calculation of the Frequencies Expected for Diseases Caused by Alterations of the Genotype

The considerations of the previous section are valid both in the case in which the species is haploid (i.e., there is only one copy of the nucleotide sequence) and when the species, like ours, is diploid (i.e., there are two copies of the nucleotide sequence).

For diploid species, a harmful mutation can be recessive or dominant. It is recessive when it carries out its harmful action only in the homozygous state (i.e., when it is present in both copies of the DNA molecule). It is dominant when it carries out its harmful action in both homozygous and heterozygous conditions (i.e., when it is present in one or both copies of the DNA molecule).

A third particular case, which is useful to consider, is when the mutation is harmful in the homozygous condition while it is advantageous in the heterozygous condition.

Now, let us consider a gene C′, which mutates to C with mutation frequency *v*, and suppose that gene C carries the disadvantage -*s*.

With this information, it is possible to calculate the equilibrium frequency of a mutation, i.e., the frequency C_e, where, at each generation, the number of new mutations is balanced by those eliminated by natural selection.

The development of the formulas that allow these calculations is detailed in "Appendix: Part A" in (Libertini et al., 2021, pp. 401–412), and it would be long and boring to repeat it here.

In short, in the first two cases mentioned above, we have:

- Recessive mutation

The equilibrium frequency of a recessive mutation is given by the formula—see also (Srb et al., 1965):

$$Ce = \sqrt{v/[s]} \tag{2.1}$$

Considering the Hardy-Weinberg formula (CC + 2 CC' + C′C′ = 1), the equilibrium frequency of the phenotype expressing the disadvantageous condition (P_e), i.e., the frequency of the homozygous condition of C (= CC = C_e^2), will be:

$$P_e = C_e^2 = v/[s] \tag{2.2}$$

- Dominant mutation

The frequency of a dominant mutation is given by the formula:

$$C_e = (1 - \sqrt{1 - 3v/[s]}/3 \approx (3v/[s])/(3+3) = 0.5v/[s] \tag{2.3}$$

Considering again the Hardy-Weinberg formula, as C is present both in CC and CC' [= C_e^2 + 2 C_e (1-C_e)], we have that:

$$P_e = C_e^2 + 2C_e(1 - C_e) = 2C_e - C_e^2 \approx 2(0.5v/[s]) - (0.5v/[s])^2 \approx v/[s] \tag{2.4}$$

Regarding genetic alterations other than mutations (i.e., changes in the number of chromosomes, losses/duplications/inversions of segments of the nucleotide sequence), it is useful to consider them as dominant mutations in which both C_e and P_e are given by the Eq. (2.2).

Figure 2.2, which is obtained from these formulas, shows that: (i) the equilibrium frequencies are related to the values of v and [*s*]; and (ii) with equal values of *v* and *s*, the phenotypic equilibrium frequencies are practically the same for the recessive and dominant mutation cases.

However, the situation immediately becomes more complex when it is considered that a gene can be altered by many different mutations. Furthermore, a metabolic pathway in which several proteins are involved is determined by many genes and, thus, can be altered by a proportionally more significant number of mutations.

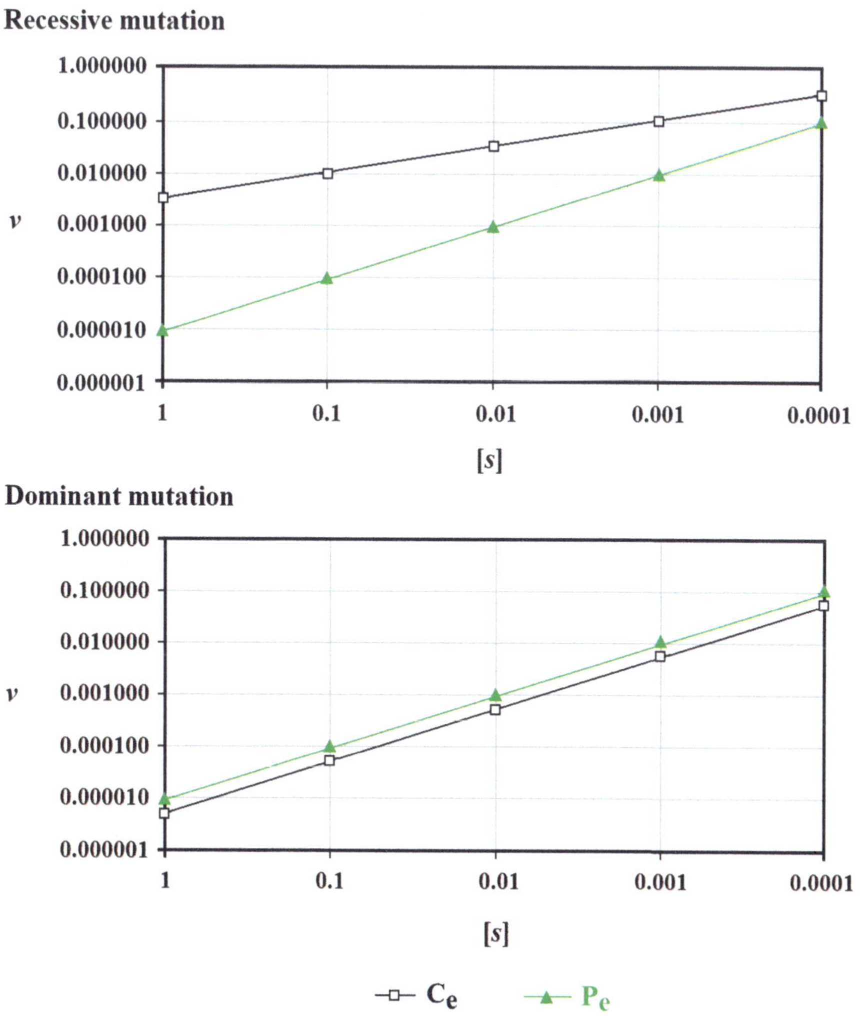

Fig. 2.2 *Top*: equilibrium frequencies of a recessive mutated gene C (C_e) and its phenotypic frequencies (P_e); *Bottom*: the same for a dominant mutated gene

Suppose that n mutations can alter the functionality of a gene or a metabolic pathway. In that case, the frequency of the phenotype in which the altered gene or metabolic pathway manifests itself will be the sum of the single equilibrium frequencies:

$$P_e = \sum_{x=1}^{n} \left(v_x / [s_x] \right) \tag{2.5}$$

If, as simplifying conditions, we set $v_x = v$; and $s_x = s$, we will have:

$$P_e = n\,v / [s] \tag{2.6}$$

A well-known example of a disease that can be caused by multiple types of mutations is cystic fibrosis (CF). For the altered protein whose harmful modification is at the origin of this disease (cystic fibrosis transmembrane conductance regulator): "there have been > 2000 CF mutations identified to date (http://www.genet.sickkids.on.ca/cftr/), over 360 are CF disease-causing variants (www.cftr2.org)." (Lee et al., 2021).

In the third case mentioned above, if C is disadvantageous in the homozygous condition ($s < 0$) and advantageous in the heterozygous condition ($s' > 0$), disregarding the mutation and reverse mutation frequencies (i.e., the mutations from C′ to C and vice versa), the equilibrium frequency will be given by the formula:

$$C_e = -\frac{2s'}{s - 4s'} = \frac{2s'}{[s] + 4s'} \tag{2.7}$$

In the previous chapter, the case of sickle cell anemia was reported. The genetic alteration of this case:

- in the homozygous state, causes a severe form of anemia;
- in the heterozygous state, there is a light form of anemia but also a resistance to malaria due to the prompt lysis of erythrocytes when infected with *Plasmodium falciparum*, the causative agent of malaria (Allison, 1954a, b; Livingstone, 1958).

Even for this third case, it is simplistic to think that a single type of mutation can cause a specific genetic alteration.

For example, another genetic alteration that also affects resistance to malaria is beta-thalassemia, characterized by a reduced or absent synthesis of well-working beta chains of hemoglobin. Beta-thalassemia is caused by various recessive mutations and cases of deletion of the beta-globin gene: "More than 200 disease-causing mutations have been so far identified. The majority of mutations are single nucleotide substitutions, deletions, or insertions of oligonucleotides leading to frameshift. Rarely, beta-thalassemia results from gross gene deletion." (Cao & Galanello, 2010).

Severe anemia manifests in the homozygous forms of such mutations or in any genetic alteration where the synthesis of beta chains is seriously compromised (thalassemia major). On the contrary, in the heterozygous form, or in any case where the defect in the synthesis of the beta chains of hemoglobin is minor, there is a mild asymptomatic microcytic anemia (thalassemia minor), which also confers certain protection against malaria. In cases where the impairment of beta chain synthesis is between those of minor and major thalassemia, there is a form of the disease of intermediate severity (thalassemia intermedia) (Origa, 2017).

The disease is widespread in populations around the Mediterranean (or among people who emigrated from these areas), particularly in coastal areas of Greece, Turkey, Cyprus, Italy (Sicily and Sardinia in particular), Malta, Corsica, North Africa, and Western Asia (Galanello & Origa, 2010), where once there were marshes that favored the spread of *Plasmodium falciparum* (Flint et al., 1998). Again, as with sickle cell anemia, opposing selective pressures either counteract or favor the spread of thalassemia genes.

In short, we can express these considerations as follows:

- The frequency of a specific dysfunction (i.e., disease) caused by one or more alterations of the genotype is directly proportional to the number and frequency of mutations and of other genetic alterations causing it and inversely proportional to the damage ($-s$) determined by the dysfunction;
- For a severe or fatal dysfunction ($-s = -1$), the equilibrium value of the phenotypic expression (P_e) will be small because v_x is, in general, very small. In practice, in the case of a dysfunction caused by n mutations, it will be:

$$Pe = \sum_{n}^{x=1} \left(v_x / 1\right) = \sum_{n}^{x=1} \left(v_x\right) \approx n\,v \tag{2.8}$$

- The equilibrium frequency can be considerable when the mutation involves an advantage in the heterozygous state or when its effect is partial. For example, the frequency of cases of beta-thalassemia carriers (thalassemia minor) reaches 14% in Cyprus and 10.3% in Sardinia (Flint et al., 1998).

However, these predictions about equilibrium frequencies refer to the frequencies at the time of conception of an individual. In the case of a genetic alteration incompatible with the viability of an embryo or a fetus, the embryo or the fetus will die before birth. So, the frequency of the genetic alteration at the moment of birth will be zero. Similarly, if a genetic alteration observable at birth in "normal" conditions is always fatal before adulthood, the equilibrium frequency will be zero in adulthood. Nevertheless, the equilibrium frequency will be as predicted in protected conditions where survival is allowed.

Now, let us consider that:

(i) the length of the human nucleotide sequence is about 3.2 billion base pairs, with about 20,000 protein-coding genes (International Human Genome Sequencing Consortium, 2004);
(ii) "Although only 2% of the genome is coding, >90% is transcribed. This transcriptional activity largely produces long noncoding RNAs (lncRNA), the functions of which have remained mostly unknown" (Chu et al., 2017). However, it is presumable that a substantial part of the non-protein-coding genome has essential regulatory functions on the overall genome functioning and protein-coding activity.

From this, it is immediately deducible that the number of genetic alterations that can alter the functioning of the genotype is enormous. So, they cause many dysfunctions (diseases) or syndromes (sets of pathological manifestations).

This general prediction is confirmed by the long list of known "rare diseases" (GARD (National Institutes of Health, National Center for Advancing Translational Sciences, Genetic and Rare Diseases Information Center), 2019):

> "What is a rare disease?
>
> The Orphan Drug Act defines a rare disease as a disease or condition that impacts less than 200,000 people in the U.S. [that is about 1 in 1650 people].
>
> Other countries may have their official definitions of a rare disease. For example, the European Union defines a disease as rare when it affects less than 1 in 2000 people.
>
> There are about 7000 known rare diseases and collectively, about 1 in 10 people (or 30 million people) in the U.S. have a rare disease." (GARD (National Institutes of Health, National Center for Advancing Translational Sciences, Genetic and Rare Diseases Information Center), 2019)

Not all rare diseases have a genetic origin. However, genetic alterations are the most frequent cause. Here are two examples:

> (1) 18 Hydroxylase deficiency
>
> Other Names: 18 alpha-hydroxylase deficiency; Aldosterone deficiency 1; Aldosterone deficiency due to defect in 18 hydroxylase; CMO 1 deficiency; Corticosterone methyloxidase type 1 deficiency
>
> Summary: CMO type I deficiency is an autosomal recessive disorder caused by a defect in the penultimate biochemical step of aldosterone biosynthesis, the 18-hydroxylation of corticosterone (B) to 18-hydroxycorticosterone (18-OHB). This enzymatic defect results in decreased aldosterone and salt-wasting. In CMO I deficiency, aldosterone is undetectable, whereas its immediate precursor, 18-OHB, is low or normal. These patients have an increased ratio of corticosterone to 18-OHB (Portrat-Doyen et al., 1998). The CYP11B2 gene product also catalyzes the final step in aldosterone biosynthesis: the 18-oxidation of 18-OHB to aldosterone. A defect in that enzymatic step results in CMO type II deficiency (610600), an allelic disorder with an overlapping phenotype but distinct biochemical features. In CMO II deficiency, aldosterone can be low or normal, but at the expense of increased secretion of 18-OHB. These patients have a low ratio of corticosterone to 18-OHB (Portrat-Doyen et al., 1998).

(2) 12q14 microdeletion syndrome

Other Names: Del(12)(q14); Deletion 12q14; Monosomy 12q14; Osteopoikilosis-short stature-intellectual disability syndrome.

Summary: 12q14 microdeletion syndrome is a genetic syndrome caused by a missing piece of chromosome 12. The signs and symptoms depend on the size of the missing piece and the genes involved, but generally include growth delay, short stature and feeding difficulties. When this syndrome is inherited, it is passed on in a dominant pattern. It can be diagnosed through a genetic test, either a chromosomal microarray or whole exome sequencing.

In many cases, the definition of a rare disease or syndrome likely includes the manifestations, more or less coincident, of multiple different genetic alterations, which are unknown. For example:

(1) Absent breasts and nipples

Other Names: Amastia; Complete absence of breasts

Summary: A rare breast malformation characterized by congenital absence of breast and nipple (amastia), or nipple or mammary gland (athelia or amazia, respectively). It can be unilateral or bilateral and may occur as an isolated malformation or be associated with a syndrome or cluster of other anomalies.

(2) Acromelic frontonasal dysostosis

Other Names: None

Summary: A rare frontonasal dysplasia characterized by distinct craniofacial (large fontanelle, hypertelorism, bifid nasal tip, nasal clefting, brachycephaly, median cleft face, carp-shaped mouth), brain (interhemispheric lipoma, agenesis of the corpus callosum), and limb (tibial hypoplasia/aplasia, club foot, symmetric preaxial polydactyly of the feet and bilateral clubbed and thickened nails of halluces) malformations as well as intellectual disability. Other manifestations sometimes reported include absent olfactory bulbs, hypopituitarism, and cryptorchidism.

Another database of rare diseases, without specific treatments due to their rarity ("orphan diseases"), reports "10,673 Clinical Entities" and "8,120 Disease gene relationships" (Orphanet, 2023).

This evidence confirms that the number of diseases and syndromes caused by genotype alterations is large and difficult to define precisely, although they surely affect a substantial part of the population (GARD (National Institutes of Health, National Center for Advancing Translational Sciences, Genetic and Rare Diseases Information Center), 2019).

At this point, describing the situation as much as possible in "normal" conditions, i.e., in the ecological niche to which our species is adapted, is appropriate. It largely coincides with the state of populations that lived or still live in primitive conditions.

It should be noted that describing such conditions does not mean an ideal description that satisfies the myth of a golden age, nor does it mean describing what happened or is happening under such conditions as a positive fact in itself.

As far as possible, the description must be objective and free from ethical judgments. Suggestions can be deduced from this description, which will then be evaluated and adopted in the forms and measures that will be deemed most appropriate after scrutiny based on the ethical standards that will be deemed appropriate.

Mutations and other genetic alterations (for simplicity, considered under the general definition of "mutations"), which, by definition, are harmful, cannot be neutral as regards the action of natural selection.

The first effect is that mutations not allowing correct embryo development can be fatal for the embryo from the first stages after conception, leading to its automatic loss.

In the modern age:

(i) Only a third, or even less, of conceptions reach the stage of live birth (Wang et al., 2003; Wilcox et al., 1988; Zinaman et al., 1996)
(ii) About 30% of conceptions are lost before implantation, i.e., in the first 2 weeks. Another 30% are lost after implantation and before the missed menstrual period, i.e., in the third or fourth week of gestation, and are defined as preclinical losses (Macklon et al., 2002).
(iii) Early clinical pregnancy loss (gestational weeks from 5 to 12) is estimated to be 15% of conceptions. Whereas loss incidence is 10% in women aged 20–24 years, it rises to 51% in women aged 40–44 years (Nybo Andersen et al., 2000). Late clinical pregnancy losses (gestational weeks from 12 to 21) are about 4% of pregnancy outcomes (Ugwumadu et al., 2003).

This situation has been described by the evaluation: "Human reproduction is characterized by its inefficiency." (Larsen et al., 2013). However, the authors do not appear to realize that this "inefficiency" is an enormous and valid defense against a large number of embryos or fetuses with some genetic defect.

In fact:

- in miscarriages, "... it has long been recognized that lethal whole chromosome abnormalities are very common in human embryos..." (Brosens et al., 2022)
- "Changes in parental (paternal and maternal) genetic material that contribute to the increased risk of miscarriage in successive pregnancies include karyotype abnormalities, recessive and dominant disease carrier status, as well as mutations in genes responsible for coagulation and the metabolism of folates. The presence of karyotype abnormalities in one of the parents is one of the most common of the known causes of recurrent miscarriages." (Kacprzak et al., 2016)

It is likely that this defense also in primitive conditions had similar effects on the number of suppressed embryos or fetuses and the effectiveness of the defense against dysfunctions caused by genetic alterations.

Other mutations do not alter the embryo so seriously, but the uterus has in some way the ability to recognize embryos with anomalies, and in such cases, spontaneous abortions occur. It is like the uterus has some way of measuring viability and recognizing embryos with anomalies, in such cases initiating spontaneous abortions. The uterus, especially the endometrium, is a likely "biosensor of embryo quality"

(Larsen et al., 2013) and when, in some way, the embryo or the fetus does not satisfy certain requirements, mechanisms are activated which determine their loss.

After birth, there is a further selection because, based on physical deformities or other anomalies, various individuals are suppressed. It has been documented that infanticide is a practice present in many primitive cultures for those born with "deformed, physical birth defects" or even for "pathological behavior" by the decision of the mother or father or the family or the "birth attendant" or the "society" (Scrimshaw, 1984, p. 445, Table I).

Infanticide for "Poor infant quality" ("Deformed or very ill") has been documented in 21 out of 37 primitive societies studied (Daly & Wilson, 1984, Table I, pp. 490–491). The study's authors clearly state the evolutionary rationale: "... choosing not to raise a deformed child obviously serves the parents' fitness interests by avoiding the squandering of prolonged effort upon a child with poor reproductive prospects." (Daly & Wilson, 1984).

The first period of life is characterized by high mortality eliminating subjects with some disease or defect or vulnerability to infection/intoxication or other damaging factors. The data reported for mortality in children aged 0–3 years and 4–14 years in a hunter-gatherer population of Paraguay (Ache people) living in primitive conditions are illustrated in Table 2.1.

The mortality rates for the Ache people—data from (Hill & Hurtado, 1996)— and for the Mucajai Yanomamo—another hunter-gatherer population living between Venezuela and Brazil, data from (Early & Peters, 1990)—, illustrated by Fig. 2.3, show for both populations a high mortality rate in the first period of life.

Table 2.1 Causes of death for Ache people in natural conditions (forest period) in children aged 0–3 years and 4–14 years

In children aged 0–3 years			In children aged 4–14 years		
Violence and accidents			*Violence and accidents*		
Homicide/neglect	52		Homicide/neglect	17	
Captured/shot by Paraguayan	21		Captured/ shot by Paraguayan	56	
Accidents	3		Accidents	11	
Total:	76	58.02%	Total:	84	84.85%
Infections/Intoxications			*Infections/Intoxications*		
Various causes	36		Various causes	11	
Total:	36	27.48%	Total:	11	11.11%
Congenital causes			*Other causes*		
Unspecified newborn death/ defective	17		Sick (unspecified)/sick in lungs	4	
Childbirth/Mother had no milk	2		Total:	4	4.04%
Total:	19	14.50%	*Total*:	99	100%
Total:	131	100%			

Data from (Hill & Hurtado, 1996), discussed in (Libertini, 2013)

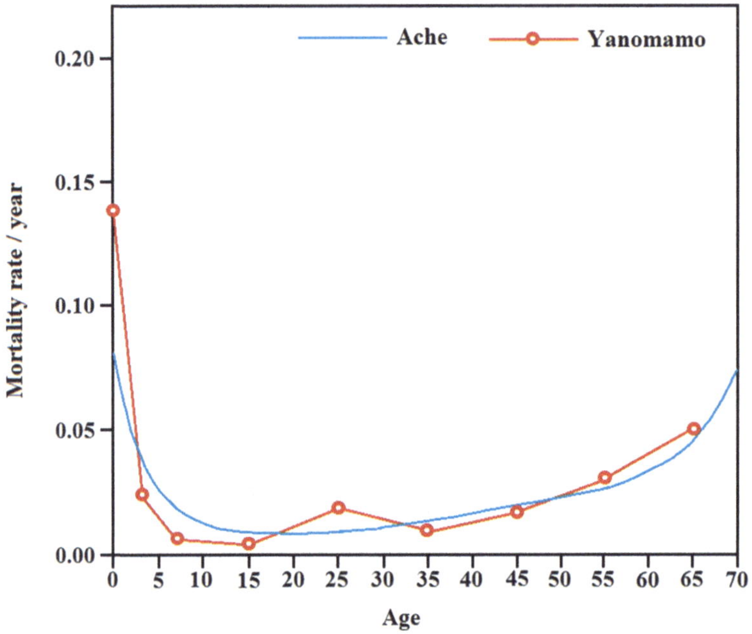

Fig. 2.3 Mortality rates of Ache and Yanomamo, both hunter-gatherer populations. Data from (Hill & Hurtado, 1996) and (Early & Peters, 1990), sex combined

2.3 Prevention and Care of the Diseases Deriving from Alterations of Genotype

Under "normal" conditions, that is to say, in the ecological niche to which we are adapted, subjects affected by genetic alterations die mostly before birth. Those living at birth, apart from the unfortunate ones eliminated by infanticides due to physical defects, for their reduced fitness are subject to higher mortality than other subjects without genetic alterations.

For ethical reasons, it is unacceptable to propose a return to the original conditions or equivalent conditions as regards the survival of subjects with genetic alterations.

It is true that even in modern conditions, most of these subjects are not in conditions to develop at the embryonic level or to survive as a fetus. Therefore, they are lost either due to their intrinsic incapacity for life or abortions spontaneously activated from the uterus. However, the number of individuals with genetic alterations is so large and varied that many survive in the early stages of life and often reach later ages.

This state of affairs entails a serious economic and social burden and constitutes a psychological torment for those who would like to do something. So, let us consider what operational and practical criteria could be useful to tackle such a large and demanding problem.

First, it is necessary to establish possible ethical criteria, which do not constitute a scientific fact but only questionable, modifiable, or acceptable choices according to a logic that goes beyond science.

As possible general ethical criteria, in the most effective ways and measures possible, medical care should try:

- to allow the maximum duration of life of affected subjects;
- to reduce their suffering;
- to limit the recurrence of new cases;
- to avoid actions that are ineffective or can be defined as forms of futile medical care.

In any case, it is imperative to avoid measures that, even in particular cases only, involve the suppression of individuals with genetic defects or that aim to pursue a population arbitrarily defined as genetically better. In this regard, a negative example, never to be forgotten, is the dramatic and criminal historical experience of Nazi ideology for which the systematic murder of individuals with physical or mental illnesses was practiced, considering them a burden to society and therefore unworthy of living, and the objective of a phantom superior race was pursued.

From these general criteria, other criteria could be derived that follow them in a more detailed form:

(1) If an embryo is unfit for implantation or development, it is useless to look for means that should avoid the natural course of the loss of the embryo;
(2) Similarly, if a fetus is completely and irreparably non-viable, it is useless to try to avoid the logical, natural course of the loss of the fetus, and indeed, it is necessary to consider the possibility of therapeutic abortion within the limits of appropriate ethical criteria that are recognized and so regulated by laws;
(3) If the fetus has some deformity or metabolic alteration that in some way can be treated even before birth, it is necessary to apply the possible suitable treatments;
(4) Individuals with genetic defects alive at birth must be assisted in the best possible way to improve the chances of survival, reduce suffering, and increase life span, avoiding futile medical care (to be precisely defined according to ethical criteria);
(5) For diseases for which it is possible to have or develop tests for early diagnosis (before or after birth), it is necessary to use or develop such tests and apply or develop appropriate treatments where possible;
(6) For severe illnesses with no possible cure diagnosed before birth, within the limits of appropriate ethical criteria that are recognized and so regulated by laws, therapeutic abortion must be considered;
(7) For diseases that can be transmitted with certainty from a parent to a child (for example, seriously harmful diseases deriving from a dominant gene or genetic

alterations of other types), procreation must be discouraged or prohibited based on ethical criteria that are recognized and regulated by laws;

(8) For diseases in which transmission is possible but not in all cases (for example, recessive mutations present in both parents), the risk of the birth of sick children (e.g., for homozygous individuals in the case of recessive mutations) must be evaluated and the possible parents must be informed. Subsequently, in the event of a pregnancy with a fetus recognized by specific tests as affected by the disease, the possibility of therapeutic abortion must be evaluated, always within the limits of appropriate ethical criteria that are recognized and regulated by laws;

(9) For individuals with genetic alterations whose survival can be allowed by appropriate protective measures (drugs, gene therapy, or other), such measures should be applied as best as possible, but procreation—if it is foreseeable that will give rise to subjects with similar problems—should be discouraged or prohibited, always within the limits of appropriate recognized ethical criteria and the related laws, and this to avoid the birth of subjects with a similar problem;

(10) Therapeutic measures suitable for treating or correcting the manifestations of genetic alterations must be actively sought, making the best use of available resources.

These measures only partially alleviate and do not solve the enormous problem of the numerous subjects suffering from genotype alterations.

Furthermore, it might seem that for this type of disease, the approach and (partial) remedies proposed by evolutionary medicine are the same as in current medicine.

However, there are some substantial differences:

- For current medicine, diseases caused by genotype alterations are a messy set of unpredictable genotype alterations without reason or logic. On the contrary, for evolutionary medicine, the presence of many random alterations of the genotype, and the resulting dysfunctions is an integral and essential part of the mechanism of evolution, and their existence as a general phenomenon is a logical prediction based on the mechanisms of evolution. A species without random modifications of the genotype, which most of the time are alterations and only in rare cases are improvements, could not evolve.
- According to evolutionary medicine, under "normal" conditions, the natural "remedy" for these alterations exists (elimination of embryos, fetuses, and, after birth, non-viable individuals or those with reduced fitness), but this "remedy" is unacceptable for ethical reasons;
- For evolutionary medicine, without prejudice to the priorities given to the protection of the life of individuals with genetic alterations, it is of fundamental importance to try to limit the spread of harmful genetic alterations by limiting or inhibiting the procreation of subjects with such alterations, within the limits of appropriate recognized ethical criteria and the resulting legislation.

References

Allison, A. C. (1954a). The distribution of the sickle-cell trait in East Africa and elsewhere, and its apparent relationship to the incidence of subtertian malaria. *Transactions of the Royal Society of Tropical Medicine and Hygiene, 48*(4), 312–318. https://doi.org/10.1016/0035-9203(54)90101-7

Allison, A. C. (1954b). Protection afforded by sickle-cell trait against subtertian malareal infection. *British Medical Journal, 1*(4857), 290–294. https://doi.org/10.1136/bmj.1.4857.290

Brookfield, J. F. (2009). Evolution and evolvability: Celebrating Darwin 200. *Biology Letters, 5*(1), 44–46. https://doi.org/10.1098/rsbl.2008.0639

Brosens, J. J., Bennett, P. R., Abrahams, V. M., Ramhorst, R., Coomarasamy, A., Quenby, S., et al. (2022). Maternal selection of human embryos in early gestation: Insights from recurrent miscarriage. *Seminars in Cell & Developmental Biology, 131*, 14–24. https://doi.org/10.1016/j.semcdb.2022.01.007

Cagan, A., Baez-Ortega, A., Brzozowska, N., Abascal, F., Coorens, T. H. H., Sanders, M. A., et al. (2022). Somatic mutation rates scale with lifespan across mammals. *Nature, 604*(7906), 517–524. https://doi.org/10.1038/s41586-022-04618-z

Cao, A., & Galanello, R. (2010). Beta-thalassemia. *Genetics in Medicine, 12*(2), 61–76. https://doi.org/10.1097/GIM.0b013e3181cd68ed

Carusillo, A., & Mussolino, C. (2020). DNA damage: From threat to treatment. *Cells, 9*(7), 1665. https://doi.org/10.3390/cells9071665

Chu, H.-P., Cifuentes-Rojas, C., Kesner, B., Aeby, E., Lee, H.-G., Wei, C., et al. (2017). TERRA RNA antagonizes ATRX and protects telomeres. *Cell, 170*(1), 86–101. https://doi.org/10.1016/j.cell.2017.06.017

Daly, M., & Wilson, M. (1984). A sociobiological analysis of human infanticide. In G. Hausfater & S. B. Hrdy (Eds.), *Infanticide: Comparative and evolutionary perspectives* (pp. 487–502). AldineTransaction. ISBN 978-0202362212.

Early, J., & Peters, J. (1990). *The population dynamics of the Mucajai Yanomamo*. Academic. ISBN 978-0122276809.

Flint, J., Harding, R. M., Boyce, A. J., & Clegg, J. B. (1998). The population genetics of the hemoglobinopathies. *Bailliere's Clinical Hematology, 11*, 1–50. https://doi.org/10.1016/S0950-3536(98)80069-3

Galanello, R., & Origa, R. (2010). Beta-thalassemia. *Orphanet Journal of Rare Diseases, 5*, 11. https://doi.org/10.1186/1750-1172-5-11

García-Rodríguez, A., Gosálvez, J., Agarwal, A., Roy, R., & Johnston, S. (2018). DNA damage and repair in human reproductive cells. *International Journal of Molecular Sciences, 20*(1), 31. https://doi.org/10.3390/ijms20010031

GARD (National Institutes of Health, National Center for Advancing Translational Sciences, Genetic and Rare Diseases Information Center). (2019). rarediseases.info.nih.gov/. Accessed 30.12.22.

Gorelick, A. N., & Naxerova, K. (2022). Mutational clocks tick differently across species. *Nature, 604*(7906), 435–436. https://doi.org/10.1038/d41586-022-00976-w

Hill, K., & Hurtado, A. M. (1996). *Ache life history*. Aldine De Gruyter. ISBN 978-0202020363.

International Human Genome Sequencing Consortium. (2004). Finishing the euchromatic sequence of the human genome. *Nature, 431*, 931–945. https://doi.org/10.1038/nature03001

Kacprzak, M., Chrzanowska, M., Skoczylas, B., Moczulska, H., Borowiec, M., & Sieroszewski, P. (2016). Genetic causes of recurrent miscarriages. *Ginekologia Polska, 87*(10), 722–726. https://doi.org/10.5603/GP.2016.0075

Lanz, M. C., Dibitetto, D., & Smolka, M. B. (2019). DNA damage kinase signaling: Checkpoint and repair at 30 years. *The EMBO Journal, 38*(18), e101801. https://doi.org/10.15252/embj.2019101801

Larsen, E. C., Christiansen, O. B., Kolte, A. M., & Macklon, N. (2013). New insights into mechanisms behind miscarriage. *BMC Medicine, 11*, 154. https://doi.org/10.1186/1741-7015-11-154

Lee, J. A., Cho, A., Huang, E. N., Xu, Y., Quach, H., Hu, J., & Wong, A. P. (2021). Gene therapy for cystic fibrosis: New tools for precision medicine. *Journal of Translational Medicine, 19*(1), 452. https://doi.org/10.1186/s12967-021-03099-4

Libertini, G. (2013). Evidence for aging theories from the study of a hunter-gatherer people (ache of Paraguay). *Biochemistry (Moscow), 78*(9), 1023–1032. https://doi.org/10.1134/S0006297913090083

Libertini, G., Corbi, G., Conti, V., Shubernetskaya, O., & Ferrara, N. (2021). *Evolutionary gerontology and geriatrics: Why and how we age*. In *Advances in studies of aging and health* (Vol. 2). Springer Nature. ISBN 978-3030737740.

Livingstone, F. B. (1958). Anthropological implications of sickle cell gene distribution in West Africa. *American Anthropologist, 60*(3), 533–562. https://doi.org/10.1525/aa.1958.60.3.02a00110

Macklon, N. S., Geraedts, J. P., & Fauser, B. C. (2002). Conception to ongoing pregnancy: The 'black box' of early pregnancy loss. *Human Reproduction Update, 8*, 333–343. https://doi.org/10.1093/humupd/8.4.333

Medawar, P. B. (1952). *An unsolved problem in biology. London: H. K. Lewis. Reprinted in: Medawar, P. B. (1957). The uniqueness of the individual*. Methuen.

Mueller, L. D. (1987). Evolution of accelerated senescence in laboratory populations of drosophila. *Proceedings of the National Academy of Sciences of the United States of America, 84*(7), 1974–1977. https://doi.org/10.1073/pnas.84.7.1974

Nybo Andersen, A. M., Wohlfahrt, J., Christens, P., Olsen, J., & Melbye, M. (2000). Maternal age and fetal loss: Population based register linkage study. *British Medical Journal, 320*, 1708–1712. https://doi.org/10.1136/bmj.320.7251.1708

Origa, R. (2017). β-Thalassemia. *Genetics in Medicine, 19*(6), 609–619. https://doi.org/10.1038/gim.2016.173

Orphanet: an online rare disease and orphan drug data base. (2023). © INSERM 1999. Available on http://www.orpha.net. Accessed from https://www.orphadata.com on 03.01.23.

Payne, J. L., & Wagner, A. (2019). The causes of evolvability and their evolution. *Nature Reviews. Genetics, 20*(1), 24–38. https://doi.org/10.1038/s41576-018-0069-z

Pigliucci, M. (2008). Is evolvability evolvable? *Nature Reviews. Genetics, 9*(1), 75–82. https://doi.org/10.1038/nrg2278

Portrat-Doyen, S., Tourniaire, J., Richard, O., Mulatero, P., Aupetit-Faisant, B., Curnow, K. M., et al. (1998). Isolated aldosterone synthase deficiency caused by simultaneous E198D and V386A mutations in the CYP11B2 gene. *The Journal of Clinical Endocrinology and Metabolism, 83*(11), 4156–4161. https://doi.org/10.1210/jcem.83.11.5258

Riederer, J. M., Tiso, S., van Eldijk, T. J. B., & Weissing, F. J. (2022). Capturing the facets of evolvability in a mechanistic framework. *Trends in Ecology and Evolution, 37*(5), 430–439. https://doi.org/10.1016/j.tree.2022.01.004

Robinson, P. S., Coorens, T. H. H., Palles, C., Mitchell, E., Abascal, F., Olafsson, S., et al. (2021). Increased somatic mutation burdens in normal human cells due to defective DNA polymerases. *Nature Genetics, 53*(10), 1434–1442. https://doi.org/10.1038/s41588-021-00930-y

Scrimshaw, S. C. M. (1984). Infanticide in human populations: Societal and individual concerns. In G. Hausfater & S. B. Hrdy (Eds.), *Infanticide: Comparative and evolutionary perspectives* (pp. 439–462). AldineTransaction. ISBN 978-0202362212.

Srb, A. M., Owen, R. D., & Edgar, R. S. (1965). *General Genetics* (2nd ed.). W. H. Freeman and Company.

Ugwumadu, A., Manyonda, I., Reid, F., & Hay, P. (2003). Effect of early oral clindamycin on late miscarriage and preterm delivery in asymptomatic women with abnormal vaginal flora and bacterial vaginosis: A randomised controlled trial. *Lancet, 361*, 983–988. https://doi.org/10.1016/S0140-6736(03)12823-1

Wang, X., Chen, C., Wang, L., Chen, D., Guang, W., & French, J. (2003). Conception, early pregnancy loss, and time to clinical pregnancy: A population-based prospective study. *Fertility and Sterility, 79*, 577–584. https://doi.org/10.1016/s0015-0282(02)04694-0

Wilcox, A. J., Weinberg, C. R., O'Connor, J. F., Baird, D. D., Schlatterer, J. P., Canfield, R. E., et al. (1988). Incidence of early loss of pregnancy. *New England Journal of Medicine, 319*, 189–194. https://doi.org/10.1056/NEJM198807283190401

Zinaman, M. J., Clegg, E. D., Brown, C. C., O'Connor, J., & Selevan, S. G. (1996). Estimates of human fertility and pregnancy loss. *Fertility and Sterility, 65*, 503–509. PMID: 8774277.

Chapter 3
Diseases Deriving from Alterations of the Ecological Niche—Part I: Not Regarding the Holobiont

The diseases deriving from alterations of the ecological niche are of great importance due to the number and incidence of troubles that fall into this group. Furthermore, the knowledge of their primary causes is essential because they are preventable by avoiding these primary causes, i.e., the alterations of the ecological niche to which the species is adapted.

In this chapter, only diseases due to alterations of the ecological niche not regarding the holobiont will be discussed. Those with alterations regarding the holobiont will be debated in the next chapter.

3.1 General Concepts About the Diseases Deriving from Alterations of the Ecological Niche

There are some general concepts regarding the primary causes of diseases deriving from alterations of the ecological niche that should be highlighted and clarified.

A species is best adapted to its ecological niche, provided that the ecological niche has not changed for too few generations for a sufficient adaptation. Optimal adaptation implies that the individuals of the species have their functions in the best possible conditions, i.e., they are in optimal health conditions.

The ecological niche of a species and the physiological mechanisms that allow the best adaptation of a species to its ecological niche constitute a complex system. A random modification in a complex system has, as the most probable consequence, excluding the cases in which the consequences are irrelevant, an alteration of the system (see Sect. 1.2.1 *Consequences of a random modification in a complex system*).

The cases in which the system constituted by a species and its ecological niche is modified by alterations of the characteristics of the species, i.e., the cases in which

G. Libertini et al., *The Essential Revolution of Evolutionary Medicine*, Advances in Studies of Aging and Health 4, https://doi.org/10.1007/978-3-031-84877-3_3

there is a random alteration of the genotype of the species, have been discussed in the previous chapter. On the contrary, the cases in which the modification concerns the ecological niche of the species are discussed in this and the next chapter.

When the ecological niche suddenly changes, a "mismatch" condition is likely to occur between the adaptation of the species and the new conditions of the ecological niche to which the species is not adapted. From this, a condition of maladaptation is derived and, therefore, impaired functionality and disease. This concept was outlined in 1983 (Libertini, 1983), and independently of this contribution, the successful definition of "mismatch" was proposed in 1988 (Eaton et al., 1988a) (s. Fig. 3.1).

The concept of ecological niche has been discussed in *Sect. 1.2.2 The ecological niche of a species*. It is essential to remember that the ecological niche is by no means limited to the physical environment in which the species lives and includes the whole set of ways in which individuals of the species experience every aspect of their existence and their relationships with other living beings.

Any modification of the ecological niche of a species must be considered presumptively as a factor negatively affecting the state of adaptation unless it is demonstrated with suitable empirical evidence that the modification of the ecological niche does not cause maladaptation and damage.

It is in no way correct to have positive or negative judgments regarding the characteristics of an ecological niche, which must be considered simply as a set of conditions to which the species is adapted. Consequently, the arbitrary evaluation of a modification of the ecological niche as an improvement in the living conditions and

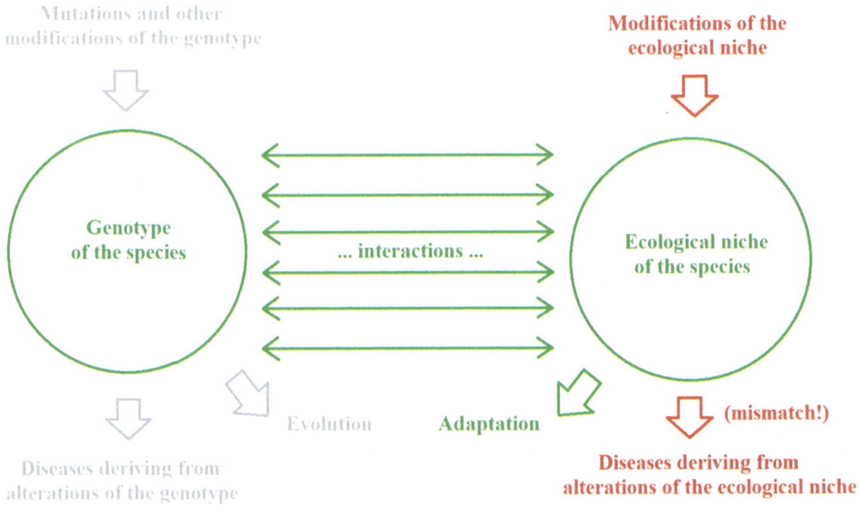

Fig. 3.1 In most cases, modifications of the ecological niche cause "mismatches" between the genotype of the species and the modified ecological niche to which the species is not adapted. This imbalance is a possible cause of diseases. However, the modified ecological niche determines the necessity of the evolution of the genotype that must adapt to the new conditions, although this may happen only in evolutionary times, i.e., within a sufficient number of generations

for the survival of the individuals of a species must be considered an unmotivated and probably erroneous prejudice. For example, suppose a species is adapted to an ecological niche in which there is reduced availability of salt. In that case, the species develops means of perception that make the salty taste particularly welcome, and individuals of the species will actively seek out salty foods. The idea that the scarcity of available salt is a negative element and that, therefore, greater availability of salt is a positive factor, as indicated by the fact that salt is pleasing to the taste, is a potentially erroneous prejudice. In the specific example, for humans, a diet rich in salt is one of the main causal factors of hypertension and consequential illnesses (Grillo et al., 2019; He et al., 2020).

Diseases caused by alterations of the ecological niche are easily identifiable, at least as a highly probable hypothesis, according to the simple criterion of disease incidence.

In fact, except in particular cases, the frequency of a disease in evolutionarily "normal" conditions is limited to a small number of cases due to specific random alterations of the genotype.

A disease caused by alterations of the ecological niche, such as in the case of a population non-living under evolutionarily "normal" conditions, will have a higher frequency, even much, than a population that lives or lived under "normal" conditions.

This criterion allows us to indicate a large number of widespread diseases as likely caused by alterations of the ecological niche.

However, it is necessary never to forget the reference to the evolutionary "normal" condition, that is to say, the ecological niche to which the species is adapted, and not to the more or less different conditions in which a modern population lives.

Let us consider the case of a study conducted on two groups of individuals living in different life conditions defined as "A" and "B" and showing different frequencies of a disease X (X_A and X_B, with $X_A > X_B$ and $\Delta x = X_A - X_B$). An interpretation could be that the lower frequency (X_B) is the expression of genetic factors, while the difference between the two frequencies (Δx) would be the expression of the different life conditions. This is a likely misleading interpretation because the difference between the two frequencies could also be exclusively the effect of alterations in living conditions that are more marked in A and less marked in B. The comparison should be made between a group of individuals who live in evolutionary "normal" living conditions, i.e., those to which the species is adapted (defined as "N"), or at least under conditions approaching those defined as "normal", and one or more groups living in different conditions ("A" "B" "C"...).

The low frequency of the disease in group N can be considered as due to genetic factors, while the higher frequency of the disease in groups A, B, C,... will presumably be attributable to harmful lifestyles, i.e., to changes in the ecological niche to which the species is adapted. For example, suppose type 2 and type 1 diabetes mellitus, hypertension, myocardial infarctions, and many other diseases are absent or very rare in populations of hunter-gatherers (i.e., in a condition near or equal to the "normal" condition), while the same diseases show a considerable frequency in populations with a modern lifestyle. In that case, the simple conclusion is that the diseases above are due to alterations of the ecological niche, except for a small portion due to genetic alterations.

Continuing on the essentiality of the reference to the evolutionary "normal" condition, in some studies on modern populations, individuals are indicated as more vulnerable to a disease based on specific genes that make them vulnerable, even defined as "pathological" genes. Consequently, the disease would be due in part to risk factors and in part to the presence of these pathological genes. These studies do not consider that the same genes in "normal" conditions are not a cause of the disease and that their "pathogenicity" appears only in conditions that are not "normal".

Let us take an extreme example considering a detention camp with extreme conditions of life where, after a certain period, only a part of the individuals survive. A careful genetic analysis of all individuals, whether survivors or not, could demonstrate that some genes are more frequent among non-surviving individuals. One misleading conclusion could be that the death of non-survivors is due both to the harsh living conditions and the presence of genes that have made these individuals more vulnerable. However, the correct conclusion is that the death of the non-surviving individuals is due only to the "non-normal" conditions of life imposed on them and not to the action of genes that would have no harmful effect under "normal" conditions.

In short, the criteria by which it is possible to identify a disease deriving from alterations of the ecological niche are the following:

1) The disease is absent or very rare in conditions of life that are evolutionarily "normal" or close to "normal";
2) The disease is frequent or very frequent in populations that live in conditions different from the "normal" ones;
3) The frequency of the disease has rapidly changed in a few years from minimum to high values, with a speed that a rapid genetic of the population change cannot justify;
4) When the alteration of the ecological niche that is the primary cause of the disease is known, the correction of the alteration reduces or eliminates the frequency of the disease.

Even where the alterations of the ecological niche as primary likely causes of disease are unknown, what is important is the difference between the frequency of cases in evolutionarily "normal" conditions and the frequency of cases of the same disease in the conditions studied.

Other important data are:

(i) the difference in the frequency of the disease in different populations. For example, the epidemiological data that: "Breast cancer incidence is more common in high-income countries (571/100 000) than in low-income countries (95/10 000)" (Kashyap et al., 2022) tells us that even if the cases in low-income countries were all due to intrinsic factors, for the cases in high-income countries no less than 83% [= (571–95)/571] are due to extrinsic factors, i.e., alterations of the ecological niche.

(ii) the difference in the disease frequency in the same population over time. For example: "Prior to the first description of hay fever in 1870 there was very little awareness of allergic disease, which is actually similar to the situation in pre-hygiene villages in Africa today.... there were no clear reports of an increase in pediatric asthma until 1970. Further the current 'epidemic' of food allergy does not appear to have started until after 1990." (Platts-Mills, 2015). This means that the current allergy epidemics are due totally or almost totally to extrinsic factors.

3.2 Possible Alterations of the Ecological Niche Not Regarding the Holobiont

The possible alterations of the ecological niche, the primary causes of many diseases, are numberless.

In the following sections of this chapter, some main types of alteration of the ecological niche not regarding the holobiont and originating possible "mismatches" will be discussed:

Sect. 3.2.1 Diet alterations;

Sect. 3.2.2 Diet alterations together with other factors;

Sect. 3.2.3 Exposure to harmful substances;

Sect. 3.2.4 Excessive noise levels;

Sect. 3.2.5 Altered exposure to light in the first years of life;

Sect. 3.2.6 Multiple carcinogenic factors;

Sect. 3.2.7 Altered social interactions;

Sect. 3.2.8 Infectious epidemics.

In the sections of the next chapter, some main types of alterations of the ecological niche regarding our holobiont and originating possible "mismatches" will be discussed:

Sect. 4.1.1 Alterations of the holobiont—A: Reduced exposure to allergens;

Sect. 4.1.2 Alterations of the holobiont—B: Elimination of intestinal parasites;

Sect. 4.1.3 Alterations of the holobiont—C: Alterations of the microbiome at birth;

Sect. 4.1.4 Alterations of the holobiont—D: Alterations of skin and mucosal microbiomes;

Sect. 4.1.5 Alterations of the holobiont—E: Alterations of the gut microbiome and obesity.

Sect. 4.1.6 Alterations of the holobiont—F: The case of *Helicobacter pylori*.

These types of alterations are only a part of the possible alterations of the ecological niche.

3.2.1 Diet Alterations

The foods eaten by modern populations are very different in type, quality, and composition from those of populations that lived (or live) in primitive conditions, i.e., similar to those of the ecological niche to which our species is adapted.

The associations between some alterations of modern diets and certain pathological states have been well-known for some time (Eaton et al., 1988b). Here are some of the main relationships:

1) *Excessive ingestion of salt → hypertension* (Grillo et al., 2019; He et al., 2020), *as well as endothelial damage and increased cardiovascular risk* (Bragulat & de la Sierra, 2002; Feng et al., 2017).

Hypertension is, in turn, associated with several diseases that are its consequences. For example, concentric left ventricular hypertrophy, diastolic dysfunction, heart failure, and progressive renal failure (Messerli et al., 2017); left atrial remodeling, atrial contractile dysfunction, and atrial fibrillation (Kockskämper & Pluteanu, 2022); and increased risk of sudden cardiac death, supraventricular and ventricular arrhythmias, stroke, and chronic kidney disease (Afzal et al., 2019).

Excessive use of salt from childhood increases the risk of hypertension later in life (Samadian et al., 2016), while a reduction in salt intake reduces the risk of hypertension (Aburto et al., 2013; Ghimire et al., 2021; He et al., 2013; Sookram et al., 2015).

2) *Excessive ingestion of unsaturated fats, caloric foods, meat with high-fat content, insufficient physical activity → overweight/obesity, type 2 diabetes mellitus (T2DM), and the so-called metabolic syndrome (overweight/obesity + type 2 diabetes + hypertension)* (Bellou et al., 2018; Lin & Li, 2021; Safaei et al., 2021; Wu et al., 2014).

T2DM and overweight/obesity are associated with an increased risk of many diseases and related complications (cardiac hypertrophy, congestive heart failure, cardiac arrhythmias with sudden death, myocardial infarction, cerebral vascular disease, other vascular diseases, etc.) with very documentation about it. E.g.: "Overweight and obesity are one of the main lifestyle illnesses that leads to further health concerns and contributes to numerous chronic diseases, including cancers, diabetes, metabolic syndrome, and cardiovascular diseases." (Safaei et al., 2021); "T2DM patients are more susceptible to different forms of both short- and long-term complications.... the complications include macrovascular diseases (hypertension, hyperlipidemia, heart attacks, coronary artery disease, strokes, cerebral vascular disease, and peripheral vascular disease), microvascular diseases (retinopathy, nephropathy, and neuropathy) and cancers." (Wu et al., 2014).

In hunter-gatherer populations, overweight/obesity, T2DM, metabolic syndrome, and insufficient physical activity are practically non-existent (Early & Peters, 1990; Eaton et al., 1988a; Hill & Hurtado, 1996).

The tendency to attribute the origins of T2DM to both dietetic and genetic factors is quite common. E.g.: "Genome-wide association studies have implicated around

250 genomic regions in predisposition to type 2 diabetes, with evidence for causal variants and genes emerging for several of these regions." (Langenberg & Lotta, 2018); "Several recent large population-based studies and their meta-analyses have identified multiple potential genetic and non-genetic biomarkers for the risk of T2D." (Laakso, 2019).

However, in these studies about correlations between T2DM and genetic factors, there is no consideration of how these genetic factors act in evolutionarily "normal" conditions, i.e., whether in such conditions, they are the cause of the disease or, on the contrary, are completely without pathogenetic effects.

Let us consider the well-documented case of the Pima, a people living in an area divided between Mexico and the state of Arizona (USA). Before Westernization, among Pima, who lived as hunter-gatherers, T2DM and obesity were unknown. However, already at the beginning of the twentieth century, some Pima were obese while diabetes was still unknown (Russell, 1908): "The traditional Pima diet was minimally processed, including wheat, maize, beans, saguaro fruit, diverse mammals, fish, birds, and worms.... As western settlers diverted water supplies, they relied on government rations, mainly sugar, white flour, lard, as well as store-bought food. Obesity rose before diabetes. In the early 1900s, some older Pima were obese but no diabetes was identified in two surveys.... By 1971, most adult Pima were obese.... In 1976, among adults over age 40 years, males averaged 122% above ideal weight, women 149%. In 1994, average adult BMI for was 35.5 for women and 30.8 for men.... Adult diabetes prevalence was <1% in 1940..., 4% in 1955..., 30% in 1963..., and 38% in 2006.... The greatest rise in diabetes occurred in the late 1950s and early 1960s, when caloric consumption was higher than most Americans (2,780 calories/day),..." (Pressler et al., 2022) (references omitted).

Nowadays, there are substantial differences in the incidence of diabetes and obesity between the Pima population living in Mexico and that living in Arizona: "The age- and sex-adjusted prevalence of type 2 diabetes in the Mexican Pima Indians (6.9%) was less than one-fifth that in the U.S. Pima Indians (38%) and similar to that of non-Pima Mexicans (2.6%). The prevalence of obesity was similar in the Mexican Pima Indians (7% in men and 20% in women) and non-Pima Mexicans (9% in men and 27% in women) but was much lower than in the U.S. Pima Indians. Levels of physical activity were much higher in both Mexican groups than in the U.S. Pima Indians." (Schulz et al., 2006).

The case of the Pima people demonstrates that if genetic factors caused, in modern lifestyles, the higher vulnerability to T2DM, in the original conditions of life of the Pima, which was essentially a hunter-gatherer people, the same genes did not in any way cause T2DM or obesity. Therefore, considering them as pathological genes because, in conditions of life to which the population is not adapted, they cause vulnerability to the development of a disease, is a misleading way of describing them. In fact, according to evolutionary medicine, it is correct to bring Westernized living conditions back to healthier living conditions as an objective alternative to classifying some genetic characteristics as pathological ("bad" genes) because they are unsuitable for tolerating Westernized living conditions without pathological consequences.

The global prevalence of obesity and diabetes has continuously increased in recent decades (see Tables 3.1 and 3.2), posing a growing threat to health.

About the significant variation in the prevalence of diabetes in various parts of the world: "The highest national prevalences - generally those in Oceania, and the Middle East and North Africa - are now five to ten times greater than the lowest prevalences, which are in some western European countries." (NCD RisC, 2016, Diabetes).

3) *Low dietary fiber intake → constipation, and increased risk of hemorrhoids, anal fissures, colon diverticulosis, appendicitis, colon and stomach carcinoma, type 2 diabetes mellitus, metabolic syndrome, and cardiovascular diseases* (Trepel, 2004)

About the relationship with constipation, an index, Healthy Eating Index (HEI), was proposed "to evaluate the overall dietary quality": "the components of HEI-2015 that is total fruits, whole fruits, total vegetables, greens and beans, whole grains, total protein foods, seafood and plant proteins, fatty acids ratio decreased the odds of constipation, and that is sodium increased the odds. Either in male or female, HEI-2015 score was negatively associated with constipation." (Liu et al., 2022).

For treating constipation, an increase in fiber intake is the first therapeutic suggestion (Bharucha & Lacy, 2020).

The relationship between dietary fiber deficiency and some diseases, such as diverticulitis, appendicitis, hernias, varicose veins, diabetes, and atherosclerosis, which are associated with high-income lifestyles, is known (Burkitt, 1971; O'Keefe, 2019).

About the relationship with atherosclerosis and related diseases, dietary fiber intake lowers blood cholesterol and appears to decrease the risk of cardiovascular diseases (Soliman, 2019).

About appendicitis, in particular:

(1) The first description of appendicitis was in 1759. In 1894, early mortality caused by appendicitis was reported to be 26% (Krzyzak & Mulrooney, 2020). In 1990, for the U.S., the lifetime risk of suffering from appendicitis was estimated to be 8.6% for males and 6.7% for females (Addiss et al., 1990).

Table 3.1 Prevalence and number of people with diabetes in WHO

	Prevalence (%)		Number (millions)	
WHO Region	1980	2014	1980	2014
Eastern Mediterranean Region	5.9	13.7	6	43
South-East Asia Region	4.1	8.6	17	96
Western Pacific Region	4.4	8.4	29	131
Region of the Americas	5	8.3	18	62
European Region	5.3	7.3	33	64
African Region	3.1	7.1	4	25
Total	4.7	8.5	108	422

(World Health Organization) regions, in 1980 and 2014, ranked by prevalence in 2014 (WHO, 2016a, Diabetes), data from (NCD RisC, 2016, Diabetes)

Table 3.2 Age-standardized prevalence (%) of obesity in 1975 and 2016 in adults and children distinguished by gender for the 30 most populous countries in the world ranked by prevalence of obesity among adult women in 2016

	1975				2016			
State	Women	Men	Girls	Boys	Women	Men	Girls	Boys
Egypt	19	6	2	1	42	24	19	16
Turkey	14	4	<1	<1	41	25	11	12
South Africa	18	2	<1	<1	41	16	13	10
USA	14	11	6	5	38	36	19	23
Mexico	14	6	2	3	34	25	13	17
Iran	11	3	<1	<1	33	20	9	11
UK	11	8	3	2	30	28	9	11
Colombia	10	3	<1	<1	28	18	6	7
Russia	21	6	1	1	28	19	4	9
Brazil	8	3	1	<1	26	19	9	13
Spain	12	8	2	4	24	25	8	13
France	11	7	3	3	22	23	7	9
Germany	10	8	3	3	21	25	7	11
Italy	11	7	4	5	20	21	10	15
Nigeria	2	<1	<1	<1	14	5	2	2
Thailand	1	<1	<1	<1	13	7	9	14
Tanzania	1	<1	<1	<1	13	4	3	2
Kenya	1	<1	<1	<1	12	3	3	1
Pakistan	2	<1	<1	<1	12	6	3	4
DR Congo	1	<1	<1	<1	10	4	3	1
Indonesia	<1	<1	<1	<1	9	5	5	7
Myanmar	<1	<1	<1	<1	8	4	2	5
Philippines	1	<1	<1	<1	8	5	3	6
China	<1	<1	<1	<1	7	6	7	15
Ethiopia	<1	<1	<1	<1	7	2	1	1
Bangladesh	<1	<1	<1	<1	5	2	2	3
India	<1	<1	<1	<1	5	3	2	2
South Korea	<1	<1	<1	1	5	5	5	13
Japan	1	<1	<1	2	4	5	2	5
Viet Nam	<1	<1	<1	<1	3	2	1	4

(adults: ≥ 20 years; children: 5–19 years; obesity in adult: body mass index or BMI ≥30 kg/m^2; obesity in children: BMI >2 SD above the median of the WHO growth reference). Data from (NCD-RisC, 2017, Adiposity). The values that have increased from 1975 to 2016 are in Italic/Bold font

(2) The global incidence of appendicitis grew by 63.55% in the period 1990–2019. In 2019, there were 17.70 million new cases of appendicitis, with an age-standardized incidence rate of 229.86 per 100,000 population (Yang et al., 2022).
(3) "Appendicitis is a global disease. The incidence of appendicitis is 7–12% in the population of the USA and Europe, and, stable in most Western countries...

Data from newly industrialized countries are sparse but suggest appendicitis is rising rapidly, although it is still lower than the reported incidence amongst patients in the developed world... In Western countries, 16% of the population undergoes appendicectomy but in Africa and Asia with a high roughage diet, the prevalence is much lower (Burkitt, 1971). Burkitt described an increase in disease in Africa associated with the adoption of a Western diet (Burkitt, 1971).... The fact that increasing the dietary fiber in modern Western society can reduce the incidence of appendicitis, is demonstrated by the substantial fall over the last 30 years in Western countries." (Weledji, 2022)

(4) "... appendicitis is a disease associated with industrialized cultures but not developing cultures (Barker & Morris, 1988; Barker et al., 1988a; Barker et al., 1988b; Bickler & DeMaio, 2008). For example, the typical incidence of childhood appendicitis in the United States appears to be about 35-fold greater than the incidence in segments of the African population that are unaffected by modern health care and sanitation practices (Bickler & DeMaio, 2008). Importantly, the incidence of appendicitis increases with the adoption of Western lifestyles in both African countries (Bickler & DeMaio, 2008) and in Europe (Barker & Morris, 1988; Barker et al., 1988a; b)." (Laurin et al., 2011)

(5) "Nowadays, dietary fibre intake in HICs [High-Income Countries] is around 15 g/day (well below the amount of fibre Burkitt advocated of >50 g/day - which is associated with diets from rural, southern and eastern sub-Saharan Africa). Since Burkitt's death in 1993, his hypothesis has been verified and extended by large-scale epidemiological studies, which have reported that fibre deficiency increases the risk of colon, liver, and breast cancer and increases all cancer mortality and death from cardiovascular, infectious, and respiratory diseases, diabetes, and all non-cardiovascular, non-cancer causes." (O'Keefe, 2019)

(6) "Epidemiological studies have consistently demonstrated the benefits of dietary fibre on gastrointestinal health through consumption of unrefined whole foods, such as wholegrains, legumes, vegetables and fruits. Mechanistic studies and clinical trials on isolated and extracted fibres have demonstrated promising regulatory effects on the gut (for example, digestion and absorption, transit time, stool formation) and microbial effects (changes in gut microbiota composition and fermentation metabolites) that have important implications for gastrointestinal disorders." (Gill et al., 2021)

4) *Excessive alcohol intake → hepatic steatosis, acute and chronic hepatitis, cirrhosis, hepatocellular carcinoma* (Adachi & Brenner, 2005; Jeon & Carr, 2020; Kim & Park, 2018; Taniai, 2020).

For the relationships between excessive alcohol intake and pancreatic damage, or acute or chronic pancreatitis, see (Pandol et al., 2010; Stram et al., 2016; Vonlaufen et al., 2007).

About the relationships between the inflammation of the esophageal and gastric mucosa and gastroesophageal reflux disease, see (Akiyama et al., 2008; Minatsuki et al., 2013; Rocco et al., 2014).

About the relationship between excessive alcohol consumption and cancer: "Approximately 4% of cancers worldwide are caused by alcohol consumption. Drinking alcohol increases the risk of several cancer types, including cancers of the upper aerodigestive tract, liver, colorectum, and breast." (Rumgay et al., 2021).

> Alcohol consumption contributes to 3 million deaths each year globally as well as to the disabilities and poor health of millions of people. Overall, harmful use of alcohol is responsible for 5.1% of the global burden of disease. Harmful use of alcohol is accountable for 7.1% and 2.2% of the global burden of disease for males and females respectively. Alcohol is the leading risk factor for premature mortality and disability among those aged 15 to 49 years, accounting for 10% of all deaths in this age group. (WHO, 2023a, Alcohol)

"Exposure to alcohol must have been rare if it occurred at all during the Paleolithic. Honey and many fruits can undergo spontaneous fermentation, so it is likely that at least some preagricultural peoples occasionally consumed alcohol beverages. In the present century, nearly half of all preliterate societies studied by anthropologists have made use of beverage alcohol; yet *none* of the groups actually producing it were hunters and gatherers." (Eaton et al., 1988a, pp. 60–61). So, it is easy to deduce that in Paleolithic peoples, the diseases caused by alcohol were inexistent. When alcohol consumption began, as our genotype was adapted to metabolize only limited quantities of alcohol, excessive alcohol consumption constituted an alteration of the ecological niche to which we are not adapted.

5) *Various dietary alterations → urolithiasis*

Urolithiasis (kidney stone disease, nephrolithiasis) was already known in ancient times, as demonstrated by the description of bladder stones by Hippocrates (Dimopoulos et al., 1980). However, its epidemic diffusion occurs in modern times:

(1) "There has been considerable increase in the incidence of idiopathic renal stone in Europe, North America, Australasia and Japan within the present century... Renal stone is rare among persons living in poor or primitive socio-economic circumstances and is very rare in African Bantu living under tribal conditions (Modlin, 1969).... To summarize, from being virtually unknown in historical times, renal stone has become significant as a common morbid condition in the affluent, westernized countries within the last 80 years whilst remaining rare in communities where the people live in primitive and poor conditions.... their immunity must be dependent on the absence of a risk factor or factors common in the environment of the affluent communities. These are most likely to be associated with dietary structure..." (Blacklock, 1981)

(2) "Kidney stone disease is a global health burden with an increasing estimated prevalence between 5-15% in various regions (Khan et al., 2017; Sorokin et al., 2017; Thongprayoon et al., 2020).... The prevalence and incidence are increasing in many countries over the last 4 decades with 2 to 4-fold increases with regional differences. These differences in reported prevalence and incidence may reflect to some extent different definitions and detection methods of stone disease. However, the increase observed globally is paralleled by an increase in risk factors for stone disease. These include, as discussed below, an increase in obesity and diabetes mellitus, consumption of diets rich in salt, animal protein,

sucrose, sugar-sweetened beverages, and global warming (Thongprayoon et al., 2020)." (Wagner, 2021)

(3) The ranges of nephrolithiasis rates for 2014 were estimated as 7–13% in North America, 5–9% in Europe, and 1–5% in Asia (Sorokin et al., 2017).

(4) As for Asia, according to a 2018 review: "There has been a stone forming belt stretching across the West Asia, Southeast Asia, South Asia as well as several developed countries, including South Korea and Japan, with prevalence ranging from 5% to 19.1%. However, it is only 1%-8% in the remaining countries. In addition, the prevalence and incidence of urolithiasis have increased in most of the countries of Asia over the last several decades." (Liu et al., 2018)

(5) Changes in the incidence of renal stones were documented in Japan and Norway during the periods 1935–1955 and 1900–1960, respectively (Andersen, 1969; Inada et al., 1958) (s. Fig. 3.2).

(6) In Japan, the prevalence of urolithiasis increased from 4.3% in 1965 to 9.0% in 2005 (+109%). In the same period, the incidence ascended from 54.2 to 114.3/100000 (+110%) (Yasui et al., 2008).

(7) In the U.S., the prevalence of kidney stones increased from 3.8% to 5.2% in 1976–1994 and then to 8.8% in 2007–2010. After a slight reduction, 8.7% in 2007–2008 and 7.2% in 2011–2012, again increased to 9.0% in 2013–2014 and to 10.1% in 2007–2016 (Stamatelou & Goldfarb, 2023).

(8) "Recent data provide evidence that the incidence of nephrolithiasis in children is rising." (Sas, 2011)

(9) "Pediatric urolithiasis has increased globally in the last few decades." (Sharma & Filler, 2010)

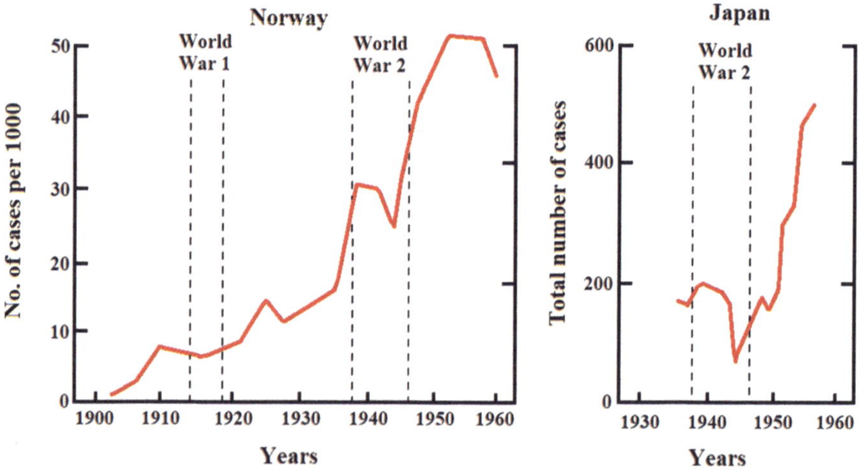

Fig. 3.2 Changes in incidence of renal stone. On the left: in Oslo City Hospital, Norway, 1900–1960, data from (Andersen, 1969); on the right: in Japan, 1935–1955, data from (Inada et al., 1958)

Risk factors for nephrolithiasis are obesity and overweight, metabolic syndrome, diabetes, hypertension, cardiovascular disease, hypertension, hypertriglyceridemia, low high-density lipoprotein, elevated fasting glucose, low intake of dietary fiber, fruit, and vegetable, higher sodium intake and lower potassium intake, high protein intake, low water intake, low dietary calcium intake (while supplemental calcium and vit. D increased the risk) (Ferraro et al., 2017, 2020; Siener, 2021; Stamatelou & Goldfarb, 2023).

> Obesity, another metabolic syndrome trait, has been associated with impaired carbohydrate tolerance, inappropriate calcium response to glucose ingestion, the increased excretion of calcium, sodium, oxalate, and uric acid, hypocitraturia, and defects in renal ammoniagenesis, all contributing to kidney stone formation (Chou et al., 2011; Maalouf et al., 2007). (Stamatelou & Goldfarb, 2023).

A systematic review and meta-analysis (Lin et al., 2020) indicated as prominent risk factors for urolithiasis body mass index, dietary sodium, fructose, meat, animal protein, and soda, while protective factors were fluid intake, alcohol, water, coffee, tea, vegetables, fruits, dietary fiber, dietary calcium, and potassium.

Vitamin D and calcium supplementation alone appeared to increase the risk of stones in observational studies but not in randomized controlled trials, while the association of Vitamin D and calcium supplementation appeared to increase the risk in both types of study, as shown in another metanalysis (Kahwati et al., 2018).

It has been known for some time that urolithiasis frequency is inversely related to dietary calcium intake (Curhan et al., 1993, 1997; Heller, 1999; Parmar, 2004), even though supplemental calcium may increase the risk (Curhan et al., 1997). Dietary calcium reduces oxalate absorption and the urinary excretion of oxalate, lowering the risk of kidney stones of calcium oxalate, the prevalent type of stones (Curhan et al., 1993). This fact "may be due to increased binding of oxalate by calcium in the gastrointestinal tract" (Curhan et al., 1993).

The intake of fiber and plant foods reduces urinary calcium excretion and thus the frequency of the stones, while carbohydrate intake has the opposite effect (Heller, 1999). In patients with recurring nephrolithiasis, hypercalciuria was found in 60.9% and hyperuricosuria in 35.8% (Levy et al., 1995).

The mechanism by which hypercalciuria causes an increased risk of renal stones is known. Some foods reduce calcium absorption, and therefore the urinary calcium (K, PO_4, fiber, Alkali Load alias fruits and vegetables), and others that have the opposite effect (supplemental Ca, Na, Mg, Carbohydrates, Acid Load alias animal flesh). The mechanisms that cause these effects are quite known (Heller, 1999). However, an increase in dietary calcium reduces oxalate absorption and oxalate excretion in the urine and thus reduces the frequency with which they form calcium oxalate stones, the most common type of kidney stones (Curhan et al., 1993).

It is essential to compare the ancestral diet with that of contemporary Western populations (Eaton et al., 2009; Konner & Eaton, 2010). In Table 3.3, this comparison is outlined, and the factors in the modern diet that increase urolithiasis risk are highlighted in italic and bold characters.

Table 3.3 Table (modified and with additions) from (Eaton et al., 2009), re-proposed with small modifications in (Konner & Eaton, 2010). The factors more increasing urolithiasis risk are highlighted in italic and bold characters and with a gray background

	Ancestral Diet (Hunter-Gatherer)	**Contemporary Western Diet**
Acid-base effect (net endogenous acid production)	Alkaline	***Acidic***
Added sugars and refined carbohydrates	Very little	***Much more***
Fiber	More	***Less***
Free water intake	More	***Less***
Fruits and vegetables	Twice as much	***Half as much***
Glycemic load	Relatively low	***High***
Protein intake	More [N.B.: lean, from game]	***Less [N.B.: fat, from breeding]***
Sodium:potassium	Much more potassium	***More sodium***
Total carbohydrate intake	Less	***More***
Antioxidant capacity	Greater	Lesser
Caloric concentration	Very little	Much
Cereal grains	Very little	A great deal
Cholesterol intake	Equal or more	Equal or less
"Dairy" products	Mother's milk only	Considerable
Dietary bulk	More	Less
Long-chain essential fatty acids	More	Less
Micronutrient intake	More	Less
Serum cholesterol-raising fat	Less	More
Soluble:insoluble	Closer to equal	Far more insoluble
Total energy intake	More	Less
Total fat intake	Equivalent	Equivalent
Total polyunsaturated fat	More	Less
ω-6:ω-3	Closer to equal	Far more ω-6

As far as protein intake is concerned, it should be noted that, in the ancestral diet, meat is low in fat, as it comes from wild animals, while in the contemporary diet, it is fatty because it comes from farmed animals. The table does not show the reduced dietary calcium intake in modern diets (a factor related to greater urolithiasis frequency). At the same time, it is high in ancestral diet due to the greater intake of fruits and vegetables.

Urolithiasis appears clearly caused by alterations of the ancestral diet, a modification of the ecological niche to which our species is not adapted.

6) *Various dietary alterations → gallstones (cholelithiasis) (→ cholecystitis, pancreatitis, and gallbladder, pancreatic, and colorectal cancer)*

"Cholelithiasis and cholecystitis were thought to be rare in Africans. Cholelithiasis is said to be an uncommon disease in tropical Africa... This is in contrast to the high incidence of gallstones in western Europe and the United States. Autopsy studies have indicated that 50% of Swedish... and 20% of American women will eventually harbor stones. Becker and Chatgidaki (Becker & Chatgidakis, 1952) in South Africa reported an incidence of 2% in a study of 4494 autopsies over a 15-year period; a total of 92 stones were found (six stones per year). Bremner (Bremner, 1971) in a review of disease pattern at Baragwanath Hospital (also in South Africa) found a four-fold increase of cholecystectomy in the decade 1959–1969; however, the incidence was still very small....

Cholelithiasis and cholecystitis are not as common in Africa as in Europe and North America. However, with the changing lifestyle and dietary habits, especially increased intake of fat and refined sugar, there will continue to be an increase in the incidence of cholelithiasis and cholecystitis in Africans. In a personal communication between Burkitt and Parnis in 1963 (Parnis, 1964), it was reported that only one case of gallbladder disease was operated upon in an African patient in Uganda over a period of 17 years. In a 5 year review from Ibadan in 1964, 35 patients - an average of seven per year - were operated upon for inflammatory disease of the gallbladder...." (Rahman, 2005).

> Gall bladder disease [GBD] seems to have been rare in Amerindians prior to World War II, based on the few systematic reports of Amerindian health. In these reports, GBD is either absent or is noted by the authors to be rare... Since about 1950, there have been many reports documenting very high rates of GBD in Amerindians across the American continents, from Alaska to parts of South America... (Weiss et al., 1984)

In modern developed countries, about 10–15% of adults suffer from gallstones (Ansaloni et al., 2016).

> Approximately 10%–20% of the national adult populations currently carry gallstones, and gallstone prevalence is rising. In addition, nearly 750,000 cholecystectomies are performed annually in the United States... Cholelithiasis is also strongly associated with gallbladder, pancreatic, and colorectal cancer occurrence. (Pak & Lindseth, 2016)

About 72% of patients with symptomatic gallstones suffer from biliary pain, cholecystitis, pancreatitis, gallstone ileus, biliary tract obstruction, and gallbladder emphysema, which may cause a rupture of the gallbladder (Gupta & Shukla, 2004).

Among patients with asymptomatic gallstones, about 10%–25% will suffer from cholecystitis, cholangitis, or pancreatitis during their lifetimes (Friedman, 1993; Halldestam et al., 2004).

Risk factors for cholesterol gallstones (apart from non-modifiable factors such as advanced age, female gender, and genetic factors) are obesity, higher BMI, chronic hyper-nutrition, high intakes of carbohydrates, high glycemic load, high caloric diet, scarcity of fibers in diet, long-chain saturated fatty acids, reduced physical activity, hyperinsulinemia, dyslipidemia, type 2 diabetes, metabolic syndrome, use of refined sugar, rapid weight loss, and therapies with estrogens. In contrast, physical activity and diets rich in unsaturated fat, fiber, ascorbic acid, calcium, fish oil, fresh fruits, and vegetables reduce the risk of forming cholesterol gallstones (Pak & Lindseth, 2016).

3.2.2 *Diet Alterations, Together with Other Factors*

Dental troubles such as caries and malocclusion, and their complications, are today among the most frequent diseases worldwide:

(1) "The Global Burden of Disease (GBD) 2010 Study produced comparable estimates of the burden of 291 diseases and injuries in 1990, 2005, and 2010. This article reports on the global burden of untreated caries, severe periodontitis, and severe tooth loss in 2010 and compares those figures with new estimates for 1990. We used disability-adjusted life-years (DALYs) and years lived with disability (YLDs) metrics to quantify the burden. Oral conditions affected 3.9 billion people, and untreated caries in permanent teeth was the most prevalent condition evaluated for the entire GBD 2010 Study (global prevalence of 35% for all ages combined)." (Marcenes et al., 2013)

(2) "The results of the global burden of disease study released by Lancet in 2017 showed that among 328 diseases, the prevalence of permanent dental caries ranked first, and the incidence ranked second (GBD 2017 Disease and Injury Incidence and Prevalence Collaborators, 2018). There are around 2.44 billion population worldwide suffering from permanent tooth decay." (Cheng et al., 2022)

(3) "Malocclusion, defined as a handicapping dento-facial anomaly by the World Health Organization, refers to abnormal occlusion and/or disturbed craniofacial relationships, which may affect esthetic appearance, function, facial harmony, and psychosocial well-being.... It is one of the most common dental problems, with high prevalence ranging from 20% to 100% reported by different researchers (Akbari et al., 2016; Alvarado et al., 2017; Mtaya et al., 2009)." (Zou et al., 2018)

(4) In the USA, in 1998: "Noticeable incisor irregularity occurs in the majority of all racial/ethnic groups, with only 35% of adults having well-aligned mandibular incisors. Irregularity is severe enough in 15% that both social acceptability and function could be affected, and major arch expansion or extraction of some teeth would be required for correction." (Proffit et al., 1998)

(5) In 2009, a study on Tanzanian children showed that 63.8 percent of the subjects had at least one type of malocclusion (Mtaya et al., 2009).

(6) In 2011, a study on Peruvian children following a modern lifestyle observed that: "The prevalence of malocclusions was 85.6%" (Aliaga-Del Castillo et al., 2011).

As for dental caries, the role of sugars, especially drinks sweetened with sugars and consumed regularly, is prominent as a sure cause of tooth decay:

(i) "Excess added sugars, particularly in the form of sugar-sweetened beverages, is a leading cause of tooth decay in US children." (Chi & Scott, 2019);

(ii) "The role of sugar (and other fermentable carbohydrates such as highly refined flour) as a risk factor in the initiation and progression of dental caries is overwhelming." (Gupta et al., 2013)

Excessive ingestion of simple and refined carbohydrates (in particular sugar) and the use of industrially refined food were already indicated as causes of tooth decay and as modifications of the diet to which our species is adapted (Eaton et al., 1988a; Price, 1939).

As for malocclusion, the presence of skeletal defects and the correlation of these defects with vitamin D3 (VD3) deficiency is evident:

> The abnormal growth of the craniofacial bone leads to skeletal and dental defects, which result in the presence of malocclusions.... In about 42.1% of patients, the presence of a skeletal defect was found, and in 46.5% of patients, dentoalveolar malocclusion. The most common defect was transverse constriction of the maxilla with a narrow upper arch (30.7%). The concentration of vitamin 25 (OH) D in the study group was on average 23.6 ± 10.5 (ng/mL). VD3 deficiency was found in 86 subjects (75.4%). Our research showed that VD3 deficiency could be... an important factor influencing maxillary development. Patients had a greater risk of a narrowed upper arch..., crowding... and crossbite.... Thus, there was a link between the deficiency of this hormone and the underdevelopment of the maxilla. (Leszczyszyn et al., 2021)

VD3 is essential for normal skeletal development and bone health. A major part of VD3 originates from exposure to ultraviolet rays of sunlight, while only a small portion depends on food intake (Saraff & Shaw, 2016). VD3 deficiency and its consequences are not a thing of the past but a serious global health problem today:

(A) "Vitamin D deficiency (serum 25-hydroxyvitamin D [25(OH)D] < 50 nmol/L or 20 ng/ml) is associated with unfavorable skeletal outcomes, including fractures and bone loss. A 25(OH)D level of >50 nmol/L or 20 ng/ml is, therefore, the primary treatment goal, although some data suggest a benefit for a higher threshold. Severe vitamin D deficiency with a 25(OH)D concentration below <30 nmol/L (or 12 ng/ml) dramatically increases the risk of excess mortality, infections, and many other diseases, and should be avoided whenever possible.... Prevalence rates of severe vitamin D deficiency, defined as 25(OH)D <30 nmol/L (or 12 ng/ml), of 5.9% (US)..., 7.4% (Canada)..., and 13% (Europe)... have been reported. Estimates of the prevalence of 25(OH)D levels <50 nmol/L (or 20 ng/ml) have been reported as 24% (US), 37% (Canada), and 40% (Europe)... This may vary by age, with lower levels in childhood and the elderly..., and also ethnicity in different regions, for example, European Caucasians show lower rates of vitamin D deficiency compared with nonwhite individuals... Worldwide, many countries report very high prevalences of low vitamin D status. 25(OH)D levels <30 nmol/L (or 12 ng/ml) in >20% of the population are common in India, Tunisia, Pakistan, and Afghanistan. For example, it has been estimated that 490 million individuals are vitamin D deficient in India..." (Amrein et al., 2020) (references omitted)

(B) "An overall pooled estimate, irrespective of age group, ethnic mix, and latitude of study populations, showed that 13.0% of the 55,844 European individuals had serum 25(OH)D concentrations <30 nmol/L on average in the year, with 17.7% and 8.3% in those sampled during the extended winter (October-March) and summer (April-November) periods, respectively. According to an alternate suggested definition of vitamin D deficiency (<50 nmol/L), the prevalence was 40.4%." (Cashman et al., 2016)

(C) The lack of VD3 due to insufficient exposure to the sun and air pollution, which reduces ultraviolet exposure and therefore the synthesis of VD3, is a problem of global importance, also widespread in tropical countries, although this may seem like a paradox (Augustine et al., 2021). Critical factors are the covering of the skin with clothing and air pollution (evaluated as a weighted annual concentration of particulate matter less than or equal in diameter to 2.5 microns, $PM_{2.5}$). For example, in Bangladesh (rural population = 66%; $PM_{2.5}$ = 83.3) 80% of the population shows VD3 < 50 nmol/l. On the contrary, in Fiji (rural population = 46%, $PM_{2.5}$ = 7) there is only 11% of the population with VD3 < 50 nmol/l (data from (Augustine et al., 2021), Table 1).

However, as already intuited by Price (see Figs. 3.3, 3.4, 3.5, 3.6 and 3.7, showing some splendid photos from his outstanding book (Price, 1939), which should be purchased and read in its entirety), for the correct development of the skull skeleton, but also of the whole skeletal system, a diet similar to that of our ancestors is necessary. The contemporary one, made up of foods industrially processed in such a way as to eliminate critical natural elements that Price himself was unable to indicate, does not satisfy these needs. A diet that reflects what we are adapted to, i.e., as close as possible to that of our hunter-gatherer ancestors, allows for the correct development of the skeletal system (of which a useful indicator is the measurement of the Bone Mineral Density, BMD):

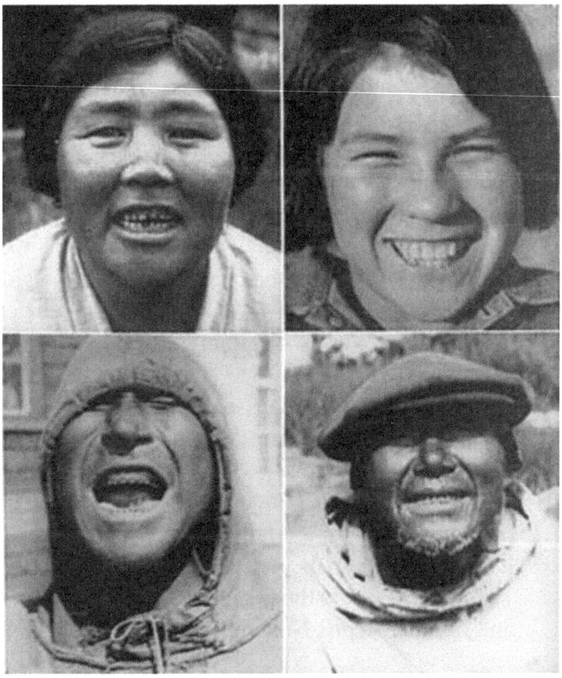

Fig. 3.3 Native Alaskan Eskimos: "broad faces and broad arches and no dental caries... Upper left, woman has a broken lower tooth. She has twenty-six children with no tooth decay" (Price, 1939, fig. 9)

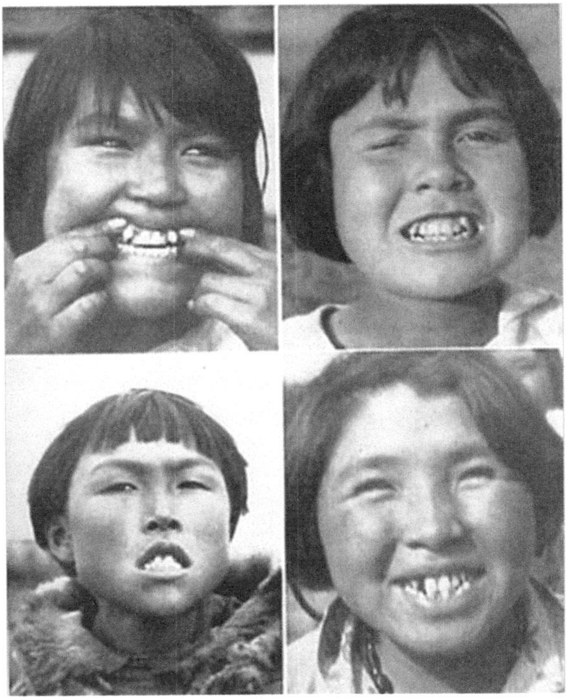

Fig. 3.4 First generation of Eskimo children raised with modern processed foods: "narrow nostrils and changed facial form" (Price, 1939, fig. 12)

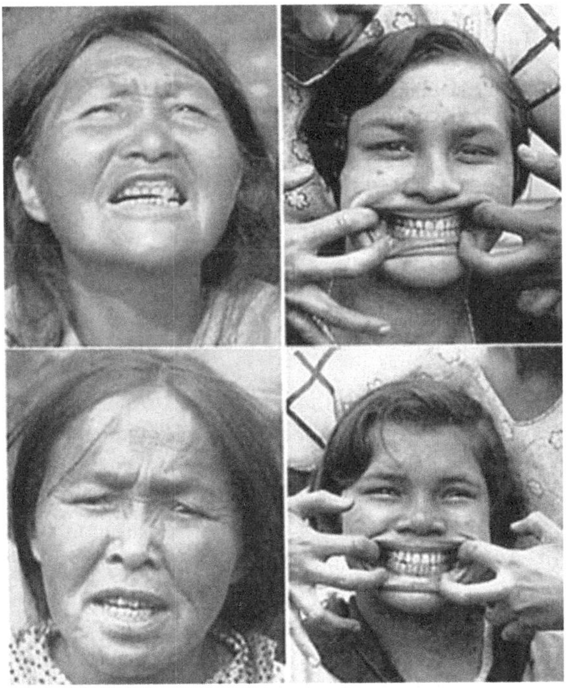

Fig. 3.5 Indians living on their native foods: "facial and dental arch form was superb with nearly complete immunity to dental caries" (Price, 1939, fig. 17)

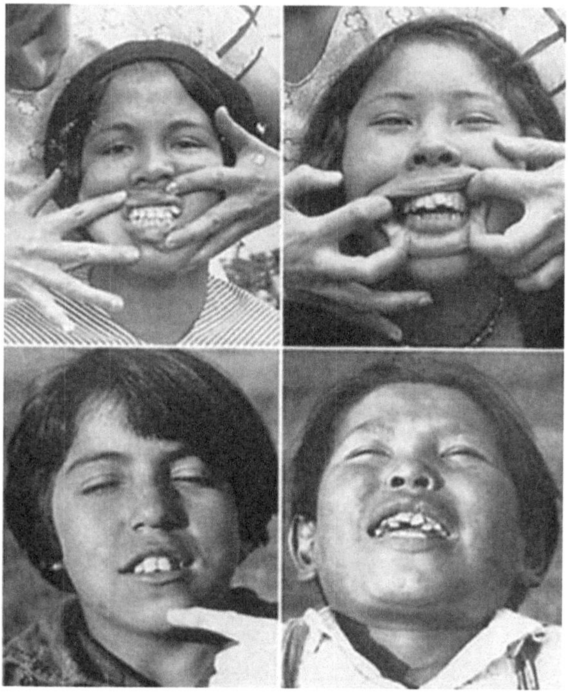

Fig. 3.6 Indians of the first generation living on modern processed foods: "deformed dental arches... narrow nostrils and crowded teeth" (Price, 1939, fig. 19)

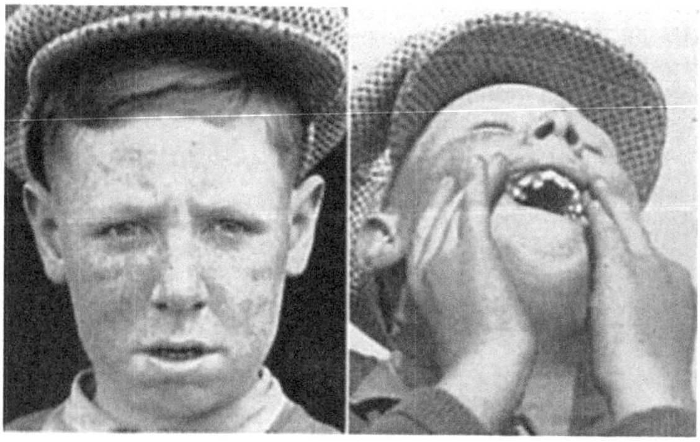

Fig. 3.7 An Alaskan white boy raised on imported foods: "lack of development of the air passages, so that he breathes through his mouth. Lack of bone development creates the crowded condition of the teeth... narrow nostrils." (Price, 1939, fig. 14)

(1) "Nutrition plays an important role, especially key vitamins D3 and K2 which are necessary for proper dentofacial development and food consistency influence on crowding and dental arches narrowing. Changes in our dentition and facial appearance are caused by changing our diet from primitive hunter gatherer to a more modern industrialized agriculture." (Tyszkowski, 2016)

(2) "Fruits and vegetables provide a multitude of micronutrients such as vitamin K, folate, magnesium, potassium as well as antioxidants such as vitamin C and carotenoids. Higher fruit and vegetable intakes have been associated with higher BMD (Prynne et al., 2006) and less BMD loss over time (Macdonald et al., 2004)." (Sahni et al., 2015)

(3) "Traditionally, nutrition research has focused on single nutrients in relation to health. Yet this approach is limited... To overcome these limitations, some investigators have proposed a "dietary patterns" approach. Results from the Framingham Study Cohorts have shown dietary patterns to be predictive of BMD among adults. In the Framingham Original Cohort six dietary patterns were identified with the greatest proportions of energy intake from the following groups: (1) meat, dairy and bread; (2) meat and sweet baked products; (3) sweet baked products; (4) alcohol; (5) candy; (6) fruit, vegetables and cereal (Tucker et al., 2002). For men, the fruit, vegetables, and cereal group had the greatest BMD at all bone sites examined." (Sahni et al., 2015)

It is useful to underline that in order to have an optimal development of the skeletal system with a high BMD and to avoid or limit osteoporosis, valid physical activity is also essential (Cosman et al., 2014; Faienza et al., 2020; Troy et al., 2018), i.e., as for hunter-gatherer populations (Eaton et al., 1988a).

The harmful consequences of the transition from an ancestral lifestyle to a modern lifestyle are also demonstrated by other studies. For example, regarding the teeth of Alaskan Eskimo children following a modern lifestyle, it was observed: "In spite of periodic dental care provided by the United States Public Health Service, these children continue to lose teeth at an alarming rate as a result of the high incidence of caries. This condition has been shown to be caused by the change from a primitive diet of meat and fat to one which incorporates the starches and refined sugars introduced by the white man.... Although previous studies of primitive Eskimos have reported practically no malocclusion, 82 per cent of the children in this study had malocclusions." (Wood, 1971).

Some examples of malocclusions are shown in Fig. 3.8.

Fig. 3.8 Same examples of malocclusions

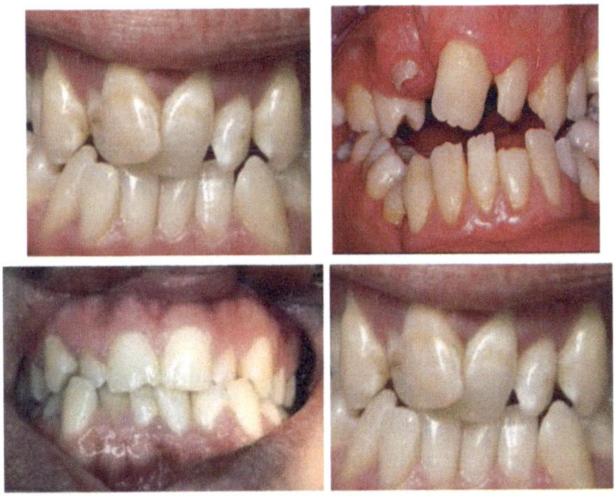

3.2.3 Exposure to Harmful Substances

For the sake of brevity, attention will be focused on just two types of common exposure to harmful substances and related diseases.

Tobacco Smoking and/or Air Pollution

The habit of smoking tobacco, non-existent in the ecological niche to which we are adapted, creates a serious mismatch with our physiology as evidenced by various diseases that are correlated with this habit:

(A) cardiovascular disease, heart failure, atrial fibrillation, and venous thromboembolism (Benowitz & Liakoni, 2022; Gallucci et al., 2020; Kondo et al., 2019; Teo & Rafiq, 2021);
(B) respiratory diseases, chronic obstructive pulmonary disease, emphysema (Lange et al., 2021; Lu et al., 2018; Sales et al., 2019);
(C) various types of cancer (Sun et al., 2020), in particular, gastrointestinal cancer (Lu et al., 2021; Scherübl, 2022), lung cancer (Barta et al., 2019; de Groot et al., 2018; Parascandola & Xiao, 2019; Thandra et al., 2021), kidney cancer (Chow et al., 2010; Harrison et al., 2021), bladder cancer (Cumberbatch et al., 2018; Ng, 2022), laryngeal carcinoma (Cavaliere et al., 2021); prostate cancer (Pernar et al., 2018); and pancreatic cancer (Hu et al., 2021);
(D) a variety of oral diseases: "increased pocket depth, loss of alveolar bone, tooth mobility, oral lesions, ulcerations, halitosis, and stained teeth" (Ahmed et al., 2021);
(E) "various neuropsychiatric diseases, such as dementia / cognitive decline, schizophrenia / psychosis, depression, anxiety disorder, and suicidal behavior" (Hahad et al., 2021);
(F) urolithiasis (Yue et al., 2022).

In short: "... tobacco use is a major risk factor for cardiovascular and respiratory diseases, over 20 different types or subtypes of cancer, and many other debilitating health conditions. Every year, more than 8 million people die from tobacco use." (WHO, 2023c, Tobacco)

The relationships between tobacco smoking (and/or air pollution) and chronic bronchitis (Viegi et al., 2006) and emphysema (Taraseviciene-Stewart & Voelkel, 2008) are well known.

It has also been demonstrated that, in a population, the increase or decrease in smoking habit, estimated on the basis of the percentage of individuals who smoke, is correlated with the increase or decrease, respectively, of the incidence of chronic bronchitis (Pelkonen et al., 2014; Vartiainen et al., 2022).

Exposure to Chemical Substances Artificially Synthesized

The ability of artificial substances to cause or aggravate allergic diseases has been known for some time. For example, a study on the types of allergic responses among employees working in the chemical industry showed that artificial chemicals are correlated with a wide variety of hypersensitivity reactions (Kirchner, 2002).

This harmful effect of substances that certainly did not exist in the ecological niche to which we are adapted has been confirmed by subsequent studies (Pawankar & Wang, 2021; Nanda et al., 2022; Sanchez Armengol et al., 2022; Yong et al., 2022).

For example: "...various environmental pollutants, such as air pollutants and chemical substances, have been shown to worsen various allergies in experimental studies. For example, diesel exhaust particles (DEPs), which are an agglomeration of particles and a wide array of chemical substances, aggravate asthma, primarily due to the principle organic chemical components of DEPs. In addition, environmental chemicals such as phthalate esters, which are commonly used as plasticizers in plastic products, also aggravate atopic dermatitis." (Takano & Inoue, 2017).

Substances present in the natural environment at very low concentrations may have a similar harmful effect, causing allergies and other diseases, when they reach higher concentrations due to industrial procedures. E. g., nickel: "Environmental pollution from nickel may be due to industry, the use of liquid and solid fuels, as well as municipal and industrial waste. Nickel contact can cause a variety of side effects on human health, such as allergy, cardiovascular and kidney diseases, lung fibrosis, lung and nasal cancer." (Genchi et al., 2020).

3.2.4 Excessive Noise Levels

Occupational noise or excessive exposure to noise from other sources → hearing loss (Chen et al., 2020; Daniel, 2007; Ding et al., 2019; Mao & Chen, 2021).

In addition to excessive noise, smoking, obesity/excess weight (i.e., a high Body Mass Index), and type 2 diabetes mellitus are positively correlated with hearing loss (Fransen et al., 2008; Samocha-Bonet et al., 2021; Yang et al., 2020).

It should be emphasized that these risk factors for hearing loss derive from conditions that do not exist in primitive conditions and, therefore, constitute alterations of the "normal" condition and likely cause of "mismatch".

"Life in Western nations is an assault on the ears... noise levels common in Western society are a major threat to hearing. About 25 million Americans have enough hearing impairment to qualify for workmen's compensation..... The acoustic environment of our ancestors was relatively noise free. Otologists (ear specialists) have studied the hearing of Maaban agriculturalists living in the southern Sudan and of San (Bushmen) in the Kalahari – people whose lives are spent amid noise levels similar to those encountered by Paleolithic humans. When the hearing of these people is contrasted with that of Westerners, marked differences in the patterns of age-related hearing loss become evident. Average seventy-year-old Maaban men have a 15 decibel (dB) hearing loss (at a standard frequency of 4000 cycles per second) when compared with fifteen-year-old Maaban boys. But, among American males, seventy-year-olds have a 65 dB loss when compared with fifteen-year-old American youths – five times as much hearing impairment! The difference between Maaban and American women is slightly less pronounced, but still dramatic: a 20 dB loss for Maabans, as opposed to a 50 dB loss for Americans! None of the Maabans or the San have presbycusis (hearing loss with age) sufficient to produce functional impairment. Before these studies were conducted, presbycusis severe enough to interfere with conversation had been considered a 'normal' concomitant of aging, but the absence of significant hearing impairment among people who, like our ancestors, live in an environment where there is little exposure to noise suggests that what we have accepted as the normal accompaniment of advancing years is just another indication of the abnormally traumatic nature of our 'civilized' environment." (Eaton et al., 1988a, pp. 57–58). For the data about Maaban and San populations, see (Rosen et al., 1962; Jarvis & Van Heerden, 1967).

Nowadays: "Deafness and hearing loss are widespread and found in every region and country. Currently more than 1.5 billion people (nearly 20% of the global population) live with hearing loss. 430 million of them have disabling hearing loss. It is expected that by 2050, there could be over 700 million people with disabling hearing loss." (WHO, 2023b, Deafness).

3.2.5 Altered Exposure to Light in the First Years of Life

Concerning the primary causes of refractive defects of vision (myopia, astigmatism, and hyperopia), alterations now so widespread as to be considered almost normal (Fig. 3.9), there is a series of facts that must be considered in order to place these disturbances in the logic of evolutionary medicine.

Absence or Rarity in People Living in Evolutionary "Normal" Conditions

The study of primitive peoples shows that these defects are very rare or nonexistent among them:

(1) As regards Australian Aborigines: "The marvelous vision of these primitive people is illustrated by the fact that they can see many stars that our race cannot see. In this connection, it is authoritatively recorded regarding the Maori of New Zealand that they can see the satellites of Jupiter which are only visible to the white man's eye with the aid of telescopes. These people prove that they can see the satellites by telling the man at the telescope when the eclipse of one of the stars occurs. It is said of these primitive Aborigines of Australia that they can see animals moving at a distance of a mile which ordinary white people cannot see at all." (Price, 1939)

Fig. 3.9 The current epidemic of refractive defects

(2) As regards Yakuts (a Siberian people): "Many travelers observed what some of them call 'telescopic' eyesight among these peoples. A Yakut distinguished with the naked eye stars in the Pleiades not usually seen without a telescope. The Yakuts say there are many stars in this group, but only seven large ones." (De Hutorowicz & Adler, 1911)

(3) As regards two hunter-gatherer populations: "Using a retinoscope and cycloplegia, Holm (Holm, 1937) refracted 2364 members (aged 20-65 years) of several hunter-gatherer tribes in Gabon (formerly French Equatorial Africa) in 1936. Of the 3624 eyes examined, only 14 were classified as myopic (nine eyes from -0.50 to 1.00 D; five eyes from -3.00 to -9.00 D), thereby yielding a myopia incidence rate of 0.4%. Similar low rates for myopia were reported by Skeller (Skeller, 1954), who refracted the eyes of 775 Angmagssalik Eskimos as part of a comprehensive anthropological study carried out in 1954. Retinoscopy in conjunction with cycloplegia demonstrated that of the 1123 eyes examined, only 13 (1.2%) were classified as myopic (nine eyes = -1.00 D; four eyes = -1.25 D)." (Cordain et al., 2002)

Current High Frequency

In modern populations, for some decades, refractive defects have been very common (Chow et al., 1990; Fredrick, 2002) and varying in frequency from population to population:

(A) "Myopia is not a simple refractive error, but an eyesight-threatening disease. There is a high prevalence of myopia, 80% to 90%, in young adults in East Asia; myopia has become the leading cause of blindness in this area. As the myopic population increases globally, the severity of its impact is predicted. Approximately one fifth of the myopic population has high myopia (≥-6 diopters), which results in irreversible vision loss such as retinal detachment, choroidal neovascularization, cataracts, glaucoma, and macular atrophy." (Wu et al., 2016a)

(B) In papers published between 2013 and 2019: "Myopia prevalence remains higher in Asia (60%) compared with Europe (40%) using cycloplegic refraction examinations. Studies reporting on non-cycloplegic measurements show exceptionally high myopia prevalence rates in school children in East Asia (73%), and high rates in North America (42%). Low prevalence under 10% was described in African and South American children." (Grzybowski et al., 2020)

(C) "The prevalence of myopia in Asia is as high as 70-90%, in Europe and America 30-40%, and in Africa 10-20 %" (Fredrick, 2002).

(D) In a study of US, Western European, and Australian adults (aged 40 years or older), the estimated crude prevalence for hyperopia (+3 D or greater) was 9.9%, 11.6%, and 5.8%, respectively; for myopia (−1 D or less), 25.4%, 26.6%, and 16.4%, respectively; for high myopia (−5 D or less), 4.5%, 4.6%, and 2.8%, respectively (Kempen et al., 2004).

(E) In one work on Singaporean adults (aged 40 to 79 years), the overall prevalence of myopia, hyperopia, astigmatism, and anisometropia was 38.7%, 28.4%, 37.8%, and 15.9%, respectively, and that of high myopia was 9.1% (Wong et al., 2000).

Strong Frequency Increase in a Few Decades

Multiple studies show a significant increase in refractive defects in a few decades:

(1) "The prevalence of myopia amongst the Chinese [of Hong Kong] has escalated in recent decades." (Edwards & Lam, 2004);

(2) In 1969, it was observed (Young et al., 1969) for recently acculturated Eskimos in Barrow (Alaska) that the incidence of myopia was a function of age: while for subjects over 41 years of age, who had lived in the traditional Eskimo mode, myopia was sporadic (1.5%), in subjects between 11 and 40 years of age myopia of at least -0.25 D had a substantial incidence (51.4%). This difference was necessarily due to the new lifestyle;

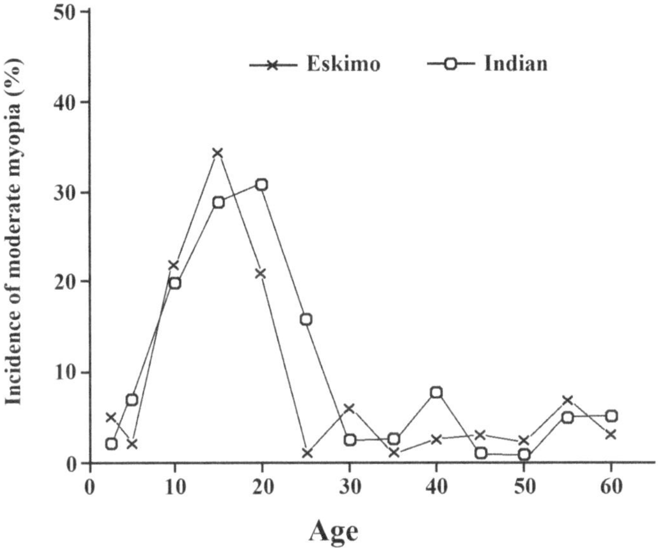

Fig. 3.10 Incidence of moderate myopia (−1.0–5.0 D) by age in Indians (n. = 844) and Eskimos (n. = 2833) of the Yukon and Northwest territories; data from (Morgan & Munro, 1973)

(3) In 1973, an analogous study (Morgan & Munro, 1973) on recently acculturated Indians and Eskimo of the Yukon and Northwest territories showed a high incidence of myopia (25–35%) in subjects aged 10–20 years, while in older subjects aged 30–60 years, the incidence was 2–7% (Fig. 3.10).

Different Frequencies in Analogous Populations Divided by Lifestyles

In Nepal, Sherpa and Tibetan children have a common ancestry but different educational and environmental conditions, i.e., rural lifestyle for Sherpa children and more rigorous schooling for Tibetan children. The prevalence of myopia in the two populations was 2.9% and 21.7%, respectively (Garner et al., 1999).

Hypotheses on Possible Alterations of the Ecological Niche Underlying Refractive Defects of Vision and on Possible Pathogenetic Mechanisms

"When the human visual system is examined from an evolutionary perspective, it becomes apparent that humans, living in the original environmental niche for which our species is genetically adapted (as hunter-gatherers), are either slightly hypermetropic or emmetropic and rarely develop myopia. Myopia occurs when novel environmental conditions associated with modern civilization are introduced into the hunter-gatherer lifestyle.... It is certainly plausible that the natural selection pressures that had previously strongly selected against myopia in hunter-gatherers may have been slightly reduced with the advent of organized society. Moreover, these selective pressures would have been almost completely eliminated with the wide-scale availability of spectacle lenses in the 200 years since the industrial revolution. However, a number of lines of evidence strongly reject the notion that this recent (in evolutionary terms) relaxation of natural selection pressures could be responsible for the high incidence of myopia in modern, technological societies." (Cordain et al., 2002)

A common hypothesis for the causes of refractive defects, in particular myopia, is excessive near work, especially using artificial lighting, an improbable condition in the ecological niche to which we are adapted (Gordon-Shaag et al., 2021; McBrien & Adams, 1997; Zylbermann et al., 1993).

Against this hypothesis: "In an earlier study of 977 school children (6-17 years of age) on the remote South Pacific island of Vanuatu, Garner et al. (Garner et al., 1985) found that only 1.3% of subjects had myopia greater than -0.25 D, despite engaging in about 8 hrs of school work per day." (Cordain et al., 2002).

Moreover, in a study on 3009 Singaporean children aged 6–72 months, i.e., before school age, the prevalence of overall myopia, high myopia (at least −6.00 D), hyperopia, astigmatism, and anisometropia were 11%, 0.2%, 1.4%, 8.6%, and 0.6%, respectively (Dirani et al., 2010). The hypothesis of excessive near work as the cause of refractive defects of vision does not explain their early onset.

Another factor, on the other hand, appears to be correlated with refractive defects.

A study was made of two homogeneous groups of 6–7-years children of Chinese ethnicity, living in Singapore and Sidney, respectively, with the same frequency of myopia in their parents: "Children in Sidney read more books per week ($P < 0.001$) and did more near-work activity ($P = 0.002$). Children in Sidney spent more time on outdoor activities (13.75 vs 3.05 hours per week) ($P < 0.001$)". The prevalence of myopia was 3.3% in Sidney and 29.1% in Singapore (Rose et al., 2008b).

This notice suggests that the critical factor is the outdoor activity, alias the exposition to natural light, a hypothesis confirmed by other studies (Dirani et al., 2009; Rose et al., 2008a).

In particular, "Higher levels of total time spent outdoors, rather than sport per se, were associated with less myopia" (Rose et al., 2008a).

In these studies, near-work activities appeared to be an independent aggravating factor but not the leading cause of myopia (Dirani et al., 2009; Rose et al., 2008a).

Study of the Mechanisms Linking Alterations of the Ecological Niche to the Pathogenesis of the Disease

It is known—from experiments on chickens—that the application (for 7 days, beginning at 5 days of age) of spherical defocus lenses or of translucent occluders to one eye, or exposure to constant light, cause astigmatism associated with myopia or hyperopia.

In control birds, astigmatism is normal at birth and disappears, or decreases, over the following days, while in chickens with visual manipulation, the percentage of astigmatism >1 D ranged from 66.7% to 100% (Kee & Deng, 2008). Similar results were obtained in monkeys (Kee et al., 2005).

In our species, too, astigmatism is regular at birth and disappears in the first years of life:

(i) "Infants have a high incidence of clinically significant astigmatism.... Of 28 children who had large amounts of astigmatism in the first year, all showed elimination or a large reduction in the amount of the cylindrical error by 4 years" (Gwiazda et al., 1984);
(ii) "Full term newborn babies are known to be on average hypermetropic at birth. Preterm babies tend to be myopic when examined at an age corresponding to term …" (Varghese et al., 2009)

Thus, at birth, eye refraction is imperfect as the image is focused either too far before the retina (hyperopia) or too far behind (myopia) and often, there is also a deformation of the image on one plane versus the other (astigmatism).

The genetic program of eye development appears to be: (i) at birth, an eye will have an imperfect vision; (ii) immediately after birth, there is modulation of the further eye development in order to achieve optimum vision ["emmetropization" (Jonas et al., 2019; Mutti et al., 2005; Troilo et al., 2019); see Fig. 3.11].

Under modern conditions of life, particularly in urban areas with high population densities, babies do not grow up outdoors with exposure to natural light. It is a widespread belief that babies should be protected as much as possible from the external environment, in particular from direct exposure to sunlight, as this could damage them in some way.

Many infants spend their first months of life almost exclusively indoors, only exposed to artificial lighting, with uneven and weak brightness, and, in all cases, under conditions quite different from those to which our species is adapted.

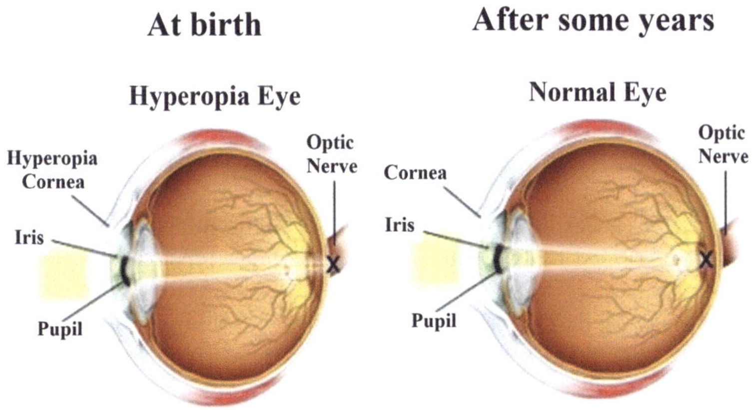

Fig. 3.11 At birth, in general, the eye is a little too short, and the image is in focus before the retina (hyperopia). Furthermore, there is often a difference in curvature between the axial planes (astigmatism). Immediately afterward, under normal lighting conditions, the eye adjusts its proportions so that the image is focused exactly on the retina (emmetropia)

For proper eye development, babies should be exposed as much as possible to natural light conditions, sunlight should not be avoided, and lighting should be as similar as possible to the original ones.

3.2.6 Multiple Carcinogenic Factors

The term "cancer" indicates a heterogeneous set of diseases characterized by uncontrolled reproduction of somatic cells. These illnesses have different causes, pathogenetic mechanisms, course, and prognosis. Neoplastic diseases have a significant incidence and an enormous social and economic impact: "The overall 0-74 years risk of developing cancer is 20.2% (22.4% in men and 18.2% in women, respectively). A total number of 18 million new cases have been diagnosed in 2018, the most frequent of which are lung (2.09 million cases), breast (2.09 million cases), and prostate (1.28 million cases) cancers.... As mortality concerns, cancer is the second worldwide cause of death (8.97 million deaths) after ischemic heart disease." (Mattiuzzi & Lippi, 2019).

The large incidence of cancer might lead to the belief that our species is intrinsically at a high risk of developing malignant pathologies. However, according to the criteria of evolutionary medicine, it is necessary to verify the incidence of this type of disease in populations living in evolutionarily "normal" conditions or at least as close as possible to such conditions.

For the Dobe !Kung people (Bushmen of the Kalahari Desert of Botswana) living in the original conditions of hunter-gatherers: "Three of the major causes of death in most modern societies – cardiovascular disease, cancer, and motor vehicle accidents – are nonexistent or minor problems for the !Kung" (Howell, 1979, p. 66). This assertion is confirmed even considering the age structure of deaths and competing risks (Truswell & Hansen, 1976).

About the Ache people, a hunter-gatherer population of Paraguay, before the contact with modern populations, i.e., in "the Forest Period", the survivors at age 70 were about 15–20%. Before this age, no case of cancer was observed. After this age, only for a few older individuals not killed by accidents or other identified causes, cancer could have been a cause of death, but no case was described (Hill & Hurtado, 1996).

The rarity of cancer in populations living according to evolutionary "normal" lifestyles is confirmed by two authoritative testimonies reported by Price (Price, 1939), one from Dr. J. R. Nimmo, the government physician in charge of Torres Strait Islands people during that period, and the other from Dr. J. Romig a surgeon of Anchorage with 36 years of experience in Alaska:

(i) Dr. Nimmo, in his 13 years with Torres Strait Islands people, "had not seen a single case of malignancy, and seen only one that he had suspected might be malignancy among the entire four thousand native populations. He stated that during this same period he had operated on several dozen malignancies for the white populations, which numbers about three hundred." (Price, 1939, p. 179);

(ii) Dr. Romig, "a surgeon of great skill and with an experience among the Eskimos and the Indians, both the primitives and the modernized … stated that in his thirty-six years of contact with these people he had never seen a case of malignant disease among the truly primitive Eskimos and Indians, although it frequently occurs when they become modernized." (Price, 1939, p. 83).

These data show that the differences between evolutionarily "normal" lifestyles and those of "modernized" populations were associated with an enormous difference in the incidence of neoplastic diseases.

According to the criteria of evolutionary medicine, when the frequency of a disease, or a set of diseases, passes from a minimum frequency (in an ecological niche that is identical or similar to that to which the species is adapted) to a high frequency, which is associated with the altered ecological niches of modern lifestyles, the increase cannot be due as primary cause to genetic alterations, as these cannot become frequent in a few years or very few generations. On the contrary, the causes must be sought in alterations of the ecological niche.

This affirmation is in accordance with the comment that "Only 5-10% of all cancer cases can be attributed to genetic defects, whereas the remaining 90-95% have their roots in the environment and lifestyle." (Anand et al., 2008).

However, in a paper from a few years ago, the authors proposed that only one-third of the cases of cancer could be attributed to environmental factors or genetic predisposition. On the contrary, most of the cases, i.e., two-thirds, must have been the consequence of unfortunate harmful somatic mutations (Tomasetti & Vogelstein, 2015). Apart from other objections to the work, a subsequent study (Wu et al., 2016b) proposed that noxious mutations could be responsible for no more than 10–30% of cancer cases, while the majority of cases (70–90%) must be caused by extrinsic factors.

Neither of these works explained how intrinsic factors, i.e., bad somatic mutations, in modern populations would cause cancer in a large percentage of cases (67% or 10–30%, respectively), while in primitive populations, these bad mutations do not cause cancer with a similar incidence. However, it is essential to consider the age structure of deaths since, in modern populations, there is a growing number of individuals who reach ages that are inexistent or very rare in primitive conditions and so are more vulnerable to the accumulation of harmful somatic mutations. At the same time, it should be considered that natural selection is capable of developing protective mechanisms against mutations, as demonstrated by recent work (Cagan et al., 2022), where an inverse correlation between the lifespan of a species and detectable somatic mutation rates per year was observed. Indeed, this more rigorous restriction of somatic mutation rates prevents, at a certain age, a greater number of cells from reaching a critical level of mutations that could trigger cancer (Gorelick & Naxerova, 2022). However, at the same time, the anti-cancer capacity of natural selection cannot have any effect for ages that are inexistent or very rare in natural conditions.

For many types of cancer, associations between modern risk factors, i.e., alterations of the ecological niche to which we are adapted, and specific types of cancer are known and documented. For example, tobacco smoking, which does not exist in the ecological niche to which we are adapted, is a major risk factor for many types of cancer. For other types of cancer, risk factors are less obvious and often multiple. In these cases, as the type of cancer under examination is nonexistent or very rare in evolutionarily "normal" life conditions, for evolutionary medicine, the primary cause is certainly one or more alterations of the ecological niche. Consequently, it is of overriding importance to search for these alterations and to look for ways to reduce or eliminate them. Where other causes are to be proposed, it is necessary to explain why the proposed causes do not cause cancer under evolutionarily "normal" conditions.

Regarding some types of cancer and the association with specific risk factors that are alterations of the ecological niche to which we are adapted, Table 3.4 shows various known correlations between some types of cancer and specific risk factors that are alterations of the ecological niche to which we are adapted.

Table 3.4 is necessarily incomplete, both for a possible insufficiency of studies regarding particular associations or for the lack of identification of the same associations, but also for two possible reasons:

(i) a specific alteration can cause a particular type of cancer with minimal frequency, and the relevant incidence of that type of cancer can also depend, at least in part, on the summation of the effects of many alterations with minimal incidence. For example, suppose many substances present in the environment or the foods or particular conditions are carcinogenic with low pathogenic incidence. In that case, we will have many associations of minimal importance, while the overall effect will be considerable without a single factor with predominant responsibility. As a specific example, there are more than 4000 components in cigarette smoke (Smith & Hansch, 2000). It is practically impossible, and perhaps not very useful, to define the pathogenic effect separately for each component on the numerous pathologies associated with tobacco smoking;
(ii) many alterations can have a synergistic pathogenic effect, and therefore, the effect of a single alteration will not be evident.

Indeed, for many types of cancer, but also other types of diseases, we will have one or more factors that are the main culprits and a remaining part of the cases that have no obvious causing factors. However, even where not all the factors responsible for a disease are known, it is essential to consider the difference between the frequency of cases in evolutionarily "normal" conditions and the frequency in the conditions studied.

Table 3.4 Known associations between specific alterations of the ecological niche and certain types of cancer

Cancer type	Association with...
Lung cancer	tobacco smoking, secondhand smoke, exposure to radon, occupational exposure to carcinogens (Barta et al., 2019; Schabath & Cote, 2019); use of coal and unprocessed biomass fuels (Kurmi et al., 2012)
Larynx cancer	tobacco smoking, alcohol abuse, exposure to asbestos, polycyclic aromatic hydrocarbons, and textile dust (Deng et al., 2020; Steuer et al., 2017)
Pancreatic cancer	diabetes mellitus, tobacco smoking, alcohol, dietary factors, pancreatitis, obesity, infection, and microbiota modifications (Hart et al., 2008; Hu et al., 2021); in the US, the estimated new cases were about 27,800 in 2001 and 57,400 in 2019 (+106%) (Hu et al., 2021)
Bladder cancer	Tobacco smoking and exposure to certain chemicals in the working and general environments (Cumberbatch et al., 2018; Janković & Radosavljević, 2007; Ng, 2022); exposure to aromatic amines, aromatic polycyclic hydrocarbons, and chlorate hydrocarbons, and water consumption with high levels of arsenic (Rozanec & Secin, 2020)
Gallbladder cancer	Gallstones, cholangitis, obesity/overweight, diabetes mellitus, consumption of sweetened beverages (Schmidt et al., 2019)
Kidney cancer	Tobacco smoking (Harrison et al., 2021); tobacco smoking, obesity/overweight, hypertension, diabetes mellitus, and an inverse correlation with physical activity (Chow et al., 2010; Lipworth et al., 2006)
Gastric cancer	*Helicobacter pylori* (see *Sect. 4.1.6 Alterations of the holobiont—F: The case of Helicobacter pylori*), tobacco smoking, alcohol abuse (Machlowska et al., 2020); roiled and charbroiled animal meats, salt-preserved foods, and smoked foods, while fruit and vegetables are protectors against gastric cancer (Kim et al., 2014)
Gastrointestinal cancer	Alcohol consumption, tobacco smoking, obesity/overweight, insufficient physical activity, and high blood cholesterol level, and inverse correlation with hypertension and diabetes (Lu et al., 2021); tobacco smoking (Scherübl, 2022)
Proximal and distal colon cancer	Low dietary fiber intake (Ma et al., 2018); red meat, particularly processed meat (Aykan, 2015; Battaglia Richi et al., 2015); red and processed meat intake, alcohol abuse, and inverse correlation with use of whole grains, vegetables, and fish intake (Vieira et al., 2017)
Prostate cancer	Obesity/overweight, tobacco smoking, low physical activity (Pernar et al., 2018)
Soft tissue sarcomas	Tobacco smoking, chronic inflammation, toxins, and genetic predisposition (a factor not due to alterations of the ecological niche) (Weskamp et al., 2022)
Liver cancer	apart from cases due to hepatitis viruses, heavy alcohol consumption, diabetes, and/or obesity (→ nonalcoholic fatty liver disease) (Kim & Park, 2018; Taniai, 2020)
Breast cancer	early menarche, late menopause, nulliparity or first childbirth after age 30, lower mean duration of breastfeeding or not breastfeeding, obesity/overweight, tobacco smoking, high dietary intake of fats or fatty foods, poor physical activity, alcohol abuse, contraception, environment toxicants, and genetic predisposition (a factor not due to alterations of the ecological niche), while there is an inverse relationship with the intake of dietary fruits, vegetables, and plant- and soy-based products (Kashyap et al., 2022; Youn & Han, 2020)

3.2.7 Altered Social Interactions

There are various psychiatric diseases that, in most cases, originate from alterations of our ecological niche, leading to a mismatch between our genetic adaptation and the new living conditions. Some of these pathologies are mentioned here.

Eating Disorders

Although cases of eating disorders have also been witnessed from ancient times, these diseases have become common from the 1960s onwards in technologically advanced countries, first in women of the wealthier classes and then spreading to other classes as well (Rosenvinge & Pettersen, 2015).

> Eating disorders are considered chronic diseases of civilization. The most studied and well known are anorexia and bulimia nervosa. Anorexia is considered one of the most common psychiatric problems of girls in puberty and adolescence. (Kontić et al., 2012)

"Over the years, the number of individuals experiencing high levels of body dissatisfaction, body distortion and eating disorders has grown in epidemic proportions (WHO, 1992; APA (American Psychiatric Association), 2000)... Eating disorders involve a serious disturbance in eating behavior – either eating too much or too little, in addition to great concern over body size and shape. The voluntary eating of smaller or larger portions of food than usual is common, but for some people this develops into a compulsion and the eating behaviors become extreme. Individuals with anorexia refuse to maintain a minimally normal body weight, carry an intense fear of gaining weight and have a distorted perception of the shape or size of their bodies. Individuals with bulimia undertake binge eating and then use compensatory methods to prevent weight gain, such as induced vomiting, excessive exercise or laxative abuse.... It is estimated that 3% of women will be affected by eating disorders in their lifetime. Approximately 0.5% to 4% of women will develop AN [Anorexia Nervosa] during their lifetime and about 1 to 4% will develop BN [Bulimia Nervosa]. AN and BN predominantly affect young women. Most vulnerable period for the occurrence of these diseases is puberty and adolescence (Crichton, 1996; APA (American Psychiatric Association), 2000)." (Kontić et al., 2012).

Anxiety Disorders

Anxiety and some phobias (e.g., for snakes/spiders/heights) are defense mechanisms that lead us to avoid dangerous situations (Nesse, 2019). Their nature as physiological defenses favored and shaped by natural selection will be discussed in *Sect. 6.3 Mental and behavioral disorders as an adaptation to particular situations*. However, these mechanisms, in modern society with its innumerable complications and stressful situations not existing in the ecological niche to which we are adapted, likely are transformed into psychiatric disorders:

(1) "Anxiety disorders are the most prevalent psychiatric disorders (with a current worldwide prevalence of 7.3% (4.8%-10.9%)... Among them, specific phobias are the most common, with a prevalence of 10.3%, then panic disorder (with or without agoraphobia) is the next most common with a prevalence of 6.0%, followed by social phobia (2.7%) and generalized anxiety disorder (2.2%). Evidence is lacking as to whether these disorders have become more prevalent in recent decades. Generally speaking, women are more prone to develop emotional disorders with an onset at adolescence; they are 1.5 to 2 times more likely than men to have an anxiety disorder..." (Thibaut, 2017)
(2) "Anxiety disorders form the most common type of mental illness.... High prevalence, chronicity, and comorbidity led WHO to rank anxiety disorders as the ninth most health-related cause of disability (GBD 2016 Disease and Injury Incidence and Prevalence Collaborators, 2017). Worldwide, anxiety disorders heavily affect patients and society, accounting for 3.3% of the global burden of disease and costing approximately €74 billion for 30 European countries (Gustavsson et al., 2011)." (Penninx et al., 2021)
(3) "... anxiety disorders very often precede the onset of other psychiatric disorders... anxiety symptoms may be a predictor of worse outcome (e.g., suicidality in patients with depression)... anxiety disorders are associated with substantial individual impairment (including reduced educational attainment, more unstable marriage, lower occupational status)..." (Stein et al., 2017)
(4) "Factors positively associated with anxiety and depressive disorders were female sex, middle age, low level of education, financial difficulty, being a housewife, and relationship problems. Arguments with husbands and relational problems with in-laws were positively associated in 3/11 studies. Those who had close confiding relationships were less likely to have anxiety and depressive disorders. Mean overall prevalence of anxiety and depressive disorders in the community population was 34% (range 29-66% for women and 10-33% for men). There were no rigorously controlled trials of treatments for these disorders.... Available evidence suggests a major social cause for anxiety and depressive disorders in Pakistan. This evidence is limited because of methodological problems, so caution must be exercised in generalising this to the whole of the population of Pakistan." (Mirza & Jenkins, 2004)

Pathologic Depression

Just as anxiety and some phobias have an adaptive value, even low mood, which is a way of saying a mild form of depression, has an adaptive value in particular situations (Nesse, 2019).

The adaptive nature of low mood will be discussed in *Sect. 6.2 Mental and behavioral disorders as an adaptation to particular situations*. Even this mechanism, in the innumerable complications and stressful situations of modern society, probably is transformed into widespread pathological conditions:

(A) "Depression is currently one of the most common psychiatric disorders and the number of patients receiving antidepressant treatment is increasing every year. Therefore, it is essential to understand the underlying mechanisms that are associated with higher prevalence of depression. The main component leading to the change in functioning, in the form of apathy, anhedonia, lack of motivation and sleep disturbances, is stress. This is the factor that in recent decades - due to the civilization speed, dynamic technological development as well as competitiveness and competition in relationships - significantly affects the psychophysical condition, which results in an increase in the prevalence of civilization diseases, including depression." (Gałecka et al., 2021);

(B) "MDE [Major Depressive Episodes] includes depressive episodes that occur as part of a bipolar disorder while MDD [Major Depressive Disorders] excludes bipolar depression. As the vast majority of lifetime MDE is MDD, the difference between the two is not of great importance when examining lifetime prevalence estimates.... Weissman et al. (Weissman et al., 1996) published the first cross-national comparison of major depression from 10 population-based surveys that administered the Diagnostic Interview Schedule... to representative community samples. DSM-III criteria were used to define depression. Lifetime prevalence estimates of MDE ranged from 1.5% (Taiwan) to 19.0% (Beirut) in these studies, with the midpoints at 9.2% (West Germany) and 9.6% (Edmonton, Canada). Twelve-month prevalence estimates ranged from 0.8% (Taiwan) to 5.8% (Christchurch, New Zealand), with the midpoints at 3.0% (US) and 4.5% (Paris). A subsequent cross-national comparison (Andrade et al., 2003) included 10 population-based studies that administered the WHO Composite International Diagnostic Interview (CIDI) for DSM-III-R and DSM-IV... Lifetime prevalence estimates of MDD ranged from 1.0% (Czech Republic) to 16.9% (US), with midpoints at 8.3% (Canada) and 9.0% (Chile). The 12-month prevalence estimates ranged from 0.3% (Czech Republic) to 10% (US), with midpoints at 4.5% (Mexico) and 5.2% (West Germany). Moussavi et al. (2007) subsequently summarized data on ICD-10 depressive episode (MDE) in the WHO World Health Survey across 60 countries. Twelve-month prevalence averaged 3.2% in participants without comorbid physical disease and 9.3% to 23.0% in participants with chronic conditions." (Kessler & Bromet, 2013)

Suicide

For the indigenous peoples Ache of Paraguay and Yanomamo of Venezuela, considered in their wildlife conditions before the contact with Researchers (until 1971 for the Ache and between 1970 and 1974 for the Yanomamo), no suicide cases were reported (Hill & Hurtado, 1996; Melancon, 1982).

In 2018, the global annual mortality rate by suicide was estimated by the WHO to be 10.7/100,000 individuals, with rates between 5.0 and 9.9 in 31 countries,

Table 3.5 Risk factors for suicide; from (Fazel & Runeson, 2020), modified

Risk factor	Strength of Association with suicide	Quality of evidence
Neuropsychiatric disorders (severe mental illness, depression, personality disorders)	Strong	High
Drug and alcohol misuse	Strong	High
Family history of suicidal behavior	Strong	High
Life events	Moderate	High
Previous suicide attempt	Moderate	High
Access to lethal means	Moderate	High
New diagnosis of terminal or chronic physical illness	Moderate	Moderate
Adverse childhood experiences	Moderate	Moderate
Effects of the media	Weak	Moderate
Socioeconomic deprivation	Weak	Low

between 10.0 and 14.9 in 22 countries, and greater than 15 in 26 countries (Bachmann, 2018).

In 2020, according to WHO, deaths caused by suicide were 703,220 and ranked 17th among the causes of death (World Cause of Death, 2020).

In 2002, suicide was the second cause of mortality in the age group 15–29 years and the third in the age group 15–44 years (Bertolote & Fleischmann, 2002).

> "Overall, suicides account for 1.4% of premature deaths worldwide.... Completed suicides are three times more common in males than females; for suicide attempts, an inverse ratio can be found. Suicide attempts are up to 30 times more common compared to suicides..." (Bachmann, 2018)

> "Suicide is a major public health concern in the United States. Between 2000 and 2018, US suicide rates increased by 35%..." (Martínez-Alés et al., 2022).

For suicide, risk factors are presented in Table 3.5 in order of strength of association with suicide and quality of evidence (Fazel & Runeson, 2020). Most risk factors are dependent on changes in living conditions compared with those of hunter-gatherer populations, i.e., derive from changes in the ecological niche.

3.2.8 *Infectious Epidemics*

The few surviving hunter-gatherer populations, which are a fairly faithful representation of the living conditions of our ancestors, indicate that our evolutionarily normal condition is a very low population density and the absence of urban agglomerations, factors that hinder the origin, spread, and persistence of an infectious epidemic: "The disease ecology of contemporary hunter-gatherers also provides a model for that of Paleolithic foragers. For example, from studies of contemporary groups we understand that the population size reached by foraging groups would be too small to support continuous transmission of many of the 'crowd' diseases. For example, infections such as influenza, measles, mumps, and smallpox require too large a host population to have been present in the Paleolithic disease-scape..." (Harper & Armelagos, 2011).

Regarding the criticality of population density, as an example, it has been estimated that measles requires a population of more than 500,000 susceptible individuals to sustain an epidemic (Black, 1966).

The transition to ever-greater demographic densities and urban concentrations, as well as increasingly rapid transfers from one community to another, have created ideal conditions for the epidemic spread of pathogenic microorganisms.

However, other factors, which also constitute alterations of the ecological niche, are often necessary for the origin of an infectious epidemic and its persistence. In particular:

(i) Transfer of the microorganism from other animals (zoonosis, spillover infection);
(ii) Concentration of excrement in confined spaces in the absence of sewers and other treatments; close coexistence with species such as mice, insects, and other animals that derive their nourishment from untreated human and animal dung; use of water polluted by such excrement;
(iii) Use of non-sterilized instruments and needles, blood transfusions, organ transplants;
(iv) Particular sexual habits that cause transmission of the infection by blood.

All this has triggered, and history bears witness to, countless infectious epidemics that have caused enormous losses of human life (Bray, 1996; Cartwright & Biddiss, 2014).

It is expected to attribute the infectious epidemics to the microbes determining the pathological manifestations. For example, the cholera epidemic is said to be caused by the *Vibrio cholerae* bacterium, the plague epidemic by the *Yersinia pestis* bacterium, the influenza epidemic by various strains of the influenza virus, and so on. This simple identification of a microbial pathogen as the cause of an infectious epidemic disregards that the primary causes are alterations of the ecological niche without which the microbe could not originate the epidemic.

Let us now briefly consider the primary causes of some of the leading infectious epidemics, always considering that for any common epidemic, other primary causes are the modern great demographic density and the speed of transfer of a pathogen from one community to another:

(1) Acquired Immunodeficiency Syndrome (AIDS) is a zoonosis caused by Human Immunodeficiency Viruses (HIV), originated from contacts with non-human primates in West-central Africa in the early twentieth century and favored in its spread by parenteral transmission and particular sexual habits (Faria et al., 2014; Sharp & Hahn, 2011). In 2020, among the causes of death worldwide, 674,662 were attributed to AIDS (World Cause of Death, 2020);

(2) Chagas disease is a zoonosis also known as American trypanosomiasis because it is largely present in Latin America. It is caused by *Trypanosoma cruzi*, a protozoan, which has, as a natural reservoir, more than 150 animal species and is spread mostly by insects of the subfamily *Triatominae*. (Rassi Jr. et al., 2010). Researchers have found, by polymerase chain reaction, the presence of *T. cruzi*'s kinetoplast DNA in mummies from Peru and Chile dating to approximately 7000 BC or about 9000 years ago (Aufderheide et al., 2004). Charles Darwin was also infected with *T. cruzi* when he landed in Chile during his famous voyage aboard the HMS Beagle (Chao et al., 2020), but the disease was identified as trypanosomiasis caused by *T. cruzi* only in 1909 (Chagas, 1909).

For a long time, *T. cruzi* infections occurred mostly in rural areas. Subsequently, as a result of migratory phenomena and increasingly widespread means of communication, the infections spread in the city centers of Latin America and later also in the USA (Chao et al., 2020).

There are about "6–8 million infected with *T. cruzi* in the Latin American and Caribbean region... With the globalization and increasing mobility during the last decade, Chagas disease is spreading to non-endemic areas, including the United States of America, Canada, and many European and some Eastern Mediterranean and Western Pacific countries..." (Liu et al., 2020) In 2020, among the causes of death worldwide, 7558 were estimated as due to Chagas disease (World Cause of Death, 2020);

(3) For an epidemic of cholera, a disease caused by the *Vibrio cholerae* bacterium, a dense population that discharges untreated excrement in waters that are then used for drinking is necessary. At this point, the germ from the intestine of an infected individual passes through the water used for drinking to other individuals, triggering the epidemic. The ingestion of about 100 million *Vibrio cholerae* bacteria is necessary to cause cholera in a healthy adult (Sack et al., 2004), because few bacteria survive the acidic barrier of the stomach and reach the small intestine where they can thrive, causing the disease (Almagro-Moreno et al., 2015). In 2015, due to the presence of the conditions above, mainly in the developing world (Reidl & Klose, 2002), about 2.8 million people worldwide suffered from cholera, with approximately 95,000 deaths/year (Ali et al., 2015);
(4) COVID-19 (SARS-CoV-2) is a zoonosis originated by a virus transmitted from bats or other wild animals in China (Andersen et al., 2020; Holmes et al., 2021);
(5) Dengue fever is caused by an RNA virus of the family *Flaviviridae*. In 2013, there were about 60 million symptomatic infections worldwide, with about 13,600 deaths (Shepard et al., 2016). The virus is primarily transmitted by *Aedes* mosquitos, in particular *A. aegypty*, and our species is its primary host (Gould & Solomon, 2008). As for malaria and yellow fever, an essential factor for the spread of the disease is the increase in population near areas and points where there are stagnant waters that allow the proliferation of mosquitoes (Normile, 2013);

(6) Hepatitis A is usually spread by drinking water or eating food contaminated by untreated excrement or following contact with infected subjects. For the spread of the disease, the existence of urban agglomerations and untreated excrement is necessary: "Outbreaks are often linked to poor sanitation, overcrowding, or food and water contamination." (Abutaleb & Kottilil, 2020);
(7) Hepatitis B is transmitted by infected blood or bodily fluids, sexual contact, or vertical transmission from mother to child (Ganem & Prince, 2004). In 2006, it was estimated that worldwide, about 360 million people suffered from chronic hepatitis B, and about 2 billion people had been infected with HBV (Shepard et al., 2006). In 2020, among the causes of death worldwide, 36,036 were attributed to Hepatitis B (World Cause of Death, 2020);
(8) "HCV [Hepatitis C] is estimated to affect approximately 3% of the population worldwide, which translates into nearly 200 million cases worldwide. There is a wide geographic variation in disease prevalence with rates of approximately 1.5% in the United States to nearly 15% in Egypt and up to 50% in certain age groups in Egypt... The unusually high rate of HCV positivity in Egypt has been traced to campaigns to administer parenteral anti-schistosomiasis treatment with inadequate needle sterilization in the 1950-1980s and subsequent spread with blood transfusions and medical and dental procedures... Hepatitis C transmission occurs with parenteral exposures, most commonly from intravenous drug use and contaminated blood transfusions with a notable exception in Egypt..." (Odenwald & Paul, 2022) In 2020, among the causes of death worldwide, 22,147 were attributed to Hepatitis C (World Cause of Death, 2020);
(9) Influenza is an infectious disease caused by viruses of the family *Orthomyxoviridae* (Influenza A to D viruses) with a remarkable ability to develop new antigenic variants by mutations and also by recombination of the antigenic characteristics of various strains infecting the same host (Dharmapalan, 2020; Krammer et al., 2018). A human or animal subject with antibodies against a previous flu strain is not protected from infection by a new strain as a consequence of this antigenic variability. Influenza viruses can infect many animals, including aquatic birds that are their primary reservoir (e.g., both avian and mammalian influenza viruses can infect pigs). When an individual is infected with more than one strain, then genetic reassortment is possible, generating new flu strains (Joseph et al., 2017). The intensive farms of pigs, chickens, and ducks are focal points where there are co-infections and generations of new influenza strains capable of triggering new epidemics on an annual basis. The trade of infected farm animals is an easy source of spreading the infection, and migratory birds can diffuse the infection across long distances. Subsequently, the infection is transmitted from man to man as influenza viruses are easily transmitted by the respiratory route. Influenza viruses have caused and are causing countless epidemics, including the Spanish flu (1918–1920) which was frightening because it had the greatest mortality in young and healthy subjects (Mills et al., 2004) and for the significant number of deaths widely varying in the assessments: between 17 million (Spreeuwenberg et al., 2018) and 100 million (Morens & Fauci, 2007);

(10) Leishmaniasis is a group of diseases caused by more than 20 species of the trypanosome genus *Leishmania*, spread by the bite of phlebotomine sandflies, genera *Phlebotomus* and *Lutzomyia* (Rawat et al., 2021; Roy et al., 2022), and occurring most frequently in tropical and subtropical regions of Africa, Asia, South and Central America, and southern Europe (Ejazi & Ali, 2013).

Reservoir hosts include humans and various animal species such as dogs, wild rodents, and hyraxes (Özbel et al., 2022; Tabbabi, 2019). Moreover, a systematic review by a literature search for the period 1990–2021 showed that: "One hundred and eighty-nine species from ten orders (i.e., Carnivora, Chiroptera, Cingulata, Didelphimorphia, Diprotodontia, Lagomorpha, Eulipotyphla, Pilosa, Primates and Rodentia) were reported to be infected" with zoonotic *Leishmania* species (Azami-Conesa et al., 2021).

Risk factors include poverty, malnutrition, labor migration and forced migration, urbanization, deforestation, or human incursion into forested areas (Grifferty et al., 2021; WHO, 2010). Moreover: "certain types of household construction materials create the optimal survival conditions for sandflies, which rest and breed in cracks or holes in walls and floors. Lack of adequate sanitation attracts wild or domestic animals and may be breeding sites for the sandflies" (Grifferty et al., 2021).

In 2012, for cases with clinical manifestations, 0.2–0.4 million visceral leishmaniasis cases and 0.7–1.2 million cutaneous leishmaniasis cases were estimated to occur each year, with 20,000–40,000 deaths per year (Alvar et al., 2012). In 2020, worldwide deaths caused by leishmaniasis were estimated to be 5780 (World Cause of Death, 2020);

(11) Leprosy, or Hansen's disease, is a long-term infection caused by *Mycobacterium leprae* (White & Franco-Paredes, 2015). The spread of leprosy occurs from person to person through the upper respiratory tract (Eichelmann et al., 2013). A contact with a subject with leprosy increases the probability of contracting the disease (Schreuder et al., 2016), although extensive contact appears necessary (Suzuki et al., 2012).

Risk factors for developing leprosy are malnutrition and illnesses, which reduce immune function (Schreuder et al., 2016), and, as poverty is a cause of malnutrition, leprosy occurs more frequently among people living in poverty (Suzuki et al., 2012). However, an estimated 95% of the world population appears to be resistant to the disease (Lockwood & Saunderson, 2012).

In India, pathological alterations in skeletal remains from the second millennium B.C. indicate that leprosy has affected humanity from ancient times (Robbins et al., 2009).

According to WHO, the number of new leprosy cases was 407,791 in 2004, 218,075 in 2011 (WHO, 2012, Leprosy), and 127,558 in 2020, of which 74.0% in India, Indonesia, and Brazil (Barbosa et al., 2022). In 2020, among the causes of death worldwide, 253 were attributed to leprosy (World Cause of Death, 2020);

(12) Malaria is a zoonosis caused by microorganisms of the *Plasmodium* group, in particular *P. falciparum*, and spread by bites of infected *Anopheles* mosquitoes

(Walter & John, 2022). Malaria caused by *P. falciparum* likely originated in West Africa from wild gorillas (or chimpanzees) (Prugnolle et al., 2011). However, the disease spread was observed only after the increase in population concurrently with the spread of agriculture (Harper & Armelagos, 2011).

In 2020, 241 million cases of malaria worldwide, with 627,000 deaths, were estimated, mainly (95% of cases and deaths) in sub-Saharan Africa (WHO, 2021, Malaria). As for dengue fever and yellow fever, the spread of the disease is determined by a high human population density near areas with stagnant waters where the proliferation of vector mosquitoes is possible. Malaria was already well known in Roman times when it was widespread in various areas of Italy, such as south Italy, Sardinia, Pontine Marshes, and coastal Etruria due to their zones with stagnant waters (Hays, 2005). Similar conditions were present in other parts of the Mediterranean basin and the world. So, the long cohabitation between man and malaria has favored, by natural selection, genetic modifications that allow greater resistance to the disease (Kariuki & Williams, 2020), in particular, thalassemia (Weatherall, 2015), G6PD deficiency (Mbanefo et al., 2017), and sickle cell disease (Allison, 2009).

In the past decades, with a significant commitment of resources, many malaria vaccines have been developed, but for the moment, none appears able to block or eradicate the disease (Bonam et al., 2021);

(13) Measles is a potentially fatal disease caused by a virus of the *Paramyxoviridae* family (Leung et al., 2018), which originated from the rinderpest virus of cattle between the eleventh and twelfth centuries in Europe (Furuse et al., 2010).

Death from measles may be caused by viral pneumonia, complicated by secondary bacterial pneumonia, post-viral encephalitis (1 in 1000 measles cases in the U.S. before the vaccine era), and, in rare cases, by subacute sclerosing panencephalitis (Porter & Goldfarb, 2019).

The disease is spread by airborne routes and is highly contagious, as the basic reproduction number (R_0) appears to be more than 18 (Guerra et al., 2017). The main risk factor is any crowding of non-immune individuals (Su et al., 2020).

In the pre-vaccine era, measles had a high morbidity of approximately 40–726 cases per 100,000 per year (Su et al., 2020) and caused an estimated minimum of one million fatalities annually (Plemper, 2020).

Fortunately, the disease is easily preventable using specific vaccines (Griffin, 2018), but the reduction in vaccine use in countries where the disease was virtually eradicated has led to a return of the disease (Porter & Goldfarb, 2019; Su et al., 2020). "In 2017, there were 109,000 predicted deaths from measles globally compared with 545,000 deaths due to measles in 2000. From 2000 to 2017, immunisation programs prevented approximately 21.1 million deaths worldwide (Dabbagh et al., 2018). After reaching a historic low of fewer than 100,000 deaths in 2016, the number of deaths due to the illness increased to over 140,000 in 2018 (Patel et al., 2019)." (Rabaan et al., 2022) In 2020, the number of deaths due to measles has been estimated to be 165,756 (World Cause of Death, 2020);

(14) For plague epidemics, *Yersinia pestis* circulates in animal reservoirs, particularly in rodents, and is transmitted by fleas that bite an infected rodent and then a person (Demeure et al., 2019). For the origin and diffusion of a plague epidemic, there must be poor hygienic conditions of the housings and the presence of mice. In these conditions, frequent in ancient and medieval times, a few infected mice and the presence of fleas carrying the germ are enough to trigger the epidemic. In some areas of the world where these conditions still exist (and elsewhere as isolated cases), there are still cases of plague today: "Between 2013 and 2018, 2,886 cases and 504 deaths have been notified to WHO (with a reported case fatality of 17.5%), of which 95% derive from sub-Saharan Africa, mainly in Madagascar and the Ituri region of the Democratic Republic of Congo... Occasional scattered cases of human plague are regularly declared in China, Mongolia, the Russian Federation, Kyrgyzstan, Peru, Bolivia, Uganda, Tanzania, and the US..." (Vallès et al., 2020);

(15) Schistosomiasis, also known as bilharzia and snail fever, is a group of diseases caused by some parasitic flatworms of the genus *Schistosoma* (*S. mansoni, japonicum, mekongi, intercalatum, guineensis, haematobium*), each distinctly present in various parts of Africa, the Caribbean, eastern South America, Southeast Asia, and Middle East (Colley et al., 2014). Schistosoma species need freshwater snails as an intermediate host and a mammalian species (e.g., *Homo s.*) as a definitive host. Eggs excreted by the mammalian host release *miracidia* that infect specific freshwater snails; then *cercariae* emerge from the snails, and the mammalians are infected by contact with fresh water contaminated with these *cercariae* (Aula et al., 2021). In Africa, schistosomiasis was present in ancient times, as shown by the description of symptoms of urinary schistosomiasis in Egyptian papyri and by eggs of *S. haematobium* in the urinary tracts of mummies (Aula et al., 2021).

In 2018, it was estimated that 250 million were infected worldwide, with more than 90% of all cases occurring in sub-Saharan Africa (McManus et al., 2018).

Schistosomiasis is generally endemic in rural communities with low incomes without access to potable water and proper sanitation. Estimates regarding the number of deaths vary. The Global Burden of Disease Study issued in 2010 estimated 12,000 direct deaths worldwide (Lozano et al., 2012) and, in 2020, among the world causes of death, 11,792 deaths were attributed to schistosomiasis (World Cause of Death, 2020).

In 2003, in sub-Saharan Africa alone, 280,000 deaths were estimated as due to schistosomiasis annually (Gryseels et al., 2006). However, the large difference between this estimate (280,000 deaths/year) and the two much smaller ones previously reported (about 12,000/year) is perhaps because the lower estimates refer to deaths directly caused by schistosomiasis. On the contrary, the higher one refers to cases of individuals weakened by both schistosomiasis and other diseases (e.g., schistosomiasis associated with HIV in Zambia (Wall et al., 2018);

(16) Sleeping sickness, or African trypanosomiasis, is a zoonosis caused by two subspecies of *Trypanosoma brucei*: *gambiense* (humans and also pigs, sheep,

dogs, goats, and many different wild animals as a reservoir host) and *rhodesiense* (wild animals as a reservoir host). It is present in some rural areas of western and central Africa and is transmitted by the bite of an infected tsetse fly (Franco et al., 2014).

The disease is estimated to have caused 34,000 deaths in 1990 and 3500 in 2015, mostly due to *T. brucei gambiense* (Lozano et al., 2012; Vos et al., 2016). In 2020, it was estimated to have caused 1616 deaths (World Cause of Death, 2020). It has been present in Africa from the times of the ancient Egyptians (Steverding, 2008). In the past, due to low demographic densities and lack of travel between indigenous people, African trypanosomiasis was limited to isolated pockets. In 1901, during the colonial administration, perhaps due to population and cattle mobility from endemic foci outwards, there was a devastating epidemic in Uganda caused by *T. brucei rhodesiense* that killed more than 250,000 people, a third of the population of the region (Fèvre et al., 2004);

(17) Smallpox (variola) was caused by variola/smallpox virus. Its spread in human populations was likely originated by probable transmission from an animal host, with the earliest evidence found in Egyptian mummies (Thèves et al., 2016). Smallpox, an eradicated infection from about 1980, is estimated to have killed up to 300 million people in the past century (Henderson, 2009, p. 12, 2011);

(18) Syphilis is an infection transmitted sexually or via vertical transmission (from mother to child) and caused by the bacterium *Treponema pallidum*, an obligate human pathogen (Peeling et al., 2017). Paleopathologists have demonstrated that syphilis was present in the Americas before European contact (Baker & Armelagos, 1988; Powell & Cook, 2005; Williams, 1932, 1936). According to the Columbian theory, the men who sailed with Christopher Columbus in 1492 brought the infection to Spain, and from there, the infection spread soon elsewhere, beginning in 1495 with a serious epidemic in Naples. The city was among the most populous European cities of the time, and, with the arrival of the soldiers of the French king Charles the VIII: "a plague broke out among the French leader's troops. When the army disbanded shortly after the campaign, the troops, composed largely of mercenaries, returned to their homes and disseminated the disease across Europe." (Maatouk & Moutran, 2014).

As the plague was attributed to the French troops, the people of Naples called it the "French disease" (Winters, 2006, p. 17). From the sixteenth to the nineteenth century, syphilis was one of the most widespread and feared diseases for its contagiousness and diffusion and for the symptoms, disability, and social stigma it entailed (de Kruif, 1932; Rayment & Sullivan, 2011). It has been estimated that in late eighteenth-century London, more than 20% of individuals aged 15 to 34 years were treated for syphilis (Szreter & Kevin, 2020).

It should be emphasized that despite the epidemiological evidence of a devastating and highly contagious disease that suddenly appeared after 1492, the hypothesis that syphilis could be a disease already existing earlier in the Old World is still strongly supported. This assumption is likely due to the evident similarities with non-sexually transmitted diseases such as yaws and bejel or endemic syphilis, also caused by varieties of *Treponema pallidum* (Baker et al., 2020).

In 1990, it has been estimated that there were 202,000 deaths caused by syphilis (Lozano et al., 2012). In 2015, about 45.4 million people had syphilis infections, which caused about 107,000 deaths (Vos et al., 2016), while, in 2020, among the causes of death worldwide, 43,427 were estimated as due to the disease (World Cause of Death, 2020).

The epidemiological characteristics of syphilis were typical of an infection to which Old World populations were not adapted, just as New World populations were inexperienced and were not adapted to those of the Old World. The changes in ecological niches caused by the contact between the two Worlds had terrible effects on both populations (see below);

(19) Tuberculosis is caused by *Mycobacterium tuberculosis* and is transmitted through the airborne route by individuals with active tuberculosis. In 2015, the WHO reported 1.4 million deaths attributable to tuberculosis and 10.4 million incident cases of active tuberculosis (WHO, 2016b, Tuberculosis). Tuberculosis has been present in humans since ancient times (Lawn & Zumla, 2011). "Tuberculosis is responsible for more deaths worldwide than any other infectious disease... *Mycobacterium tuberculosis* has no environmental niche and does not naturally infect animals. Thus, in nature, it is essentially an obligate human pathogen... the majority of people who acquire aerosol infection go straight to 'latent TB', with no clinical symptoms." (Guinn & Rubin, 2017) Most patients infected with *M. tuberculosis* become healthy carriers of the microbe, while in general, those who develop a clinically evident disease are weakened by some other condition. In fact, patients affected by diabetes mellitus develop active tuberculosis at higher rates than nondiabetic patients (Dooley & Chaisson, 2009); "Nutritional deficiencies appear to predispose to infection, and strikingly, those who have their stomach removed (gastrectomy) or who have malabsorption of multiple nutrients are at particular risk of TB..." (Guinn & Rubin, 2017);

(20) Yellow fever (YF) is caused by an arbovirus of the family *Flaviviridae*, transmitted primarily by mosquitoes of various species, which carry the virus after receiving it by biting infected humans or non-human primates (Gianchecchi et al., 2022): "... globalization, the continuous population growth, urbanization associated with inadequate public health infrastructures and climate changes constitute important promoting factors for the spread of not only new infectious agents but also well-known pathogens, including YF virus (Brent et al., 2018)." (Gianchecchi et al., 2022) As for malaria and dengue fever, an essential factor for the spread of the disease is the increase in population near areas and points where there are stagnant waters, allowing the proliferation of mosquitoes.

This list is partial and intended to represent only a general and non-exhaustive discussion of the primary causes of the reported epidemics. Table 3.6 briefly shows the main primary causes of the epidemics discussed.

Table 3.6 Some infectious epidemic and their main primary causes

Epidemic	Demographic density and rapid transfer of infected individuals	Zoonosis/ spillover infection	Altered or poor hygienic conditions	Other (s. Notes)
AIDS	Y	Y		1
Chagas' disease	Y	Y		
Cholera	Y		Y	
Covid-19	Y	Y		
Hepatitis A	Y		Y	
Hepatitis B	Y			2
Hepatitis C	Y			3
Influenza	Y	Y		4
Dengue fever	Y			5
Leishmaniasis	Y	Y		6
Leprosy	Y			
Malaria	Y	Y		5
Measles	Y	Y		
Plague pest	Y	Y	Y	
Schistosomiasis	Y		Y	
Sleeping sickness	Y	Y		
Smallpox	Y	Y		
Syphilis	Y			7
Tuberculosis	Y			8
Yellow fever	Y	Y		5

Notes: 1: Sexual and parenteral transmission, 2: Parenteral transmission and vertical transmission (from mother to child), 3: Parenteral transmission and medical treatment with inadequate needle sterilization, 4: Formation of new strains in intensive breeding, 5: Population living in areas with stagnant waters, 6: Migrations and deforestations, 7: Sexual and vertical transmission (from mother to child), 8: Conditions in which the organism is weakened by other causes

Over the centuries, due to their serious consequences, the more ancient of these epidemics have led to strong selective pressures with progressively greater resistance to their pathogens. A tragic demonstration of the greater resistance of populations subjected for a long time to such selective pressures, compared with populations without this experience, can be found in the consequences of contacts between the two types of populations.

In fact, when a population that has lived in isolation for millennia is contacted by individuals of a non-isolated population, there is high mortality among the individuals of the first population because it is unable to sustain infections quite trivial in non-isolated populations. E.g., for the Ache, a hunter-gatherer people of Paraguay, in the transition from isolated living conditions (the "forest period") when they were contacted by the researchers who first studied them, there was high mortality from infectious diseases unknown to them until then (Hill & Hurtado, 1996).

However, the most resounding example is the consequence of the first contacts between populations of the Old World, strongly selected by millennia of epidemics,

Table 3.7 Estimates of the depopulation of the Americas after 1492, data from Table 1 in (Koch et al., 2019)

	Pongratz et al. (2008)	Klein Goldewijk et al. (2010)	Kaplan et al. (2011)	Koch et al. (2019)
Population at 1500 A.D. (million)	41.1	41.2	60	60.5 (44.8–78.2)
Population at 1600 A.D. (million)	14.2	9.7	6	6.1 (4.5–7.8)
Depopulation (%)	65%	76%	90%	90%

with the inhabitants of the New World. The latter, having much less experience with epidemic diseases due to their relatively smaller demographic density and total population, were heavily decimated several times by the diseases brought by the former. According to recent estimates, comparing the population of the Americas in 1492 and 1600, "55 million indigenous people died following the European conquest of the Americas beginning in 1492", and there was a 90% population decline (Koch et al., 2019). "Existing evidence suggests that the indigenous population collapse was primarily caused by the introduction of pathogens unknown to the American continent ("virgin soil epidemics") together with warfare and slavery... Part of a wider Columbian Exchange of once-separate continental fauna and flora, these epidemics were introduced by European settlers and African slaves and were passed on to an indigenous population that had not been previously exposed to these pathogens and therefore did not initially possess suitable antibodies... Such diseases included smallpox, measles, influenza, the bubonic plague, and later malaria, diphtheria, typhus, and cholera..." (Koch et al., 2019) (s. Table 3.7).

An exception to the rule of disease transmission from the Old to the New World was syphilis, which existed in the Americas before 1492 and then spread to Europe and elsewhere (s. above the sub-section dedicated to Syphilis).

References

Aburto, N. J., Ziolkovska, A., Hooper, L., Elliott, P., Cappuccio, F. P., & Meerpohl, J. J. (2013). Effect of lower sodium intake on health: systematic review and meta-analyses. *British Medical Journal (BMJ), 346*, f1326. https://doi.org/10.1136/bmj.f1326

Abutaleb, A., & Kottilil, S. (2020). Hepatitis A: Epidemiology, Natural History, Unusual Clinical Manifestations, and Prevention. *Gastroenterology Clinics of North America, 49*(2), 191–199. https://doi.org/10.1016/j.gtc.2020.01.002

Adachi, M., & Brenner, D. A. (2005). Clinical syndromes of alcoholic liver disease. *Digestive Diseases, 23*(3–4), 255–263. https://doi.org/10.1159/000090173

Addiss, D. G., Shaffer, N., Fowler, B. S., & Tauxe, R. V. (1990). The epidemiology of appendicitis and appendectomy in the United States. *American Journal of Epidemiology, 132*(5), 910–925. https://doi.org/10.1093/oxfordjournals.aje.a115734

Afzal, M. R., Savona, S., Mohamed, O., Mohamed-Osman, A., & Kalbfleisch, S. J. (2019). Hypertension and Arrhythmias. *Heart Failure Clinics, 15*(4), 543–550. https://doi.org/10.1016/j.hfc.2019.06.011

Ahmed, N., Arshad, S., Basheer, S. N., Karobari, M. I., Marya, A., Marya, C. M., et al. (2021). Smoking a Dangerous Addiction: A Systematic Review on an Underrated Risk Factor for Oral Diseases. *International Journal of Environmental Research and Public Health, 18*(21), 11003. https://doi.org/10.3390/ijerph182111003

Akbari, M., Lankarani, K. B., Honarvar, B., Tabrizi, R., Mirhadi, H., & Moosazadeh, M. (2016). Prevalence of malocclusion among Iranian children: A systematic review and meta-analysis. *Dental Research Journal (Isfahan), 13*, 387–395. https://doi.org/10.4103/1735-3327.192269

Akiyama, T., Inamori, M., Iida, H., Mawatari, H., Endo, H., Hosono, K., et al. (2008). Alcohol consumption is associated with an increased risk of erosive esophagitis and Barrett's epithelium in Japanese men. *Biomed Central Gastroenterology, 8*, 58. https://doi.org/10.1186/1471-230X-8-58

Ali, M., Nelson, A. R., Lopez, A. L., & Sack, D. A. (2015). Updated Global Burden of Cholera in Endemic Countries. *PLoOS Neglected Tropical Diseases, 9*(6), e0003832. https://doi.org/10.1371/journal.pntd.0003832

Aliaga-Del Castillo, A., Mattos-Vela, M. A., Aliaga-Del Castillo, R., & Del Castillo-Mendoza, C. (2011). Maloclusiones en niños y adolescentes de caseríos y comunidades nativas de la Amazonía de Ucayali, Perú [Malocclusions in children and adolescents from villages and native communities in the Ucayali Amazon region in Peru] (article in Spanish). *Revista Peruana de Medicina Experimental y Salud Pública, 28*(1), 87–91. https://doi.org/10.1590/s1726-46342011000100014

Allison, A. C. (2009). Genetic control of resistance to human malaria. *Current Opinion in Immunology, 21*(5), 499–505. https://doi.org/10.1016/j.coi.2009.04.001

Almagro-Moreno, S., Pruss, K., & Taylor, R. K. (2015). Intestinal Colonization Dynamics of Vibrio cholerae. *PLoS Pathogens, 11*(5), e1004787. https://doi.org/10.1371/journal.ppat.1004787

Alvar, J., Vélez, I. D., Bern, C., Herrero, M., Desjeux, P., Cano, J., WHO Leishmaniasis Control Team, et al. (2012). Leishmaniasis worldwide and global estimates of its incidence. *PLoS One, 7*(5), e35671. https://doi.org/10.1371/journal.pone.0035671

Alvarado, K., López, L., Hanke, R., Picón, F., & Rivas-Tumanyan, S. (2017). Prevalence of Malocclusion and Distribution of Occlusal Characteristics in 13- to 18-year-old Adolescents Attending Selected High Schools in the Municipality of San Juan, PR (2012-2013). *Puerto Rico Health Sciences Journal, 36*(2), 61–66. PMID: 28622400.

Amrein, K., Scherkl, M., Hoffmann, M., Neuwersch-Sommeregger, S., Köstenberger, M., Tmava Berisha, A., et al. (2020). Vitamin D deficiency 2.0: an update on the current status worldwide. *European Journal of Clinical Nutrition, 74*(11), 1498–1513. https://doi.org/10.1038/s41430-020-0558-y

Anand, P., Kunnumakkara, A. B., Sundaram, C., Harikumar, K. B., Tharakan, S. T., Lai, O. S., et al. (2008). Cancer is a preventable disease that requires major lifestyle changes. *Pharmaceutical Research, 25*(9), 2097–2116. https://doi.org/10.1007/s11095-008-9661-9

Andersen, D. A. (1969). Historical and geographical differences in the pattern of incidence of urinary stones considered in relation to aetiological factors. In A. Hodgkinson & B. E. C. Nordin (Eds.), *Proceedings of the International Symposium on Renal Stone Research* (pp. 7–32). Churchill.

Andersen, K. G., Rambaut, A., Lipkin, W. I., Holmes, E. C., & Garry, R. F. (2020). The proximal origin of SARS-CoV-2. *Nature Medicine, 26*(4), 450–452. https://doi.org/10.1038/s41591-020-0820-9

Andrade, L., Caraveo-Anduaga, J. J., Berglund, P., Bijl, R. V., De Graaf, R., Vollebergh, W., et al. (2003). The epidemiology of major depressive episodes: results from the International Consortium of Psychiatric Epidemiology (ICPE) Surveys. *International Journal of Methods in Psychiatric Research, 12*(1), 3–21. https://doi.org/10.1002/mpr.138. Erratum in: *International Journal of Methods in Psychiatric Research*, 2003, 12(3), 165.

Ansaloni, L., Pisano, M., Coccolini, F., Peitzmann, A. B., Fingerhut, A., Catena, F., et al. (2016). 2016 WSES guidelines on acute calculous cholecystitis. *World Journal of Emergency Surgery, 11*, 25. https://doi.org/10.1186/s13017-016-0082-5

APA (American Psychiatric Association). (2000). *Diagnostic and statistical manual of mental disorders* (4th ed. (DSM-IV)). American Psychiatric Association.

Aufderheide, A. C., Salo, W., Madden, M., Streitz, J., Buikstra, J., Guhl, F., et al. (2004). A 9,000-year record of Chagas' disease. *Proceedings of the National Academy of Sciences of the United States of America, 101*(2004), 2034–2039. https://doi.org/10.1073/pnas.0307312101

Augustine, L. F., Nair, K. M., & Kulkarni, B. (2021). Sun exposure as a strategy for acquiring vitamin D in developing countries of tropical region: Challenges & way forward. *The Indian Journal of Medical Research, 154*(3), 423–432. https://doi.org/10.4103/ijmr.IJMR_1244_18

Aula, O. P., McManus, D. P., Jones, M. K., & Gordon, C. A. (2021). Schistosomiasis with a Focus on Africa. *Tropical Medicine and Infectious Disease, 6*(3), 109. https://doi.org/10.3390/tropicalmed6030109

Aykan, N. F. (2015). Red Meat and Colorectal Cancer. *Oncology Reviews, 9*(1), 288. https://doi.org/10.4081/oncol.2015.288

Azami-Conesa, I., Gómez-Muñoz, M. T., & Martínez-Díaz, R. A. (2021). A Systematic Review (1990-2021) of Wild Animals Infected with Zoonotic Leishmania. *Microorganisms, 9*(5), 1101. https://doi.org/10.3390/microorganisms9051101

Bachmann, S. (2018). Epidemiology of suicide and the psychiatric perspective. *International Journal of Environmental Research and Public Health, 15*(7), 1425. https://doi.org/10.3390/ijerph15071425

Baker, B. J., & Armelagos, G. J. (1988). The origin and antiquity of syphilis: Paleopathological diagnosis and interpretation. *Current Anthropology, 29*, 703–738. https://doi.org/10.1086/203691

Baker, B. J., Crane-Kramer, G., Dee, M. W., Gregoricka, L. A., Henneberg, M., Lee, C., et al. (2020). Advancing the Understanding of Treponemal Disease in the Past and Present. *Yearbook of Physical Anthropology, 171*, 5–41. https://doi.org/10.1002/ajpa.23988

Barbosa, C. C., Bezerra, G. S. N., Xavier, A. T., Albuquerque, M. F. P. M., Bonfim, C. V. D., Medeiros, Z. M., & Souza, W. V. (2022). Systematic Review of Survival Analysis in Leprosy Studies-Including the Following Outcomes: Relapse, Impairment of Nerve Function, Reactions and Physical Disability. *International Journal of Environmental Research and Public Health, 19*(19), 12155. https://doi.org/10.3390/ijerph191912155

Barker, D. J., & Morris, J. (1988). Acute appendicitis, bathrooms, and diet in Britain and Ireland. *British Medical Journal (Clinical research edition), 296*(6627), 953–955. https://doi.org/10.1136/bmj.296.6627.953

Barker, D. J., Morris, J. A., Simmonds, S. J., & Oliver, R. H. (1988a). Appendicitis epidemic following introduction of piped water to Anglesey. *Journal of Epidemiology and Community Health, 42*(2), 144–148. https://doi.org/10.1136/jech.42.2.144

Barker, D. J., Osmond, C., Golding, J., & Wadsworth, M. E. (1988b). Acute appendicitis and bathrooms in three samples of British children. *British Medical Journal (Clinical Research Edition), 296*(6627), 956–958. https://doi.org/10.1136/bmj.296.6627.956

Barta, J. A., Powell, C. A., & Wisnivesky, J. P. (2019). Global epidemiology of lung Cancer. *Annals of Global Health, 85*(1), 8. https://doi.org/10.5334/aogh.2419

Battaglia Richi, E., Baumer, B., Conrad, B., Darioli, R., Schmid, A., & Keller, U. (2015). Health risks associated with meat consumption: A review of epidemiological studies. *International Journal for Vitamin and Nutrition Research, 85*(1–2), 70–78. https://doi.org/10.1024/0300-9831/a000224

Becker, B. J. P., & Chatgidakis, C. B. (1952). Carcinoma of the gallbladder and cholelithiasis on the Witwatersrand, an autopsy survey of the racial incidence. *South African Journal of Clinical Science, 3*, 13–22. PMID: 14930982.

Bellou, V., Belbasis, L., Tzoulaki, I., & Evangelou, E. (2018). Risk factors for type 2 diabetes mellitus: An exposure-wide umbrella review of meta-analyses. *PLoS One, 13*(3), e0194127. https://doi.org/10.1371/journal.pone.0194127

Benowitz, N. L., & Liakoni, E. (2022). Tobacco use disorder and cardiovascular health. *Addiction, 117*(4), 1128–1138. https://doi.org/10.1111/add.15703

Bertolote, J. M., & Fleischmann, A. (2002). A global perspective in the epidemiology of suicide. *Suicidologi, 7*, 6–8. https://doi.org/10.5617/suicidologi.2330
Bharucha, A. E., & Lacy, B. E. (2020). Mechanisms, evaluation, and management of chronic constipation. *Gastroenterology, 158*(5), 1232–1249.e3. https://doi.org/10.1053/j.gastro.2019.12.034
Bickler, S. W., & DeMaio, A. (2008). Western diseases: current concepts and implications for pediatric surgery research and practice. *Pediatric Surgery International, 24*(3), 251–255. https://doi.org/10.1007/s00383-007-2095-3
Black, F. L. (1966). Measles endemicity in insular populations: critical community size and its evolutionary implication. *Journal of Theoretical Biology, 11*(2), 207–211. https://doi.org/10.1016/0022-5193(66)90161-5
Blacklock, N. (1981). Renal stone. In H. C. Trowell & D. P. Burkitt (Eds.), *Western diseases, their emergence and prevention* (pp. 60–70). Edward Arnold. ISBN 978-0674950207.
Bonam, S. R., Rénia, L., Tadepalli, G., Bayry, J., & Kumar, H. M. S. (2021). Plasmodium falciparum Malaria Vaccines and Vaccine Adjuvants. *Vaccines (Basel), 9*(10), 1072. https://doi.org/10.3390/vaccines9101072
Bragulat, E., & de la Sierra, A. (2002). Salt intake, endothelial dysfunction, and salt-sensitive hypertension. *Journal of Clinical Hypertension (Greenwich, Conn.), 4*(1), 41–46. https://doi.org/10.1111/j.1524-6175.2002.00503.x
Bray, R. S. (1996). *Armies of pestilence. The impact of pndemics on history*. James Clarke &. ISBN 978-0718829490.
Bremner, C. G. (1971). The changing pattern of disease seen in Baragwanath Hospital. *South African Journal of Surgery, 9*, 127–131. PMID: 5138517.
Brent, S. E., Watts, A., Cetron, M., German, M., Kraemer, M. U., Bogoch, I. I., et al. (2018). International travel between global urban centres vulnerable to yellow fever transmission. *Bulletin of the World Health Organization, 96*(5), 343–354B. https://doi.org/10.2471/BLT.17.205658
Burkitt, D. P. (1971). The aetiology of appendicitis. *The British Journal of Surgery, 58*(9), 695–699. https://doi.org/10.1002/bjs.1800580916
Cagan, A., Baez-Ortega, A., Brzozowska, N., Abascal, F., Coorens, T. H. H., Sanders, M. A., et al. (2022). Somatic mutation rates scale with lifespan across mammals. *Nature, 604*(7906), 517–524. https://doi.org/10.1038/s41586-022-04618-z
Cartwright, F. F., & Biddiss, M. (2014). *Disease & History* (3rd ed.). Thistle Publishing. ISBN 978-1910198230.
Cashman, K. D., Dowling, K. G., Škrabáková, Z., Gonzalez-Gross, M., Valtueña, J., De Henauw, S., et al. (2016). Vitamin D deficiency in Europe: pandemic? *The American Journal of Clinical Nutrition, 103*(4), 1033–1044. https://doi.org/10.3945/ajcn.115.120873
Cavaliere, M., Bisogno, A., Scarpa, A., D'Urso, A., Marra, P., Colacurcio, V., et al. (2021). Biomarkers of laryngeal squamous cell carcinoma: a review. *Annals of Diagnostic Pathology, 54*, 151787. https://doi.org/10.1016/j.anndiagpath.2021.151787
Chagas, C. (1909). Nova tripanozomiase humana. Estudos sobre a morfolojía e o ciclo evolutivo de Schizotrypanum cruzi n. gen., n. sp., ajente etiolójico de nova entidade morbida do homen (in Portuguese). *Memórias do Instituto Oswaldo Cruz, 1*, 159–218.
Chao, C., Leone, J. L., & Vigliano, C. A. (2020). Chagas disease: Historic perspective. *Biochimica et biophysica acta. Molecular Basis of Disease, 1866*(5), 165689. https://doi.org/10.1016/j.bbadis.2020.165689
Chen, K. H., Su, S. B., & Chen, K. T. (2020). An overview of occupational noise-induced hearing loss among workers: epidemiology, pathogenesis, and preventive measures. *Environmental Health and Preventive Medicine, 25*(1), 65. https://doi.org/10.1186/s12199-020-00906-0
Cheng, L., Zhang, L., Yue, L., Ling, J., Fan, M., Yang, D., et al. (2022). Expert consensus on dental caries management. *International Journal of Oral Science, 14*(1), 17. https://doi.org/10.1038/s41368-022-00167-3
Chi, D. L., & Scott, J. M. (2019). Added sugar and dental caries in children: A scientific update and future steps. *Dental Clinics of North America, 63*(1), 17–33. https://doi.org/10.1016/j.cden.2018.08.003

Chou, Y. H., Su, C. M., Li, C. C., Liu, C. C., Liu, M. E., Wu, W. J., & Juan, Y. S. (2011). Difference in urinary stone components between obese and non-obese patients. *Urological Research, 39*(4), 283–287. https://doi.org/10.1007/s00240-010-0344-8

Chow, Y. C., Dhillon, B., Chew, P. T., & Chew, S. J. (1990). Refractive errors in Singapore medical students. *Singapore Medical Journal, 31*, 472–473. PMID: 2259947.

Chow, W. H., Dong, L. M., & Devesa, S. S. (2010). Epidemiology and risk factors for kidney cancer. *Nature Reviews. Urology, 7*(5), 245–257. https://doi.org/10.1038/nrurol.2010.46

Colley, D. G., Bustinduy, A. L., Secor, W. E., & King, C. H. (2014). Human schistosomiasis. *Lancet, 383*(9936), 2253–2264. https://doi.org/10.1016/s0140-6736(13)61949-2

Cordain, L., Eaton, S. B., Brand Miller, J. B., Lindeberg, S., & Jensen, C. (2002). An evolutionary analysis of the aetiology and pathogenesis of juvenile-onset myopia. *Acta Ophthalmologica Scandinavica, 80*(2), 125–135. https://doi.org/10.1034/j.1600-0420.2002.800203.x

Cosman, F., de Beur, S. J., LeBoff, M. S., Lewiecki, E. M., Tanner, B., Randall, S., Lindsay, R., & National Osteoporosis Foundation. (2014). Clinician's Guide to Prevention and Treatment of Osteoporosis. *Osteoporos International, 25*(10), 2359–2381. https://doi.org/10.1007/s00198-014-2794-2

Crichton, P. (1996). Were the Roman emperors Claudius and Vitellius bulimic? *The International Journal of Eating Disorders, 19*(2), 203–207. https://doi.org/10.1002/(SICI)1098-108X(199603)19:2<203::AID-EAT11>3.0.CO;2-V

Cumberbatch, M. G. K., Jubber, I., Black, P. C., Esperto, F., Figueroa, J. D., Kamat, A. M., et al. (2018). Epidemiology of Bladder Cancer: A Systematic Review and Contemporary Update of Risk Factors in 2018. *European Urology, 74*(6), 784–795. https://doi.org/10.1016/j.eururo.2018.09.001

Curhan, G. C., Willett, W. C., Rimm, E. B., & Stampfer, M. J. (1993). A prospective study of dietary calcium and other nutrients and the risk of symptomatic kidney stones. *The New England Journal of Medicine, 328*(12), 833–838. https://doi.org/10.1056/NEJM199303253281203

Curhan, G. C., Willett, W. C., Speizer, F. E., Spiegelman, D., & Stampfer, M. J. (1997). Comparison of dietary calcium with supplemental calcium and other nutrients as factors affecting the risk for kidney stones in women. *Annals of Internal Medicine, 126*(7), 497–504. https://doi.org/10.7326/0003-4819-126-7-199704010-00001

Dabbagh, A., Laws, R. L., Steulet, C., Dumolard, L., Mulders, M. N., Kretsinger, K., et al. (2018). Progress toward regional measles elimination - worldwide, 2000-2017. *MMWR. Morbidity and Mortality Weekly Report, 67*(47), 1323–1329. https://doi.org/10.15585/mmwr.mm6747a6

Daniel, E. (2007). Noise and hearing loss: a review. *The Journal of School Health, 77*(5), 225–231. https://doi.org/10.1111/j.1746-1561.2007.00197.x

de Groot, P. M., Wu, C. C., Carter, B. W., & Munden, R. F. (2018). The epidemiology of lung cancer. *Translational Lung Cancer Research, 7*(3), 220–233. https://doi.org/10.21037/tlcr.2018.05.06

De Hutorowicz, H., & Adler, B. F. (1911). Maps of Primitive Peoples. *Bulletin of the American Geographical Society, 43*, 669–679.

de Kruif, P. (1932). Chapter 7: Schaudinn: The Pale Horror. In *Men Against Death*. Harcourt, Brace.

Demeure, C. E., Dussurget, O., Mas Fiol, G., Le Guern, A. S., Savin, C., & Pizarro-Cerdá, J. (2019). Yersinia pestis and plague: an updated view on evolution, virulence determinants, immune subversion, vaccination, and diagnostics. *Genes and Immunity, 20*(5), 357–370. https://doi.org/10.1038/s41435-019-0065-0

Deng, Y., Wang, M., Zhou, L., Zheng, Y., Li, N., Tian, T., et al. (2020). Global burden of larynx cancer, 1990-2017: estimates from the global burden of disease 2017 study. *Aging (Albany NY), 12*(3), 2545–2583. https://doi.org/10.18632/aging.102762

Dharmapalan, D. (2020). Influenza. *Indian Journal of Pediatrics, 87*(10), 828–832. https://doi.org/10.1007/s12098-020-03214-1

Dimopoulos, C., Gialas, A., Likourinas, M., Androutsos, G., & Kostakopoulos, A. (1980). Hippocrates: founder and pioneer of urology. *British Journal of Urology, 52*(2), 73–74. https://doi.org/10.1111/j.1464-410x

Ding, T., Yan, A., & Liu, K. (2019). What is noise-induced hearing loss? *British Journal of Hospital Medicine (London), 80*(9), 525–529. https://doi.org/10.12968/hmed.2019.80.9.525

Dirani, M., Tong, L., Gazzard, G., Zhang, X., Chia, A., Young, T. L., et al. (2009). Outdoor activity and myopia in Singapore teenage children. *The British Journal of Ophthalmology, 93*, 997–1000. https://doi.org/10.1136/bjo.2008.150979

Dirani, M., Chan, Y. H., Gazzard, G., Hornbeak, D. M., Leo, S. W., Selvaraj, P., et al. (2010). Prevalence of refractive error in Singaporean Chinese children: the strabismus, amblyopia, and refractive error in young Singaporean Children (STARS) study. *Investigative Ophthalmology & Visual Science, 51*(3), 1348–1355. https://doi.org/10.1167/iovs.09-3587

Dooley, K. E., & Chaisson, R. E. (2009). Tuberculosis and diabetes mellitus: convergence of two epidemics. *The Lancet. Infectious Diseases, 9*, 737–746. https://doi.org/10.1016/S1473-3099(09)70282-8

Early, J., & Peters, J. (1990). *The population dynamics of the Mucajai Yanomamo*. Academic. ISBN 978-0122276809.

Eaton, S. B., Shostak, M., & Konner, M. (1988a). *The paleolithic prescription: a program of diet and exercise and a design for living*. Harper and Row. ISBN 978-0060916354.

Eaton, S. B., Konner, M., & Shostak, M. (1988b). Stone agers in the fast lane: Chronic degenerative diseases in evolutionary perspective. *The American Journal of Medicine, 84*(4), 739–749. https://doi.org/10.1016/0002-9343(88)90113-1

Eaton, S. B., Konner, M. J., & Cordain, L. (2009). Diet-dependent acid load, Paleolithic nutrition, and evolutionary health promotion. *The American Journal of Clinical Nutrition, 91*(2), 295–297. https://doi.org/10.3945/ajcn.2009.29058

Edwards, M. H., & Lam, C. S. (2004). The epidemiology of myopia in Hong Kong. *Annals of the Academy of Medicine, Singapore, 33*, 34–38. PMID: 15008559.

Eichelmann, K., González González, S. E., Salas-Alanis, J. C., & Ocampo-Candiani, J. (2013). Leprosy. An update: definition, pathogenesis, classification, diagnosis, and treatment. *Actas Dermo-Sifiliográficas, 104*(7), 554–563. https://doi.org/10.1016/j.adengl.2012.03.028

Ejazi, S. A., & Ali, N. (2013). Developments in diagnosis and treatment of visceral leishmaniasis during the last decade and future prospects. *Expert Review of Anti-Infective Therapy, 11*(1), 79–98. https://doi.org/10.1586/eri.12.148

Faienza, M. F., Lassandro, G., Chiarito, M., Valente, F., Ciaccia, L., & Giordano, P. (2020). How physical activity across the lifespan can reduce the impact of bone ageing: A literature review. *International Journal of Environmental Research and Public Health, 17*(6), 1862. https://doi.org/10.3390/ijerph17061862

Faria, N. R., Rambaut, A., Suchard, M. A., Baele, G., Bedford, T., Ward, M. J., et al. (2014). The early spread and epidemic ignition of HIV-1 in human populations. *Science, 346*(6205), 56–61. https://doi.org/10.1126/science.1256739

Fazel, S., & Runeson, B. (2020). Suicide. *The New England Journal of Medicine, 382*(3), 266–274. https://doi.org/10.1056/NEJMra1902944. Erratum in: N Engl J Med. 2020, 382(11), 1078.

Feng, W., Dell'Italia, L. J., & Sanders, P. W. (2017). Novel Paradigms of Salt and Hypertension. *Journal of the American Society of Nephrology, 28*(5), 1362–1369. https://doi.org/10.1681/ASN.2016080927

Ferraro, P. M., Taylor, E. N., Gambaro, G., & Curhan, G. C. (2017). Dietary and lifestyle risk factors associated with incident kidney stones in men and women. *The Journal of Urology, 198*(4), 858–863. https://doi.org/10.1016/j.juro.2017.03.124

Ferraro, P. M., Bargagli, M., Trinchieri, A., & Gambaro, G. (2020). Risk of kidney stones: Influence of dietary factors, dietary patterns, and vegetarian-vegan diets. *Nutrients, 12*(3), 779. https://doi.org/10.3390/nu12030779

Fèvre, E. M., Coleman, P. G., Welburn, S. C., & Maudlin, I. (2004). Reanalyzing the 1900-1920 sleeping sickness epidemic in Uganda. *Emerging Infectious Diseases, 10*(4), 567–573. https://doi.org/10.3201/eid1004.020626

Franco, J. R., Simarro, P. P., Diarra, A., & Jannin, J. G. (2014). Epidemiology of human African trypanosomiasis. *Clinical Epidemiology, 6*, 257–275. https://doi.org/10.2147/CLEP.S39728

Fransen, E., Topsakal, V., Hendrickx, J. J., Van Laer, L., Huyghe, J. R., Van Eyken, E., et al. (2008). Occupational noise, smoking, and a high Body Mass Index are risk factors for age-related hearing impairment and moderate alcohol consumption is protective: a European population-based multicenter study. *Journal of the Association for Research in Otolaryngology, 9*(3), 264–276. https://doi.org/10.1007/s10162-008-0123-1

Fredrick, D. R. (2002). Myopia. *British Medical Journal (BMJ), 324*, 1195–1199. https://doi.org/10.1136/bmj.324.7347.1195

Friedman, G. D. (1993). Natural history of asymptomatic and symptomatic gallstones. *American Journal of Surgery, 165*(4), 399–404. https://doi.org/10.1016/s0002-9610(05)80930-4

Furuse, Y., Suzuki, A., & Oshitani, H. (2010). Origin of measles virus: divergence from rinderpest virus between the 11th and 12th centuries. *Virology Journal, 7*, 52. https://doi.org/10.1186/1743-422X-7-52

Gałecka, M., Bliźniewska-Kowalska, K., Maes, M., Su, K. P., & Gałecki, P. (2021). Update on the neurodevelopmental theory of depression: is there any 'unconscious code'? *Pharmacol Reports, 73*(2), 346–356. https://doi.org/10.1007/s43440-020-00202-2

Gallucci, G., Tartarone, A., Lerose, R., Lalinga, A. V., & Capobianco, A. M. (2020). Cardiovascular risk of smoking and benefits of smoking cessation. *Journal of Thoracic Diseases, 12*(7), 3866–3876. https://doi.org/10.21037/jtd.2020.02.47

Ganem, D., & Prince, A. M. (2004). Hepatitis B virus infection—natural history and clinical consequences. *The New England Journal of Medicine, 350*(11), 1118–1129. https://doi.org/10.1056/NEJMra031087. Erratum in: *The New England Journal of Medicine*, 2004, 351(12), 351.

Garner, L. F., Kinnear, R. F., Klinger, J. D., & McKellar, M. J. (1985). Prevalence of myopia in school children in Vanuatu. *Acta Ophthalmologica (Copenh), 63*(3), 323–326. https://doi.org/10.1111/j.1755-3768.1985.tb06814.x

Garner, L. F., Owens, H., Kinnear, R. F., & Frith, M. J. (1999). Prevalence of myopia in Sherpa and Tibetan children in Nepal. *Optometry and Vision Science, 76*, 282–285. https://doi.org/10.1097/00006324-199905000-00014

GBD 2016 Disease and Injury Incidence and Prevalence Collaborators. (2017). Global, regional, and national incidence, prevalence, and years lived with disability for 328 diseases and injuries for 195 countries, 1990-2016: a systematic analysis for the Global Burden of Disease Study 2016. *Lancet, 390*(10100), 1211–1259. https://doi.org/10.1016/S0140-6736(17)32154-2. Erratum in: *Lancet*, 2017, 390(10106), e38.

GBD 2017 Disease and Injury Incidence and Prevalence Collaborators. (2018). Global, regional, and national incidence, prevalence, and years lived with disability for 354 diseases and injuries for 195 countries and territories, 1990-2017: a systematic analysis for the Global Burden of Disease Study 2017. *Lancet, 218, 392*(10159), 1789–1858. https://doi.org/10.1016/S0140-6736(18)32279-7. Erratum in: *Lancet*, 2019, 393(10190), e44.

Genchi, G., Carocci, A., Lauria, G., Sinicropi, M. S., & Catalano, A. (2020). Nickel: Human health and environmental toxicology. *International Journal of Environmental Research and Public Health, 17*(3), 679. https://doi.org/10.3390/ijerph17030679

Ghimire, K., Mishra, S. R., Satheesh, G., Neupane, D., Sharma, A., Panda, R., et al. (2021). Salt intake and salt-reduction strategies in South Asia: From evidence to action. *Journal of Clinical Hypertension (Greenwich, Conn.), 23*(10), 1815–1829. https://doi.org/10.1111/jch.14365

Gianchecchi, E., Cianchi, V., Torelli, A., & Montomoli, E. (2022). Yellow fever: Origin, epidemiology, preventive strategies and future prospects. *Vaccines (Basel), 10*(3), 372. https://doi.org/10.3390/vaccines10030372

Gill, S. K., Rossi, M., Bajka, B., & Whelan, K. (2021). Dietary fibre in gastrointestinal health and disease. *Nature Reviews. Gastroenterology & Hepatology, 18*(2), 101–116. https://doi.org/10.1038/s41575-020-00375-4

Gordon-Shaag, A., Shneor, E., Doron, R., Levine, J., & Ostrin, L. A. (2021). Environmental and behavioral factors with refractive error in Israeli boys. *Optometry and Vision Science, 98*(8), 959–970. https://doi.org/10.1097/OPX.0000000000001755

Gorelick, A. N., & Naxerova, K. (2022). Mutational clocks tick differently across species. *Nature, 604*(7906), 435–436. https://doi.org/10.1038/d41586-022-00976-w
Gould, E. A., & Solomon, T. (2008). Pathogenic flaviviruses. *Lancet, 371*(9611), 500–509. https://doi.org/10.1016/S0140-6736(08)60238-X
Grifferty, G., Shirley, H., McGloin, J., Kahn, J., Orriols, A., & Wamai, R. (2021). Vulnerabilities to and the socioeconomic and psychosocial impacts of the leishmaniases: A review. *Research and Reports in Tropical Medicine, 12*, 135–151. https://doi.org/10.2147/RRTM.S278138
Griffin, D. E. (2018). Measles Vaccine. *Viral Immunology, 31*(2), 86–95. https://doi.org/10.1089/vim.2017.0143
Grillo, A., Salvi, L., Coruzzi, P., Salvi, P., & Parati, G. (2019). Sodium intake and hypertension. *Nutrients, 1*(9), 1970. https://doi.org/10.3390/nu11091970
Gryseels, B., Polman, K., Clerinx, J., & Kestens, L. (2006). Human schistosomiasis. *Lancet, 368*(9541), 1106–1118. https://doi.org/10.1016/S0140-6736(06)69440-3
Grzybowski, A., Kanclerz, P., Tsubota, K., Lanca, C., & Saw, S. M. (2020). A review on the epidemiology of myopia in school children worldwide. *Biomed Central Ophthalmology, 20*(1), 27. https://doi.org/10.1186/s12886-019-1220-0
Guerra, F. M., Bolotin, S., Lim, G., Heffernan, J., Deeks, S. L., Li, Y., et al. (2017). The basic reproduction number (R0) of measles: a systematic review. *The Lancet. Infectious Diseases, 17*(12), e420–e428. https://doi.org/10.1016/S1473-3099(17)30307-9
Guinn, K. M., & Rubin, E. J. (2017). Tuberculosis: Just the FAQs. *MBio, 8*(6), e01910–e01917. https://doi.org/10.1128/mBio.01910-17
Gupta, S. K., & Shukla, V. K. (2004). Silent gallstones: A therapeutic dilemma. *Tropical Gastroenterology, 25*(2), 65–68. PMID: 15471318.
Gupta, P., Gupta, N., Pawar, A. P., Birajdar, S. S., Natt, A. S., & Singh, H. P. (2013). Role of sugar and sugar substitutes in dental caries: A review. *ISRN Dentistry, 2013*, 519421. https://doi.org/10.1155/2013/519421
Gustavsson, A., Svensson, M., Jacobi, F., Allgulander, C., Alonso, J., Beghi, E., CDBE2010Study Group, et al. (2011). Cost of disorders of the brain in Europe 2010. *European Neuropsychopharmacology, 21*(10), 718–779. https://doi.org/10.1016/j.euroneuro.2011.08.008
Gwiazda, J., Scheiman, M., Mohindra, I., & Held, R. (1984). Astigmatism in children: Changes in axis and amount from birth to six years. *Investigative Ophthalmology & Visual Science, 25*, 88–92. PMID: 6698734.
Hahad, O., Daiber, A., Michal, M., Kuntic, M., Lieb, K., Beutel, M., & Münzel, T. (2021). Smoking and neuropsychiatric disease-associations and underlying mechanisms. *International Journal of Molecular Science, 22*(14), 7272. https://doi.org/10.3390/ijms22147272
Halldestam, I., Enell, E. L., Kullman, E., & Borch, K. (2004). Development of symptoms and complications in individuals with asymptomatic gallstones. *The British Journal of Surgery, 91*(6), 734–738. https://doi.org/10.1002/bjs.4547
Harper, K., & Armelagos, G. (2011). The changing disease-scape in the third epidemiological transition. *International Journal of Environmental Research and Public Health, 7*(2), 675–697. https://doi.org/10.3390/ijerph7020675
Harrison, H., Thompson, R. E., Lin, Z., Rossi, S. H., Stewart, G. D., Griffin, S. J., & Usher-Smith, J. A. (2021). Risk prediction models for kidney cancer: A systematic review. *European Urology Focus, 7*(6), 1380–1390. https://doi.org/10.1016/j.euf.2020.06.024
Hart, A. R., Kennedy, H., & Harvey, I. (2008). Pancreatic cancer: A review of the evidence on causation. *Clinical Gastroenterology and Hepatology, 6*(3), 275–282. https://doi.org/10.1016/j.cgh.2007.12.041
Hays, J. N. (2005). *Epidemics and pandemics: Their impacts on human history*. ABC-CLIO. ISBN 978-1851096589.
He, F. J., Li, J., & Macgregor, G. A. (2013). Effect of longer term modest salt reduction on blood pressure: Cochrane systematic review and meta-analysis of randomised trials. *British Medical Journal (BMJ), 346*, f1325. https://doi.org/10.1136/bmj.f1325

He, F. J., Tan, M., Ma, Y., & MacGregor, G. A. (2020). Salt reduction to prevent hypertension and cardiovascular disease: JACC state-of-the-art review. *Journal of the American College of Cardiology, 75*(6), 632–647. https://doi.org/10.1016/j.jacc.2019.11.055

Heller, H. J. (1999). The role of calcium in the prevention of kidney stones. *Journal of the American College of Nutritions, 18*(5 Suppl), 373S–378S. https://doi.org/10.1080/07315724.1999.10718901

Henderson, D. (2009). *Smallpox: The death of a disease*. Prometheus Books. ISBN 978-1615922307.

Henderson, D. A. (2011). The eradication of smallpox—an overview of the past, present, and future. *Vaccine, 29*(Suppl 4), D7–D9. https://doi.org/10.1016/j.vaccine.2011.06.080

Hill, K., & Hurtado, A. M. (1996). *Ache life history*. Aldine De Gruyter. ISBN 978-0202020363.

Holm, S. (1937). The ocular refraction state of the Palae-Negroids in Gabon, French Equatorial Africa. *Acta Ophthalmologica, Suppl 13*, 1–299.

Holmes, E. C., Goldstein, S. A., Rasmussen, A. L., Robertson, D. L., Crits-Christoph, A., Wertheim, J. O., et al. (2021). The origins of SARS-CoV-2: A critical review. *Cell, 184*(19), 4848–4856. https://doi.org/10.1016/j.cell.2021.08.017

Howell, N. (1979). *Demography of the Dobe! Kung*. Academic. ISBN 978-0202306496.

Hu, J. X., Zhao, C. F., Chen, W. B., Liu, Q. C., Li, Q. W., Lin, Y. Y., & Gao, F. (2021). Pancreatic cancer: A review of epidemiology, trend, and risk factors. *World Journal of Gastroenterology, 27*(27), 4298–4321. https://doi.org/10.3748/wjg.v27.i27.4298

Inada, T., Miyazaki, S., Omori, T., Nihira, H., & Hino, T. (1958). Statistical study on urolithiasis in Japan. *Urologia Internationalis (Basel), 7*, 150–165. https://doi.org/10.1159/000277382

Janković, S., & Radosavljević, V. (2007). Risk factors for bladder cancer. *Tumori, 93*(1), 4–12. https://doi.org/10.1177/030089160709300102

Jarvis, J. F., & Van Heerden, H. G. (1967). The Acuity of Hearing in the Kalahari Bushmen: A Pilot Survey. *The Journal of Laryngology & Otology, 81*(1), 63–68. https://doi.org/10.1017/S0022215100066779

Jeon, S., & Carr, R. (2020). Alcohol effects on hepatic lipid metabolism. *Journal of Lipid Research, 61*(4), 470–479. https://doi.org/10.1194/jlr.R119000547

Jonas, J. B., Ohno-Matsui, K., & Panda-Jonas, S. (2019). Myopia: Anatomic changes and consequences for its etiology. *Asia-Pacific Journal of Ophthalmology, 8*(5), 355–359. https://doi.org/10.1097/01.APO.0000578944.25956.8b

Joseph, U., Su, Y. C., Vijaykrishna, D., & Smith, G. J. (2017). The ecology and adaptive evolution of influenza A interspecies transmission. *Influenza and Other Respiratory Viruses, 11*(1), 74–84. https://doi.org/10.1111/irv.12412

Kahwati, L. C., Weber, R. P., Pan, H., Gourlay, M., LeBlanc, E., Coker-Schwimmer, M., et al. (2018). Vitamin D, calcium, or combined supplementation for the primary prevention of fractures in community-dwelling adults: evidence report and systematic review for the US preventive services task force. *JAMA, 319*(15), 1600–1612. https://doi.org/10.1001/jama.2017.21640

Kaplan, J. O., Krumhardt, K. M., Ellis, E. C., Ruddiman, W. F., Lemmen, C., & Klein Goldewijk, K. (2011). Holocene carbon emissions as a result of anthropogenic land cover change. *The Holocene, 21*(5), 775–791. https://doi.org/10.1177/0959683610386983

Kariuki, S. N., & Williams, T. N. (2020). Human genetics and malaria resistance. *Human Genetics, 139*(6–7), 801–811. https://doi.org/10.1007/s00439-020-02142-6

Kashyap, D., Pal, D., Sharma, R., Garg, V. K., Goel, N., Koundal, D., et al. (2022). Global increase in breast cancer incidence: Risk factors and preventive measures. *BioMed Research International, 2022*, 9605439. https://doi.org/10.1155/2022/9605439

Kee, C. S., & Deng, L. (2008). Astigmatism associated with experimentally induced myopia or hyperopia in chickens. *Investigative Ophthalmology and Vision Science, 49*, 858–867. https://doi.org/10.1167/iovs.06-1370

Kee, C. S., Hung, L. F., Qiao-Grider, Y., Ramamirtham, R., & Smith, E. L., 3rd. (2005). Astigmatism in monkeys with experimentally induced myopia or hyperopia. *Optometry and Vision Science, 82*, 248–260. https://doi.org/10.1097/01.opx.0000159357.61498.6b

Kempen, J. H., Mitchell, P., Lee, K. E., Tielsch, J. M., Broman, A. T., Taylor, H. R., Eye Diseases Prevalence Research Group, et al. (2004). The prevalence of refractive errors among adults in the United States, Western Europe, and Australia. *Archives of Ophthalmology, 122*(4), 495–505. https://doi.org/10.1001/archopht.122.4.495

Kessler, R. C., & Bromet, E. J. (2013). The epidemiology of depression across cultures. *Annual Review of Public Health, 34*, 119–138. https://doi.org/10.1146/annurev-publhealth-031912-114409

Khan, S. R., Pearle, M. S., Robertson, W. G., Gambaro, G., Canales, B. K., Doizi, S., et al. (2017). Kidney stones. *Nature Reviews. Disease Primers, 3*, 17001. https://doi.org/10.1038/nrdp.2017.1

Kim, B. H., & Park, J. W. (2018). Epidemiology of liver cancer in South Korea. *Clinical and Molecular Hepatology, 24*(1), 1–9. https://doi.org/10.3350/cmh.2017.0112

Kim, J., Cho, Y. A., Choi, W. J., & Jeong, S. H. (2014). Gene-diet interactions in gastric cancer risk: a systematic review. *World Journal of Gastroenterology, 20*(28), 9600–9610. https://doi.org/10.3748/wjg.v20.i28.9600

Kirchner, D. B. (2002). The spectrum of allergic disease in the chemical industry. *International Archives of Occupational and Environmental Health, 75*(Suppl), S107–S112. https://doi.org/10.1007/s00420-002-0345-1

Klein Goldewijk, K., Beusen, A., & Janssen, P. (2010). Long-term dynamic modeling of global population and built-up area in a spatially explicit way: hyde 3.1. *The Holocene, 20*(4), 565–573. https://doi.org/10.1177/0959683609356587

Koch, A., Brierley, C., Maslin, M. M., & Lewis, S. L. (2019). Earth system impacts of the European arrival and Great Dying in the Americas after 1492. *Quaternary Science Reviews, 207*, 13–36. https://doi.org/10.1016/j.quascirev.2018.12.004

Kockskämper, J., & Pluteanu, F. (2022). Left Atrial Myocardium in arterial hypertension. *Cells, 11*(19), 3157. https://doi.org/10.3390/cells11193157

Kondo, T., Nakano, Y., Adachi, S., & Murohara, T. (2019). Effects of tobacco smoking on cardiovascular disease. *Circulation Journal, 83*(10), 1980–1985. https://doi.org/10.1253/circj.CJ-19-0323

Konner, M., & Eaton, S. B. (2010). Paleolithic nutrition: Twenty-five years later. *Nutrition in Clinical. Practice, 25*, 594–602. https://doi.org/10.1177/0884533610385702

Kontić, O., Vasiljević, N., Trisović, M., Jorga, J., Lakić, A., & Gasić, M. J. (2012). Eating disorders. *Srpski Arhiv Za Celokupno Lekarstvo (Serbian Archives of Medicine), 140*(9–10), 673–678. https://doi.org/10.2298/SARH1210673K

Krammer, F., Smith, G. J., Fouchier, R. A., Peiris, M., Kedzierska, K., Doherty, P. C., et al. (2018). Influenza. *Nature Reviews. Disease Primers, 4*(1), 3. https://doi.org/10.1038/s41572-018-0002-y

Krzyzak, M., & Mulrooney, S. M. (2020). Acute Appendicitis Review: Background, Epidemiology, Diagnosis, and Treatment. *Cureus, 12*(6), e8562. https://doi.org/10.7759/cureus.8562

Kurmi, O. P., Arya, P. H., Lam, K. B., Sorahan, T., & Ayres, J. G. (2012). Lung cancer risk and solid fuel smoke exposure: A systematic review and meta-analysis. *The European Respiratory Journal, 40*(5), 1228–1237. https://doi.org/10.1183/09031936.00099511

Laakso, M. (2019). Biomarkers for type 2 diabetes. *Molecular Metabolism, 27*, S139–S146. https://doi.org/10.1016/j.molmet.2019.06.016

Lange, P., Ahmed, E., Lahmar, Z. M., Martinez, F. J., & Bourdin, A. (2021). Natural history and mechanisms of COPD. *Respirology, 26*(4), 298–321. https://doi.org/10.1111/resp.14007

Langenberg, C., & Lotta, L. A. (2018). Genomic insights into the causes of type 2 diabetes. *Lancet, 391*, 2463–2474. https://doi.org/10.1016/S0140-6736(18)31132-2

Laurin, M., Everett, M. L., & Parker, W. (2011). The cecal appendix: one more immune component with a function disturbed by post-industrial culture. *Anatomical Record (Hoboken, N.J.), 294*(4), 567–579. https://doi.org/10.1002/ar.21357

Lawn, S. D., & Zumla, A. I. (2011). Tuberculosis. *Lancet, 378*(9785), 57–72. https://doi.org/10.1016/S0140-6736(10)62173-3

Leszczyszyn, A., Hnitecka, S., & Dominiak, M. (2021). Could Vitamin D3 Deficiency Influence Malocclusion Development? *Nutrients, 13*(6), 2122. https://doi.org/10.3390/nu13062122

Leung, A. K., Hon, K. L., Leong, K. F., & Sergi, C. M. (2018). Measles: A disease often forgotten but not gone. *Hong Kong Medical Journal, 24*(5), 512–520. https://doi.org/10.12809/hkmj187470

Levy, F. L., Adams-Huet, B., & Pak, C. Y. (1995). Ambulatory evaluation of nephrolithiasis: An update of a 1980 protocol. *The American Journal of Medicine, 98*(1), 50–59. https://doi.org/10.1016/S0002-9343(99)80080-1

Libertini, G. (1983). *Ragionamenti evoluzionistici [book in Italian].* Società Editrice Napoletana. English edition (2011): *Evolutionary arguments on aging, disease, and other topics.* Crownsville (MD): Azinet Press. ISBN 978-0978870928.

Lin, X., & Li, H. (2021). Obesity: Epidemiology, Pathophysiology, and Therapeutics. *Frontiers in Endocrinology (Lausanne), 12*, 706978. https://doi.org/10.3389/fendo.2021.706978

Lin, B. B., Lin, M. E., Huang, R. H., Hong, Y. K., Lin, B. L., & He, X. J. (2020). Dietary and lifestyle factors for primary prevention of nephrolithiasis: a systematic review and meta-analysis. *Biomed Central Nephrology, 21*(1), 267. https://doi.org/10.1186/s12882-020-01925-3

Lipworth, L., Tarone, R. E., & McLaughlin, J. K. (2006). The epidemiology of renal cell carcinoma. *The Journal of Urology, 176*(6 Pt 1), 2353–2358. https://doi.org/10.1016/j.juro.2006.07.130

Liu, Y., Chen, Y., Liao, B., Luo, D., Wang, K., Li, H., & Zeng, G. (2018). Epidemiology of urolithiasis in Asia. *Asian Journal of Urology, 5*(4), 205–214. https://doi.org/10.1016/j.ajur.2018.08.007

Liu, Q., Chen, J., & Zhou, X. N. (2020). Preparedness for Chagas disease spreading worldwide. *Infectious Diseases of Poverty, 9*(1), 44. https://doi.org/10.1186/s40249-020-00658-7

Liu, Q., Kang, Y., & Yan, J. (2022). Association between overall dietary quality and constipation in American adults: A cross-sectional study. *Biomed Central Public Health, 22*, 1971. https://doi.org/10.1186/s12889-022-14360-w

Lockwood, D. N., & Saunderson, P. R. (2012). Nerve damage in leprosy: a continuing challenge to scientists, clinicians and service providers. *International Health, 4*(2), 77–85. https://doi.org/10.1016/j.inhe.2011.09.006

Lozano, R., Naghavi, M., Foreman, K., Lim, S., Shibuya, K., Aboyans, V., et al. (2012). Global and regional mortality from 235 causes of death for 20 age groups in 1990 and 2010: A systematic analysis for the Global Burden of Disease Study 2010. *Lancet, 380*(9859), 2095–2128. https://doi.org/10.1016/S0140-6736(12)61728-0

Lu, Q., Gottlieb, E., & Rounds, S. (2018). Effects of cigarette smoke on pulmonary endothelial cells. *American Journal of Physiology. Lung Cellular and Molecular Physiology, 314*(5), L743–L756. https://doi.org/10.1152/ajplung.00373.2017

Lu, L., Mullins, C. S., Schafmayer, C., Zeißig, S., & Linnebacher, M. (2021). A global assessment of recent trends in gastrointestinal cancer and lifestyle-associated risk factors. *Cancer Communications (London, England), 41*(11), 1137–1151. https://doi.org/10.1002/cac2.12220

Ma, Y., Hu, M., Zhou, L., Ling, S., Li, Y., Kong, B., & Huang, P. (2018). Dietary fiber intake and risks of proximal and distal colon cancers: A meta-analysis. *Medicine, 97*(36), e11678. https://doi.org/10.1097/MD.0000000000011678

Maalouf, N. M., Cameron, M. A., Moe, O. W., Adams-Huet, B., & Sakhaee, K. (2007). Low urine pH: A novel feature of the metabolic syndrome. *Clinical Journal of the American Society of Nephrology, 2*(5), 883–888. https://doi.org/10.2215/CJN.00670207

Maatouk, I., & Moutran, R. (2014). History of syphilis: between poetry and medicine. *Sexual Medicine, 11*(1), 307–310. https://doi.org/10.1111/jsm.12354

Macdonald, H. M., New, S. A., Golden, M. H., Campbell, M. K., & Reid, D. M. (2004). Nutritional associations with bone loss during the menopausal transition: Evidence of a beneficial effect of calcium, alcohol, and fruit and vegetable nutrients and of a detrimental effect of fatty acids. *The American Journal of Clinical Nutrition, 79*(1), 155–165. https://doi.org/10.1093/ajcn/79.1.155

Machlowska, J., Baj, J., Sitarz, M., Maciejewski, R., & Sitarz, R. (2020). Gastric cancer: Epidemiology, risk factors, classification, genomic characteristics and treatment strategies. *International Journal of Molecular Science, 21*(11), 4012. https://doi.org/10.3390/ijms21114012

Mao, H., & Chen, Y. (2021). Noise-induced hearing loss: Updates on molecular targets and potential interventions. *Neural Plasticity, 2021*, 4784385. https://doi.org/10.1155/2021/4784385

Marcenes, W., Kassebaum, N. J., Bernabé, E., Flaxman, A., Naghavi, M., Lopez, A., & Murray, C. J. (2013). Global burden of oral conditions in 1990-2010: A systematic analysis. *Journal of Dental Research, 92*(7), 592–597. https://doi.org/10.1177/0022034513490168

Martínez-Alés, G., Jiang, T., Keyes, K. M., & Gradus, J. L. (2022). The recent rise of suicide mortality in the United States. *Annual Review of Public Health, 43*, 99–116. https://doi.org/10.1146/annurev-publhealth-051920-123206

Mattiuzzi, C., & Lippi, G. (2019). Current cancer epidemiology. *Journal of Epidemiology and Global Health, 9*(4), 217–222. https://doi.org/10.2991/jegh.k.191008.001

Mbanefo, E. C., Ahmed, A. M., Titouna, A., Elmaraezy, A., Trang, N. T., Phuoc Long, N., et al. (2017). Association of glucose-6-phosphate dehydrogenase deficiency and malaria: A systematic review and meta-analysis. *Scientific Reports, 7*, 45963. https://doi.org/10.1038/srep45963

McBrien, N. A., & Adams, D. W. (1997). A longitudinal investigation of adult-onset and adult-progression of myopia in an occupational group. Refractive and biometric findings. *Investigative Ophthalmology, 38*, 321–333. PMID: 9040464.

McManus, D. P., Dunne, D. W., Sacko, M., Utzinger, J., Vennervald, B. J., & Zhou, X. N. (2018). Schistosomiasis. *Nature Reviews. Disease Primers, 4*, 13. https://doi.org/10.1038/s41572-018-0013-8

Melancon, T. (1982). *Marriage and Reproduction among the Yanomamo Indians of Venezuela*. Ph. D. Dissertation, Pennsylvania State University.

Messerli, F. H., Rimoldi, S. F., & Bangalore, S. (2017). The Transition From Hypertension to Heart Failure: Contemporary Update. *JACC. Heart Failure, 5*(8), 543–551. https://doi.org/10.1016/j.jchf.2017.04.012. Erratum in: *JACC. Heart Failure*, 2017, 5(12):948.

Mills, C. E., Robins, J. M., & Lipsitch, M. (2004). Transmissibility of 1918 pandemic influenza. *Nature, 432*(7019), 904–906. https://doi.org/10.1038/nature03063

Minatsuki, C., Yamamichi, N., Shimamoto, T., Kakimoto, H., Takahashi, Y., Fujishiro, M., et al. (2013). Background factors of reflux esophagitis and non-erosive reflux disease: a cross-sectional study of 10,837 subjects in Japan. *PLoS One, 8*, e69891. https://doi.org/10.1371/journal.pone.0069891

Mirza, I., & Jenkins, R. (2004). Risk factors, prevalence, and treatment of anxiety and depressive disorders in Pakistan: systematic review. *British Medical Journal (BMJ), 328*(7443), 794. https://doi.org/10.1136/bmj.328.7443.794

Modlin, M. (1969). Renal calculus in the Republic of South Africa. In *Proceeding of the renal stone research symposium* (pp. 49–58). Churchill.

Morens, D. M., & Fauci, A. S. (2007). The 1918 influenza pandemic: insights for the 21st century. *The Journal of Infectious Diseases, 195*(7), 1018–1028. https://doi.org/10.1086/511989

Morgan, R. W., & Munro, M. (1973). Refractive problems in Northern natives. *Canadian Journal of Ophthalmology, 8*, 226–228. PMID: 4707204.

Moussavi, S., Chatterji, S., Verdes, E., Tandon, A., Patel, V., & Ustun, B. (2007). Depression, chronic diseases, and decrements in health: results from the World Health Surveys. *Lancet, 370*(9590), 851–858. https://doi.org/10.1016/S0140-6736(07)61415-9

Mtaya, M., Brudvik, P., & Astrom, A. N. (2009). Prevalence of malocclusion and its relationship with socio-demographic factors, dental caries, and oral hygiene in 12- to 14-year-old Tanzanian schoolchildren. *European Journal of Orthodontics, 31*, 467–476. https://doi.org/10.1093/ejo/cjn125

Mutti, D. O., Mitchell, G. L., Jones, L. A., Friedman, N. E., Frane, S. L., Lin, W. K., et al. (2005). Axial growth and changes in lenticular and corneal power during emmetropization in infants. *Investigative Ophthalmology and Vision Science, 46*(9), 3074–3080. https://doi.org/10.1167/iovs.04-1040

Nanda, A., Mustafa, S. S., Castillo, M., & Bernstein, J. A. (2022). Air pollution effects in allergies and asthma. *Immunology and Allergy Clinics of North America, 42*(4), 801–815. https://doi.org/10.1016/j.iac.2022.06.004

NCD Risk Factor Collaboration (NCD-RisC), Adiposity. (2017). http://ncdrisc.org/data-downloads-adiposity.html. Retrieved on 24.01.23.

NCD Risk Factor Collaboration (NCD-RisC), Diabetes. (2016). Worldwide trends in diabetes since 1980: a pooled analysis of 751 population-based studies with 4.4 million participants. *Lancet, 387*(10027), 1513–1530. https://doi.org/10.1016/S0140-6736(16)00618-8

Nesse, R. M. (2019). *Good Reasons for Bad Feelings*. Dutton. ISBN 978-1101985687.

Ng, K. L. (2022). The Etiology of Bladder Cancer. In N. Barber & A. Ali (Eds.), *Urologic Cancers*. Exon Publications. 2022 Sep 12. Chapter 3. https://www.ncbi.nlm.nih.gov/books/NBK585966/

Normile, D. (2013). Tropical medicine. Surprising new dengue virus throws a spanner in disease control efforts. *Science, 342*(6157), 415. https://doi.org/10.1126/science.342.6157.415

Odenwald, M. A., & Paul, S. (2022). Viral hepatitis: Past, present, and future. *World Journal of Gastroenterology, 28*(14), 1405–1429. https://doi.org/10.3748/wjg.v28.i14.1405

O'Keefe, S. J. (2019). The association between dietary fibre deficiency and high-income lifestyle-associated diseases: Burkitt's hypothesis revisited. *The Lancet. Gastroenterology & Hepatology, 4*(12), 984–996. https://doi.org/10.1016/S2468-1253(19)30257-2

Özbel, Y., Töz, S., Muñoz, C., Ortuño, M., Jumakanova, Z., Pérez-Cutillas, P., et al. (2022). The current epidemiology of leishmaniasis in Turkey, Azerbaijan and Georgia and implications for disease emergence in European countries. *Zoonoses and Public Health, 69*(5), 395–407. https://doi.org/10.1111/zph.12977

Pak, M., & Lindseth, G. (2016). Risk factors for cholelithiasis. *Gastroenterol Nursing, 39*(4), 297–309. https://doi.org/10.1097/SGA.0000000000000235

Pandol, S. J., Gorelick, F. S., Gerloff, A., & Lugea, A. (2010). Alcohol abuse, endoplasmic reticulum stress and pancreatitis. *Digestive Disease, 28*(6), 776–782. https://doi.org/10.1159/000327212

Parascandola, M., & Xiao, L. (2019). Tobacco and the lung cancer epidemic in China. *Cancer Translational Lung Research, 8*(Suppl 1), S21–S30. https://doi.org/10.21037/tlcr.2019.03.12

Parmar, M. S. (2004). Kidney stones. *BMJ, 328*(7453), 1420–1424. https://doi.org/10.1136/bmj.328.7453.1420

Parnis, R. O. (1964). Gall bladder disease in Nigeria: A five-year review. *Transactions of the Royal Society of Tropical Medicine and Hygiene, 58*(5), 437–440. https://doi.org/10.1016/0035-9203(64)90092-6

Patel, M. K., Dumolard, L., Nedelec, Y., Sodha, S. V., Steulet, C., Gacic-Dobo, M., et al. (2019). Progress toward regional measles elimination - Worldwide, 2000-2018. *MMWR. Morbidity and Mortality Weekly Report, 68*(48), 1105–1111. https://doi.org/10.15585/mmwr.mm6848a1

Pawankar, R., & Wang, J. Y. (2021). Climate change, air pollution, and biodiversity in Asia pacific and impact on respiratory allergies. *Immunology and Allergy Clinics of North America, 41*(1), 63–71. https://doi.org/10.1016/j.iac.2020.09.008

Peeling, R. W., Mabey, D., Kamb, M. L., Chen, X. S., Radolf, J. D., & Benzaken, A. S. (2017). Syphilis. *Nature Reviews. Disease Primers, 3*, 17073. https://doi.org/10.1038/nrdp.2017.73

Pelkonen, M. K., Notkola, I. L., Laatikainen, T. K., & Koskela, H. O. (2014). Twenty-five year trends in prevalence of chronic bronchitis and the trends in relation to smoking. *Respiratory Medicine, 108*(11), 1633–1640. https://doi.org/10.1016/j.rmed.2014.08.007

Penninx, B. W., Pine, D. S., Holmes, E. A., & Reif, A. (2021). Anxiety disorders. *Lancet, 397*(10277), 914–927. https://doi.org/10.1016/S0140-6736(21)00359-7. Erratum in: *Lancet*, 2021, 397(10277), 880.

Pernar, C. H., Ebot, E. M., Wilson, K. M., & Mucci, L. A. (2018). The epidemiology of prostate cancer. *Cold Spring Harbor Perspectives in Medicine, 8*(12), a030361. https://doi.org/10.1101/cshperspect.a030361

Platts-Mills, T. A. E. (2015). The allergy epidemics: 1870-2010. *The Journal of Allergy and Clinical Immunology, 136*(1), 3–13. https://doi.org/10.1016/j.jaci.2015.03.048

Plemper, R. K. (2020). Measles resurgence and drug development. *Current Opinion in Virology, 41*, 8–17. https://doi.org/10.1016/j.coviro.2020.02.007
Pongratz, J., Reick, C., Raddatz, T., & Claussen, M. (2008). A reconstruction of global agricultural areas and land cover for the last millennium. *Global Biogeochemical Cycles, 22*, GB3018. https://doi.org/10.1029/2007GB003153
Porter, A., & Goldfarb, J. (2019). Measles: A dangerous vaccine-preventable disease returns. *Cleveland Clinical Journal of Medicine, 86*(6), 393–398. https://doi.org/10.3949/ccjm.86a.19065
Powell, M. L., & Cook, D. C. (2005). *The myth of Syphilis: The natural history of treponematosis in North America*. University Press of Florida. ISBN 978-0813027944.
Pressler, M., Devinsky, J., Duster, M., Lee, J. H., Glick, C. S., Wiener, S., et al. (2022). Dietary transitions and health outcomes in four populations - Systematic review. *Frontiers in Nutrition, 9*, 748305. https://doi.org/10.3389/fnut.2022.748305
Price, W. A. (1939). *Nutrition and physical degeneration*. Paul B. Hoeber. [Other editions in 1945, 1970, 1989, 1997, 2004, 2006, 2008, 2011; the last edition – ISBN 978-0916764203 - has been used].
Proffit, W. R., Fields, H. W., Jr., & Moray, L. J. (1998). Prevalence of malocclusion and orthodontic treatment need in the United States: estimates from the NHANES III survey. *The International Journal of Adult Orthodontics and Orthognathic Surgery, 13*(2), 97–106. PMID: 9743642.
Prugnolle, F., Durand, P., Ollomo, B., Duval, L., Ariey, F., Arnathau, C., et al. (2011). A fresh look at the origin of Plasmodium falciparum, the most malignant malaria agent. *PLoS Pathogens, 7*(2), e1001283. https://doi.org/10.1371/journal.ppat.1001283
Prynne, C. J., Mishra, G. D., O'Connell, M. A., Muniz, G., Laskey, M. A., Yan, L., et al. (2006). Fruit and vegetable intakes and bone mineral status: a cross sectional study in 5 age and sex cohorts. *The American Journal of Clinical Nutrition, 83*(6), 1420–1428. https://doi.org/10.1093/ajcn/83.6.1420
Rabaan, A. A., Mutair, A. A., Alhumaid, S., Garout, M., Alsubki, R. A., Alshahrani, F. S., et al. (2022). Updates on Measles Incidence and Eradication: Emphasis on the Immunological Aspects of Measles Infection. *Medicina (Kaunas, Lithuania), 58*(5), 680. https://doi.org/10.3390/medicina58050680
Rahman, G. A. (2005). Cholelithiasis and cholecystitis: Changing prevalence in an African community. *Journal of the National Medical Association, 97*(11), 1534–1538. PMID: 16334500.
Rassi, A., Jr., Rassi, A., & Marin-Neto, J. A. (2010). Chagas disease. *Lancet, 375*(9723), 1388–1402. https://doi.org/10.1016/S0140-6736(10)60061-X
Rawat, A., Roy, M., Jyoti, A., Kaushik, S., Verma, K., & Srivastava, V. K. (2021). Cysteine proteases: Battling pathogenic parasitic protozoans with omnipresent enzymes. *Microbiological Research, 249*, 126784. https://doi.org/10.1016/j.micres.2021.126784
Rayment, M., & Sullivan, A. K. (2011). "He who knows syphilis knows medicine" - The return of an old friend. *British Journal of Cardiology, 18*(2), 56–58.
Reidl, J., & Klose, K. E. (2002). Vibrio cholerae and cholera: Out of the water and into the host. *FEMS Microbiology Reviews, 26*(2), 125–139. https://doi.org/10.1111/j.1574-6976.2002.tb00605.x
Robbins, G., Tripathy, V. M., Misra, V. N., Mohanty, R. K., Shinde, V. S., Gray, K. M., & Schug, M. D. (2009). Ancient skeletal evidence for leprosy in India (2000 B.C.). *PLoS One, 4*(5), e5669. https://doi.org/10.1371/journal.pone.0005669
Rocco, A., Compare, D., Angrisani, D., Sanduzzi Zamparelli, M., & Nardone, G. (2014). Alcoholic disease: Liver and beyond. *World Journal of Gastroenterolgy, 20*(40), 14652–14659. https://doi.org/10.3748/wjg.v20.i40.14652
Rose, K. A., Morgan, I. G., Ip, J., Kifley, A., Huynh, S., Smith, W., & Mitchell, P. (2008a). Outdoor activity reduces the prevalence of myopia in children. *Ophthalmology, 115*, 1279–1285. https://doi.org/10.1016/j.ophtha.2007.12.019
Rose, K. A., Morgan, I. G., Smith, W., Burlutsky, G., Mitchell, P., & Saw, S.-M. (2008b). Myopia, lifestyle, and schooling in students of Chinese ethnicity in Singapore and Sydney. *Archives of Ophthalmology, 126*, 527–530. https://doi.org/10.1001/archopht.126.4.527

Rosen, S., Bergman, M., Plester, D., El-Mofty, A., & Satti, M. H. (1962). LXII Presbycusis Study of a Relatively Noise-Free Population in The Sudan. *Annals of Otology, Rhinology & Laryngology, 71*(3), 727–743. https://doi.org/10.1177/000348946207100313

Rosenvinge, J. H., & Pettersen, G. (2015). Epidemiology of eating disorders, part I: introduction to the series and a historical panorama. *Advances in Eating Disorders, 3*(1), 76–90. https://doi.org/10.1080/21662630.2014.898206

Roy, M., Rawat, A., Kaushik, S., Jyoti, A., & Srivastava, V. K. (2022). Endogenous cysteine protease inhibitors in upmost pathogenic parasitic protozoa. *Microbiological Research, 261*, 127061. https://doi.org/10.1016/j.micres.2022.127061

Rozanec, J. J., & Secin, F. P. (2020). Epidemiología, etiología, prevención del cáncer vesical [Epidemiology, etiology and prevention of bladder cancer. In Spanish]. *Archivos Españoles de Urología, 73*(10), 872–878. PMID: 33269706.

Rumgay, H., Murphy, N., Ferrari, P., & Soerjomataram, I. (2021). Alcohol and Cancer: Epidemiology and Biological Mechanisms. *Nutrients, 13*(9), 3173. https://doi.org/10.3390/nu13093173

Russell, F. (1908). *The Pima Indians*. Bureau of American Ethnology.

Sack, D. A., Sack, R. B., Nair, G. B., & Siddique, A. K. (2004). Cholera. *Lancet, 363*(9404), 223–233. https://doi.org/10.1016/S0140-6736(03)15328-7

Safaei, M., Sundararajan, E. A., Driss, M., Boulila, W., & Shapi'i, A. (2021). A systematic literature review on obesity: Understanding the causes & consequences of obesity and reviewing various machine learning approaches used to predict obesity. *Computers in Biology and Medicine, 136*, 104754. https://doi.org/10.1016/j.compbiomed.2021.104754

Sahni, S., Mangano, K. M., McLean, R. R., Hannan, M. T., & Kiel, D. P. (2015). Dietary approaches for bone health: Lessons from the Framingham osteoporosis study. *Current Osteoporosis Reports, 13*(4), 245–255. https://doi.org/10.1007/s11914-015-0272-1

Sales, M. P. U., Araújo, A. J., Chatkin, J. M., Godoy, I., Pereira, L. F. F., Castellano, M. V. C. O., et al. (2019). Update on the approach to smoking in patients with respiratory diseases. *Jornal Brasileiro de Pneumologia, 45*(3), e20180314. https://doi.org/10.1590/1806-3713/e20180314

Samadian, F., Dalili, N., & Jamalian, A. (2016). Lifestyle modifications to prevent and control hypertension. *Iranian Journal of Kidney Disease, 10*(5), 237–263. PMID: 27721223.

Samocha-Bonet, D., Wu, B., & Ryugo, D. K. (2021). Diabetes mellitus and hearing loss: A review. *Ageing Research Reviews, 71*, 101423. https://doi.org/10.1016/j.arr.2021.101423

Sanchez Armengol, E., Blanka Kerezsi, A., & Laffleur, F. (2022). Allergies caused by textiles: Control, research and future perspective in the medical field. *International Immunopharmacology, 110*, 109043. https://doi.org/10.1016/j.intimp.2022.109043

Saraff, V., & Shaw, N. (2016). Sunshine and vitamin D. *Archives of Disease in Childhood, 101*(2), 190–192. https://doi.org/10.1136/archdischild-2014-307214

Sas, D. J. (2011). An update on the changing epidemiology and metabolic risk factors in pediatric kidney stone disease. *Clinical Journal of the American Society of Nephrology, 6*(8), 2062–2068. https://doi.org/10.2215/CJN.11191210

Schabath, M. B., & Cote, M. L. (2019). Cancer progress and priorities: Lung cancer. *Cancer Epidemiology, Biomarkers & Prevention, 28*(10), 1563–1579. https://doi.org/10.1158/1055-9965.EPI-19-0221

Scherübl, H. (2022). Tobacco smoking and gastrointestinal cancer risk. *Visceral Medicine, 38*(3), 217–222. https://doi.org/10.1159/000523668

Schmidt, M. A., Marcano-Bonilla, L., & Roberts, L. R. (2019). Gallbladder cancer: Epidemiology and genetic risk associations. *Chinese Clinical Oncology, 8*(4), 31. https://doi.org/10.21037/cco.2019.08.13

Schreuder, P. A., Noto, S., & Richardus, J. H. (2016). Epidemiologic trends of leprosy for the 21st century. *Clinics in Dermatology, 34*(1), 24–31. https://doi.org/10.1016/j.clindermatol.2015.11.001

Schulz, L. O., Bennett, P. H., Ravussin, E., Kidd, J. R., Kidd, K. K., Esparza, J., & Valencia, M. E. (2006). Effects of traditional and western environments on prevalence of type 2 diabetes in Pima Indians in Mexico and the U.S. *Diabetes Care, 29*(8), 1866–1871. https://doi.org/10.2337/dc06-0138

Sharma, A. P., & Filler, G. (2010). Epidemiology of pediatric urolithiasis. *Indian Journal of Urology, 26*(4), 516–522. https://doi.org/10.4103/0970-1591.74450

Sharp, P. M., & Hahn, B. H. (2011). Origins of HIV and the AIDS Pandemic. *Cold Spring Harbor Perspectives in Medicine, 1*(1), a006841. https://doi.org/10.1101/cshperspect.a006841

Shepard, C. W., Simard, E. P., Finelli, L., Fiore, A. E., & Bell, B. P. (2006). Hepatitis B virus infection: Epidemiology and vaccination. *Epidemiologic Reviews, 28*, 112–125. https://doi.org/10.1093/epirev/mxj009

Shepard, D. S., Undurraga, E. A., Halasa, Y. A., & Stanaway, J. D. (2016). The global economic burden of dengue: a systematic analysis. *The Lancet. Infectious Diseases, 16*(8), 935–941. https://doi.org/10.1016/S1473-3099(16)00146-8

Siener, R. (2021). Nutrition and kidney stone disease. *Nutrients, 13*(6), 1917. https://doi.org/10.3390/nu13061917

Skeller, E. (1954). Anthropological and ophthalmological studies on the Angmagssalik Eskimos. In *Meddelelser om Grønland* (Vol. 107, pp. 167–211). C. A. Reitzel.

Smith, C. J., & Hansch, C. (2000). The relative toxicity of compounds in mainstream cigarette smoke condensate. *Food and Chemial Toxicology, 38*, 637–646. https://doi.org/10.1016/S0278-6915(00)00051-X

Soliman, G. A. (2019). Dietary fiber, atherosclerosis, and cardiovascular disease. *Nutrients, 11*(5), 1155. https://doi.org/10.3390/nu11051155

Sookram, C., Munodawafa, D., Phori, P. M., Varenne, B., & Alisalad, A. (2015). WHO's supported interventions on salt intake reduction in the sub-Saharan Africa region. *Cardiovascular Diagnosis and Therapy, 5*(3), 186–190. https://doi.org/10.3978/j.issn.2223-3652.2015.04.04

Sorokin, I., Mamoulakis, C., Miyazawa, K., Rodgers, A., Talati, J., & Lotan, Y. (2017). Epidemiology of stone disease across the world. *World Journal of Urology, 35*(9), 1301–1320. https://doi.org/10.1007/s00345-017-2008-6

Spreeuwenberg, P., Kroneman, M., & Paget, J. (2018). Reassessing the Global Mortality Burden of the 1918 influenza pandemic. *American Journal of Epidemiology, 187*(12), 2561–2567. https://doi.org/10.1093/aje/kwy191

Stamatelou, K., & Goldfarb, D. S. (2023). Epidemiology of kidney stones. *Healthcare (Basel), 11*(3), 424. https://doi.org/10.3390/healthcare11030424

Stein, D. J., Scott, K. M., de Jonge, P., & Kessler, R. C. (2017). Epidemiology of anxiety disorders: From surveys to nosology and back. *Dialogues in Clinical Neuroscience, 19*(2), 127–136. https://doi.org/10.31887/DCNS.2017.19.2/dstein

Steuer, C. E., El-Deiry, M., Parks, J. R., Higgins, K. A., & Saba, N. F. (2017). An update on larynx cancer. *CA: a Cancer Journal for Clinicians, 67*(1), 31–50. https://doi.org/10.3322/caac.21386

Steverding, D. (2008). The history of African trypanosomiasis. *Parasites & Vectors, 1*(1), 3. https://doi.org/10.1186/1756-3305-1-3

Stram, M., Liu, S., & Singhi, A. D. (2016). Chronic pancreatitis. *Surgical Pathology Clinics, 9*(4), 643–659. https://doi.org/10.1016/j.path.2016.05.008

Su, S. B., Chang, H. L., & Chen, A. K. (2020). Current status of mumps virus infection: Epidemiology, pathogenesis, and vaccine. *International Journal of Environmental Research and Public Health, 17*(5), 1686. https://doi.org/10.3390/ijerph17051686

Sun, D., Li, H., Cao, M., He, S., Lei, L., Peng, J., & Chen, W. (2020). Cancer burden in China: Trends, risk factors and prevention. *Cancer Biology & Medicine, 17*(4), 879–895. https://doi.org/10.20892/j.issn.2095-3941.2020.0387

Suzuki, K., Akama, T., Kawashima, A., Yoshihara, A., Yotsu, R. R., & Ishii, N. (2012). Current status of leprosy: Epidemiology, basic science and clinical perspectives. *The Journal of Dermatology, 39*(2), 121–129. https://doi.org/10.1111/j.1346-8138.2011.01370.x

Szreter, S., & Kevin, S. (2020). The pox in Boswell's London: An estimate of the extent of syphilis infection in the metropolis in the 1770s. *The Economic History Review, 74*(2), 372–399. https://doi.org/10.1111/ehr.13000

Tabbabi, A. (2019). Review of Leishmaniasis in the Middle East and North Africa. *African Health Sciences, 19*(1), 1329–1337. https://doi.org/10.4314/ahs.v19i1.4

Takano, H., & Inoue, K. I. (2017). Environmental pollution and allergies. *Journal of Toxicologic Pathology, 30*(3), 193–199. https://doi.org/10.1293/tox.2017-0028
Taniai, M. (2020). Alcohol and hepatocarcinogenesis. *Clinical and Molecular Hepatology, 26*(4), 736–741. https://doi.org/10.3350/cmh.2020.0203
Taraseviciene-Stewart, L., & Voelkel, N. F. (2008). Molecular pathogenesis of emphysema. *The Journal of Clinical Investigation, 118*(2), 394–402. https://doi.org/10.1172/JCI31811
Teo, K. K., & Rafiq, T. (2021). Cardiovascular risk factors and prevention: A perspective from developing countries. *The Canadian Journal of Cardiology, 37*(5), 733–743. https://doi.org/10.1016/j.cjca.2021.02.009
Thandra, K. C., Barsouk, A., Saginala, K., Aluru, J. S., & Barsouk, A. (2021). Epidemiology of lung cancer. *Contemporary Oncology (Poznań, Poland), 25*(1), 45–52. https://doi.org/10.5114/wo.2021.103829
Thèves, C., Crubézy, E., & Biagini, P. (2016). History of smallpox and its spread in human populations. *Microbiology Spectrum, 4*(4). https://doi.org/10.1128/microbiolspec.PoH-0004-2014
Thibaut, F. (2017). Anxiety disorders: A review of current literature. *Dialogues in Clinical Neuroscience, 19*(2), 87–88. https://doi.org/10.31887/DCNS.2017.19.2/fthibaut
Thongprayoon, C., Krambeck, A. E., & Rule, A. D. (2020). Determining the true burden of kidney stone disease. *Nature Reviews. Nephrology, 16*(12), 736–746. https://doi.org/10.1038/s41581-020-0320-7
Tomasetti, C., & Vogelstein, B. (2015). Variation in cancer risk among tissues can be explained by the number of stem cell divisions. *Science, 347*(6217), 78–81. https://doi.org/10.1126/science.1260825
Trepel, F. (2004). Dietary fibre: more than a matter of dietetics. II. Preventative and therapeutic uses, article in German. *Wiener Klinische Wochenschrift, 116*(15–16), 511–522. https://doi.org/10.1007/bf03217703
Troilo, D., Smith, E. L., 3rd, Nickla, D. L., Ashby, R., Tkatchenko, A. V., Ostrin, L. A., et al. (2019). IMI - Report on experimental models of emmetropization and myopia. *Investigative Ophthalmology & Visual Science, 60*(3), M31–M88. https://doi.org/10.1167/iovs.18-25967
Troy, K. L., Mancuso, M. E., Butler, T. A., & Johnson, J. E. (2018). Exercise early and often: Effects of physical activity and exercise on women's bone health. *International Journal of Environmental Research and Public Health, 15*(5), 878. https://doi.org/10.3390/ijerph15050878
Truswell, A. S., & Hansen, J. D. L. (1976). Medical research among the !Kung. In R. B. Lee & I. DeVore (Eds.), *Kalahari hunter-gatherers*. Harvard University Press. ISBN 978-0674499850.
Tucker, K. L., Chen, H., Hannan, M. T., Cupples, L. A., Wilson, P. W., Felson, D., & Kiel, D. P. (2002). Bone mineral density and dietary patterns in older adults: the Framingham Osteoporosis Study. *The American Journal of Clinical Nutrition, 76*(1), 245–252. https://doi.org/10.1093/ajcn/76.1.245
Tyszkowski, M. (2016). Is there a relationship between nutrition, facial development, and crowding of the teeth? *International Journal of Orthodontics (Milwaukee, Wisconsin), 27*(1), 15–17. PMID: 27319035.
Vallès, X., Stenseth, N. C., Demeure, C., Horby, P., Mead, P. S., Cabanillas, O., et al. (2020). Human plague: An old scourge that needs new answers. *PLoS Neglected Tropical Diseases, 14*(8), e0008251. https://doi.org/10.1371/journal.pntd.0008251
Varghese, R. M., Sreenivas, V., Mammen Puliye, J., & Varughese, S. (2009). Refractive status at birth: Its relation to newborn physical parameters at birth and gestational age. *PLoS One, 4*(2), e4469. https://doi.org/10.1371/journal.pone.0004469
Vartiainen, V. A., Jousilahti, P., Laatikainen, T., & Vartiainen, E. (2022). Contribution of smoking change to 45-year trend in prevalence of chronic bronchitis in Finland. *Scandinavian Journal of Public Health, 19*, 14034948221104351. https://doi.org/10.1177/14034948221104351
Viegi, G., Maio, S., Pistelli, F., Baldacci, S., & Carrozzi, L. (2006). Epidemiology of chronic obstructive pulmonary disease: Health effects of air pollution. *Respirology, 11*(5), 523–532. https://doi.org/10.1111/j.1440-1843.2006.00886.x

Vieira, A. R., Abar, L., Chan, D. S. M., Vingeliene, S., Polemiti, E., Stevens, C., et al. (2017). Foods and beverages and colorectal cancer risk: a systematic review and meta-analysis of cohort studies, an update of the evidence of the WCRF-AICR Continuous Update Project. *Annals of Oncology, 28*(8), 1788–1802. https://doi.org/10.1093/annonc/mdx171

Vonlaufen, A., Wilson, J. S., Pirola, R. C., & Apte, M. V. (2007). Role of alcohol metabolism in chronic pancreatitis. *Alcohol Research & Health, 30*(1), 48–54. PMID: 17718401.

Vos, T., Allen, C., Arora, M., Barber, R. M., Bhutta, Z. A., Brown, A., GBD 2015 Mortality and Causes of Death Collaborators, et al. (2016). Global, regional, and national life expectancy, all-cause mortality, and cause-specific mortality for 249 causes of death, 1980-2015: a systematic analysis for the Global Burden of Disease Study 2015. *Lancet, 388*(10053), 1459–1544. https://doi.org/10.1016/s0140-6736(16)31012-1

Wagner, C. A. (2021). Etiopathogenic factors of urolithiasis. *Archivos Españoles de Urología, 74*(1), 16–23. https://www.aeurologia.com/EN/Y2021/V74/I1/16

Wall, K. M., Kilembe, W., Vwalika, B., Dinh, C., Livingston, P., Lee, Y. M., et al. (2018). Schistosomiasis is associated with incident HIV transmission and death in Zambia. *PLoS Neglected Tropical Diseases, 12*(12), e0006902. https://doi.org/10.1371/journal.pntd.0006902

Walter, K., & John, C. C. (2022). Malaria. *JAMA, 327*(6), 597. https://doi.org/10.1001/jama.2021.21468

Weatherall, D. J. (2015). The Thalassemias: Disorders of globin synthesis. In *Williams Hematology* (9th ed.). McGraw Hill Professional. ISBN 978-0071833011.

Weiss, K. M., Ferrell, R. E., Hanis, C. L., & Styne, P. N. (1984). Genetics and epidemiology of gallbladder disease in New World native peoples. *American Journal of Human Genetics, 36*(6), 1259–1278. PMID: 6517051.

Weissman, M. M., Bland, R. C., Canino, G. J., Faravelli, C., Greenwald, S., Hwu, H. G., et al. (1996). Cross-national epidemiology of major depression and bipolar disorder. *JAMA, 276*(4), 293–299. PMID: 8656541.

Weledji, E. P. (2022). Appendicitis: Epidemiology, evaluation, and controversy in management. In E. P. Weledji (Ed.), *Appendicitis-causes and treatments*. IntechOpen. https://www.intechopen.com/chapters/1132565

Weskamp, P., Ufton, D., Drysch, M., Wagner, J. M., Dadras, M., Lehnhardt, M., et al. (2022). Risk factors for occurrence and relapse of soft tissue sarcoma. *Cancers (Basel), 14*(5), 1273. https://doi.org/10.3390/cancers14051273

White, C., & Franco-Paredes, C. (2015). Leprosy in the 21st century. *Clinical Microbiology Reviews, 28*(1), 80–94. https://doi.org/10.1128/CMR.00079-13

WHO. (1992). *The ICD-X classification of mental and behavioral disorders: Diagnostic criteria for research*. World Health Organization.

WHO. (2010). *Control of the leishmaniases: Report of a meeting of the WHO Expert Committee on the Control of Leishmaniases, Geneva, 22–26 March 2010*. World Health Organization. ISBN 978-9241209496.

WHO. (2012). Leprosy. Global leprosy situation, 2012. *Weekly Epidemiological Record - Relevé épidémiologique hebdomadaire, 87*(34), 317–328.

WHO. (2016a). *Diabetes*. Global report on diabetes. https://www.who.int/publications/i/item/9789241565257, accessed on 24.01.23. ISBN 978-9241565257.

WHO. (2016b). *Tuberculosis. Global tuberculosis report 2016*. World Health Organization.

WHO. (2021). *Malaria. World Malaria Report 2021*. World Health Organization. ISBN 978-9240040496.

WHO. (2023a). *Alcohol*. https://www.who.int/health-topics/alcohol#tab=tab_1, accessed on 14.01.23.

WHO. (2023b). *Deafness. Deafness and Hearing Loss*. https://www.who.int/health-topics/hearing-loss#tab=tab_2, accessed on 14.01.23.

WHO. (2023c). *Tobacco*. https://www.who.int/health-topics/tobacco#tab=tab_1, accessed on 16.01.23.

Williams, H. U. (1932). The origin and antiquity of syphilis: The evidence from diseased bones, a review, with some new material from America. *Archives of Pathology, 13*(779–814), 931–983.
Williams, H. U. (1936). The origin of syphilis: Evidence from diseased bones: A supplementary report. *Archives of Dermatology and Syphilology, 33*(5), 783–787. https://doi.org/10.1001/archderm.1936.01470110003001
Winters, A. (2006). *Syphilis*. Rosen Pub. Group. ISBN 978-1404209060.
Wong, T. Y., Foster, P. J., Hee, J., Ng, T. P., Tielsch, J. M., Chew, S. J., et al. (2000). Prevalence and risk factors for refractive errors in an adult Chinese population in Singapore. *Investigative Ophthalmology & Visual Science, 41*(9), 2486–2494. PMID: 10937558.
Wood, B. F. (1971). Malocclusion in the modern Alaskan Eskimo. *American Journal of Orthodontics, 60*(4), 344–354. https://doi.org/10.1016/0002-9416(71)90147-3
World Cause of Death. (2020). (*Data from WHO*), www.worldlifeexpectancy.com/world-rankings-total-deaths. accessed 15.12.22.
Wu, Y., Ding, Y., Tanaka, Y., & Zhang, W. (2014). Risk factors contributing to type 2 diabetes and recent advances in the treatment and prevention. *International Journal of Medical Sciences, 11*(11), 1185–1200. https://doi.org/10.7150/ijms.10001
Wu, P. C., Huang, H. M., Yu, H. J., Fang, P. C., & Chen, C. T. (2016a). Epidemiology of Myopia. *Asia-Pacific Journal of Ophthalmology (Philadelphia), 5*(6), 386–393. https://doi.org/10.1097/APO.0000000000000236
Wu, S., Powers, S., Zhu, W., & Hannun, Y. A. (2016b). Substantial contribution of extrinsic risk factors to cancer development. *Nature, 529*, 43–47. https://doi.org/10.1038/nature16166
Yang, J. R., Hidayat, K., Chen, C. L., Li, Y. H., Xu, J. Y., & Qin, L. Q. (2020). Body mass index, waist circumference, and risk of hearing loss: A meta-analysis and systematic review of observational study. *Environmental Health and Preventive Medicine, 25*(1), 25. https://doi.org/10.1186/s12199-020-00862-9
Yang, Y., Guo, C., Gu, Z., Hua, J., Zhang, J., Qian, S., & Shi, J. (2022). The Global Burden of appendicitis in 204 countries and territories from 1990 to 2019. *Clinical Epidemiology, 14*, 1487–1499. https://doi.org/10.2147/CLEP.S376665
Yasui, T., Iguchi, M., Suzuki, S., & Kohri, K. (2008). Prevalence and epidemiological characteristics of urolithiasis in Japan: National trends between 1965 and 2005. *Urology, 71*(2), 209–213. https://doi.org/10.1016/j.urology.2007.09.034
Yong, S. B., Gau, S. Y., Guo, Y. C., & Wei, J. C. (2022). Allergy from perspective of environmental pollution effects: From an aspect of atopic dermatitis, immune system, and atmospheric hazards-a narrative review of current evidences. *Environmental Science and Pollution Research International, 29*(38), 57091–57101. https://doi.org/10.1007/s11356-022-21582-3
Youn, H. J., & Han, W. (2020). A review of the epidemiology of breast cancer in Asia: Focus on risk factors. *Asian Pacific Journal of Cancer Prevention, 21*(4), 867–880. https://doi.org/10.31557/APJCP.2020.21.4.867
Young, F. A., Leary, G. A., Baldwin, W. R., West, D. C., Box, R. A., Harris, E., & Johnson, C. (1969). The transmission of refractive errors within Eskimo families. *American Journal of Optometry and Archives of American Academy of Optometry, 46*, 676–685. https://doi.org/10.1097/00006324-196909000-00005
Yue, L., Pai, Q., Wu, X., & Zhang, J. (2022). Smoking and risk of urolithiasis: Meta-analysis of observational studies. *Frontiers in Public Health, 10*, 816756. https://doi.org/10.3389/fpubh.2022.816756
Zou, J., Meng, M., Law, C. S., Rao, Y., & Zhou, X. (2018). Common dental diseases in children and malocclusion. *International Journal of Oral Science, 10*(1), 7. https://doi.org/10.1038/s41368-018-0012-3
Zylbermann, R., Landau, D., & Berson, D. (1993). The influence of study habits on myopia in Jewish teenagers. *Journal of Pediatric Ophthalmology and Strabismus, 30*, 319–322. https://doi.org/10.3928/0191-3913-19930901-12

Chapter 4
Diseases Deriving from Alterations of the Ecological Niche—Part II: Regarding the Holobiont

4.1 Possible Alterations of the Ecological Niche Regarding the Holobiont

In this chapter, the diseases deriving from alterations of the ecological niche and regarding our holobiont are discussed.

4.1.1 Alterations of the Holobiont—A: Reduced Exposure to Allergens

Allergic diseases comprise a group of chronic diseases caused by alterations in the immune system (atopic dermatitis, allergic rhinitis, bronchial asthma, food allergy, and acute urticaria).

These diseases: (i) were non-existent or very rare in the rather recent past; (ii) have become more and more frequent, and now are among the most frequent types of disease; (iii) have shown and show increased frequencies in a few decades; and (iv) have diverse frequencies in parts of the same population with different lifestyles.

Absence or Rarity in People Living in Evolutionary "Normal" Conditions

A study about Hiwi of southwestern Venezuela raised the question of "why this and other lowland South American Indian groups do not express the allergic response at levels seen in industrialized contexts. Allergic sensitization among the Hiwi appears to be negligible. This absence occurs in the context of high exposure to macroparasites (mainly hookworm), nutritional stress, frequent and prolonged breastfeeding,

G. Libertini et al., *The Essential Revolution of Evolutionary Medicine*, Advances in Studies of Aging and Health 4, https://doi.org/10.1007/978-3-031-84877-3_4

low indoor allergen deposition, and few hours spent per day indoors." (Hurtado et al., 1997).

A similar absence of allergic phenomena was observed in the study of the Ache, a hunter-gatherer people of Paraguay (Hill & Hurtado, 1996).

Current High Frequency

About the current high frequency of allergic disease:

(1) "Worldwide, respiratory allergic diseases alone, namely asthma and allergic rhinitis, affect nearly 700 million subjects. It has been clearly reported that the prevalence of these diseases has been on the rise for the last few decades, but at different rates, in various areas of the world." (Hendaus et al., 2016);
(2) "The atopic diseases - atopic dermatitis, asthma, and hay fever - pose a great burden to the individual and society, not least, since these diseases have reached epidemic proportions during the past decades in industrialized and, more recently, in developing countries. Whereas the prevalence of the atopic diseases now seems to have reached a plateau in many Western countries, they are still on the increase in the developing world.... About 20% of all children develop symptoms of atopic dermatitis at some point in their lives." (Thomsen, 2015);
(3) "Allergic rhinitis represents a global health concern where it affects approximately 400 million people worldwide... Allergic rhinitis prevalence has increased significantly since the 1990s. It is reported to affect approximately 25 and 40% of children and adult globally, respectively." (Nur Husna et al., 2022);
(4) In Asia, allergic rhinitis affects a large population, ranging from 27% in South Korea (An et al., 2015) to 32% in the United Arab Emirates (Alsowaidi et al., 2010);
(5) In Sweden, in 2006–2007, the prevalence of self-reported allergic rhinitis was 28.0%... and 33.6% in subjects 30–40 years old (Eriksson et al., 2012);
(6) "Allergic rhinitis affects 20 to 30% of adults in both the United States and Europe and perhaps a somewhat higher percentage of children." (Hoyte & Nelson, 2018);
(7) "In the US, a population-based cross-sectional prevalence survey of over 50,000 households published in 2018 estimated that IgE-mediated food allergy is likely to affect approximately 1 in 10 adults and 1 in 12 children.... over 10% of the US population is likely to suffer from at least one IgE-mediated food allergy - with even more individuals reporting current food allergy in the absence of convincingly IgE-mediated symptoms." (Warren et al., 2020);
(8) "Estimated convincing food allergy prevalence among US adults was 10.8%, although 19.0% of adults self-reported a food allergy. The most common allergies were shellfish, milk, peanut, tree nut, and fin fish. Among food-allergic adults, 51.1% experienced a severe food allergy reaction, 45.3% were allergic to multiple foods, and 48.0% developed food allergies as an adult." (Gupta et al., 2019)

Strong Frequency Increase in a Few Decades

Due to the enormous diffusion of allergic diseases today, it is difficult to believe that these diseases have been identified and have spread in different ways and times, starting only about a century and a half ago, even if it is probable that sporadic cases have occurred in previous times.

The first convincing description of hay fever dates back to 1828. However, the first scientific publication about it dates back to 1870 by Blackley, an English scholar of grass pollen, and by Wigweed, an American scholar of ragweed pollen. In those years, hay fever was the only recognized allergic disease. Clear reports of an increase in pediatric asthma cases did not occur until 1970, while the food allergy epidemic was not described until 1990 (Platts-Mills, 2015).

The rapid increase and different spread times of allergic diseases are illustrated in Fig. 4.1.

By 1900, hay fever was well known, as it had spread in the cities, especially among the more affluent classes, and there were such severe cases that "Sites at which patients with hay fever could go during the season to avoid exposure to pollen were identified" (Platts-Mills, 2015). Furthermore, in those years, injections of pollen extract began to be tested to stimulate immunity against the presumed pollen toxins. However, in 1946, in cities like New York, the hay fever problem became so severe that a ragweed eradication campaign was launched (Platts-Mills, 2015). During the 1960s, the reported cases of asthma became more frequent, but only in 1969 a substantial increase in the disease from 1958 to 1968 was reported, and a correlation between the disease and the positivity of skin tests to dust mites was observed (Smith et al., 1969).

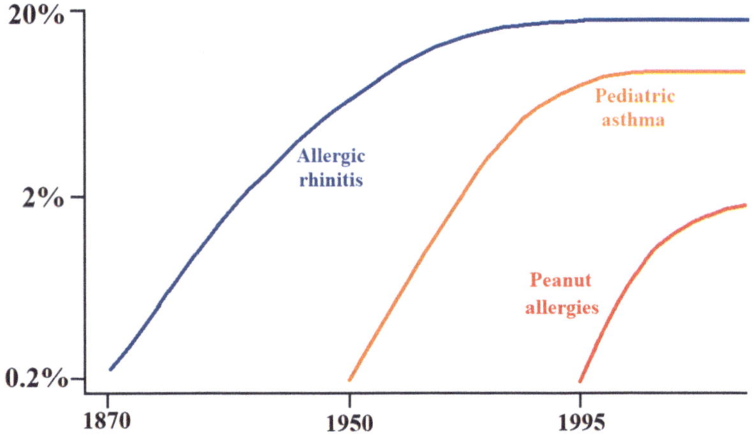

Fig. 4.1 Some crucial dates relating to the spread of allergic diseases, mainly in the USA and England: (i) In 1870, the definition of hay fever, with a subsequent substantial increase up to 20% in 1990 when a plateau was reached; (ii) From 1950, a strong increase in cases of pediatric asthma until 1995 when a plateau is reached; and (iii) Since 1995 onset of peanut allergy, data from (Platts-Mills, 2015)

In subsequent years, an increasing prevalence of asthma was reported in several countries, but: "Although it is well known today that asthma has increased in all Western countries, it might be forgotten that this did not become clear until 1990. In that year, data on asthma among recruits to the Finnish and Swedish armies came out, showing a progressive increase over 20 years.... Allergic or anaphylactic reactions to peanuts and other foods have been recognized for many years. However, starting about 20 years ago, most clinics in the United States and United Kingdom observed an increase in the number of cases. Furthermore, it was clear that, at least for peanut, the titers of IgE antibodies to the relevant proteins were often very high. The observations made in clinics have been confirmed in population-based birth cohorts. This epidemic cannot possibly be ascribed to the changes in water quality that occurred 70 years earlier or to changes in physical activity that started at least 30 years earlier" (Platts-Mills, 2015).

> The rise in prevalence of symptoms in many centres is concerning, but the absence of increases in prevalence of asthma symptoms for centres with existing high prevalence in the older age-group is reassuring. (Asher et al., 2006).

Allergic diseases are in constant increase globally and no longer a rare fact in Africa (Ait-Khaled et al., 2007; Kung et al., 2014): "The consensus is that allergic diseases are increasing in Africa." (Mvoundza Ndjindji & Djoba Siawaya, 2022); "... allergic manifestations such as asthma and atopic dermatitis continue to rise in the higher-income African countries.... Food allergy is definitely an emerging disease in Africa..." (Kung et al., 2014).

The increase in asthma frequency in some western states from 1952 to 1982 is illustrated by Fig. 4.2.

Different Frequencies in the Same Population Divided by Lifestyles

Apart from the differences of the past mentioned above in the frequency of hay fever between urban populations, especially among the wealthier social classes, compared to the populations of rural centers (Platts-Mills, 2015), the differences in hay fever frequencies in two European populations, each divided into two parts living in different conditions, are documented:

(1) *West and East Germany in the first years after reunification*

A first study (von Mutius et al., 1992) compared Munich (West Germany) with Leipzig (East Germany), while another work (Nicolai & von Mutius, 1996) compared Munich with Leipzig and Halle (East Germany). The two East German cities were heavily polluted as a consequence of industrial emissions and coal burning for home heating. In contrast, Munich had lower industrial emissions but pollution due to heavy car traffic. For all cities, similar incidences of bronchitis and asthma were found in the children studied. At the same time, hay fever was much less frequent in the cities of East Germany despite the heavy environmental pollution. Similarly, the skin test reactivity to aeroallergens was considerably lower in East Germany. These

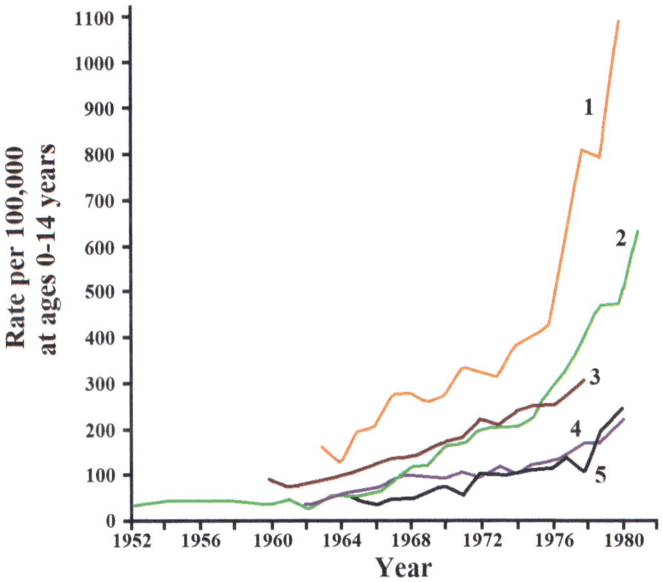

Fig. 4.2 Asthma increase in Queensland - Australia (1), New Zealand (2), Canada (3), England and Wales (4), USA (5), from 1952 to 1982; data from (Mitchell, 1985)

data confirmed the importance of environmental pollution as a cause of bronchitis and asthma while indicating factors associated with Western lifestyle and not environmental pollution as the primary causes of hay fever.

(2) *Finnish Karelia and Russian Karelia*

These two areas are inhabited mainly by the same Finnish-speaking population but living in two different economic and social conditions. In the Finnish part, the economic conditions are much better; there is less crowding in the houses, a lower infection rate, and drinking water in optimal hygienic conditions is used. In the Russian part, the conditions are the opposite; water taken from Lake Ladoga with minimal treatment is used as drinking water, and there is more significant contact with cows and chickens. However, in the Finnish part, allergies were much more frequent, as well as the frequency of two types of autoimmune diseases such as celiac disease and type 1 diabetes (Vartiainen et al., 2002; Kondrashova et al., 2005, 2008; von Hertzen et al., 2007; Laatikainen et al., 2011).

Some Data About the Pathophysiology of Allergic Diseases

In general, the onset of allergic diseases follows a chronological order defined as an "atopic march". In fact, we have two pairs of allergic diseases, the first including atopic dermatitis (AD) and food allergy (FA), the other comprising allergic asthma (AA) and allergic rhinitis (AR). The allergic diseases of the first pair (AD and FA) develop first. Then, they tend to subside while the allergic diseases of the second pair (AS and AR) grow in frequency and intensity (s. Fig. 4.3) (Hill et al., 2016; Hill & Spergel, 2018; Tsuge et al., 2021).

The later development time of specific IgE for inhalant allergens compared to that of specific IgE for food and other atopic allergens provides a possible explanation for the delayed age of onset of the second pair of allergic diseases compared to the first (Kulig et al., 1999).

Primary Cause of the Epidemic of Allergy Diseases

The absence or rarity of allergic diseases in people living in evolutionary "normal" conditions in contrast to the current high incidence, the notable increase of their incidence in a few decades, and the different frequencies in parts of the same population divided by lifestyles, are all elements that indicate one or more modifications of the ecological niche to which we are adapted as the primary cause of allergic diseases.

In 1989, Strachan proposed as a possible cause the reduced exposure to bacterial allergens due to better hygienic conditions and, therefore, a lower frequency of infections (Strachan, 1989), a possibility later defined as the "hygiene hypothesis" (Strachan, 2000). In fact, Strachan suggested a modification of our ecological niche as a cause of allergic diseases.

Subsequently, this hypothesis was integrated with the so-called "old friends hypothesis" proposed in 2004 (Rook et al., 2004). Indeed, this hypothesis highlights the co-evolution of the human immune system with those of microorganisms (bacteria, monocellular fungi, viruses, etc.) and macro-organisms, such as parasitic

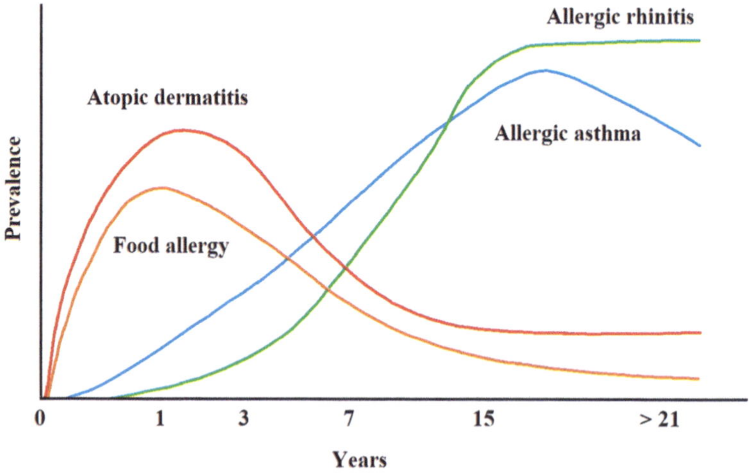

Fig. 4.3 The "atopic march". Outline of the timing of onset of the four types of allergic diseases; from (Tsuge et al., 2021), modified and adapted

helminths, which have constant relations with our species. Furthermore, it was proposed that the relations with these organisms were indispensable for the normal development of the immune system (Rook et al., 2004; Rook & Brunet, 2005).

> For example, a study in Gabon found that school children diagnosed with schistosomiasis, caused by infection with helminth parasites from the *Schistosoma* genus, exhibited lower levels of allergen reactivity than their uninfected classmates.... Since then, additional studies have highlighted this seemingly protective effect of helminths in many mouse models of allergic diseases... There is also experimental animal model evidence suggesting the ability of helminths to ameliorate symptoms of T1D [type 1 diabetes] and colitis... Non-obese diabetic (NOD) mice spontaneously develop T1D, which is significantly inhibited when they are infected with *H. polygyrus* or the filarial nematode *Litomosoides sigmodontis*.... Helminth infection has also been shown to reduce inflammation in murine models of colitis.... (Stiemsma et al., 2015)

If populations using clean water and uncontaminated food, with decreased exposure to farm animals and soil bacteria and with helminth eradication are defined as "posthygiene societies", and those without these characteristics are considered "prehygiene societies", then "In prehygiene villages children and adults generally have increased total serum IgE levels and low-titer IgE antibodies to mite and other allergens and, if tested, often have IgE antibodies to parasite-related antigens, such as *Ascaris* species or the tick-related galactose-α-1,3-galactose (Endara et al., 2015; Perzanowski et al., 2002; Stevens et al., 2011). In posthygiene societies mean total IgE levels are lower, and specific IgE antibodies to common inhalant allergens are common and often present in high titers. In addition, the presence of high-titer IgE antibodies correlates with both hay fever and asthma (Ronmark et al., 2009; Soto-Quiros et al., 2012). In some studies, mean total serum IgE levels among nonatopic subjects are as low as 20 IU/mL. In posthygiene communities, asthma is often severe, and many children require regular treatment. By contrast, in prehygiene communities, although wheezing is not uncommon, it is generally not severe, and treatment is unusual (Perzanowski et al., 2002; Scrivener et al., 2001)." (Platts-Mills, 2015).

Overall, these observations appear to confirm the thesis that the strong increase and the great incidence of allergic diseases have as their primary cause the drastic alterations of our holobiont caused by the modified relationships with microorganisms and also macro-organisms with which we have been co-evolving since time immemorial. It should be noted that the primary cause of allergic diseases, at least in part, coincides with the primary cause of autoimmune diseases, which will be discussed in the next section.

4.1.2 Alterations of the Holobiont—B: Elimination of Intestinal Parasites

The definition of autoimmune disease encompasses a large group of diseases with an immune-mediated attack on the body's own organs.

Indeed, any organ of the body can be attacked, and the symptoms, evolution, and prognosis of autoimmune diseases are extremely varied. As a consequence of these differences, the common pathogenetic aspects have long been underestimated, and the "Autoimmune diseases have conventionally been considered as distinct disorders, likely as a consequence of their being treated by separate medical specialties based on organ system of involvement." (Cooper et al., 2009).

This state of affairs, the low incidence of these diseases in the fairly recent past, and the possible difficulty of recognizing a disease as autoimmune have hindered the elaboration of overall statistics from which the incidence and importance of this group of diseases could be understood.

The long list of autoimmune diseases according to the Autoimmune Association (Autoimmune Association, 2023b—Disease-information) is reported in Table 4.1.

An accurate list of autoimmune disorders is reported in the Autoimmune Registry (Autoimmune Registry, 2023) with a notation of whether the evidence for the nature of the disorder as an autoimmune disease is strong, moderate, or unconfirmed. After opportune reworking, data obtained from this source, are shown in Tables 4.2, 4.3, 4.4 and 4.5. According to this Registry, in the US, the cases of autoimmune diseases with the level of evidence defined as strong or moderate are between 32.2 and 43.1 million; cases of diseases defined as unconfirmed (i.e., a disease with *no known cause* but sharing features common with autoimmune diseases, and autoimmunity suspected in at least one scientific article as a cause, or it is comorbid with confirmed autoimmune diseases) are between 33.9 and 75.3 million; cases of troubles defined as symptoms (i.e., occurring commonly among people with autoimmune diseases) are between 12.9 and 93.3 million.

Even if there are many cases in which several autoimmune diseases and "symptoms" coexist in the same patient, a fact that reduces the overall number of subjects affected by one or more autoimmune diseases, the diffusion and impact of autoimmune diseases certainly appear considerable and such as to define them as a type of pathology of primary importance.

Total of US cases (Min-Max) = 32,184,498–43,110,193

Same images regarding autoimmune diseases are shown in Figs. 4.4, 4.5 and 4.6.

Absence or Rarity in People Living in Evolutionary "Normal" Conditions

In some studies, no cases of autoimmune diseases have been described in populations living in the hunter-gatherer condition, i.e., in the condition as close as possible to the ecological niche to which our species is adapted (Early & Peters, 1990; Hill & Hurtado, 1996). This lack of data may be due to the smallness of the populations studied and the difficulty of identifying cases of such diseases. More reliable assessments are obtained using data from African populations or other areas of the world before the growing westernization of lifestyles. As a rule, these data relate to specific autoimmune diseases and not to all of these diseases:

Table 4.1 List of autoimmune diseases reported by Autoimmune Association

Achalasia; Addison's disease; Adult Still's disease; Agammaglobulinemia; Alopecia areata; Amyloidosis; Ankylosing spondylitis; Anti-GBM/Anti-TBM nephritis; Antiphospholipid syndrome; Autoimmune angioedema; Autoimmune dysautonomia; Autoimmune encephalitis; Autoimmune hepatitis; Autoimmune inner ear disease (AIED); Autoimmune myocarditis; Autoimmune oophoritis; Autoimmune orchitis; Autoimmune pancreatitis; Autoimmune retinopathy; Autoimmune urticaria; Axonal & neuronal neuropathy (AMAN); Baló disease; Behcet's disease; Benign mucosal pemphigoid (Mucous membrane pemphigoid); Bullous pemphigoid; Castleman disease (CD); Celiac disease; Chagas disease; Chronic inflammatory demyelinating polyneuropathy (CIDP); Chronic recurrent multifocal osteomyelitis (CRMO); Churg-Strauss syndrome (CSS) or Eosinophilic granulomatosis (EGPA); Cicatricial pemphigoid; Cogan's syndrome; Cold agglutinin disease; Complex regional pain syndrome (formerly known as reflex sympathetic dystrophy); Congenital heart block; Coxsackie myocarditis; CREST syndrome; Crohn's disease; Dermatitis herpetiformis; Dermatomyositis; Devic's disease (neuromyelitis optica); Discoid lupus; Dressler's syndrome; Endometriosis; Eosinophilic esophagitis (EoE); Eosinophilic fasciitis; Erythema nodosum; Essential mixed cryoglobulinemia; Evans syndrome; Fibromyalgia; Fibrosing alveolitis; Giant cell arteritis (temporal arteritis); Giant cell myocarditis; Glomerulonephritis; Goodpasture's syndrome; Granulomatosis with polyangiitis; Graves' disease; Guillain-Barre syndrome; Hashimoto's thyroiditis; Hemolytic anemia; Henoch-Schonlein purpura (HSP); Herpes gestationis or pemphigoid gestationis (PG); Hidradenitis suppurativa (HS) (Acne inversa); IgA nephropathy; IgG4-related sclerosing disease; Immune thrombocytopenic purpura (ITP); Inclusion body myositis (IBM); Interstitial cystitis (IC); Juvenile arthritis; Juvenile diabetes (Type 1 diabetes); Juvenile myositis (JM); Kawasaki disease; Lambert-Eaton syndrome; Lichen planus; Lichen sclerosus; Ligneous conjunctivitis; Linear IgA disease (LAD); Lupus; Lyme disease chronic; Meniere's disease; Microscopic polyangiitis (MPA); Mixed connective tissue disease (MCTD); Mucha-Habermann disease; Multifocal motor neuropathy (MMN) or MMNCB; Multiple sclerosis; Myasthenia gravis; Myelin oligodendrocyte glycoprotein antibody disorder; Myositis; Narcolepsy; Neonatal lupus; Neuromyelitis optica/devic disease; Neutropenia; Ocular cicatricial pemphigoid; Optic neuritis; Palindromic rheumatism (PR); PANDAS (Pediatric autoimmune neuropsychiatric disorders associated with streptococcus infections); Paraneoplastic cerebellar degeneration (PCD); Paroxysmal nocturnal hemoglobinuria (PNH); Pars planitis (peripheral uveitis); Parsonage-Turner syndrome; Pemphigus; Peripheral neuropathy; Perivenous encephalomyelitis; Pernicious anemia (PA); POEMS syndrome; Polyarteritis nodosa; Polyglandular syndromes type I, II, III; Polymyalgia rheumatica; Polymyositis; Postmyocardial infarction syndrome; Postpericardiotomy syndrome; Primary biliary cholangitis; Primary sclerosing cholangitis; Progesterone dermatitis; Progressive hemifacial atrophy (PHA) Parry Romberg syndrome; Psoriasis; Psoriatic arthritis; Pure red cell aplasia (PRCA); Pyoderma gangrenosum; Raynaud's phenomenon; Reactive arthritis; Relapsing polychondritis; Restless legs syndrome (RLS); Retroperitoneal fibrosis; Rheumatic fever; Rheumatoid arthritis; Sarcoidosis; Schmidt syndrome or Autoimmune polyendocrine syndrome type II; Scleritis; Scleroderma; Sjögren's Disease; Stiff person syndrome (SPS); Susac's syndrome; Sympathetic ophthalmia (SO); Takayasu's arteritis; Temporal arteritis/giant cell arteritis; Thrombocytopenic purpura (TTP); Thrombotic thrombocytopenic purpura (Ttp); Thyroid eye disease (Ted); Tolosa-Hunt syndrome (THS); Transverse myelitis; Type 1 diabetes; Ulcerative colitis (UC); Undifferentiated connective tissue disease (UCTD); Uveitis; Vasculitis; Vitiligo; Vogt-Koyanagi-Harada disease; Warm autoimmune hemolytic anemia

Table 4.2 List of autoimmune diseases with strong or moderate levels of evidence (Autoimmune Registry, 2023)

Diseases with the level of evidence: Strong (S) or Moderate (M)	US cases Min–Max
Psoriasis (subtype: Guttate psoriasis) (M)	8,320,000–11,840,000
Celiac disease (adult form of celiac disease, acquired celiac disease) (S)	2,400,000–3200,000
Autoimmune urticaria (S)	2560,000–2560,000
Rheumatoid arthritis (S)	1,056,000–2,528,000
Crohn's disease (S)	1,918,400–2,464,000
Ulcerative colitis (S)	1,569,600–2,016,000
Graves' disease (autoimmune thyrotoxicosis) (S)	2,012,800–2,012,800
Autoimmune thyroiditis (Hashimoto's thyroiditis) (S)	1664,000–1,856,000
Idiopathic pulmonary fibrosis (M)	647,040–1,582,400
Pernicious anemia (S)	1,580,000–1,580,000
Sjogren's disease (S)	480,000–1,376,000
Diabetes mellitus type 1 (S)	832,000–1,088,000
Ankylosing spondylitis (M)	832,000–832,000
Polymyalgia rheumatica (S)	708,288–816,000
Psoriatic arthritis (M)	192,000–800,000
Multiple sclerosis (disseminated sclerosis, disseminated encephalomyelitis; subtypes: Baló concentric sclerosis, primary progressive multiple sclerosis, Schilder disease, disseminated sclerosis) (S)	757,926–762,846
Alopecia (areata, totalis, universalis) (S)	672,000–672,000
Vitiligo (M)	320,000–640,000
Graves' ophthalmopathy (S)	416,000–464,000
Reactive arthritis (subtype: Reiter's syndrome) (M)	249,920–339,840
Autoimmune uveitis (M)	330,240–330,240
Giant cell arteritis (cranial arteritis, temporal arteritis) (S)	175,168–276,352
Cutaneous lupus erythematosus (subtypes: Chronic discoid lupus erythematosus, discoid lupus erythematosus, discoid lupus erythematosus of the eyelid, discoid lupus erythematosus of the oral mucosa, lupus erythematosus tumidus, subacute cutaneous lupus erythematosus) (M)	224,000–262,400
Systemic lupus Erythematosus (lupus) (S)	208,960–259,200
Episcleritis (M)	168,320–168,320
Antiphospholipid syndrome (anticardiolipin antibody syndrome, Hughes syndrome) (S)	128,000–160,000
Lichen sclerosus (M)	160,000–160,000
Primary idiopathic dilated cardiomyopathy (S)	116,800–116,800
Kawasaki's disease (lymph node syndrome, mucocutaneous lymph node syndrome) (S)	111,680–111,680
Systemic scleroderma (CREST syndrome) (S)	48,000–96,000
Primary biliary cholangitis (primary biliary cirrhosis) (S)	93,760–93,760
Post-myocardial infarction syndrome (subtypes: Post-cardiotomy syndrome, post-pericardiotomy pericarditis, subacute bacterial endocarditis, Dressler's syndrome) (S)	89,000–89,000

(continued)

Table 4.2 (continued)

Diseases with the level of evidence: Strong (S) or Moderate (M)	US cases Min–Max
Autoimmune myocarditis (acute rheumatic myocarditis) (M)	67,200–83,200
Bullous pemphigoid (S)	82,880–82,880
Evans syndrome (S)	55,680–82,560
Autoimmune gastritis (S)	76,800–76,800
Interstitial cystitis (M)	33,920–72,000
Addison's disease (subtype: Autoimmune adrenalitis) (S)	13,344–70,720
Polymyositis (S)	16,000–70,400
Lupus nephritis (SLE glomerulonephritis syndrome) (S)	66,908–69,043
Dermatomyositis (S)	68,544–68,544
Myasthenia gravis (S)	48,000–64,000
Anti-neutrophil cytoplasmic antibody-associated vasculitis (Churg-Strauss syndrome, Wegener's granulomatosis; subtypes: Eosinophilic granulomatosis with polyangiitis, granulomatosis with polyangiitis) (S)	14,720–58,880
Autoimmune hepatitis (S)	54,080–54,080
Myositis (S)	41,088–49,472
Autoimmune inner ear disease (subtype: Meniere's disease) (M)	48,000–48,000
Narcolepsy with cataplexy (M)	44,800–44,800
Autoimmune encephalitis (M)	43,840–43,840
Warm autoimmune hemolytic anemia (autoimmune hemolytic anemia) (S)	36,300–42,900
Microscopic polyangiitis (microscopic polyarteritis) (S)	18,560–41,280
Juvenile arthritis (juvenile idiopathic arthritis, juvenile rheumatoid arthritis) (S)	28,934–37,148
Dermatitis herpetiformis (S)	24,000–35,840
Felty syndrome (S)	32,000–32,000
Autoimmune thrombocytopenic purpura (immune thrombocytopenia) (S)	14,400–30,400
Lupus vasculitis (S)	25,379–26,188
Purpura rheumatica (anaphylactoid purpura, Henoch-Schonlein purpura; subtype: Immunoglobulin A vasculitis) (M)	4210–24,660
Primary sclerosing cholangitis (S)	15,424–23,520
Pemphigus vulgaris (S)	15,680–17,280
Scleritis (M)	16,640–16,640
Behcet's disease (subtype: Hughes-Stovin syndrome) (M)	16,640–16,640
Rheumatic fever (S)	16,000–16,000
Chronic inflammatory demyelinating polyneuropathy (subtypes: Relapsing polyneuropathy, chronic inflammatory demyelinating polyneuritis, chronic inflammatory demyelinating polyneuropathy, Miller-fisher syndrome) (S)	6304–15,264
Autoimmune pancreatitis (S)	14,720–14,720
IgA nephropathy (M)	9920–14,400
Neuromyelitis optica (Devic's disease) (S)	1664–14,080
Antisynthetase syndrome (S)	9600–12,800

(continued)

Table 4.2 (continued)

Diseases with the level of evidence: Strong (S) or Moderate (M)	US cases Min–Max
Myelin oligodendrocyte glycoprotein disease (M)	5120–10,944
Mixed connective tissue disease (S)	7872–10,824
Morphea (localized scleroderma) (M)	1280–8640
Guillain-Barre syndrome (subtypes: Acute motor axonal neuropathy, progressive inflammatory neuropathy) (S)	1984–8512
Juvenile dermatomyositis (S)	7040–7040
Encephalopathy associated with autoimmune thyroid disease (Hashimoto's encephalopathy, steroid-responsive encephalopathy associated with autoimmune thyroiditis) (M)	6720–6720
Intermediate uveitis (pars planitis, peripheral uveitis) (M)	4480–6400
Cold agglutinin disease (S)	5184–5184
Retroperitoneal fibrosis (M)	4480–4480
Paraneoplastic cerebellar degeneration (S)	3915–3915
Cryptogenic organizing pneumonia (M)	3520–3520
Rheumatoid vasculitis (angioneurotic edema) (S)	2038–2613
IgG4-related diseases (hyper-IgG4 disease, systemic IgG4 related plasmocytic syndrome) (S)	2560–2560
Autoimmune neutropenia (S)	1600–1600
Takayasu arteritis (M)	1504–1504
Relapsing polychondritis (chronic atrophic polychondritis, generalized chondromalacia, Meyenburg-Altherr-Uehlinger syndrome, systemic chondromalacia) (S)	1120–1120

(A) "There is some doubt about whether multiple sclerosis occurs in indigenous Africans (Dean, 1967; Gelfand, 1957; Haddock, 1965; Trowell, 1960). During the period 1952–69 we have encountered only two cases..." [in two main hospitals of Kenya] (Foster & Harries, 1970);

(B) "M.S. [multiple sclerosis] has not, however, been reported among the Australian aborigines and is apparently uncommon among the Maoris of New Zealand.... No single patient with multiple sclerosis has yet been found among South Africa's 11 million Bantu. The disease does occur among the Coloured and Asian South African, but it is very uncommon..... For all White South Africans the average annual incidence was 0.6 per 100,000." (Dean, 1967);

(C) "... fibrocystic disease of the pancreas is of the utmost rarity in Africans, coeliac disease is probably equally rare and no case of Crohn's disease has yet been conclusively demonstrated." (Trowell, 1960);

(D) In the University College Hospital (UCH) of Ibadan (Nigeria), in 10 years, there were only 42 admissions for rheumatoid arthritis among 100,000 patients. A population survey carried out in two nearby villages found only two mild cases of polyarthritis "among 571 subjects studied clinically, radiologically and serologically.... Systemic collagen diseases, Hashimoto's thyroiditis, myxoedema, pernicious anemia and Addison's disease were all found to be extremely unusual causes of admission to this hospital; thyrotoxicosis,

Table 4.3 List of autoimmune diseases with the level of evidence "unconfirmed" (Autoimmune Registry, 2023)

Diseases with the level of evidence = Unconfirmed (U), i.e., a disease with *no known cause* but sharing features common with autoimmune diseases, and autoimmunity suspected in at least one scientific article as a cause, or it is comorbid with confirmed autoimmune diseases	US cases Min–Max
Long COVID (U)	25,000,000–50,000,000
Myalgic encephalomyelitis (chronic fatigue syndrome) (U)	736,000–13,856,000
Fibromyalgia (U)	4,544,000–6,624,000
Hidradenitis suppurativa (U)	1,696,000–2,688,000
Endometriosis (U)	787,500–825,000
Pediatric autoimmune neuropsychiatric disorder associated with streptococcus (U)	370,000–370,000
Microscopic colitis (U)	329,600–329,600
Sarcoidosis (U)	188,928–192,576
Complex regional pain syndrome (reflex neurovascular dystrophy, reflex sympathetic dystrophy; subtype: Amplified musculoskeletal pain syndrome) (U)	65,824–65,824
Eosinophilic esophagitis (U)	32,000–182,400
Pityriasis lichenoides et varioliformis acuta (Mucha-Habermann disease) (U)	96,590–96,590
Pyoderma gangrenosum (U)	17,920–19,520
Erythema nodosum (U)	3200–16,000
Leukocytoclastic vasculitis (urticarial vasculitis) (U)	14,400–14,400
Eosinophilic fasciitis (Shulman's syndrome) (U)	3200–6720
Inclusion body myositis (U)	3200–3200

Total of US cases Min - Max = 33,888,362–75,289,830

myasthenia gravis and ulcerative colitis were seen only slightly more frequently." (Greenwood et al., 1970);

(E) In general, autoimmune diseases were reported as rare in tropical rural populations of Africa (Clark et al., 1996; Adebajo, 1997);

(F) The clinical study of 7966 hospital patients and the finding that autoantibodies were uncommon suggested that autoimmune diseases were rare in Ethiopia. "Thus, both clinical and serological studies confirm that previously held views concerning the rarity of autoimmune diseases in many parts of Africa are also true in Ethiopia." (Tsega et al., 1980)

Current High Frequency

> Historically, autoimmune diseases were considered to be rare but, through rigorous epidemiological studies, have now been shown to affect 3–5% of the population... (Wang et al., 2015)

In a 1997 review aimed at determining the frequency of autoimmune diseases, 24 autoimmune diseases were considered as reported in studies covering the previous

Table 4.4 Autoimmune diseases with US cases <1000 (Autoimmune Registry, 2023)

Acquired hemophilia (M); Acute disseminated encephalomyelitis (Hurst's disease, perivenous encephalomyelitis, Weston-Hurst syndrome; subtypes: acute hemorrhagic encephalomyelitis, acute hemorrhagic leukoencephalitis, acute necrotizing hemorrhagic leukoencephalitis, non-vasculitic autoimmune meningoencephalitis) (M); Adiposis dolorosa (Dercum's disease) (U); Adult-onset Still's disease (S); Anti-Glomerular Basement Membrane nephritis (Goodpasture disease) (S); Anti-n-methyl-D-aspartate receptor encephalitis (M); Anti-Sperm Antibodies (S); Aplastic anemia (S); Autoimmune Angioedema (U); Autoimmune enteropathy (S); Autoimmune lymphoproliferative syndrome (S); Autoimmune oophoritis (S); Autoimmune orchitis (S); Autoimmune polyendocrine syndrome type 1 (autoimmune polyendocrinopathy-candidiasis-ectodermal dystrophy, polyglandular autoimmune syndrome 1, Whitaker syndrome) (S); Autoimmune polyendocrine syndrome type 2 (polyglandular autoimmune syndrome 2) (S); Autoimmune polyendocrine syndrome type 3 (polyglandular autoimmune syndrome 3) (S); Autoimmune progesterone dermatitis (U); Autoimmune retinopathy (U); Bickerstaff's encephalitis (S); Chronic Lyme disease (U); Cicatricial pemphigoid (benign mucosal pemphigoid; subtype: ocular cicatricial pemphigoid (S); Cogan syndrome (S); Enthesitis-related arthritis (S); Epidermolysis bullosa acquisita (S); Essential mixed cryoglobulinemia (S); Gestational pemphigoid (Herpes gestationis) (S); Lambert-Eaton myasthenic syndrome (S); Lichen planus (U); Ligneous conjunctivitis (U); Linear IgA disease (M); Mooren's ulcer (U); Neuromyotonia (subtype: Isaac's syndrome) (M); Opsoclonus myoclonus syndrome (M); Palindromic rheumatism (Hench-Rosenberg syndrome) (U); Paroxysmal nocturnal hemoglobinuria (U); Parry Romberg syndrome (U); POEMS syndrome (U); Polyarteritis nodosa (Kussmaul-Maier disease, panarteritis nodosa, periarteritis nodosa) (S); Pure red cell aplasia (S); Retinocochleocerebral vasculopathy (Susac's syndrome) (U); Rheumatic chorea (Sydenham chorea) (U); Schnitzler syndrome (U); Stiff person syndrome (Stiff man syndrome) (M); Sympathetic ophthalmia (Sympathetic uveitis) (S); Tolosa-Hunt syndrome (U)

Table 4.5 "Symptoms" related to autoimmune diseases (Autoimmune Registry, 2023)

Symptoms (i.e., they occur commonly among people with autoimmune diseases)	US cases Min–Max
Secondary Raynaud's phenomenon	6,656,000–27,872,000
Restless leg syndrome	12,800,000–92,800,000
Undifferentiated connective tissue disease	44,800–448,000
Transverse myelitis	7872–7872

Total of US cases Min-Max = 12,852,672–93,255,872

30 years, and the prevalence of these diseases was estimated to be 3.2% (Jacobson et al., 1997).

A 2007 review regarding Denmark's population and the prevalence of 31 autoimmune diseases based on data obtained from the national hospitalization registry reported a prevalence of 5.3% (Eaton et al., 2007).

A third review, carried out in 2009 (Cooper et al., 2009), largely based on data from the period 1989–2008, highlighted that, for certain autoimmune diseases with low hospitalization rates (e.g., alopecia, vitiligo, psoriasis, celiac disease, hyperthyroidism, hypothyroidism), there was a significant underestimation. E.g., one study conducted on over 50,000 adults in the Netherlands showed that the prevalence of celiac disease was 0.016% for clinically diagnosed cases and 0.35% for unrecognized cases (Schweizer et al., 2004). However, applying correction factors to

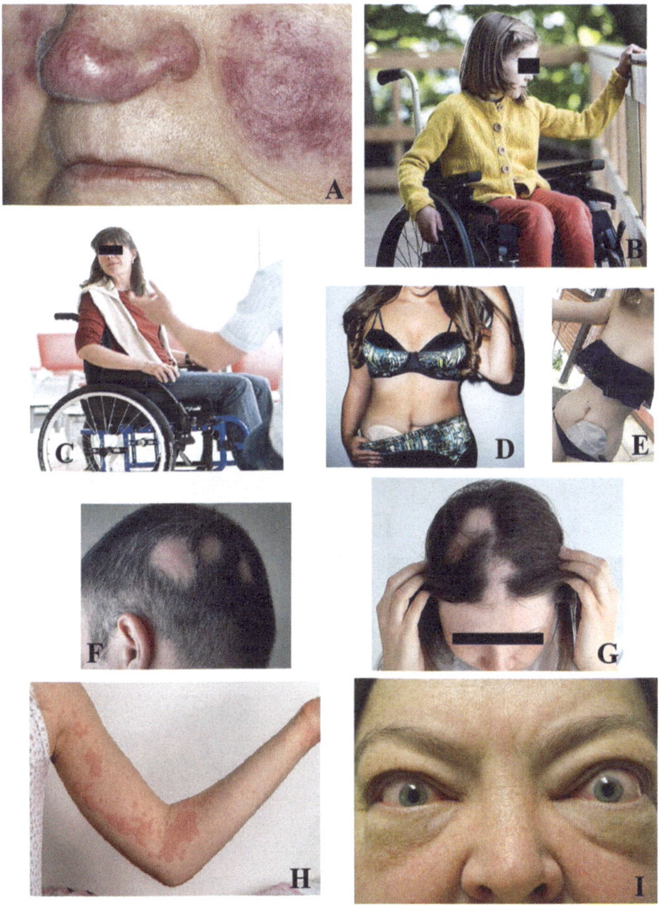

Fig. 4.4 Some images of autoimmune diseases: (**a**) lupus; (**b**), (**c**) multiple sclerosis; (**d**), (**e**) Crohn's disease; (**f**), (**g**) alopecia areata; (**h**) autoimmune urticaria; (**i**) Graves' disease

account for undiagnosed and / or nonhospitalized cases, a prevalence between 7.6% and 9.4% was estimated.

> The incidence of T1D [Type 1 diabetes] is 5–10, 10–20 and <1 per 100,000 person-years in populations from Europe, the USA and China, respectively. (Wang et al., 2015)

> Collectively, autoimmune diseases affect some 20 million Americans, predominantly women. Many of these diseases are increasing in frequency in industrialized countries. (Rose, 2016)

According to the Autoimmune Association, in the US, today more than 24 million people are affected by autoimmune diseases; about 80% are women, and 5–10% of Americans suffer from more than one autoimmune disease (Autoimmune Association, 2023a—About autoimmunity).

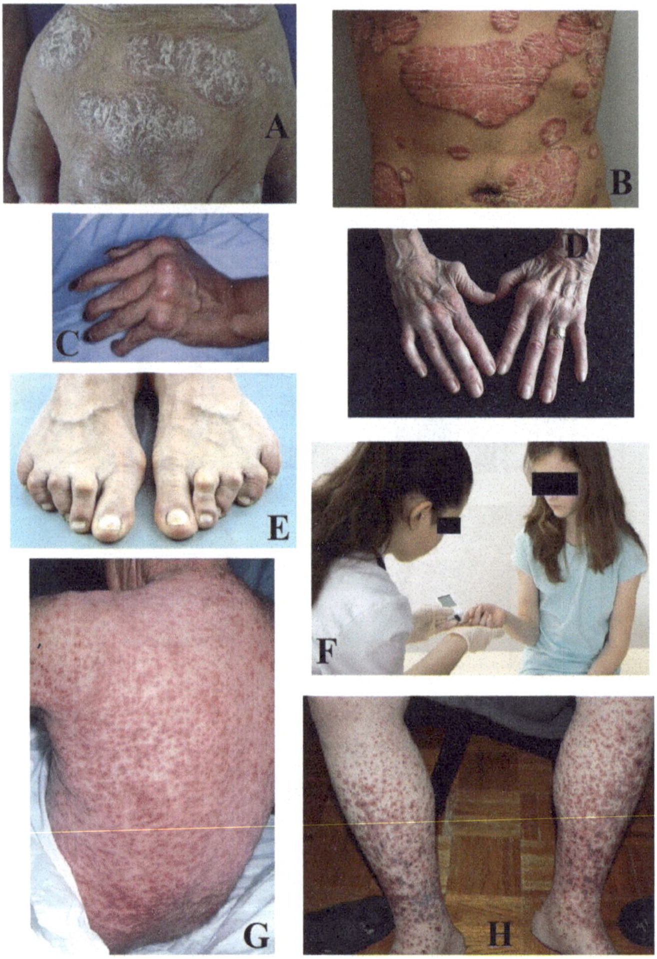

Fig. 4.5 Other images of autoimmune diseases: (**a**), (**b**) psoriasis; (**c**) psoriatic arthritis; (**d**), (**e**) rheumatoid arthritis; (**f**) type 1 diabetes; (**g**), (**h**) idiopathic thrombocytopenic purpura

Strong Frequency Increase in a Few Decades

Between 1990 and 2011, the incidence rate of celiac disease in the UK increased from 5.2 per 100,000 to 19.1 per 100,000 person-years (West et al., 2014).

A 2015 review showed that in the previous 30 years, there had been a worldwide net % increase / year incidence and prevalence of autoimmune diseases, equal to 19.1 ± 43.1 and 12.5 ± 7.9, respectively (Lerner et al., 2015).

From 1960 to 1996, about the incidence of type 1 diabetes, 37 studies in 27 countries were carried out. The pooled data showed an overall increase in incidence of 3.0% per year. The increase was statistically significant in 24 of the 37 populations, including those with high incidence (>14.6 per 100,000 a year). However, the

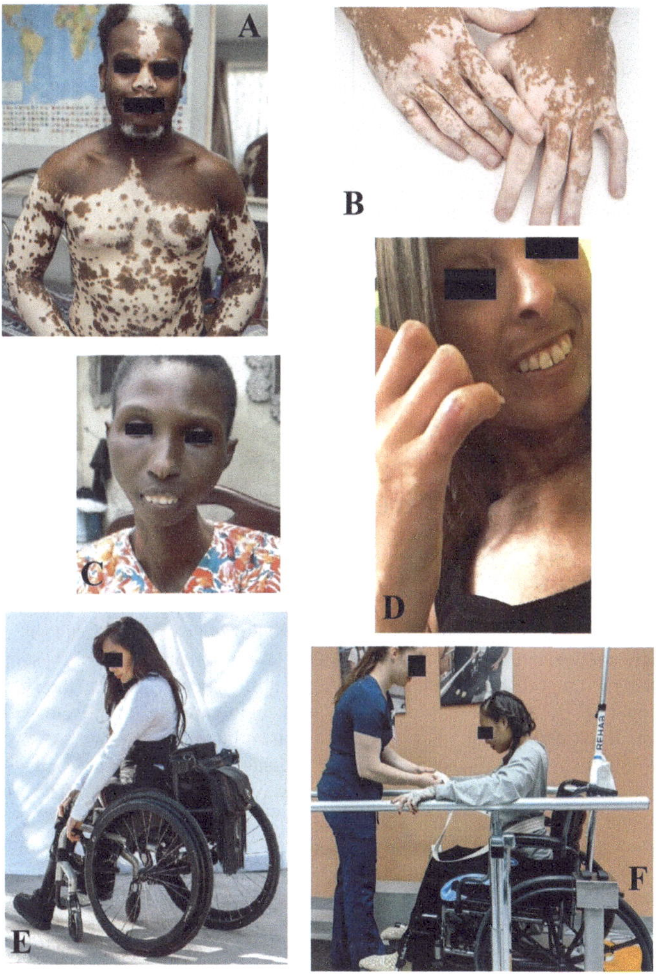

Fig. 4.6 Other images of autoimmune diseases: (**a**), (**b**) vitiligo; (**c**), (**d**) scleroderma; (**e**), (**f**) transverse myelitis

relative increase was steeper in populations with a lower incidence. The data showed that type 1 diabetes incidence was increasing worldwide, both in low and high-incidence populations. It was predicted by 2010 that there would be an incidence over 30/100000/year in many populations and 50/100.000/year in Finland (Onkamo et al., 1999).

In Finland, the incidence among children under the age of 15 years has increased from 12 to 65 new cases/100,000/year in the period 1980–2011 (Harjutsalo et al., 2013).

> I founded and continue to direct the Johns Hopkins Transverse Myelitis (TM) Center, the only center in the world dedicated to developing new therapies for this paralyzing

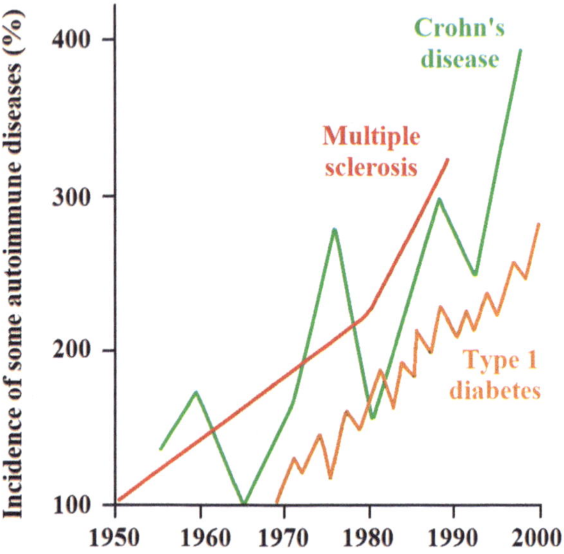

Fig. 4.7 The steep increase in the incidence of multiple sclerosis, type 1 diabetes, and Crohn's disease in developed countries in the period 1950–2000; data from (Farrokhyar et al., 2001; Pugliatti et al., 2001; Tuomilehto et al., 1999) reported in (Bach, 2002)

> autoimmune disorder. Increasingly, I see that more and more patients are being felled by this devastating disorder. Infants as young as five months old can get TM and some are left permanently paralyzed and dependent upon a ventilator to breathe. But this is supposed to be a rare disorder, reportedly affecting only one in a million people. Prior to the 1950s, there were a grand total of four cases reported in the medical literature. Currently, my colleagues at the Johns Hopkins Hospital and I hear about or treat hundreds of new cases every year. In the multiple sclerosis clinic, where I also see patients, the number of cases likewise continues to climb.... Autoimmune diseases have not always been this common. The prevalence of autoimmune diseases like systemic lupus erythematosus or lupus, multiple sclerosis, and type 1 diabetes is on the rise. In some cases, autoimmune diseases are three times more common now than they were several decades ago. These changes are not due to increased recognition of these disorders or altered diagnostic criteria. Rather, more people are getting autoimmune disorders than ever before.... The numbers are staggering: one in twelve Americans... will develop an autoimmune disorder. And since it is clear that not every patient with an autoimmune disease is correctly diagnosed, the prevalence is certainly higher than that. The American Heart Association estimates that by comparison, only one in twenty Americans will have coronary heart disease. Similarly, according to the National Center for Health Statistics, one in fourteen American adults will have cancer at some time in their life. This means that an American is more likely to get an autoimmune disease than either cancer or heart disease. (Kerr, 2009)

Figure 4.7 shows the strong increase in the incidence of three autoimmune diseases (multiple sclerosis, Crohn's disease, and type 1 diabetes) in the period 1950–2000.

Different Frequencies in the Same Population Divided by Lifestyles

There is the previously described case of Finnish Karelia and Russian Karelia, where the inhabitants, both Finnish-speaking populations, have remarkably different lifestyles. In Russian Karelia, with higher rates of infections and greater contact with animals such as cows and chickens, the frequencies of celiac disease and type 1 diabetes, both autoimmune diseases, were much lower than in Finnish Karelia (in Russian part of Karelia one-sixth of the cases of type 1 diabetes and one-fifth of cases of celiac disease of the Finnish part) (Kondrashova et al., 2005; Kondrashova et al., 2008).

Possible Alterations of the Ecological Niche As Causes of Autoimmune Diseases

Multiple elements indicate that autoimmune diseases largely depend on one or more alterations of the ecological niche to which we are adapted as a primary cause: (i) the considerable frequency of autoimmune diseases in current living conditions as opposed to the rarity or absence of such diseases in life conditions closer to the original ones to which our species is adapted; (ii) the rapid increase in the frequency of such diseases; and (iii) the frequency differences between two parts of the same population living in different conditions. These facts cannot be explained as an effect of harmful genes which, among other things, should be absent in certain conditions, present in others, and rapidly appear in a few years.

With regard to the factors that could be the cause of autoimmune diseases, in authoritative reviews, e.g.: (Rose, 2016; Wang et al., 2015), apart from the indication of genetic factors as origins of these diseases, specific alterations as possible causes are not proposed.

The main obstacle arises from considering as a subject of possible harmful effects the living being understood as an isolated organism and not as the host of a holobiont. On the contrary, if attention is paid to the equilibria of an extremely complex whole such as our holobiont, it is easy to find that serious alterations of these equilibria are plausible primary causes of autoimmune diseases.

Already in 1970, studying the frequencies of some autoimmune diseases as a cause of hospital admission in Nigeria, it was observed the rarity of diseases such as rheumatoid arthritis, systemic collagen diseases, Hashimoto's thyroiditis, myxedema, pernicious anemia, Addison's disease, thyrotoxicosis, myasthenia gravis, and ulcerative colitis. At the same time, however, a high incidence of positive tests for rheumatoid factor was found in apparently healthy populations (Greenwood, 1968; Greenwood et al., 1970). As a deduction from these facts, it was suggested that: "In many parts of tropical Africa a large proportion of the population shows a number of immunological changes which are related to the prevalence of parasitic infections. In the same areas diseases of the autoimmune group are uncommon. Consideration of these two sets of observations led one of us to suggest that the host response to multiple parasitic infections might in some way interfere with the abnormal immunological processes involved in production of diseases of the autoimmune group (Greenwood, 1968)." (Greenwood et al., 1970).

Indeed:

(A) "Worm parasites have co-evolved with the mammalian immune system for many millions of years and during this time, they have developed extremely effective strategies to modulate and evade host defences and so maintain their evolutionary fitness. It is therefore reasonable to conclude that the human immune system has been shaped by its relationship with parasitic worms and this may be a necessary requirement for maintaining our immunological health." (Zaccone et al., 2006);

(B) "... the modern-day lack of exposure to helminths due to our hygienic practices is an important environmental factor contributing to IBD [Inflammatory Bowel Disease]. Until modern times, nearly all children and most adults harboured intestinal helminths. Helminths and the immune system of *Homo sapiens* co-evolved in close proximity over many 1000s of years. Helminths regulate their host's immune system and prevent excessive inflammatory responses, which could underlie the mechanism of protection." (Weinstock et al., 2004);

(C) "Before the 1940s, many children and adults in the USA carried helminths. Worm carriage was particularly common in rural areas of the South and in indigent populations of major cities (Elliott et al., 2000). In the USA and Europe, helminthic colonisation has steadily declined. They are found in recent immigrants from less developed countries (Salas et al., 1990) and in economically disadvantaged populations living in underserved regions of the USA such as some Indian reservations (Healy et al., 1969). These groups are at low risk for IBD. There is an inverse relationship between the frequency of worm colonisation and the prevalence of CD [Crohn's disease]. There is more CD in urban versus rural populations, in northern versus southern regions of the USA and Europe, and developed versus less developed countries. The opposite is true for worm carriage." (Weinstock et al., 2004);

(D) "Concurrent infection with *S.* [*Schistosoma*] *mansoni* significantly attenuates TNBS [2,4,6-Trinitrobenzene sulphonic acid] induced colitis in the rat." (Moreels et al., 2004);

(E) "... there is an immunological basis to suspect deworming as a risk factor. People bearing helminths display dampened immune responses to unrelated concurrent antigenic exposures (Borkow et al., 2000; Elliott et al., 2000; Sabin et al., 1996).... Interactions between these parasites [helminths] and their hosts are complex and multifaceted as would be expected for such a successful co-evolutionary process that leads to 'peaceful' coexistence. Helminths not only trigger Th2 responses, which help to limit worm number in the host, they also promote production of powerful immunomodulatory molecules such as IL-1048 and TGF-β, and 'regulatory' T cells (Doetze et al., 2000).... There is now substantial human epidemiological data and several animal studies supporting the hypothesis that helminths protect the host from immunological disease. For instance, people colonised with helminths have high serum levels of IL-10, which may protect them from atopy (van den Biggelaar et al., 2000).... natural exposure to helminths may guard people from developing IBD and other immunological diseases through induction of IL-4, IL-10, TGF-β, regulatory T cells, or perhaps by other means." (Weinstock et al., 2004);

(F) In a study on Rhesus macaques (*Macaca mulatta*) about alopecia, trouble correlated with chronic dermal inflammatory infiltrates for which analogous human disorders are classified as autoimmune diseases, it was found that when comparing animals born indoors at the New England Primate Research Center with animals born outdoors infected with lung mites, the outdoor animals "had reduced dermal inflammatory cell infiltrates and a lower incidence of alopecia" (Kramer et al., 2010);

(G) Domesticated animals (horses, dogs, cats) or animals held in captivity in zoos (chimpanzees, gorillas, baboons, rhesus monkeys, cotton-top tamarins) which, as a rule, are routinely dewormed in supposed defense of their health, suffer from allergic diseases and autoimmune diseases that are similar to human ones (Velasquez-Manoff, 2012, pp. 299–302);

(H) "Epidemiological studies suggest that autoimmune diseases, such as multiple sclerosis (MS), are less frequent in individuals who are helminth carriers. This observation has been tested in murine models of colitis, MS, type 1 diabetes and asthma. In each case, mice colonized with helminths show protection from disease. This apparent down-modulation of inflammatory response resulting from helminth infection has triggered interest in exploring the potential clinical efficacy of controlled helminth infection in patients suffering from autoimmune diseases." (Correale, 2014);

(I) "Epidemiologic data and animal experimentation suggest that elimination of helminths contributes to the increasing prevalence of some immune-mediated diseases in regions with ever-improving sanitation. Diseases of increasing frequency include ulcerative colitis, Crohn's disease, Type 1 diabetes (T1D), multiple sclerosis (MS), rheumatoid arthritis (RA), asthma, food allergy and perhaps others." (Weinstock & Elliott, 2014)

Based on these facts, there are growing efforts to counter autoimmune diseases through the use of helminths (helminth therapy) (Arai & Lopes, 2022; Lothstein & Gause, 2021; Shi et al., 2022; Shields & Cooper, 2022), not always with clear positive results. However, for helminthic therapy, some concepts, which are a direct derivation from the above, should be considered:

(1) An autoimmune disease is not a disease without a cause. Evidence suggests that autoimmune diseases have as a primary cause the loss of intestinal parasites as part of a severe alteration of the holobiont;
(2) Therefore, it is rational to prevent the alteration of the holobiont by avoiding the eradication of the parasites;
(3) If the eradication of the parasites has already taken place, the alteration of the holobiont must be corrected as best and soon as possible by reinserting the helminths;
(4) This should be achieved by restoring parasites to which our species has long been adapted (e.g., *Enterobius vermicularis*, pinworm) and not trying to use parasites for which adaptation is inexistent or incomplete (e.g., parasites adapted

to other species—as *Thichuris suis*, pig whipworm, and *Hymenolepis diminuta*, rat tapeworm cysticerci—or parasites common in tropical environments for which populations living for millennia in other habitats have little or no evolutionary experience—as *Necator americanus*);

(5) Reintegration of parasites should occur as early as possible in life when the immune system is developing. Contact between host and parasite occurs under evolutionarily "natural" conditions at an early age, and it is by no means guaranteed that contact at later (not "normal") ages will lead to a complete correction of the abnormal development of the immune system.

4.1.3 Alterations of the Holobiont—C: Alterations of the Microbiome at Birth

Autism spectrum disorders (ASD, or, for brevity, autism) are a group of neurodevelopmental disorders (autistic disorder, Asperger syndrome, pervasive developmental disorder—not otherwise specified or PDD-NOS) (McPartland & Volkmar, 2012), characterized mainly by the altered capacity of social communication and stereotyped or repetitive behaviors, and also by "deficits in social-emotional reciprocity... deficits in nonverbal communicative behaviors... deficits in developing, maintaining, and understanding relationships... stereotyped or repetitive motor movements... insistence on sameness, inflexible adherence to routines, or ritualized patterns of verbal or nonverbal behavior... highly restricted, fixated interests... hyper- or hypo reactivity to sensory input..." (Sanchack & Thomas, 2016).

However, because of the variety of symptoms and manifestations and due to the different evaluation criteria for the definition of autistic subjects, it is not easy to clearly individuate autistic subjects and to get sure statistics (Chiarotti & Venerosi, 2020).

In 1943, there was the first description of cases of autistic children (Kanner, 1943), while the first cases of those who would later be defined as affected by Asperger syndrome were described in 1944 (Asperger, 1944).

For several years, autism was considered a rare disease. In fact, before 1966, a survey of all subjects aged 8–10 in the County of Middlesex identified 35 autistic cases with a prevalence of 4.5 per 10,000 (0.045%) (Lotter, 1966).

In 1988, it was observed: "Epidemiologic findings have shown that autism is a rare disorder with a prevalence of two to five per 10,000", i.e., a prevalence from 0.02% to 0.05% (Smalley et al., 1988).

In 1999, a review that considered 23 epidemiological surveys published between 1966 and 1998 on autism prevalence highlighted a median prevalence estimate of 5.2/10,000 (0,052%) (Fombonne, 1999).

In 2020, a systematic review of studies on autism prevalence in Asia showed a large variability of prevalence estimates across countries, ranging from low values (e.g., Iran, 2012, 0.063%; India, 2017, 0.153%) to larger values (e.g., Lebanon, 2016, 1.53%; South Korea, 2011, 2.64%) (Qiu et al., 2020).

In 2022, another systematic review highlighted that the results of autism incidence in 99 estimates from 71 studies from 2012 onwards indicated a global median prevalence of autism of 100/100,000 (1%) with a range from 1.09/10,000 (0.0109%) to 436.0/10,000 (4.36%), with great differences between various areas of the world (Zeidan et al., 2022).

In the US, for children aged 8 years, the frequency of autism increased from 1 in 150 in 2002 (0.66%) to 1 in 54 in 2016 (1.85%), with an increase of 181% in 14 years (s. Table 4.6) (Chiarotti & Venerosi, 2020).

In some rare cases, autistic subjects, together with severe behavioral alterations and deficits, show surprising qualities that have created false myths around the extent of the damage and the consequences of autism (see Fig. 4.8). In reality, as autism itself is not a life-threatening disorder and autistic individuals need lifelong care and assistance, the social and economic costs of autism are staggering: "Few US estimates of the economic burden of autism spectrum disorders (ASD) are available and none provide estimates for 2015 and 2025. We forecast annual direct medical, direct non-medical, and productivity costs combined will be $268 billion (range $162-$367 billion; 0.884–2.009 % of GDP) for 2015 and $461 billion (range $276-$1011 billion; 0.982–3.600 % of GDP) for 2025. These 2015 figures are on a par with recent estimates for diabetes..." (Leigh & Du, 2015) (s. Figure 4.9).

Many autism cases are related to known genetic alterations (Genovese & Butler, 2020; Styles et al., 2020). "However, genetic factors might account for only 10% to 20% of the observed ASD cases, and the concordance rate of monozygotic twins is less than 100% (77% for male pairs and 50% for female pairs), suggesting an important contribution of environmental elements in ASD development." (Berding & Donovan, 2016)

"Although ASD has been shown to be highly heritable (recent estimates 38–54%), several meta-analyses have highlighted that nongenetic prenatal causes of ASD exist, opening the door for further studies to investigate such mechanisms (Hallmayer et al., 2011). Approximately 10% of ASD cases are linked to disorders of genetic etiology, such as fragile X syndrome, tuberous sclerosis, and Rett disorder. Supporting the idea of heterogeneity of ASD, single genetic mutations account for only 1–2% of ASD cases (Abrahams & Geschwind, 2008), with the majority of cases remaining idiopathic." (Madore et al., 2016).

In the vast scientific literature on autism, there is a tendency to attribute its causes to genetic alterations. Sometimes, the novelty of identifying genes and genomic

Table 4.6 Increase of autism frequency in the US from 2002 to 2016, data from Table 5 in (Chiarotti & Venerosi, 2020)

Year	2002	2006	2008	2010	2012	2014	2016
Frequency (in %)	0.66	0.9	1.13	1.47	1.46	1.68	1.85
Incidence	1 in 150	1 in 110	1 in 88	1 in 68	1 in 68	1 in 59	1 in 54
Ratio M/F	4.2:1	4.5:1	4.6:1	4.5:1	4.5:1	4.0:1	4.3:1

Fig. 4.8 Films such as *Rain Man*, the cases of "idiot savant", the television series *The Good Doctor*, and the cases of intelligent individuals with Asperger syndrome (e.g., Elon Musk) have created the myth that the autistic subject, in addition to various alterations of behavior, often has some mental capacities of a genius. However, the autistic subject, as a rule, has both neurological problems or deficits and alterations in behavior and capacity for human interactions. This causes the need for continuous assistance and hampers the development of an independent life. In figure: (**a**): Freddie Highmore in *The Good Doctor*; (**b**): Stephen Wiltshire, a famous autistic savant; (**c**): Tom Cruise and Dustin Hoffman in *Rain Man*; (**d**): Elon Musk (Asperger syndrome)

regions that contribute to the genesis of autism is highlighted (Manoli & State, 2021), even going so far as to describe autism as "one of the most highly heritable disorders" (Thapar & Rutter, 2021).

Certainly, as in every biological phenomenon, genes are important, but specific hypothetical "bad" genes do not explain the great increase in the frequency of the disease. If a disease's frequency varies enormously both in a few years and among various populations, according to the principles of evolutionary medicine, a good portion of cases must be due to significant variations in the ecological niche and it is not among the genes that the main primary origin of autism must be defined. (Libertini et al., 2021).

Various factors appear to be correlated with the onset of the syndrome, but none appears to be decisive (Atladóttir et al., 2012; Hamad et al., 2019; Hodges et al., 2020; Wang et al., 2023).

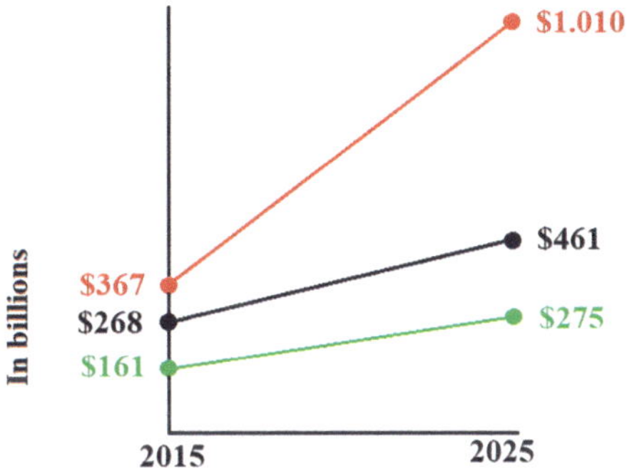

Fig. 4.9 The projected annual cost of autism assuming a prevalence of 1.1%. Alternative projections are in different colors. The costs include productivity loss for people with autism, productivity loss for parents, non-medical services, special education, residential care, and accommodation. Data from (Leigh & Du, 2015)

Perhaps the mistake is in considering the individual as an isolated entity while neglecting that he is part of something much more complex, i.e., the holobiont.

In fact:

(i) Gastrointestinal symptoms are four times more prevalent in children with autism compared to the normal population (Marler et al., 2017). A wide range of gastrointestinal symptoms are found in autistic children (constipation, diarrhea, abdominal pain, bloating, reflux, vomiting, gaseousness, foul-smelling stools, and food allergies) (Berding & Donovan, 2016; Buie, 2015; Fulceri et al., 2016; Marler et al., 2017; Santocchi et al., 2016);
(ii) There is comorbity between gastrointestinal disorders and autism spectrum disorders and their gravity (Madra et al., 2021);
(iii) In autistic children, there are alterations of the gut microbiota (Finegold et al., 2017; Kang et al., 2013) and a lower bacterial diversity (Kang et al., 2013);
(iv) In germ-free mice, structural changes in the brain (Martin & Mayer, 2017) and less sociable behaviors (Mayer et al., 2015; Vuong et al., 2017) indicate a functional connection between the microbiota and the brain.

These data (and others reported in the following references) have led to the likely hypothesis of a microbiota-gut-brain axis, strongly influenced by the intestinal microbiota, which is essential for the correct development of neurological functions and whose alteration is a likely cause of autism (Alam et al., 2017; Diaz Heijtz, 2016; Dinan & Cryan, 2017; Fattorusso et al., 2019; Fung et al., 2017; Li et al., 2017; Martin & Mayer, 2017; Principi & Esposito, 2016; Srikantha & Mohajeri, 2019; Zhang et al., 2015).

About the possible alterations of the holobiont (i.e., in this case, of the microbiota-gut-brain axis), which can be related to autism:

(1) Before, during, and after pregnancy, the holobiont of the mother and that of her child are severely altered. From her childhood, the mother's bacterial ecosystems are again and again altered by the frequent use of antibiotics;
(2) The further use of antibiotics during pregnancy increases the risk of autism (Atladóttir et al., 2012; Hamad et al., 2019);
(3) The microbiota of human infants derives from the mother's fecal, vaginal, and skin microbiota, in particular during the natural delivery (Palmer et al., 2007). Childbirth under aseptic conditions or by Caesarean section hinders the normal colonization of the germ-free bowel of the baby;
(4) After birth, the newborn is often separated from the mother and taken to a nursery. Unfortunately, without continuous contact with the mother, the bacterial ecosystems of the newborn are abnormally colonized;
(5) Breastfeeding is also often missed because it is mistakenly believed that infant formula can safely replace breast milk. However, in 2018, in the milk of the mother, over 70 different types of special oligosaccharides (Human Milk Oligosaccharides, HMOs) stimulating the growth of beneficial intestinal bacteria were known (Wu et al., 2010; Wu et al., 2011). In the following years, the number of identified HMOs increased to about 200 (Urashima et al., 2018). They show multiple functions related to the gut microbiome (Salminen et al., 2020). If the mother produces these HMOs, which, among other things, promote the growth of particular bacteria, somehow, they must be useful to the newborn. Otherwise, their existence would not be justifiable in evolutionary terms;
(6) In the following years, the child is "defended" by repeated antibiotic treatments. In developed countries, a child undergoes about 10–12 courses of antibiotics before the age of 18 years (Sharland et al., 2007), and these "curative" treatments "shift the gut microbiota to long-term alternative dysbiotic states, which may promote the development and aggravation of disease" (Lange et al., 2016). In short, the holobiont of a newborn is often strongly altered, and this alteration continues and increases in subsequent times. This discrepancy has also led to proposals to evaluate the use of probiotics to correct the imbalance of the intestinal microbiota to treat autism (Santocchi et al., 2016);
(7) The origin of autism in the gestational period, in particular as a consequence of a breaking of the equilibrium of maternal-fetal binomial, is a well-known hypothesis and has been re-proposed in very recent times (Botelho et al., 2024).

In short, various alterations of our ecological niche, regarding both maternal holobiont and baby/child holobiont, cause substantial alterations of the gut microbiota, which is a fundamental part of our holobiont, hence alterations of the gut-brain microbiota axis and autism. The study of the possible biochemical mechanisms underlying autism caused by the imbalance of the intestinal microbiota is being studied (Srikantha & Mohajeri, 2019).

As an alternative hypothesis, an inflammatory or even autoimmune genesis for autism has been proposed (Gottfried et al., 2015). Indeed, in the brain of autistic subjects, it has been observed altered systemic immunity (Goines & Van de Water,

2010; Mead & Ashwood, 2015), activated microglia in the dorsolateral prefrontal cortex (Morgan et al., 2010), activation of microglia and an increase in density in post-mortem studies (Morgan et al., 2010; Tetreault et al., 2012), and increased levels or activity of proinflammatory factors (cytokines and chemokines) in nervous tissue and cerebrospinal fluid (Ashwood et al., 2011; Morgan et al., 2010; Vargas et al., 2005).

However, as: "Cytokines and chemokines modulate brain function, as well as systemic and CNS responses to infection, injury, and inflammation (Besedovsky & del Rey, 1996). In fact, cytokines, such as TNF-α, IL-1β, IL-6, and TGF-β family, are able to modulate neuronal activity (Onore et al., 2011) and IL-6 promotes oligodendrocyte survival (Vitkovic et al., 2000)." (Gottfried et al., 2015), the cerebral inflammatory phenomena described for autism would be an effect of the alteration of the gut-brain microbiota axis.

4.1.4 Alterations of the Holobiont—D: Alterations of Skin and Mucosal Microbiomes

The human skin hosts complex ecosystems in which there are bacteria, fungi, viruses, archaea, and mites (Carmona-Cruz et al., 2022; Lunjani et al., 2019):

(i) Among the bacteria, there are *Actinobacteria*, *Firmicutes*, *Proteobacteria*, and *Bacteroidetes* (Byrd et al., 2018; Carmona-Cruz et al., 2022; McLoughlin et al., 2022);
(ii) In particular areas, there are *Archaea* (Koskinen et al., 2017), which "have been identified as a substantial component, or even as keystone species, in complex microbiomes - in the environment or accompanying a holobiont." (Moissl-Eichinger et al., 2018). Although *Archaea* are mostly present in the gut, where they are pivotal species (Koskinen et al., 2017), *Methanobrevibacter oralis* is present on the oral mucosa (in subgingival plaque) (Ferrari et al., 1994), and *Thaumarchaeota* on the skin (Probst et al., 2013);
(iii) Among the fungal species, there are *Malassezia, Cryptococcus, Rhodotorula, Aspergillus*, and *Epicoccum* spp., among which *Malassezia* spp. are the most common (Byrd et al., 2018; Lunjani et al., 2019; Rozas et al., 2021);
(iv) Among the skin viruses, there are human papillomavirus and mainly phages of the bacterial flora inhabiting the skin (Boxberger et al., 2021);
(v) Other normal components of the skin are two species of tiny mites (*Demodex folliculorum* and *Demodex brevis*), which live in the pilosebaceous follicles (Boxberger et al., 2021; Forton & De Maertelaer, 2021).

These brief notes give an idea of the complexity of the microbiota of skin and external mucous membranes and, in the context of our holobiont that includes our organism as host (or main biont), of the complexity of relationships between the various bionts, host included.

The formation of microbiota colonization begins at birth and is influenced, among other things, by the route of delivery (Chu et al., 2017). Afterward, the composition of the microbiota is certainly determined by evolutionarily "normal" factors such as skin or mucosal site, gender, age, geographical location, climate, and seasonality, but also by exposure to products absent in the "normal" ecological niche, which more or less seriously alter the composition of the microbiota. Products such as soaps, detergents, mouthwashes, toothpaste, skin disinfectants, moisturizers, lotions, shampoos, shower foams, deodorants, perfumes, cosmetics, and bactericidal creams certainly alter the composition of the microbiota, so indirectly favoring the proliferation of species that are more resistant to these new factors. For example, regarding cosmetics, see (Pinto et al., 2021): "... there is growing evidence that pollutants, phthalates, and other chemicals, also those included in cosmetic products can act as endocrine disruptors, interfering with the skin neuroendocrine system. It has also recently been observed that the application of cosmetic products to the skin can alter the balance of the skin microbiota, compromising the eubiosis of the skin, mucous membranes, and the scalp. Indeed, the topical application of cosmetic products such as hygiene products, moisturizers, anti-aging oily products, soaps, shampoos, lotions can alter the lipid film that covers the skin and affect the diversity of the skin's resident microflora (Holland & Bojar, 2002; Wallen-Russell, 2019)".

These alterations of skin and mucous microbiota are not neutral for the health of the host.

If we consider the components of skin and mucous membrane microbiota as elements in their own right, this might seem irrelevant to the health of the host. However, if we correctly consider these microbiotas as part of the holobiont of which the host is the main component, the alterations of these microbiota are unlikely to be indifferent to the health of the host. Indeed, this can be deduced from the concept that modifications of biological complex systems have as more probable consequences, alterations of physiological conditions and, therefore, diseases. In fact, the microbiotas of skin and mucous membranes perform some very important functions:

(i) their presence hinders the establishment of possible pathogenic species;
(ii) the balanced presence of the various species of a microbiota contrasts the possible prevalence of a species that could prove to be pathogenic; and.
(iii) there is a continuous interaction between host immune defenses, which shape microbiota composition, and important actions of microbiota on the development and modulation of host immune defenses: "the skin immune system should be considered a collective mixture of elements from the host and microbes acting in a mutualistic relationship." (Sanford & Gallo, 2013)

Figures 4.10 and 4.11 show some types of diseases due to likely alterations of skin microbiomes.

"Imbalances in the skin microbiota composition (dysbiosis) are associated with several skin conditions, either pathological such as eczema, acne, allergies or

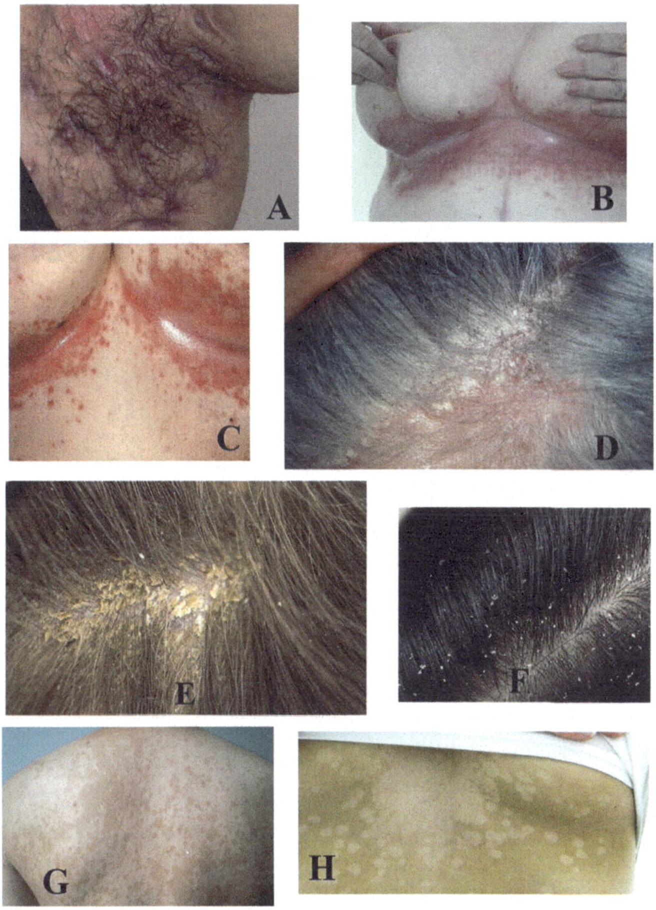

Fig. 4.10 Some dermatological diseases caused by alterations in the skin's microbiome. (**a**) armpit hidradenitis; (**b**), (**c**) submammary mycosis; (**d**), (**e**) and (**f**): seborrheic dermatitis and dandruff; (**g**), (**h**) pityriasis versicolor

dandruff or non-pathological such as sensitive skin, irritated skin or dry skin." (Sfriso et al., 2020).

These diseases are probably determined by the same general mechanism. The primary cause is a significant alteration of the skin microbiome in an area of the body surface. The alteration may be determined by the use of bactericidal and germicidal products (→ various types of pityriasis) or of antibiotics and/or corticosteroids (e.g., → oral candidiasis), or by particular conditions such as friction caused by adjacent skin surfaces, moisture, and sweat build-up trapped in particular areas causing maceration of the stratum corneum and epidermis (e.g., diapers in neonates → perineal candidiasis), poor ventilation of particular areas (e.g., safety shoes → finger intertrigo) (Nobles & Miller, 2022). This results in the proliferation of species

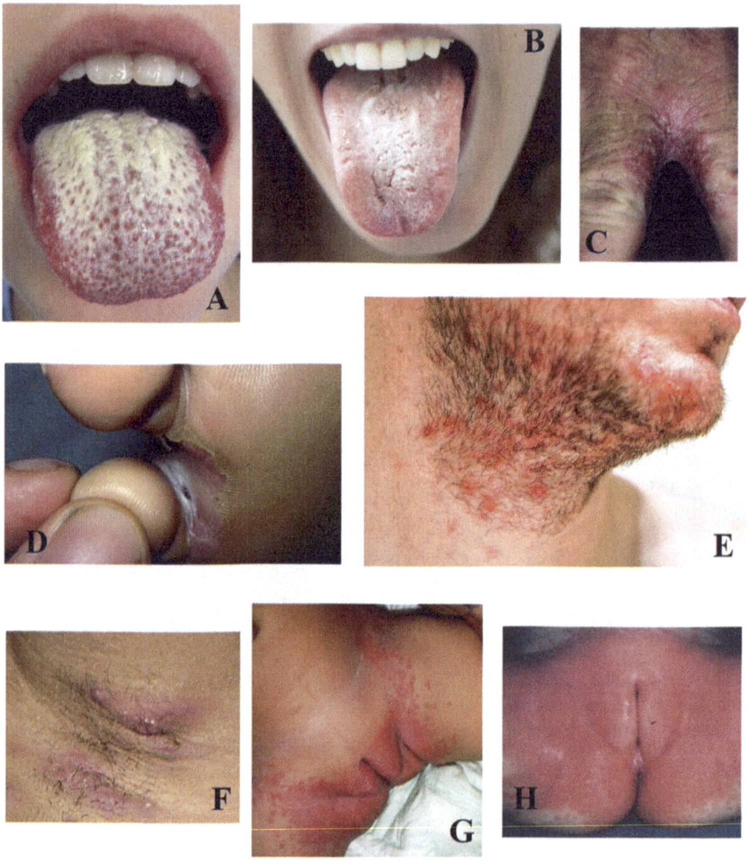

Fig. 4.11 Other dermatological diseases caused by alterations of the skin's microbiome: (**a**), (**b**) oral candidiasis; (**c**), (**d**) finger intertrigo; (**e**) seborrheic dermatitis; (**f**) hidradenitis suppurativa; (**g**), (**h**) perineal candidiasis

already existing in the normal microbiome that are more resistant to the germicidal product or adapted to particular conditions.

Let us consider dandruff and seborrheic dermatitis (SD), which can be considered two different forms of the same disease affecting the seborrheic areas of the skin. Dandruff, the mildest form, affects the scalp, while SD affects both the scalp and other seborrheic areas (e.g., → armpit hidradenitis). At least 50 million Americans suffer from dandruff, while the incidence of SD is 1–3% of the adult population (Borda & Wikramanayake, 2015).

The correlation of dandruff with alterations of the microbiome is known: "Several scalp microbiome studies from different populations have revealed the association of dandruff with bacterial and fungal dysbiosis. However, the functional role of scalp microbiota in scalp disorders and health remains scarcely explored.... *Propionibacterium acnes* and *Staphylococcus epidermidis* emerged as the core bacterial species, where the former was associated with a healthy scalp and the latter with dandruff scalp. Along with the commonly occurring *Malassezia* species (*M. restricta* and *M. globosa*) on the scalp, a strikingly high association of dandruff with yet uncharacterized *Malassezia* species was observed in the core mycobiome.... The results point toward a new potential role of bacterial commensals in maintaining the scalp nutrient homoeostasis and highlights an important and yet unknown role of the scalp microbiome, similar to the gut microbiome." (Saxena et al., 2018).

Despite the evident importance of the microbiome in the genesis of dandruff and SD, the relationship between the use of shampoos or other germicidal products and the alteration of the microbiome is not investigated in the reviews above (Borda & Wikramanayake, 2015; Saxena et al., 2018).

Furthermore, regarding the diseases caused by alterations of skin microbiomes, there are no studies comparing modern populations with populations living in conditions similar to the ecological niche to which we are adapted.

4.1.5 Alterations of the Holobiont—F: Alterations of the Gut Microbiome and Obesity

Extensive documentation shows the relationship between obesity and particular alterations of the gut microbiome, in particular, due to antibiotic treatment:

(1) "... an increased Firmicutes/Bacteroidetes ratio at the phylum level is an important feature of the gut microbiota in obesity. The family Christensenellaceae and the genera Methanobacteriales, Lactobacillus, Bifidobacteria, and Akkermansia are usually considered as probiotics, and their relative abundance is often inversely associated with obesity." (Liu et al., 2021);
(2) "... comparing the gut microbiota of obese and lean people revealed higher Firmicutes and lower Bacteroidetes proportion in obesity, and an inverse profile was found in people after 1-year diet therapy (Ley et al., 2006) and gastric bypass (Zhang et al., 2009)." (Tseng & Wu, 2019);
(3) The onset of obesity has been detected after treatment with antibiotics for endocarditis (Thuny et al., 2010) or eradication of *H. Pylori* (Francois et al., 2011);
(4) Childhood obesity appears related to the use of antibiotics in infancy (Bailey et al., 2014; Scott et al., 2016);
(5) In mice, early exposure to subtherapeutic antibiotic doses was related to alterations of gut microbiome and increased adiposity (Cho et al., 2012);
(6) In mice, fecal transplantation from an overweight adult twin to germ-free mice led to rapid increases in body and fat mass. On the contrary, germ-free mice receiving fecal transplantation from a lean twin maintained normal weight (Ridaura et al., 2013).

4.1.6 Alterations of the Holobiont—F: The Case of **Helicobacter Pylori**

In 1984, a particular type of bacterium, *Helicobacter pylori*, was demonstrated to live and thrive in the stomach lining, even though, until then, the acidity of this environment was considered a factor that made life impossible life for bacteria (Marshall & Warren, 1984). The presence of the bacterium was associated with gastritis and stomach cancer, and the importance of the discovery was sanctioned with the awarding of the Nobel Prize for Medicine in 2005: "Marshall and Warren went on to show that *H. pylori* is found in almost all individuals with gastritis, peptic ulcers, and duodenal ulcers. Indeed, more than 80% of stomach ulcers and 90% of duodenal ulcers are known to be caused by *H. pylori*. Furthermore, it became clear that stomach cancer can result in infected individuals where *H. pylori* causes widespread inflammation of the stomach lining..." (Mégraud, 2005).

Since the bacterium was clearly identified as a cause of even fatal diseases but could be eliminated by the use of antibiotics, from then on, the therapeutic indication was the eradication of the bacterium. At the same time, the continuous and growing use of antibiotics for other pathologies contributed to increasingly reducing the presence of the bacterium: "In the early twentieth century, *Helicobacter pylori* was the dominant microbe in the stomachs of almost all people. By the turn of the twenty-first century, fewer than 6% of children in the United States, Sweden and Germany were carrying the organism. Other factors may be at play in this disappearance, but antibiotics may be a culprit. For example, a single course of amoxicillin or a macrolide antibiotic, most commonly used to treat middle-ear or respiratory infections in children, may also eradicate *H. pylori* in 20–50% of cases." (Blaser, 2011).

The fact that smoking aggravated the damage caused by *H. pylori* (Halter & Brignoli, 1998; Parasher & Eastwood, 2000) did not contradict this view.

However, doubts began to arise about the role of *H. pylori* and its pathogenicity. First of all, in animal models, it was highlighted that the presence of *H. pylori* conferred protection against allergic and chronic inflammatory disorders (Arnold et al., 2011; Arnold et al., 2012; Kyburz et al., 2017; Kyburz et al., 2019). In humans, the absence of *H. pylori* is associated with a higher incidence of asthma, hay fever, and skin allergies in childhood (Amberbir et al., 2011; Chen & Blaser, 2007).

It was also observed that individuals with *H. pylori* showed greater resistance against developing active tuberculosis (Perry et al., 2010) and diarrheal diseases (Cohen et al., 2012).

Moreover, there was a bigger and contradictory problem: "as *H. pylori* has disappeared from people's stomachs, there has been also an increase in gastroesophageal reflux, and its correlated diseases such as Barrett's oesophagus and oesophageal cancer." (Blaser, 2011).

The epidemiology of gastroesophageal reflux disease (GERD) "is linked to the Western lifestyle, obesity, and the demise of *Helicobacter pylori*." (Katzka & Kahrilas, 2020). GERD can cause low- or high-grade esophagitis, regurgitations, reflux chest pain, Barrett's esophagus, laryngopharyngeal reflux, and "a fivefold risk of developing esophageal adenocarcinoma" (Katzka & Kahrilas, 2020).

Thus, the presence of *H. pylori* caused gastritis and stomach cancer risk but protected against esophagitis and esophageal cancer risk (as well as beneficial effects on the immune system), while the absence of the bacterium caused the opposite.

However, additional data made the picture more contradictory and confusing.

In 1992, Holcombe highlighted a phenomenon that he called "The African Enigma". He observed that in sub-Saharan Africa, most people acquired the bacterium early in life. However, the incidence of stomach cancer was remarkably low: "Gastritis is very common throughout Africa and shows a strong correlation with *H. pylori* infection... Gastric ulcer is a rare disease in Africa... In northern Nigeria duodenal ulcer is uncommon, yet gastric ulcer is six times less common... In Nigeria *H. pylori* gastritis is common, yet gastric cancer is uncommon" (Holcombe, 1992).

Similar observations were made for Asia ("Asian enigma") (Singh & Ghoshal, 2006) and India ("Indian enigma") (Misra et al., 2014).

Additional data were provided by a study on the number of gastric cancers, gastric ulcers, and duodenal ulcer cases from 1800 to 1920 in four hospitals in Scotland and three in the US (Sonnenberg & Baron, 2010). At the beginning of the nineteenth century, these diseases were a rarity or not reported. Subsequently, the cases of gastric cancer and gastric ulcer gradually rose from the mid-nineteenth century, while the cases of duodenal ulcer began to rise from the end of the nineteenth century. In subsequent studies, the same pattern was shown for France, Germany, Netherlands, Scotland, Spain, Sweden, Denmark, England & Wales, Italy, Japan, and Switzerland (Sonnenberg, 2010, 2011).

In fact, as antibiotics did not yet exist and presumably the presence of *H. pylori* was widespread, the situation in Europe in the first half of the nineteenth century was similar to that of Africa and India, justifying the definition of "European enigma" (Velasquez-Manoff, 2012, pp. 157–158).

The drastic change between Europe in the first half of the nineteenth century (similar to that of Africa and India today) and Europe in the second half of the nineteenth century and later required an explanation that could not be in the simple alternative of the presence or absence of *H. pylori* (Velasquez-Manoff, 2012, p. 158).

In 2000, it was shown that mice infected with a helicobacter species (*Helicobacter felis*), the equivalent for mice of human *H. pylori*, developed severe gastritis and histological alterations that were similar to premalignant lesions. The same mice also infected with the rodent worm *Heligmosomoides polygirus* did not show these alterations despite hosting a larger colony of *H. felis* than mice without the worms (Fox et al., 2000).

A few years later, two human populations were compared, the first living in Tumaco on the coast of Colombia and the other in Pasto in the interior of the country. Both populations showed the presence of *H. pylori* in 95% of the individuals. However, the children of Tumaco had twice as many worms as the Pasto population, and gastric cancer was much less frequent in the Tumaco population (Whary et al., 2005). A subsequent study (Ek et al., 2012) conducted on the same populations showed that the Tumaco population, compared to the other population, was much more frequently positive for *Ascaris lumbricoides*, a helminth, and *Toxoplasma gondii*, an obligate intracellular parasitic protozoan, and the authors concluded that these "infections may impact inflammatory responses to *H. pylori* and partially explain differences in gastric cancer risk in Colombia."

Other studies on different populations gave similar results: "Nearly all Tanzanianas had the bacterium, for example, but inhabitants of the country's highlands, and denizens of the slopes of Mount Kilimanjaro where worms were less prevalent, had much more stomach cancer. Likewise, across the Japanese archipelago, gastric cancer incidence increased with latitude, again the inverse of worm prevalence." (Velasquez-Manoff, 2012, p. 159).

Therefore, considering other data relating to the beneficial effects of the presence of *H. pylori*, for example, the protection against allergic diseases, chronic inflammatory disorders, active forms of tuberculosis, and diarrheal diseases (for the references, see above), the role of *H. pylori*, but also that of certain worm infestations, should be completely reconsidered.

In the traditional view, in both cases, there are pathogenic factors (presence of *H. pylori* or worm infestation) that must be removed as much as possible. On the contrary, for evolutionary medicine, it is necessary to consider the holobiont and the complicated relationships between the various bionts, the host included, which compose it. From this point of view, the pathogenetic factor, the primary cause, is constituted by the alterations of the balance between the various bionts, which trigger pathologies that are absent when the balance is maintained.

The relationships between *H. pylori*, intestinal worms, the health status of the host, and other infections are summarized (of course, incompletely) and represented in Fig. 4.12 and Table 4.7.

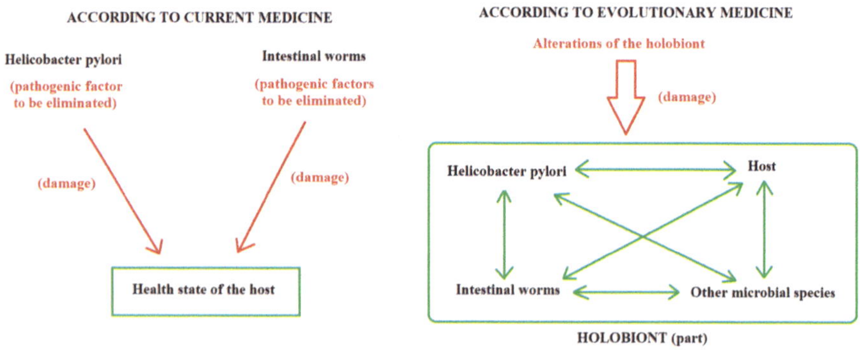

Fig. 4.12 *H. pylori* as a pathogenic factor according to the different interpretations of current medicine and evolutionary medicine

Table 4.7 Relations between infection by *H. pylori*, infestation by helminths, and some pathological states

Condition	The rule in:	Effects
Presence of both *H. Pylori* and helminths	Hunter-gatherer populations (evolutionary "normal" condition) Many modern African and Indian populations European people in the first half of the nineteenth century	Asymptomatic gastritis, no increased incidence of gastric and esophageal cancer; no gastric or duodenal ulcer; protection against allergic and chronic inflammatory disorders; resistance to active tuberculosis and diarrheal diseases
Presence of *H. Pylori* but not of helminths	European people in the second half of the nineteenth century	Increased incidence of gastric ulcer, gastric cancer, and duodenal ulcer
Absence of both *H. Pylori* and helminths	Modern civilized populations	Increased incidence of gastroesophageal reflux, esophagitis, Barrett's esophagus, esophageal cancer, asthma, hay fever, and skin allergies, reduced protection against allergic and chronic inflammatory disorders; reduced resistance to active tuberculosis and diarrheal diseases

References

Abrahams, B. S., & Geschwind, D. H. (2008). Advances in autism genetics: On the threshold of a new neurobiology. *Nature Reviews. Genetics, 9*(5), 341–355. https://doi.org/10.1038/nrg2346. Erratum in: *Nature Reviews. Genetics* 2008, 9(6), 493.

Adebajo, A. O. (1997). Low frequency of autoimmune disease in tropical Africa. *Lancet, 349*(9048), 361–362. https://doi.org/10.1016/s0140-6736(05)62867-x

Ait-Khaled, N., Odhiambo, J., Pearce, N., Adjoh, K. S., Maesano, I. A., Benhabyles, B., et al. (2007). Prevalence of symptoms of asthma, rhinitis and eczema in 13- to 14-year-old children in Africa: The international study of asthma and allergies in childhood phase III. *Allergy, 62*(3), 247–258. https://doi.org/10.1111/j.1398-9995.2007.01325.x

Alam, R., Abdolmaleky, H. M., & Zhou, J. R. (2017). Microbiome, inflammation, epigenetic alterations, and mental diseases. *American Journal of Medical Genetics. Part B, Neuropsychiatric Genetics, 174*(6), 651–660. https://doi.org/10.1002/ajmg.b.32567

Alsowaidi, S., Abdulle, A., Shehab, A., Zuberbier, T., & Bernsen, R. (2010). Allergic rhinitis: Prevalence and possible risk factors in a gulf Arab population. *Allergy, 65*(2), 208–212. https://doi.org/10.1111/j.1398-9995.2009.02123.x

Amberbir, A., Medhin, G., Erku, W., Alem, A., Simms, R., Robinson, K., et al. (2011). Effects of helicobacter pylori, geohelminth infection and selected commensal bacteria on the risk of allergic disease and sensitization in 3-year-old Ethiopian children. *Clinical and Experimental Allergy, 41*(10), 1422–1430. https://doi.org/10.1111/j.1365-2222.2011.03831.x

An, S. Y., Choi, H. G., Kim, S. W., Park, B., Lee, J. S., Jang, J. H., & Sung, M. W. (2015). Analysis of various risk factors predisposing subjects to allergic rhinitis. *Asian Pacific Journal of Allergy and Immunology, 33*(2), 143–151. https://doi.org/10.12932/AP0554.33.2.2015

Arai, T., & Lopes, F. (2022). Potential of human helminth therapy for resolution of inflammatory bowel disease: The future ahead. *Experimental Parasitology, 232*, 108189. https://

doi.org/10.1016/j.exppara.2021.108189. Erratum in: *Experimental Parasitology*, 2021, 234, 108218.

Arnold, I. C., Dehzad, N., Reuter, S., Martin, H., Becher, B., Taube, C., & Müller, A. (2011). Helicobacter pylori infection prevents allergic asthma in mouse models through the induction of regulatory T cells. *The Journal of Clinical Investigation, 121*(8), 3088–3093. https://doi.org/10.1172/JCI45041

Arnold, I. C., Hitzler, I., & Müller, A. (2012). The immunomodulatory properties of helicobacter pylori confer protection against allergic and chronic inflammatory disorders. *Frontiers in Cellular and Infection Microbiology, 2*, 10. https://doi.org/10.3389/fcimb.2012.00010

Asher, M. I., Montefort, S., Björkstén, B., Lai, C. K., Strachan, D. P., Weiland, S. K., Williams, H., & ISAAC Phase Three Study Group. (2006). Worldwide time trends in the prevalence of symptoms of asthma, allergic rhinoconjunctivitis, and eczema in childhood: ISAAC phases one and three repeat multicountry cross-sectional surveys. *Lancet, 368*(9537), 733–743. https://doi.org/10.1016/S0140-6736(06)69283-0. Erratum in: *Lancet*, 2007, 370(9593), 1128.

Ashwood, P., Krakowiak, P., Hertz-Picciotto, I., Hansen, R., Pessah, I., & Van de Water, J. (2011). Elevated plasma cytokines in autism spectrum disorders provide evidence of immune dysfunction and are associated with impaired behavioral outcome. *Brain, Behavior, and Immunity, 25*(1), 40–45. https://doi.org/10.1016/j.bbi.2010.08.003

Asperger, H. (1944). Die 'Autistischen Psychopathen' im Kindesalter. *Archiv für Psychiatrie und Nervenkrankheiten, 117*, 76–136.

Atladóttir, H. Ó., Henriksen, T. B., Schendel, D. E., & Parner, E. T. (2012). Autism after infection, febrile episodes, and antibiotic use during pregnancy: An exploratory study. *Pediatrics, 130*(6), e1447–e1454. https://doi.org/10.1542/peds.2012-1107

Autoimmune Association. (2023a). *About autoimmunity*. https://autoimmune.org/resource-center/about-autoimmunity/. Accessed on 07.04.23.

Autoimmune Association. (2023b). *Disease information*. https://autoimmune.org/disease-information/. Accessed on 07.04.23.

Autoimmune Registry. (2023). https://www.autoimmuneregistry.org/autoimmune-diseases. Accessed on 07.04.23.

Bach, J. F. (2002). The effect of infections on susceptibility to autoimmune and allergic diseases. *New England Journal of Medicine, 347*(12), 911–920. https://doi.org/10.1056/NEJMra020100

Bailey, L. C., Forrest, C. B., Zhang, P., Richards, T. M., Livshits, A., & DeRusso, P. A. (2014). Association of antibiotics in infancy with early childhood obesity. *JAMA Pediatrics, 168*(11), 1063–1069. https://doi.org/10.1001/jamapediatrics.2014.1539

Berding, K., & Donovan, S. M. (2016). Microbiome and nutrition in autism spectrum disorder: Current knowledge and research needs. *Nutrition Reviews, 74*(12), 723–736. https://doi.org/10.1093/nutrit/nuw048

Besedovsky, H. O., & del Rey, A. (1996). Immune-neuro-endocrine interactions: Facts and hypotheses. *Endocrine Reviews, 17*, 64–102. https://doi.org/10.1210/edrv-17-1-64

Blaser, M. J. (2011). Antibiotic overuse: Stop the killing of beneficial bacteria. *Nature, 476*(7361), 393–394. https://doi.org/10.1038/476393a

Borda, L. J., & Wikramanayake, T. C. (2015). Seborrheic dermatitis and dandruff: A comprehensive review. *Journal of Clinical and Investigative Dermatology, 3*(2), 10. https://doi.org/10.13188/2373-1044.1000019

Borkow, G., Leng, Q., Weisman, Z., Stein, M., Galai, N., Kalinkovich, A., & Bentwich, Z. (2000). Chronic immune activation associated with intestinal helminth infections results in impaired signal transduction and anergy. *The Journal of Clinical Investigation, 106*(8), 1053–1060. https://doi.org/10.1172/JCI10182

Botelho, R. M., Silva, A. L. M., & Borbely, A. U. (2024). The autism Spectrum disorder and its possible origins in pregnancy. *International Journal of Environmental Research and Public Health, 21*(3), 244. https://doi.org/10.3390/ijerph21030244

Boxberger, M., Cenizo, V., Cassir, N., & La Scola, B. (2021). Challenges in exploring and manipulating the human skin microbiome. *Microbiome, 9*(1), 125. https://doi.org/10.1186/s40168-021-01062-5

Buie, T. (2015). Potential etiologic factors of microbiome disruption in autism. *Clinical Therapeutics, 37*(5), 976–983. https://doi.org/10.1016/j.clinthera.2015.04.001

Byrd, A. L., Belkaid, Y., & Segre, J. A. (2018). The human skin microbiome. *Nature Reviews. Microbiology, 16*(3), 143–155. https://doi.org/10.1038/nrmicro.2017.157

Carmona-Cruz, S., Orozco-Covarrubias, L., & Sáez-de-Ocariz, M. (2022). The human skin microbiome in selected cutaneous diseases. *Frontiers in Cellular and Infection Microbiology, 12*, 834135. https://doi.org/10.3389/fcimb.2022.834135

Chen, Y., & Blaser, M. J. (2007). Inverse associations of helicobacter pylori with asthma and allergy. *Archives of Internal Medicine, 167*(8), 821–827. https://doi.org/10.1001/archinte.167.8.821

Chiarotti, F., & Venerosi, A. (2020). Epidemiology of autism Spectrum disorders: A review of worldwide prevalence estimates since 2014. *Brain Sciences, 10*(5), 274. https://doi.org/10.3390/brainsci10050274

Cho, I., Yamanishi, S., Cox, L., Methé, B. A., Zavadil, J., Li, K., et al. (2012). Antibiotics in early life alter the murine colonic microbiome and adiposity. *Nature, 488*(7413), 621–626. https://doi.org/10.1038/nature11400

Chu, D. M., Ma, J., Prince, A. L., Antony, K. M., Seferovic, M. D., & Aagaard, K. M. (2017). Maturation of the infant microbiome community structure and function across multiple body sites and in relation to mode of delivery. *Nature Medicine, 23*(3), 314–326. https://doi.org/10.1038/nm.4272

Clark, I. A., Al-Yaman, F. M., Cowden, W. B., & Rockett, K. A. (1996). Does malarial tolerance, through nitric oxide, explain the low incidence of autoimmune disease in tropical Africa? *Lancet, 348*(9040), 1492–1494. https://doi.org/10.1016/s0140-6736(96)07342-4

Cohen, D., Shoham, O., Orr, N., & Muhsen, K. (2012). An inverse and independent association between helicobacter pylori infection and the incidence of shigellosis and other diarrheal diseases. *Clinical Infectious Diseases, 54*(4), e35–e42. https://doi.org/10.1093/cid/cir916

Cooper, G. S., Bynum, M. L., & Somers, E. C. (2009). Recent insights in the epidemiology of autoimmune diseases: Improved prevalence estimates and understanding of clustering of diseases. *Journal of Autoimmunity, 33*(3–4), 197–207. https://doi.org/10.1016/j.jaut.2009.09.008

Correale, J. (2014). Helminth/parasite treatment of multiple sclerosis. *Current Treatment Options in Neurology, 16*(6), 296. https://doi.org/10.1007/s11940-014-0296-3

Dean, G. (1967). Annual incidence, prevalence, and mortality of multiple sclerosis in white south-African-born and in white immigrants to South Africa. *British Medical Journal, 2*(5554), 724–730. https://doi.org/10.1136/bmj.2.5554.724

Diaz Heijtz, R. (2016). Fetal, neonatal, and infant microbiome: Perturbations and subsequent effects on brain development and behavior. *Seminars in Fetal & Neonatal Medicine, 21*(6), 410–417. https://doi.org/10.1016/j.siny.2016.04.012

Dinan, T. G., & Cryan, J. F. (2017). Gut instincts: Microbiota as a key regulator of brain development, ageing and neurodegeneration. *The Journal of Physiology, 595*(2), 489–503. https://doi.org/10.1113/JP273106

Doetze, A., Satoguina, J., Burchard, G., Rau, T., Löliger, C., Fleischer, B., & Hoerauf, A. (2000). Antigen-specific cellular hyporesponsiveness in a chronic human helminth infection is mediated by T(h)3/T(r)1-type cytokines IL-10 and transforming growth factor-beta but not by a T(h)1 to T(h)2 shift. *International Immunology, 12*(5), 623–630. https://doi.org/10.1093/intimm/12.5.623

Early, J., & Peters, J. (1990). *The population dynamics of the Mucajai Yanomamo*. Academic. ISBN 978-0122276809.

Eaton, W. W., Rose, N. R., Kalaydjian, A., Pedersen, M. G., & Mortensen, P. B. (2007). Epidemiology of autoimmune diseases in Denmark. *Journal of Autoimmunity, 29*(1), 1–9. https://doi.org/10.1016/j.jaut.2007.05.002

Ek, C., Whary, M. T., Ihrig, M., Bravo, L. E., Correa, P., & Fox, J. G. (2012). Serologic evidence that ascaris and toxoplasma infections impact inflammatory responses to helicobacter pylori in Colombians. *Helicobacter, 17*(2), 107–115. https://doi.org/10.1111/j.1523-5378.2011.00916.x

Elliott, D. E., Urban, J. F., Jr., Argo, C. K., & Weinstock, J. V. (2000). Does the failure to acquire helminthic parasites predispose to Crohn's disease? *The FASEB Journal, 14*(12), 1848–1855. https://doi.org/10.1096/fj.99-0885hyp
Endara, P., Vaca, M., Platts-Mills, T. A., Workman, L., Chico, M. E., Barreto, M. L., et al. (2015). Effect of urban vs. rural residence on the association between atopy and wheeze in Latin America: Findings from a case-control analysis. *Clinical and Experimental Allergy, 45*(2), 438–447. https://doi.org/10.1111/cea.12399
Eriksson, J., Ekerljung, L., Rönmark, E., Dahlén, B., Ahlstedt, S., Dahlén, S. E., & Lundbäck, B. (2012). Update of prevalence of self-reported allergic rhinitis and chronic nasal symptoms among adults in Sweden. *The Clinical Respiratory Journal, 6*(3), 159–168. https://doi.org/10.1111/j.1752-699X.2011.00269.x
Farrokhyar, F., Swarbrick, E. T., & Irvine, E. J. (2001). A critical review of epidemiological studies in inflammatory bowel disease. *Scandinavian Journal of Gastroenterology, 36*(1), 2–15. https://doi.org/10.1080/00365520150218002. Erratum in: *Scandinavian Journal of Gastroenterology*, 2001, 36(6), facing 561.
Fattorusso, A., Di Genova, L., Dell'Isola, G. B., Mencaroni, E., & Esposito, S. (2019). Autism Spectrum disorders and the gut microbiota. *Nutrients, 11*(3), 521. https://doi.org/10.3390/nu11030521
Ferrari, A., Brusa, T., Rutili, A., Canzi, E., & Biavati, B. (1994). Isolation and characterization of Methanobrevibacter oralis sp. nov. *Current Microbiology, 29*, 7–12.
Finegold, S. M., Summanen, P. H., Downes, J., Corbett, K., & Komoriya, T. (2017). Detection of Clostridium perfringens toxin genes in the gut microbiota of autistic children. *Anaerobe, 45*, 133–137. https://doi.org/10.1016/j.anaerobe.2017.02.008
Fombonne, E. (1999). The epidemiology of autism: A review. *Psychological Medicine, 29*(4), 769–786. https://doi.org/10.1017/s0033291799008508
Forton, F. M. N., & De Maertelaer, V. (2021). Which factors influence Demodex proliferation? A retrospective pilot study highlighting a possible role of subtle immune variations and sebaceous gland status. *The Journal of Dermatology, 48*(8), 1210–1220. https://doi.org/10.1111/1346-8138.15910
Foster, R. M., & Harries, J. R. (1970). Multiple sclerosis in the African. *British Medical Journal, 3*(5723), 628. https://doi.org/10.1136/bmj.3.5723.628
Fox, J. G., Beck, P., Dangler, C. A., Whary, M. T., Wang, T. C., Shi, H. N., & Nagler-Anderson, C. (2000). Concurrent enteric helminth infection modulates inflammation and gastric immune responses and reduces helicobacter-induced gastric atrophy. *Nature Medicine, 6*, 536–542. https://doi.org/10.1038/75015
Francois, F., Roper, J., Joseph, N., Pei, Z., Chhada, A., Shak, J. R., et al. (2011). The effect of H. Pylori eradication on meal-associated changes in plasma ghrelin and leptin. *BMC Gastroenterology, 11*, 37. https://doi.org/10.1186/1471-230X-11-37
Fulceri, F., Morelli, M., Santocchi, E., Cena, H., Del Bianco, T., Narzisi, A., et al. (2016). Gastrointestinal symptoms and behavioral problems in preschoolers with autism Spectrum disorder. *Digestive and Liver Diseases, 48*(3), 248–254. https://doi.org/10.1016/j.dld.2015.11.026
Fung, T. C., Olson, C. A., & Hsiao, E. Y. (2017). Interactions between the microbiota, immune and nervous systems in health and disease. *Nature Neuroscience, 20*(2), 145–155. https://doi.org/10.1038/nn.4476
Gelfand, M. (1957). *The sick African. A clinical study* (3rd ed.). Juta. ISBN 978-0718602147.
Genovese, A., & Butler, M. G. (2020). Clinical assessment, genetics, and treatment approaches in autism Spectrum disorder (ASD). *International Journal of Molecular Sciences, 21*(13), 4726. https://doi.org/10.3390/ijms21134726
Goines, P., & Van de Water, J. (2010). The immune system's role in the biology of autism. *Current Opinion in Neurology, 23*(2), 111–117. https://doi.org/10.1097/WCO.0b013e3283373514
Gottfried, C., Bambini-Junior, V., Francis, F., Riesgo, R., & Savino, W. (2015). The impact of Neuroimmune alterations in autism Spectrum disorder. *Frontiers in Psychiatry, 6*, 121. https://doi.org/10.3389/fpsyt.2015.00121

Greenwood, B. M. (1968). Autoimmune disease and parasitic infections in Nigerians. *Lancet, 292*(7564), 380–382. https://doi.org/10.1016/S0140-6736(68)90595-3

Greenwood, B. M., Herrick, E. M., & Voller, A. (1970). Can parasitic infection suppress autoimmune disease? *Proceedings of the Royal Society of Medicine, 63*(1), 19–20. https://www.ncbi.nlm.nih.gov/pmc/articles/PMC1810949/

Gupta, R. S., Warren, C. M., Smith, B. M., Jiang, J., Blumenstock, J. A., Davis, M. M., Schleimer, R. P., & Nadeau, K. C. (2019). Prevalence and severity of food allergies among US adults. *JAMA Network Open, 2*(1), e185630. https://doi.org/10.1001/jamanetworkopen.2018.5630

Haddock, D. R. W. (1965). Neurological disorders in Tanzania. *Journal of Tropical Medicine and Hygiene, 68*, 161–166. PMID: 14331720.

Hallmayer, J., Cleveland, S., Torres, A., Phillips, J., Cohen, B., Torigoe, T., et al. (2011). Genetic heritability and shared environmental factors among twin pairs with autism. *Archives of General Psychiatry, 68*(11), 1095–1102. https://doi.org/10.1001/archgenpsychiatry.2011.76

Halter, F., & Brignoli, R. (1998). *Helicobacter pylori* and smoking: Two additive risk factors for organic dyspepsia. *The Yale Journal of Biology and Medicine, 71*(2), 91–99. www.ncbi.nlm.nih.gov/pmc/articles/PMC2578888/

Hamad, A. F., Alessi-Severini, S., Mahmud, S. M., Brownell, M., & Kuo, I. F. (2019). Prenatal antibiotics exposure and the risk of autism spectrum disorders: A population-based cohort study. *PLoS One, 14*(8), e0221921. https://doi.org/10.1371/journal.pone.0221921

Harjutsalo, V., Sund, R., Knip, M., & Groop, P. H. (2013). Incidence of type 1 diabetes in Finland. *JAMA, 310*(4), 427–428. https://doi.org/10.1001/jama.2013.8399

Healy, G. R., Gleason, N. N., Bokat, R., Pond, H., & Roper, M. (1969). Prevalence of ascariasis and amebiasis in Cherokee Indian school children. *Public Health Reports, 84*, 907–914. PMID: 4980739.

Hendaus, M. A., Jomha, F. A., & Ehlayel, M. (2016). Allergic diseases among children: Nutritional prevention and intervention. *Therapeutics and Clinical Risk Management, 12*, 361–372. https://doi.org/10.2147/TCRM.S98100

Hill, K., & Hurtado, A. M. (1996). *Ache life history*. Aldine De Gruyter. ISBN 978-0202020363.

Hill, D. A., & Spergel, J. M. (2018). The atopic march: Critical evidence and clinical relevance. *Annals of Allergy, Asthma & Immunology, 120*(2), 131–137. https://doi.org/10.1016/j.anai.2017.10.037. Erratum in: *Annals of allergy, asthma & immunology*, 2018, Mar 9. PMID: 29413336.

Hill, D. A., Grundmeier, R. W., Ram, G., & Spergel, J. M. (2016). The epidemiologic characteristics of healthcare provider-diagnosed eczema, asthma, allergic rhinitis, and food allergy in children: A retrospective cohort study. *BioMed Central Pediatrics, 16*, 133. https://doi.org/10.1186/s12887-016-0673-z

Hodges, H., Fealko, C., & Soares, N. (2020). Autism spectrum disorder: Definition, epidemiology, causes, and clinical evaluation. *Translational Pediatrics, 9*(Suppl 1), S55–S65. https://doi.org/10.21037/tp.2019.09.09

Holcombe, C. (1992). Helicobacter pylori: The African enigma. *Gut, 33*(4), 429–431. https://doi.org/10.1136/gut.33.4.429

Holland, K. T., & Bojar, R. A. (2002). Cosmetics: What is their influence on the skin microflora? *American Journal of Clinical Dermatology, 3*(7), 445–449. https://doi.org/10.2165/00128071-200203070-00001

Hoyte, F. C. L., & Nelson, H. S. (2018). Recent advances in allergic rhinitis. *F1000Research, 7*.:F1000 Faculty Rev-1333. https://doi.org/10.12688/f1000research.15367.1

Hurtado, A. M., Hill, K., de Hurtado, I. A., & Rodriguez, S. (1997). The evolutionary context of chronic allergic conditions: The Hiwi of Venezuela. *Human Nature (Hawthorne, N.Y.), 8*(1), 51–75. https://doi.org/10.1007/s12110-997-1004-8

Jacobson, D. L., Gange, S. J., Rose, N. R., & Graham, N. M. (1997). Epidemiology and estimated population burden of selected autoimmune diseases in the United States. *Clinical Immunology and Immunopathology, 84*(3), 223–243. https://doi.org/10.1006/clin.1997.4412

Kang, D. W., Park, J. G., Ilhan, Z. E., Wallstrom, G., Labaer, J., Adams, J. B., & Krajmalnik-Brown, R. (2013). Reduced incidence of Prevotella and other fermenters in intestinal microflora of autistic children. *PLoS One, 8*(7), e68322. https://doi.org/10.1371/journal.pone.0068322
Kanner, L. (1943). Autistic disturbances of affective contact. *The Nervous Child, 2*(3), 217–250. Kanner-Autistic-Disturbances-of-Affective-Contact-1943-vooiwn.pdf.
Katzka, D. A., & Kahrilas, P. J. (2020). Advances in the diagnosis and management of gastroesophageal reflux disease. *British Medical Journal, 371*, m3786. https://doi.org/10.1136/bmj.m3786
Kerr, D. (2009). Foreword. In D. J. Nakazawa (Ed.), *The autoimmune epidemic*. Touchstone Book. ISBN 978-0743277761.
Kondrashova, A., Reunanen, A., Romanov, A., Karvonen, A., Viskari, H., Vesikari, T., et al. (2005). A six-fold gradient in the incidence of type 1 diabetes at the eastern border of Finland. *Annals of Medicine, 37*(1), 67–72. https://doi.org/10.1080/07853890410018952
Kondrashova, A., Mustalahti, K., Kaukinen, K., Viskari, H., Volodicheva, V., Haapala, A. M., Epivir Study Group, et al. (2008). Lower economic status and inferior hygienic environment may protect against celiac disease. *Annals of Medicine, 40*(3), 223–231. https://doi.org/10.1080/07853890701678689
Koskinen, K., Pausan, M. R., Perras, A. K., Beck, M., Bang, C., Mora, M., et al. (2017). First insights into the diverse human Archaeome: Specific detection of Archaea in the gastrointestinal tract, lung, and nose and on skin. *MBio, 8*(6), e00824–e00817. https://doi.org/10.1128/mBio.00824-17
Kramer, J., Fahey, M., Santos, R., Carville, A., Wachtman, L., & Mansfield, K. (2010). Alopecia in rhesus macaques correlates with immunophenotypic alterations in dermal inflammatory infiltrates consistent with hypersensitivity etiology. *Journal of Medical Primatology, 39*(2), 112–122. https://doi.org/10.1111/j.1600-0684.2010.00402.x
Kulig, M., Bergmann, R., Klettke, U., Wahn, V., Tacke, U., & Wahn, U. (1999). Natural course of sensitization to food and inhalant allergens during the first 6 years of life. *The Journal of Allergy and Clinical Immunology, 103*(6), 1173–1179. https://doi.org/10.1016/s0091-6749(99)70195-8
Kung, S. J., Steenhoff, A. P., & Gray, C. (2014). Food allergy in Africa: Myth or reality? *Clinical Reviews in Allergy & Immunology, 46*(3), 241–249. https://doi.org/10.1007/s12016-012-8341-z
Kyburz, A., Urban, S., Altobelli, A., Floess, S., Huehn, J., Cover, T. L., & Müller, A. (2017). Helicobacter pylori and its secreted immunomodulator VacA protect against anaphylaxis in experimental models of food allergy. *Clinical and Experimental Allergy, 47*(10), 1331–1341. https://doi.org/10.1111/cea.12996
Kyburz, A., Fallegger, A., Zhang, X., Altobelli, A., Artola-Boran, M., Borbet, T., et al. (2019). Transmaternal helicobacter pylori exposure reduces allergic airway inflammation in offspring through regulatory T cells. *The Journal of Allergy and Clinical Immunology, 143*(4), 1496–1512.e11. https://doi.org/10.1016/j.jaci.2018.07.046
Laatikainen, T., von Hertzen, L., Koskinen, J. P., Mäkelä, M. J., Jousilahti, P., Kosunen, T. U., et al. (2011). Allergy gap between Finnish and Russian Karelia on increase. *Allergy, 66*(7), 886–892. https://doi.org/10.1111/j.1398-9995.2010.02533.x
Lange, K., Buerger, M., Stallmach, A., & Bruns, T. (2016). Effects of antibiotics on gut microbiota. *Digestive diseases (Basel, Switzerland), 34*(3), 260–268. https://doi.org/10.1159/000443360
Leigh, J. P., & Du, J. (2015). Brief report: Forecasting the economic burden of autism in 2015 and 2025 in the United States. *Journal of Autism and Developmental Disorders, 45*(12), 4135–4139. https://doi.org/10.1007/s10803-015-2521-7
Lerner, A., Jeremias, P., & Matthias, T. (2015). The world incidence and prevalence of autoimmune diseases is increasing. *International Journal of Celiac Disease, 3*(4), 151–155. https://doi.org/10.12691/ijcd-3-4-8
Ley, R. E., Turnbaugh, P. J., Klein, S., & Gordon, J. I. (2006). Microbial ecology: Human gut microbes associated with obesity. *Nature, 444*(7122), 1022–1023. https://doi.org/10.1038/4441022a
Li, Q., Han, Y., Dy, A. B. C., & Hagerman, R. J. (2017). The gut microbiota and autism Spectrum disorders. *Frontiers in Cellular Neuroscience, 11*, 120. https://doi.org/10.3389/fncel.2017.00120

Libertini, G., Corbi, G., Conti, V., Shubernetskaya, O., & Ferrara, N. (2021). Evolutionary gerontology and geriatrics: Why and how we age. In *Advances in studies of aging and health* (Vol. 2). Springer Nature. ISBN 978-3030737740.

Liu, B. N., Liu, X. T., Liang, Z. H., & Wang, J. H. (2021). Gut microbiota in obesity. *World Journal of Gastroenterology, 27*(25), 3837–3850. https://doi.org/10.3748/wjg.v27.i25.3837

Lothstein, K. E., & Gause, W. C. (2021). Mining Helminths for novel therapeutics. *Trends in Molecular Medicine, 27*(4), 345–364. https://doi.org/10.1016/j.molmed.2020.12.010

Lotter, V. (1966). Epidemiology of autistic conditions in young children: 1. Prevalence. *Social Psychiatry, 1*, 124–135. https://doi.org/10.1007/BF00584048

Lunjani, N., Hlela, C., & O'Mahony, L. (2019). Microbiome and skin biology. *Current Opinion in Allergy and Clinical Immunology, 19*(4), 328–333. https://doi.org/10.1097/ACI.0000000000000542

Madore, C., Leyrolle, Q., Lacabanne, C., Benmamar-Badel, A., Joffre, C., Nadjar, A., & Layé, S. (2016). Neuroinflammation in autism: Plausible role of maternal inflammation, dietary omega 3, and microbiota. *Neural Plasticity, 2016*, 3597209. https://doi.org/10.1155/2016/3597209

Madra, M., Ringel, R., & Margolis, K. G. (2021). Gastrointestinal issues and autism Spectrum disorder. *The Psychiatric Clinics of North America, 44*(1), 69–81. https://doi.org/10.1016/j.psc.2020.11.006

Manoli, D. S., & State, M. W. (2021). Autism Spectrum disorder genetics and the search for pathological mechanisms. *The American Journal of Psychiatry, 178*(1), 30–38. https://doi.org/10.1176/appi.ajp.2020.20111608

Marler, S., Ferguson, B. J., Lee, E. B., Peters, B., Williams, K. C., McDonnell, E., et al. (2017). Association of Rigid-Compulsive Behavior with functional constipation in autism Spectrum disorder. *Journal of Autism and Developmental Disorders, 47*(6), 1673–1681. https://doi.org/10.1007/s10803-017-3084-6

Marshall, B. J., & Warren, J. R. (1984). Unidentified curved bacilli in the stomach of patients with gastritis and peptic ulceration. *Lancet, 323*(8390), 1311–1315. https://doi.org/10.1016/s0140-6736(84)91816-6

Martin, C. R., & Mayer, E. A. (2017). Gut-brain Axis and behavior. *Nestlé Nutrition Institute Workshop Series, 88*, 45–53. https://doi.org/10.1159/000461732

Mayer, E. A., Tillisch, K., & Gupta, A. (2015). Gut/brain axis and the microbiota. *The Journal of Clinical Investigation, 125*(3), 926–938. https://doi.org/10.1172/JCI76304

McLoughlin, I. J., Wright, E. M., Tagg, J. R., Jain, R., & Hale, J. D. F. (2022). Skin microbiome-the next frontier for probiotic intervention. *Probiotics and Antimicrobial Proteins, 14*(4), 630–647. https://doi.org/10.1007/s12602-021-09824-1

McPartland, J., & Volkmar, F. R. (2012). Autism and related disorders. *Handbook of Clinical Neurology, 106*, 407–418. https://doi.org/10.1016/B978-0-444-52002-9.00023-1

Mead, J., & Ashwood, P. (2015). Evidence supporting an altered immune response in ASD. *Immunology Letters, 163*(1), 49–55. https://doi.org/10.1016/j.imlet.2014.11.006

Mégraud, F. (2005). A humble bacterium sweeps this year's Nobel prize. *Cell, 123*(6), 975–976. https://doi.org/10.1016/j.cell.2005.11.032

Misra, V., Pandey, R., Misra, S. P., & Dwivedi, M. (2014). Helicobacter pylori and gastric cancer: Indian enigma. *World Journal of Gastroenterology, 20*(6), 1503–1509. https://doi.org/10.3748/wjg.v20.i6.1503

Mitchell, E. (1985). International trends in hospital admission rates for asthma. *Archive of Disease in Childhood, 60*, 376–378. https://doi.org/10.1136/adc.60.4.376

Moissl-Eichinger, C., Pausan, M., Taffner, J., Berg, G., Bang, C., & Schmitz, R. A. (2018). Archaea are interactive components of complex microbiomes. *Trends in Microbiology, 26*(1), 70–85. https://doi.org/10.1016/j.tim.2017.07.004

Moreels, T. G., Nieuwendijk, R. J., De Man, J. G., De Winter, B. Y., Herman, A. G., Van Marck, E. A., & Pelckmans, P. A. (2004). Concurrent infection with *Schistosoma mansoni* attenuates inflammation induced changes in colonic morphology, cytokine levels, and smooth muscle

contractility of trinitrobenzene sulphonic acid induced colitis in rats. *Gut, 53*(1), 99–107. https://doi.org/10.1136/gut.53.1.99
Morgan, J. T., Chana, G., Pardo, C. A., Achim, C., Semendeferi, K., Buckwalter, J., et al. (2010). Microglial activation and increased microglial density observed in the dorsolateral prefrontal cortex in autism. *Biological Psychiatry, 68*(4), 368–376. https://doi.org/10.1016/j.biopsych.2010.05.024
Mvoundza Ndjindji, O., & Djoba Siawaya, J. F. (2022). Mapping allergic diseases in sub-Saharan Africa. *Frontiers in Allergy, 3*, 850291. https://doi.org/10.3389/falgy.2022.850291
Nicolai, T., & von Mutius, E. (1996). Respiratory hypersensitivity and environmental factors: East and West Germany. *Toxicology Letters, 86*(2–3), 105–113. https://doi.org/10.1016/0378-4274(96)03681-8
Nobles, T., & Miller, R. A. (2022). Intertrigo. In *StatPearls*. StatPearls Publishing. PMID: 30285384.
Nur Husna, S. M., Tan, H. T., Md Shukri, N., Mohd Ashari, N. S., & Wong, K. K. (2022). Allergic rhinitis: A clinical and pathophysiological overview. *Frontiers in Medicine (Lausanne), 9*, 874114. https://doi.org/10.3389/fmed.2022.874114
Onkamo, P., Väänänen, S., Karvonen, M., & Tuomilehto, J. (1999). Worldwide increase in incidence of type I diabetes—The analysis of the data on published incidence trends. *Diabetologia, 42*(12), 1395–1403. https://doi.org/10.1007/s001250051309. Erratum in: *Diabetologia*, 2000, 43(5), 685.
Onore, C., Careaga, M., & Ashwood, P. (2011). The role of immune dysfunction in the pathophysiology of autism. *Brain, Behavior, and Immunity, 26*, 383–392. https://doi.org/10.1016/j.bbi.2011.08.007
Palmer, C., Bik, E. M., DiGiulio, D. B., Relman, D. A., & Brown, P. O. (2007). Development of the human infant intestinal microbiota. *PLoS Biology, 5*(7), e177. https://doi.org/10.1371/journal.pbio.0050177
Parasher, G., & Eastwood, G. L. (2000). Smoking and peptic ulcer in the helicobacter pylori era. *European Journal of Gastroenterology & Hepatology, 12*, 843–553. https://doi.org/10.1097/00042737-200012080-00003
Perry, S., de Jong, B. C., Solnick, J. V., de la Luz Sanchez, M., Yang, S., Lin, P. L., et al. (2010). Infection with helicobacter pylori is associated with protection against tuberculosis. *PLoS One, 5*(1), e8804. https://doi.org/10.1371/journal.pone.0008804
Perzanowski, M. S., Ng'ang'a, L. W., Carter, M. C., Odhiambo, J., Ngari, P., Vaughan, J. W., et al. (2002). Atopy, asthma, and antibodies to Ascaris among rural and urban children in Kenya. *The Journal of Pediatrics, 140*, 582–588. https://doi.org/10.1067/mpd.2002.122937
Pinto, D., Ciardiello, T., Franzoni, M., Pasini, F., Giuliani, G., & Rinaldi, F. (2021). Effect of commonly used cosmetic preservatives on skin resident microflora dynamics. *Scientific Reports, 11*, 8695. https://doi.org/10.1038/s41598-021-88072-3
Platts-Mills, T. A. E. (2015). The allergy epidemics: 1870-2010. *The Journal of Allergy and Clinical Immunology, 136*(1), 3–13. https://doi.org/10.1016/j.jaci.2015.03.048
Principi, N., & Esposito, S. (2016). Gut microbiota and central nervous system development. *The Journal of Infection, 73*(6), 536–546. https://doi.org/10.1016/j.jinf.2016.09.010
Probst, A. J., Auerbach, A. K., & Moissl-Eichinger, C. (2013). Archaea on human skin. *PLoS One, 8*(6), e65388. https://doi.org/10.1371/journal.pone.0065388
Pugliatti, M., Sotgiu, S., Solinas, G., Castiglia, P., Pirastru, M. I., Murgia, B., et al. (2001). Multiple sclerosis epidemiology in Sardinia: Evidence for a true increasing risk. *Acta Neurologica Scandinavica, 103*(1), 20–26. https://doi.org/10.1034/j.1600-0404.2001.00207.x
Qiu, S., Lu, Y., Li, Y., Shi, J., Cui, H., Gu, Y., et al. (2020). Prevalence of autism spectrum disorder in Asia: A systematic review and meta-analysis. *Psychiatry Research, 284*, 112679. https://doi.org/10.1016/j.psychres.2019.112679
Ridaura, V. K., Faith, J. J., Rey, F. E., Cheng, J., Duncan, A. E., Kau, A. L., et al. (2013). Gut microbiota from twins discordant for obesity modulate metabolism in mice. *Science, 341*(6150), 1241214. https://doi.org/10.1126/science.1241214

Ronmark, E., Bjerg, A., Perzanowski, M., Platts-Mills, T., & Lundback, B. (2009). Major increase in allergic sensitization in schoolchildren from 1996 to 2006 in northern Sweden. *The Journal of Allergy and Clinical Immunology, 124*, 357–363., 63.E1-15. https://doi.org/10.1016/j.jaci.2009.05.011

Rook, G. A., & Brunet, L. R. (2005). Microbes, immunoregulation, and the gut. *Gut, 54*(3), 317–320. https://doi.org/10.1136/gut.2004.053785

Rook, G. A., Adams, V., Hunt, J., Palmer, R., Martinelli, R., & Brunet, L. R. (2004). Mycobacteria and other environmental organisms as immunomodulators for immunoregulatory disorders. *Springer Seminars in Immunopathology, 25*(3–4), 237–255. https://doi.org/10.1007/s00281-003-0148-9

Rose, N. R. (2016). Prediction and prevention of autoimmune disease in the 21st century: A review and preview. *American Journal of Epidemiology, 183*(5), 403–406. https://doi.org/10.1093/aje/kwv292

Rozas, M., Hart de Ruijter, A., Fabrega, M. J., Zorgani, A., Guell, M., Paetzold, B., & Brillet, F. (2021). From Dysbiosis to healthy skin: Major contributions of Cutibacterium acnes to skin homeostasis. *Microorganisms, 9*(3), 628. https://doi.org/10.3390/microorganisms9030628

Sabin, E. A., Araujo, M. I., Carvalho, E. M., & Pearce, E. J. (1996). Impairment of tetanus toxoid-specific Th1-like immune responses in humans infected with Schistosoma mansoni. *The Journal of Infectious Diseases, 173*, 269–272. https://doi.org/10.1093/infdis/173.1.269

Salas, S. D., Heifetz, R., & Barrett-Connor, E. (1990). Intestinal parasites in central American immigrants in the United States. *Archives of Internal Medicine, 150*, 1514–1516. PMID: 2369250.

Salminen, S., Stahl, B., Vinderola, G., & Szajewska, H. (2020). Infant formula supplemented with biotics: Current knowledge and future perspectives. *Nutrients, 12*(7), 1952. https://doi.org/10.3390/nu12071952

Sanchack, K. E., & Thomas, C. A. (2016). Autism Spectrum disorder: Primary care principles. *American Family Physician, 94*(12), 972–979. PMID: 28075089.

Sanford, J. A., & Gallo, R. L. (2013). Functions of the skin microbiota in health and disease. *Seminars in Immunology, 25*(5), 370–377. https://doi.org/10.1016/j.smim.2013.09.005

Santocchi, E., Guiducci, L., Fulceri, F., Billeci, L., Buzzigoli, E., Apicella, F., et al. (2016). Gut to brain interaction in autism Spectrum disorders: A randomized controlled trial on the role of probiotics on clinical, biochemical and neurophysiological parameters. *BioMed Central Psychiatry, 16*, 183. https://doi.org/10.1186/s12888-016-0887-5

Saxena, R., Mittal, P., Clavaud, C., Dhakan, D. B., Hegde, P., Veeranagaiah, M. M., et al. (2018). Comparison of healthy and dandruff scalp microbiome reveals the role of commensals in scalp health. *Frontiers in Cellular and Infection Microbiology, 8*, 346. https://doi.org/10.3389/fcimb.2018.00346

Schweizer, J. J., von Blomberg, B. M., Bueno-de Mesquita, H. B., & Mearin, M. L. (2004). Coeliac disease in The Netherlands. *Scandinavian Journal of Gastroenterology, 39*(4), 359–364. https://doi.org/10.1080/00365520310008503

Scott, F. I., Horton, D. B., Mamtani, R., Haynes, K., Goldberg, D. S., Lee, D. Y., & Lewis, J. D. (2016). Administration of Antibiotics to children before age 2 years increases risk for childhood obesity. *Gastroenterology, 151*(1), 120–129.e5. https://doi.org/10.1053/j.gastro.2016.03.006

Scrivener, S., Yemaneberhan, H., Zebenigus, M., Tilahun, D., Girma, S., Ali, S., et al. (2001). Independent effects of intestinal parasite infection and domestic allergen exposure on risk of wheeze in Ethiopia: A nested case-control study. *Lancet, 358*, 1493–1499. https://doi.org/10.1016/S0140-6736(01)06579-5

Sfriso, R., Egert, M., Gempeler, M., Voegeli, R., & Campiche, R. (2020). Revealing the secret life of skin - with the microbiome you never walk alone. *International Journal of Cosmetic Science, 42*(2), 116–126. https://doi.org/10.1111/ics.12594

Sharland, M., Paediatric, S. A. C. A. R., & Subgroup. (2007). The use of antibacterials in children: A report of the specialist advisory committee on antimicrobial resistance (SACAR) Paediatric

Subgroup. *The Journal pf Antimicrobic Chemotherapy, 60*(Suppl 1), i15–i26. https://doi.org/10.1093/jac/dkm153

Shi, W., Xu, N., Wang, X., Vallée, I., Liu, M., & Liu, X. (2022). Helminth therapy for immune-mediated inflammatory diseases: Current and future perspectives. *Journal of Inflammation Research, 15*, 475–491. https://doi.org/10.2147/JIR.S348079

Shields, V. E., & Cooper, J. (2022). Use of helminth therapy for management of ulcerative colitis and Crohn's disease: A systematic review. *Parasitology, 149*(2), 145–154. https://doi.org/10.1017/S0031182021001670

Singh, K., & Ghoshal, U. C. (2006). Causal role of helicobacter pylori infection in gastric cancer: An Asian enigma. *World Journal of Gastroenterology, 12*(9), 1346–1351. https://doi.org/10.3748/wjg.v12.i9.1346

Smalley, S. L., Asarnow, R. F., & Spence, M. A. (1988). Autism and genetics. A decade of research. *Archives of General Psychiatry, 45*(10), 953–961. https://doi.org/10.1001/archpsyc.1988.01800340081013

Smith, J. M., Disney, M. E., Williams, J. D., & Goels, Z. A. (1969). Clinical significance of skin reactions to mite extracts in children with asthma. *British Medical Journal, 2*, 723–726. https://doi.org/10.1136/bmj.2.5659.723

Sonnenberg, A. (2010). Differences in the birth-cohort patterns of gastric cancer and peptic ulcer. *Gut, 59*, 736–743. https://doi.org/10.1136/gut.2009.195008

Sonnenberg, A. (2011). Time trends of mortality from gastric cancer in Europe. *Digestive Diseases and Sciences, 56*, 112–1118. https://doi.org/10.1007/s10620-010-1553-2

Sonnenberg, A., & Baron, J. H. (2010). Rising trends of gastric cancer and peptic ulcer in the 19th century. *Alimentary Pharmacology & Therapeutics, 32*(7), 901–907. https://doi.org/10.1111/j.1365-2036.2010.04413.x

Soto-Quiros, M., Avila, L., Platts-Mills, T. A., Hunt, J. F., Erdman, D. D., Carper, H., et al. (2012). High titers of IgE antibody to dust mite allergen and risk for wheezing among asthmatic children infected with rhinovirus. *The Journal of Allergy and Clinical Immunology, 129*(6), 1499–1505.E5. https://doi.org/10.1016/j.jaci.2012.03.040

Srikantha, P., & Mohajeri, H. (2019). The possible role of the microbiota-gut-brain-Axis in autism Spectrum disorder. *International Journal of Molecular Sciences, 20*(9), 2115. https://doi.org/10.3390/ijms20092115

Stevens, W., Addo-Yobo, E., Roper, J., Woodcock, A., James, H., Platts-Mills, T., et al. (2011). Differences in both prevalence and titre of specific immunoglobulin E among children with asthma in affluent and poor communities within a large town in Ghana. *Clinical and Experimental Allergy, 41*, 1587–1594. https://doi.org/10.1111/j.1365-2222.2011.03832.x

Stiemsma, L. T., Reynolds, L. A., Turvey, S. E., & Finlay, B. B. (2015). The hygiene hypothesis: Current perspectives and future therapies. *Immuno Targets and Therapy, 4*, 143–157. https://doi.org/10.2147/ITT.S61528

Strachan, D. P. (1989). Hay fever, hygiene, and household size. *British Medical Journal, 299*(6710), 1259–1260. https://doi.org/10.1136/bmj.299.6710.1259

Strachan, D. P. (2000). Family size, infection and atopy: The first decade of the "hygiene hypothesis". *Thorax, 55*(Suppl 1), S2–S10. https://doi.org/10.1136/thorax.55.suppl_1.s2

Styles, M., Alsharshani, D., Samara, M., Alsharshani, M., Khattab, A., Qoronfleh, M. W., & Al-Dewik, N. I. (2020). Risk factors, diagnosis, prognosis and treatment of autism. *Frontiers in Bioscience (Landmark Edition), 25*(9), 1682–1717. https://doi.org/10.2741/4873

Tetreault, N. A., Hakeem, A. Y., Jiang, S., Williams, B. A., Allman, E., Wold, B. J., & Allman, J. M. (2012). Microglia in the cerebral cortex in autism. *Journal of Autism and Developmental Disorders, 42*(12), 2569–2584. https://doi.org/10.1007/s10803-012-1513-0

Thapar, A., & Rutter, M. (2021). Genetic advances in autism. *Journal of Autism and Developmental Disorders, 51*(12), 4321–4332. https://doi.org/10.1007/s10803-020-04685-z

Thomsen, S. F. (2015). Epidemiology and natural history of atopic diseases. *European Clinical Respiratory Journal, 2*. https://doi.org/10.3402/ecrj.v2.24642

Thuny, F., Richet, H., Casalta, J. P., Angelakis, E., Habib, G., & Raoult, D. (2010). Vancomycin treatment of infective endocarditis is linked with recently acquired obesity. *PLoS One, 5*(2), e9074. https://doi.org/10.1371/journal.pone.0009074

Trowell, H. C. (1960). *Non-infective disease in Africa. The peculiarities of medical non-infective diseases in the indigenous inhabitants of Africa south of the Sahara*. Edward Arnold.

Tsega, E., Choremi, H., Bottazzo, G. F., & Doniach, D. (1980). Prevalence of autoimmune diseases and autoantibodies in Ethiopia. *Tropical and Geographical Medicine, 32*(3), 231–236. PMID: 7193928.

Tseng, C. H., & Wu, C. Y. (2019). The gut microbiome in obesity. *Journal of the Formosan Medical Association, 118*(Suppl 1), S3–S9. https://doi.org/10.1016/j.jfma.2018.07.009

Tsuge, M., Ikeda, M., Matsumoto, N., Yorifuji, T., & Tsukahara, H. (2021). Current insights into atopic march. *Children (Basel, Switzerland), 8*(11), 1067. https://doi.org/10.3390/children8111067

Tuomilehto, J., Karvonen, M., Pitkäniemi, J., Virtala, E., Kohtamäki, K., Toivanen, L., & Tuomilehto-Wolf, E. (1999). Record-high incidence of type I (insulin-dependent) diabetes mellitus in Finnish children. The Finnish childhood type I diabetes registry group. *Diabetologia, 42*(6), 655–660. https://doi.org/10.1007/s001250051212

Urashima, T., Hirabayashi, J., Sato, S., & Kobata, A. (2018). Human Milk oligosaccharides as essential tools for basic and application studies on Galectins. *Trends in Glycoscience and Glycotechnology, 30*, SJ11–SJ24. https://doi.org/10.4052/tigg.1734.1SJ

van den Biggelaar, A. H., van Ree, R., Rodrigues, L. C., Lell, B., Deelder, A. M., Kremsner, P. G., & Yazdanbakhsh, M. (2000). Decreased atopy in children infected with Schistosoma haematobium: A role for parasite-induced interleukin-10. *Lancet, 356*(9243), 1723–1727. https://doi.org/10.1016/S0140-6736(00)03206-2

Vargas, D. L., Nascimbene, C., Krishnan, C., Zimmerman, A. W., & Pardo, C. A. (2005). Neuroglial activation and neuroinflammation in the brain of patients with autism. *Annals of Neurology, 57*(1), 67–81. https://doi.org/10.1002/ana.20315

Vartiainen, E., Petäys, T., Haahtela, T., Jousilahti, P., & Pekkanen, J. (2002). Allergic diseases, skin prick test responses, and IgE levels in North Karelia, Finland, and the Republic of Karelia, Russia. *The Journal of Allergy and Clinical Immunology, 109*(4), 643–648. https://doi.org/10.1067/mai.2002.123307

Velasquez-Manoff, M. (2012). *An epidemic of absence: A new way of understanding allergies and autoimmune diseases*. Simon and Schuster. ISBN 978-1439199381.

Vitkovic, L., Konsman, J. P., Bockaert, J., Dantzer, R., Homburger, V., & Jacque, C. (2000). Cytokine signals propagate through the brain. *Molecular Psychiatry, 5*, 604–615. https://doi.org/10.1038/sj.mp.4000813

von Hertzen, L., Laatikainen, T., Pitkänen, T., Vlasoff, T., Mäkelä, M. J., Vartiainen, E., & Haahtela, T. (2007). Microbial content of drinking water in Finnish and Russian Karelia - implications for atopy prevalence. *Allergy, 62*(3), 288–292. https://doi.org/10.1111/j.1398-9995.2006.01281.x

von Mutius, E., Fritzsch, C., Weiland, S. K., Röll, G., & Magnussen, H. (1992). Prevalence of asthma and allergic disorders among children in united Germany: A descriptive comparison. *British Medical Journal, 305*(6866), 1395–1399. https://doi.org/10.1136/bmj.305.6866.1395

Vuong, H. E., Yano, J. M., Fung, T. C., & Hsiao, E. Y. (2017). The microbiome and host behavior. *Annual Review of Neuroscience, 40*, 21–49. https://doi.org/10.1146/annurev-neuro-072116-031347

Wallen-Russell, C. (2019). The role of every-day cosmetics in altering the skin microbiome: A study using biodiversity. *Cosmetics, 6*, 2. https://doi.org/10.3390/cosmetics6010002

Wang, L., Wang, F. S., & Gershwin, M. E. (2015). Human autoimmune diseases: a comprehensive update. *Journal of Internal Medicine, 278*(4), 369–395. https://doi.org/10.1111/joim.12395

Wang, L., Wang, B., Wu, C., Wang, J., & Sun, M. (2023). Autism Spectrum disorder: Neurodevelopmental risk factors, biological mechanism, and precision therapy. *International Journal of Molecular Sciences, 24*(3), 1819. https://doi.org/10.3390/ijms24031819

Warren, C. M., Jiang, J., & Gupta, R. S. (2020). Epidemiology and burden of food allergy. *Current Allergy and Asthma Reports, 20*(2), 6. https://doi.org/10.1007/s11882-020-0898-7

Weinstock, J. V., & Elliott, D. E. (2014). Helminth infections decrease host susceptibility to immune-mediated diseases. *Journal of Immunology, 193*(7), 3239–3247. https://doi.org/10.4049/jimmunol.1400927

Weinstock, J. V., Summers, R., & Elliott, D. E. (2004). Helminths and harmony. *Gut, 53*(1), 7–9. https://doi.org/10.1136/gut.53.1.7

West, J., Fleming, K. M., Tata, L. J., Card, T. R., & Crooks, C. J. (2014). Incidence and prevalence of celiac disease and dermatitis herpetiformis in the UK over two decades: Population-based study. *The American Journal of Gastroenterology, 109*(5), 757–768. https://doi.org/10.1038/ajg.2014.55

Whary, M. T., Sundina, N., Bravo, L. E., Correa, P., Quinones, F., Caro, F., & Fox, J. G. (2005). Intestinal helminthiasis in Colombian children promotes a Th2 response to helicobacter pylori: Possible implications for gastric carcinogenesis. *Cancer Epidemiology, Biomarkers & Prevention, 14*(6), 1464–1469. https://doi.org/10.1158/1055-9965.EPI-05-0095

Wu, S., Tao, N., German, J. B., Grimm, R., & Lebrilla, C. B. (2010). Development of an annotated library of neutral human milk oligosaccharides. *Journal of Proteome Research, 9*(8), 4138–4151. https://doi.org/10.1021/pr100362f

Wu, S., Grimm, R., German, J. B., & Lebrilla, C. B. (2011). Annotation and structural analysis of sialylated human milk oligosaccharides. *Journal of Proteome Research, 10*(2), 856–868. https://doi.org/10.1021/pr101006u

Zaccone, P., Fehervari, Z., Phillips, J. M., Dunne, D. W., & Cooke, A. (2006). Parasitic worms and inflammatory diseases. *Parasite Immunology, 28*(10), 515–523. https://doi.org/10.1111/j.1365-3024.2006.00879.x

Zeidan, J., Fombonne, E., Scorah, J., Ibrahim, A., Durkin, M. S., Saxena, S., et al. (2022). Global prevalence of autism: A systematic review update. *Autism Research, 15*(5), 778–790. https://doi.org/10.1002/aur.2696

Zhang, H., DiBaise, J. K., Zuccolo, A., Kudrna, D., Braidotti, M., Yu, Y., et al. (2009). Human gut microbiota in obesity and after gastric bypass. *Proceedings of the National Academy of Sciences of the United States of America, 106*(7), 2365–2370. https://doi.org/10.1073/pnas.0812600106

Zhang, Y. J., Li, S., Gan, R. Y., Zhou, T., Xu, D. P., & Li, H. B. (2015). Impacts of gut bacteria on human health and diseases. *International Journal of Molecular Sciences, 16*(4), 7493–7519. https://doi.org/10.3390/ijms16047493

Chapter 5
Diseases Deriving from Relations with Other Living Beings and from Conditions Beyond the Adaptation Range

In this chapter, two different categories of diseases are discussed:

- those deriving from infections, infestations, and other damages caused by other species, highlighting the cases where the role played by alterations of the ecological niche is predominant in the relationship with other species;
- those deriving from conditions beyond the adaptation range of our species.

5.1 Infections, Infestations, and Other Damages Caused by Other Species

Natural selection, facing the challenges posed by an inanimate agent, shapes a species to minimize any harm or even maximize benefits concerning that agent.

In principle, natural selection can arrive at a particular adaptation that nullifies the possible damage of the inanimate agent, disregarding the alterations caused by harmful mutations that compromise the particular adaptation in the affected individuals.

Conversely, in the relationship between two species, natural selection acts on both species, and it is not certain or surely expected that an adaptation is reached for one of the two species which involves the complete cancellation of the damage caused by the other species.

Let us consider the case of a pair of species in which one is a predator of the other, which is the prey of the first:

- If the predator species is too efficient in its predation, the other species tends to become extinct, and this can lead to the extinction of the predator species as well in the absence of alternative food sources for it;

G. Libertini et al., *The Essential Revolution of Evolutionary Medicine*, Advances in Studies of Aging and Health 4, https://doi.org/10.1007/978-3-031-84877-3_5

- If the prey species is too efficient at escaping predation, the other species tends to become extinct. However, in this case, natural selection in favor of the defense mechanisms of the prey becomes less intense, the defense mechanisms become less effective, the rate of predation increases, and the predatory species manages to survive.

In general, in these cases, adult and healthy individuals of the prey species succeed in escaping predation. Therefore, the predator species is clever at prevailing only over the most vulnerable subjects, i.e., subjects in the development phases, or that are sick or weakened for any reason, or elderly and so with a decline of the functions. At the same time, in the predatory species, only individuals in a state of health and not too weakened by aging are able to successfully prey and survive.

This example does not want to debate the complex relationships of the prey-predator species pair, also because often the situation is more complex as there are more predatory species for each prey species and, likewise, more prey species for each predatory species.

The example only wants to show that a fundamental limit of the capacity of natural selection on species A is constituted by natural selection operating on species B, C,... X, which are in relation to or compete with species A.

Limiting to our own species, innumerable relationships, both with the numberless species that are part of the holobiont of which we are the host species and with the numberless other species that are not part of our holobiont, are certainly relevant.

As possible occasions of harm (or of no harm) to individuals of our species, there are some general types of cases:

(A) Species that are part of other holobionts and that fail to take root in our organism either because they are incompatible with our physiology or because they are not specifically adapted. Such events sometimes happen with the help of competing species that already exist in our holobiont, and which contrast their invasion. In these cases, as a rule, we do not suffer damage, i.e., we have no disease as a result.

(B) Species that suddenly (in evolutionary times) pass from another species to ours and which can take root and proliferate, causing diseases.

This occurrence is defined as zoonosis or spillover infection.

Examples of such cases are infections causing AIDS, Chagas' disease, Covid-19, plague, and influenza. However, other conditions are necessary. Important general factors are the population density and the speed of transfer of infected individuals, which allow the infection to spread within a large community and to pass from one community to another. Other factors are decisive (s. the specific Sect. *3.2.8 Infectious epidemics*). In particular:

- for AIDS, after contact with non-human primates that are natural hosts of Human Immunodeficiency Viruses (HIV), particular sexual behaviors that allow the transfer of organic fluids or the use of needles and other materials that allow parenteral infection;

- for Chagas' disease, living in areas where there are species that are natural reservoirs of *Trypanosoma cruzi*;
- for influenza, the existence of intensive breeding of certain animals which become hotbeds of recombination of influenza viruses;
- for the plague, originated from *Yersinia pestis* in animal reservoirs, the spread favored by poor hygienic conditions of the housings and the presence of mice in the houses;
- for COVID-19, there were contacts with bats or other wild animals that are natural hosts of the SARS-CoV-2 virus variants not habitual to humans.

All these cases fall into the category of diseases caused by modifications of the ecological niche, as our species, which in these cases is a victim of zoonoses, has no evolutionary experience of the disease and, therefore, has no specific adaptation.

(C) Zoonoses that originated centuries or millennia ago and for which an adaptation is underway.

This type of event can be considered as the development over time of the previous one. For these cases, the population density and transfer speed of infected individuals are also critical factors.

Examples of these infections are measles (originated from rinderpest virus of cattle between the eleventh and twelfth centuries), smallpox (originated by likely transmission from an animal host, perhaps cattle, in ancient times), malaria (originated from wild animals such as gorillas or chimpanzees) (s. the specific Sect. *3.2.8 Infectious epidemics*). For measles and smallpox, progressive adaptations have increased the ability to resist these diseases. The use of specific vaccines for measles has dramatically slowed the spread of this infection, while in the case of smallpox, it has completely eradicated the virus that caused it. Instead, valid vaccines for malaria are not yet available (Bonam et al., 2021). In the search for an adaptation, natural selection has favored genetic modifications, such as those that cause thalassemia, G6PD deficiency, and sickle cell disease, which allow greater resistance to the disease but also cause other severe diseases in some individuals with specific genetic modifications (Kariuki & Williams, 2020).

These diseases, together with others, when with the discovery of the New World touched populations that for them had no experience, and so no specific adaptation, had disruptive effects causing the estimated loss of 90% of the indigenous American population over a century (Koch et al., 2019) (s. Sect. 3.2.8 *Infectious epidemics*).

(D) Species with which we have lived since ancestral times and which are often a stable part of our holobiont.

Numerous species have been causing infections since ancient times, for which there is an equally ancient adaptation. In general, in healthy subjects, these species cause mild and easily overcome diseases (for example, pharyngitis, diarrhea, benign skin diseases caused by viruses or bacteria that heal spontaneously without problems).

Moreover, each human holobiont is made up of innumerable species that, as a rule, do not cause damage to the host species and indeed often have advantages as they:

- hinder the taking root of other potentially harmful species;
- are integrated into the physiology of the host and, in some cases, are essential for the development of the host (s. *Chap. 4—Diseases deriving from alterations of the ecological niche—Part II: Regarding the holobiont*).

In general, the species that are part of the holobiont, although they do not cause disease (unless there are particular alterations of the ecological niche), they can be an important part of a pathogenetic mechanism in subjects that are weakened due to various causes (genetic alterations, wounds, starvation, aging). For example, a person who is significantly weakened by old age or other causes may develop terminal bronchopneumonia by action of microbes that are harmless to non-weakened persons. In these cases, the primary cause of death is the one that caused the weakening and not the germ by itself (or group of germs) that causes bronchopneumonia.

(E) Cases in which man is the prey or the victim of attacks by individuals of other species.

There are still cases in which a man is prey to other species or, in any case, dies or is injured because of attacks by individuals of other species. It is necessary to exclude from this category the cases in which the animals do not kill or damage humans directly and are only vectors transmitting agents causing a disease that is damaging or even deadly (e.g., mosquitoes that are malaria vectors). We also exclude cases where an animal unintentionally causes an accident resulting in the death of human beings (for example, deer or other animals causing road accidents either by being run over or by sudden maneuvers performed to avoid them).

Table 5.1 reports some approximate and incomplete data, with only indicative value, regarding the deaths caused by attacks of individuals of other species. Snakes, dogs, and scorpions represent a danger that should not be disregarded, while the number of deaths caused by various wild species appears very limited.

In the 2020 World Cause of Death list (World Cause of Death, 2020), this type of death is not reported as a separate item. However, in this list, considering only the deaths caused by snakes, dogs, and scorpions and accepting the estimated number of deaths (78,250–135,000), this cause of death would be between 50th and 57th place.

The cases of diseases or deaths caused by violence or war are not considered in the category *Infections, infestations, and other damages caused by other species* because the cause is by individuals of the same species, and so are included in the category *Diseases deriving from conditions beyond adaptation range* (see Sect. 5.2).

Table 5.1 Deaths caused by attacks of individuals of other species

Specie	CNET (Learish, 2016)	Business Insider (Ramsey, 2016)	Bill Gates (Gates, 2014)
Snakes	50,000	100,000	50,000
Dogs (1)	25,000	35,000	25,000
Scorpions	3250		
Crocodiles	1000	1000	1000
Hippopotamuses	500	500	500
Elephants	500	500	100
Lions	250	22+	100
African cape buffalo	200		
Bees	53		
Jellyfish	40		
Ants	30		
Leopards	29		
Horses	20		
Wolves	10	10	10
Sharks	5	6	10
Alligators	1		

Note 1: In many cases, the attacking dog is affected by rabies, and this often causes the death of the attacked subject

5.2 Diseases Deriving from Conditions Beyond the Adaptation Range

The capability of natural selection in developing sophisticated characteristics of living organisms challenges the imagination and planning abilities of the most daring minds.

The incredible complexity and functionality of organs such as the human brain, the specialized organ that allows echolocation in bats, and the flight apparatus in birds are just some of the countless examples that show the extraordinary power of evolution shaped by the engine of natural selection.

However, for each species, evolution has obligatory limits in the range of adaptation.

Some limits derive from the specialization of the species and not from an impossibility of evolution to overcome them. Suppose a species is beautifully modeled for one type of capability. In this case, it may be completely unsuitable for other capabilities, just as a Formula One racing car is no match for a speedboat racing and vice versa. Specialization in one function often results in little or no capacity in other functions. Therefore, if individuals of a species adapted to a certain function are faced with situations outside their range of adaptation, it is easy for them to be damaged, resulting in disease or death. For example, the horse is splendid in its running abilities, but it is totally incapable of flying and, falling from even a small height, risks being seriously injured or killed.

Then, there are conditions to which no living species is adapted or can resist. For example, the intense heat of a fire or the effects of a lightning strike are conditions for which it appears impossible for an organism to resist.

Aside from these extreme conditions, each species has endurance limits, dictated by its adaptation, beyond which serious injury or death occurs.

E.g., for humans, there are limits of resistance to traumas, burns, wounds, poisonous or harmful substances, anoxia, cold, and so on, beyond which damage or death occurs.

Table 5.2 shows the number of deaths worldwide in 2020 attributed by the World Health Organization (WHO) to various causes (road traffic accidents, falls, other injuries, violence, drownings, fires, poisonings, and war), which may fall within the definition of "conditions beyond adaptation range".

However, it must be considered that for some of these deaths, the primary cause is different:

(1) If a subject, following the consumption of alcohol or drugs, has a reduced reactivity and awareness of his own actions and consequently is injured or dies for one of the aforementioned events, the primary cause should be considered the abuse of a substance that alters his psychophysical capacities, which is undoubtedly an alteration of the ecological niche. In particular:

  – "Alcohol-attributable injury accounts for around one-tenth of the total impact of alcohol on health (9.9% and 12.6% in low- and high-income countries, respectively) (Institute of Health Metrics and Evaluation, 2021). Males (90%) and young people aged 15–39 years (40%) dominate alcohol-attributable injury deaths (Institute of Health Metrics and Evaluation, 2021). Impacts on health systems are considerable, with alcohol contributing to between 5% and 40% of all emergency department (ED) injury presentations across 27 countries (Ye et al., 2018)." (Chikritzhs & Livingston, 2021);

Table 5.2 World Cause of Death (2020)

Rank	Cause of death	No. of deaths	% of total no. of deaths
11	Road traffic accidents	1,282,150	2.56
18	Falls	684,277	1.37
22	Other injuries	631,108	1.26
25	Violence	474,691	0.95
39	Drownings	235,642	0.47
52	Fires	114,144	0.23
56	Poisonings	84,278	0.17
58	War	72,457	0.14
	Total:	3,597,690	7.19
	Death from other causes:	46,452,393	92.81
	Total no. of deaths:	50,050,083	100.00

- "Alcohol-impaired driving crashes account for approximately one third of all crash fatalities in the United States.... An estimated 4.2 million adults reported at least one alcohol-impaired driving episode in the preceding 30 days, resulting in an estimated 121 million episodes... annually" (Jewett et al., 2015);
- There is a relationship between road traffic accidents with consequent injuries and the use of various substances that alter psychophysical functions (including marijuana, cocaine, various opiates—also on medical prescription such as painkillers -, amphetamines, PCP). However, it appears difficult to quantify the entity of the damage caused (Naumann et al., 2020).

(2) If an elderly subject, due to his reduced psychomotor skills, dies from one of the aforementioned "conditions beyond adaptation range", in cases in which a non-elderly subject without the impairments would not have incurred the same conditions or would have overcome them without damage, aging should be considered as the primary cause. For example, for an elderly person who slips, breaks a femur, and then dies of complications, the series of events that leads to illness (fracture of the femur) and then to death has as its primary cause the vulnerability of old age, considering that for a young it is probable that this would not have occurred. In particular:

- "Indeed, more than one out of four Americans over the age of 65 experience a fall annually (Bergen et al., 2016). According to Centers for Disease Control (CDC) data, this ultimately results to over 30 million falls, seven million injuries, three million emergency room visits, 800,000 hospitalizations, and 38,000 deaths each year (Bergen et al., 2016; Burns & Kakara, 2018; CDC, Injury Prevention and Control, 2019)." (Luebbert et al., 2022).
- "Almost 28% of all traumas that result in mortalities happen among people over 65 years old. Injuries are ranked as fifth reason of mortalities in this age group (Siram et al., 2011; Stevenson, 2004). Road Traffic Injuries (RTIs) bear high rate of mortalities and injuries for all age groups... One of the most important age groups in this regard are old people (Abou-Raya & ElMeguid, 2009; Etehad et al., 2015; Kaimila et al., 2013; Orriols et al., 2013; Sadeghi-Bazargani et al., 2018).... Due to musculoskeletal problems and slow activities and reactions the elderly people cannot save themselves when a car accident happens. On the other hand, due to chronic diseases like osteoporosis the rate of bone fracture and hospitalization increases which itself augments the risk of mortalities among elderly people..." (Azami-Aghdash et al., 2018).

However, the consequences of aging as a cause of mortality will be discussed in *Chap. 8—Physiological phenomena that cause troubles and suffering or even death—Part III: Slow phenoptosis or aging.*

As regards the prevention and treatment of this category of diseases, everything seems similar to what is already commonly accepted and practiced in current medicine.

Prevention must be achieved by avoiding or limiting in every possible and feasible way the conditions that cause the diseases mentioned earlier and the resulting deaths. In cases where such cases nevertheless occur, it is certainly necessary to practice all available treatments in the best possible way.

However, it must always be remembered that: (i) it is not possible to remedy fatal cases, and (ii) treatments that fully restore previous health conditions are often impossible, which is a rational stimulus to seek prevention as much as possible.

References

Abou-Raya, S., & ElMeguid, L. A. (2009). Road traffic accidents and the elderly. *Geriatrics & Gerontology International, 9*(3), 290–297. https://doi.org/10.1111/j.1447-0594.2009.00535.x

Azami-Aghdash, S., Aghaei, M. H., & Sadeghi-Bazarghani, H. (2018). Epidemiology of road traffic injuries among elderly people; a systematic review and meta-analysis. *Bulletin of Emergency and Trauma, 6*(4), 279–291. https://doi.org/10.29252/beat-060403

Bergen, G., Stevens, M. R., & Burns, E. R. (2016). Falls and fall injuries among adults aged ≥65 years - United States, 2014. *MMWR. Morbidity and Mortality Weekly Report, 65*(37), 993–998. https://doi.org/10.15585/mmwr.mm6537a2

Bonam, S. R., Rénia, L., Tadepalli, G., Bayry, J., & Kumar, H. M. S. (2021). Plasmodium falciparum Malaria Vaccines and Vaccine Adjuvants. *Vaccines (Basel), 9*(10), 1072. https://doi.org/10.3390/vaccines9101072

Burns, E., & Kakara, R. (2018). Deaths from falls among persons aged ≥65 years - United States, 2007-2016. *MMWR. Morbidity and Mortality Weekly Report, 67*(18), 509–514. https://doi.org/10.15585/mmwr.mm6718a1

CDC (Centers for Diseases Control and Prevention), Injury Prevention and Control. (2019). *WISQARS (Web-based Injury Statistics Query and Reporting System)*. Accessed on 07.04.22. https://www.cdc.gov/injury/wisqars/index.html

Chikritzhs, T., & Livingston, M. (2021). Alcohol and the risk of injury. *Nutrients, 13*(8), 2777. https://doi.org/10.3390/nu13082777

Etehad, H., Yousefzadeh-Chabok, S., Davoudi-Kiakalaye, A., Moghadam Dehnadi, A., Hemati, H., & Mohtasham-Amiri, Z. (2015). Impact of road traffic accidents on the elderly. *Archives of Gerontology and Geriatrics, 61*(3), 489–493. https://doi.org/10.1016/j.archger.2015.08.008

Gates, B. (2014). *Gates Notes*. https://www.gatesnotes.com/Health/Most-Lethal-Animal-Mosquito-Week. Retrieved April 7, 2023.

Institute of Health Metrics and Evaluation. (2021). *Global Health data exchange (GHDx)*. University of Washington.

Jewett, A., Shults, R. A., Banerjee, T., & Bergen, G. (2015). Alcohol-impaired driving among adults - United States, 2012. *MMWR. Morbidity and Mortality Weekly Report, 64*(30), 814–817. https://doi.org/10.15585/mmwr.mm6430a2

Kaimila, B., Yamashina, H., Arai, A., & Tamashiro, H. (2013). Road traffic crashes and fatalities in Japan 2000-2010 with special reference to the elderly road user. *Traffic Injury Prevention, 14*(8), 777–781. https://doi.org/10.1080/15389588.2013.774085

Kariuki, S. N., & Williams, T. N. (2020). Human genetics and malaria resistance. *Human Genetics, 139*(6–7), 801–811. https://doi.org/10.1007/s00439-020-02142-6

Koch, A., Brierley, C., Maslin, M. M., & Lewis, S. L. (2019). Earth system impacts of the European arrival and great dying in the Americas after 1492. *Quaternary Science Reviews, 207*, 13–36. https://doi.org/10.1016/j.quascirev.2018.12.004

Learish, J. (2016). *The 24 deadliest animals on Earth, ranked. CNET*. Archived from the original on March 24, 2019. https://www.cnet.com/pictures/the-24-deadliest-animals-on-earth-ranked/24/. Retrieved April 70, 2023.

Luebbert, S., Christensen, W., Finkel, C., & Worsowicz, G. (2022). Falls in senior adults: Demographics, cost, risk stratification, and evaluation. *Missouri Medicine, 119*(2), 158–163. PMID: 36036034.

Naumann, R. B., Shiue, K., Hezaveh, A. M., Marshall, S. W., & Cherry, C. R. (2020). Connections between opioids and road injury: Linkage of prescription monitoring and crash databases. *American Journal of Preventive Medicine, 58*(3), 461–466. https://doi.org/10.1016/j.amepre.2019.10.006

Orriols, L., Wilchesky, M., Lagarde, E., & Suissa, S. (2013). Prescription of antidepressants and the risk of road traffic crash in the elderly: A case-crossover study. *British Journal of Clinical Pharmacology, 76*(5), 810–815. https://doi.org/10.1111/bcp.12090

Ramsey, L. (2016). These are the world's deadliest animals. *Business Insider*. Archived from the original on June 3, 2021. https://www.businessinsider.com/worlds-deadliest-animals-2016-9?r=US&IR=T. Retrieved April 7, 2023.

Sadeghi-Bazargani, H., Samadirad, B., & Moslemi, F. (2018). A decade of road traffic fatalities among the elderly in north-West Iran. *BioMed Central Public Health, 18*(1), 111. https://doi.org/10.1186/s12889-017-4976-2

Siram, S. M., Sonaike, V., Bolorunduro, O. B., Greene, W. R., Gerald, S. Z., et al. (2011). Does the pattern of injury in elderly pedestrian trauma mirror that of the younger pedestrian? *The Journal of Surgery Research, 167*(1), 14–18. https://doi.org/10.1016/j.jss.2010.10.007

Stevenson, J. (2004). When the trauma patient is elderly. *Journal of Perianesthesia Nursing, 19*(6), 392–400. https://doi.org/10.1016/j.jopan.2004.08.010

World Cause of Death. (2020). *(Data from WHO)*, www.worldlifeexpectancy.com/world-rankings-total-deaths. Accessed 15.12.22.

Ye, Y., Shield, K., Cherpitel, C. J., Manthey, J., Korcha, R., & Rehm, J. (2018). Estimating alcohol-attributable fractions for injuries based on data from emergency department and observational studies: A comparison of two methods. *Addiction, 114*, 462–470. https://doi.org/10.1111/add.14477

Chapter 6
Physiological Phenomena that Cause Troubles and Suffering or Even Death: Part I: Non-phenoptotic Phenomena

A general definition of disease could be any phenomenon causing trouble, suffering of any kind, and even death.

However, a number of phenomena would fall within this definition while they are certainly useful, i.e., adaptive, physiological phenomena, and so cannot be considered disease.

Such phenomena are divided into two groups in this book.

The first group, which is the subject of the exposition of this chapter, includes all phenomena not falling under the definition of "phenoptosis".

The phenomena of the second group belong to those defined as "phenoptosis" and will be discussed in the following *Chap.* 7 *-Physiological phenomena that cause troubles and suffering or even death—Part II: Phenoptotic phenomena different from slow phenoptosis*, and *Chap.* 8*—Physiological phenomena that cause troubles and suffering or even death—Part III: Slow phenoptosis or aging.*

6.1 Defenses against Infections, Toxic Substances, and Other Noxious Factors

The organism is shaped by natural selection to defend itself against any inanimate or animate agent or event that harms or may harm it, thus causing disease. Some defense mechanisms can cause trouble or pain. It is important to distinguish between diseases and these defense mechanisms, which may erroneously be considered and treated as a disease.

This distinction is beautifully outlined in the book by Randolph Nesse and George Williams, which is regarded as a meaningful step in the birth of Evolutionary Medicine (Nesse & Williams, 1994): "A fair skinned person with severe pneumonia may take on a dusky hue and have a deep cough. These two signs of pneumonia

G. Libertini et al., *The Essential Revolution of Evolutionary Medicine*, Advances in Studies of Aging and Health 4, https://doi.org/10.1007/978-3-031-84877-3_6

represent entirely different categories, one a manifestation of a defect, the other a defence. The skin is blue because hemoglobin is darker in color when it lacks oxygen.... A cough, on the other hand, is a defence. It results from a complex mechanism designed specifically to expel foreign material in the respiratory tract.... The distinction between defenses and defects is not merely of academic interest. For someone who is sick it can be crucial. Correcting a defect is almost always a good thing.... But eliminating a defense by blocking it can be catastrophic. Cut the wire that indicates a low fuel supply, and you are more likely to run out of gas. Block your cough excessively, and you may die of pneumonia." (Nesse & Williams, 1994, pp. 8–9).

Now, let us consider phenomena that are not diseases but defenses against a harmful agent or event:

1. **Cough**

The cough is a protective reflex characterized by a sudden expulsion of air through the mouth to eliminate fluids, mucus, or even foreign bodies from the upper airways (Sharma et al., 2023).

A cough frequency of about 18.6/day has been found in healthy subjects (Yousaf et al., 2013). Therefore, cough is a usual physiological means of keeping the upper airways clear in healthy subjects. Moreover, coughing is vital to avoid suffocation or lung infections when something liquid or solid penetrates or threatens to penetrate the larynx or trachea. The crossing between the upper respiratory tract and the upper digestive tract is a critical anatomical feature, particularly for infants in whom neuromotor coordination and discriminating abilities are still developing: "Accidental aspiration of an element into airways is a widespread clinical scenario among children under 3 years, predominantly males. Moreover, it is the leading cause of infantile deaths and the fourth one among preschool children.... 80% of cases are below 3 years, of which the peak frequency occurs in 1-2 years age group." (Salih et al., 2016).

Elderly subjects are another age group at risk due to the decline of neuromotor coordination and discriminating abilities, i.e., for causes mirroring those in very young subjects. A study showed that, in 2014 and in Japan, out of a total number of accidental deaths equal to 39,029, those caused by airway obstruction for the ingestion of foreign bodies (9806, 25.1%) exceeded those due to traffic accidents (5717, 14.6%) and those associated with domestic drowning or immersion in bathtub (5362, 13.7%), the last two types of causes mainly in the elderly population (Hanada, 2016).

However, it is presumable that many possible cases of suffocation due to the penetration of foreign bodies into the airways are resolved before any medical intervention by timely coughing which manages to expel the foreign body.

Cough is a common symptom of respiratory diseases, particularly infections of the respiratory tract and lungs, and other diseases (e.g., gastroesophageal reflux disease, cancer, obstructive sleep apnea, congestive heart failure). Although cough is a symptom and not a disease, it has possible complications as, in vulnerable subjects, severe bouts of cough can cause syncope, headache, rib fracture, vomiting, sweating, and urinary incontinence (Sharma et al., 2023).

Since cough, in addition to clearing the respiratory tract, also spreads microbes in the environment when there is a respiratory tract infection, it has been proposed that somehow the microbes manipulate the physiology of infected subjects by increasing the cough (Nesse & Williams, 1994). A possible case in support of this hypothesis is pertussis, or whooping cough, where the bacterium *Bordetella pertussis*, the pathogenic agent of the disease, produces the pertussis toxin, which causes hypersensitivity to histamine and is the likely cause of the severe and prolonged cough that characterizes the disease (Locht & Antoine, 2021).

2. **Sneeze**

Sneeze (or sternutation) is a reflex of sudden expulsion of air from the lungs through the nose and the mouth. A sneeze tries to free the nasal and oral cavities from mucus or other substances irritating the mucous membrane. Sneezes are also important because they try to expel from the nasal and oral cavities any foreign bodies or even insects that have penetrated or are about to penetrate them. Sneeze is a protective reflex also triggered by dust, powders, air pollution, and dry air (Songu & Önerci, 2023).

In general, a sneeze is more violent than a cough and is more effective in spreading pathogens by air: "… sneezing represents the most forceful and common mechanism to spread infectious droplets during viral respiratory infections. Studies indicate that a single sneeze can create 40,000 virus-containing droplets that reach a radius of 7 to 8 meters and suspend in the air for up to 10 min … In contrast, a cough produces 3,000 droplets at most, which is about the same number as talking continuously for 5 min …" (Li et al., 2021).

Sneezing is a common symptom of upper respiratory tract infections. Like coughing, it has been hypothesized that may be a "host manipulation" of the infectious agent to allow its diffusion into the environment: "By subtle chemical influence a parasite may gain some control over the machinery of the host's body and cause that machinery to serve the interests of parasite than host.... Perhaps the most important human examples of manipulation by pathogens are the sneezing, coughing, vomiting, and diarrhea triggered by bacteria and viruses." (Nesse & Williams, 1994, pp. 46–47).

Sneezing can also be a symptom of a disease such as allergic rhinitis (Small et al., 2018).

Furthermore, sneezes can also be triggered by the so-called photic sneeze reflex, or ACHOO (Autosomal Dominant Compelling Helio-Ophthalmic Outburst) syndrome, which is shown by about 25% of the population as an autosomal dominant character (Dean, 2012; Bobba et al., 2019): "About one in four individuals who already have a prickling sensation in their nose will sneeze in response to sunlight, but 'pure' photic sneezing is far less common (Breitenbach et al., 1993)." (Dean, 2012) The high frequency of the trait should be accounted for by some unknown advantage of sneezing under reflex-triggering conditions.

Considering vulnerable individuals, "sneezing has also been reported to result in rib fractures in osteoporotic patients, spontaneous abortion, ruptured intracranial aneurysm, aortic dissection, intervertebral disc prolapse leading to quadriplegia, and death" (Bobba et al., 2019).

3. **Pain**

Pain is a set of unpleasant sensations, which can be of various types and intensities and trigger reflexes and complex reactions to avoid or limit it. Physiological pain is defined as nociceptive pain and is caused by the perception of actual or potential tissue damage (e.g., excessive heat or cold, crushing, tearing, shearing of part of the body, effect of chemical agents). Suppose the pain arises acutely due to the sudden action of an agent. In that case, it immediately activates reflexes and actions that try to avoid the cause of the damage (for example, hand contact with too hot water triggers the retraction of the part). When the pain is chronic, it can be a signal to induce rest of the part to avoid further damage (for example, pain caused by a muscle strain or tear induces to avoid loads for the affected muscle because this would worsen the damage) (Steeds, 2009; Sneddon, 2018).

Pain is unpleasant, but it is a feeling essential to avoid or limit damage from countless causes. In fact, in the rare cases caused by various genetic defects in which there is congenital insensitivity to pain and nociception, the affected subjects undergo various damages that progressively alter their health conditions and reduce their survival capabilities (Weisman et al., 2019; Schon et al., 2018).

Leprosy and diabetes are two diseases that can cause insensitivity to pain and nociception, and even in these cases, there is progressive damage to the insensitive parts, confirming the physiological importance of nociceptive pain (Boulton, 2012).

Quite different is the case in which the pain is a consequence of diseases:

(i) a symptom of countless diseases, e.g., cancer (Scarborough & Smith, 2018);
(ii) neuropathic pain due to diseases affecting parts of the somatosensory system (Treede et al., 2008);
(iii) somatoform pain (psychogenic pain or psychalgia) due to mental, emotional, or behavioral factors (Landa et al., 2012; Egloff et al., 2014);
(iv) or disease per se, as impaired or exaggerated pain perception without tissue damage that could account for the pain, i.e., nociplastic pain (Trouvin & Perrot, 2019; Nijs et al., 2021).

4. **Itch**

Itch (or pruritus) is a sensation that causes the desire or reflex to scratch. There is an overlap of receptor systems between pain and itch. However, opioid analgesics stimulate itch and relieve pain, while opioid antagonists show antipruritic efficacy. Moreover, pain caused by scratching soothes the itch (Ikoma et al., 2006; Cevikbas & Lerner, 2020).

The fundamental difference between itch and pain is in their functional significance. Pain stimulates a powerful and immediate withdrawal reflex and aims to avoid a cause of damage. On the other hand, itch stimulates a scratch reflex, which is not as immediate or powerful as that caused by pain, and aims to remove sources of irritation and damage from the skin, such as arthropods, or to calm their effects. The itching sensation, which often also includes burning, stinging, and tingling sensations, stimulates scratching, which in turn causes inflammation, vasodilation, and edema. In addition, "scratching damages the epidermal barrier and facilitates an appropriate physiological cascade associated with sensing the insult" (Cevikbas & Lerner, 2020).

All of this indicates that itching is an essential component of an important defensive mechanism. Otherwise, there are numerous cases in which itching is one of the symptoms of a disease and has no adaptive significance (e.g., atopic dermatitis, bullous pemphigoid, cholestasis, cutaneous lupus erythematosus, cutaneous T-cell lymphoma, dermatitis herpetiformis, dermatomyositis, epidermolysis bullosa pruriginosa, lichen planus, neuropathic pruritus, nonmelanoma skin cancer, pityriasis rubra pilaris, porphyria cutanea tarda, psoriasis, psychogenic pruritus, sarcoidosis, uremia, urticaria) (Roh et al., 2022).

5. **Nausea** and **vomiting**

Vomiting (or emesis) is the involuntary, forceful expulsion through the mouth of the contents of the stomach and sometimes of the first tract of the intestine. The expulsion may be so strong that it also occurs through the nose. Before vomiting, there is nausea, which, however, does not always lead to vomiting and retching, "where the content of gastrointestinal tract is forced into the esophagus, without expulsion of the vomitus" (Zhong et al., 2021). In general, vomiting is an organism's defense against foods or substances introduced into the stomach and interpreted as harmful by the physiological mechanisms of the organism. For example, vomiting can be triggered by food poisoning, overeating, alcohol, or indigestion of food (Zhong et al., 2021).

Vomiting is also the symptom of many diseases and particular conditions (e.g., gastritis, cholecystitis, pancreatitis, appendicitis, hepatitis, food allergies, pyloric stenosis, bowel obstruction, acute abdomen and/or peritonitis, cerebral concussion, cerebral hemorrhage, migraine, brain tumors, intracranial hypertension, hydrocephalus, hypercalcemia, uremia, bulimia nervosa, overexposure to ionizing radiation, vertigo, Ménière's disease, celiac disease, opioids, chemotherapy drugs, altitude sickness, motion sickness, intense anxiety) (Goldman & Schafer, 2019; Loscalzo et al., 2022).

Vomiting as a symptom does not appear to have any adaptive value, so it is useful and opportune to use drugs to limit the activation of this reflex. On the contrary, vomiting as a defense has an important adaptive value, and therefore, in such cases, blocking the vomiting reflex does not appear rational and can be harmful.

About nausea and vomiting in pregnancy, see below in *§6.2 Troubles related to pregnancy.*

6. **Fever**

Fever (or pyrexia) is the condition of a temperature above the normal range determined by an upregulation of the body's temperature. Fever is present in many infections and represents a general defense of the organism in case of attack by pathogenic organisms (Kluger et al., 1996; Harden et al., 2015).

However, fever also occurs in cases of inflammation due to tissue damage (for example, a wound, a bone, muscle, or cartilage lesion) or other irritants (Harden et al., 2015).

A feverish state is generally accompanied by the so-called sickness behavior, which is characterized by "anorexia, fatigue, loss of interest in usual daily activities, social withdrawal, listlessness or malaise, hyperalgesia, sleep disturbances and cognitive dysfunction" (Harden et al., 2015), and malaise, depression, anxiety (Kelley et al., 2003).

Regarding the usefulness of fever in case of infection: "The induction of fever in endothermic (warm-blooded) animals occurs at a high metabolic cost such that a 1°C rise in body temperature requires a 10-12.5% increase in metabolic rate (Kluger, 1979). There is mounting evidence that the increase of 1 to 4°C in core body temperature that occurs during fever is associated with improved survival and resolution of many infections. For example, the use of antipyretic drugs to diminish fever correlates with a 5% increase in mortality in human populations infected with influenza virus and negatively affects patient outcomes in the intensive care unit (Ryan & Levy, 2003; Schulman et al., 2005; Earn et al., 2014)." (Evans et al., 2015).

The increase in metabolic activity allows for optimal activation of the body's defenses. Furthermore, the sickness behavior reduces the other activities of the body by concentrating all the resources for the needs required by the state of inflammation. Overall, fever and related manifestations appear as complex reactions having the adaptive meaning of an optimal response to the pathological condition. Therefore, in these cases, treating the fever as something to be fought at all costs does not appear rational or useful except in cases where the fever is deemed excessive and causes too much discomfort.

On the contrary, cases in which fever is a symptom of other diseases—e.g., cancer in the absence of a concomitant infection (Pasikhova et al., 2017)—and does not appear to have an adaptive value, are different and, if this is considered useful, need treatment.

A different condition is that of hyperthermia (or overheating), which is neither a symptom nor a synonym of fever but a dangerous increase in body temperature caused by excessive exposure to the sun or conditions of excessive environmental temperature and humidity (sunstroke, heat stroke) (Leon & Bouchama, 2015). This pathological condition must be treated immediately as potentially lethal (Gaudio & Grissom, 2016).

7. **Diarrhea**

Diarrhea (or diarrhoea) is an abnormal frequent emission of liquid or loose or semisolid feces. The most common causes of diarrhea are intestinal infections due to viruses (Bányai et al., 2018), bacteria (Fleckenstein et al., 2021), or parasites (de Truchis & de Truchis, 2007).

Acute diarrhea can be triggered by foodborne bacteria, intake of medicines (e.g., antibiotics, laxative abuse, sorbitol, colchicine, cardiac antidysrhythmic, nonsteroidal anti-inflammatory drugs, chemotherapeutics, antacids), and ingestion of toxic substances (e.g., organophosphates, poisonous mushrooms, arsenic, ciguatera or scombroid) (Graves, 2013).

Non-infectious causes of chronic diarrhea are lactose intolerance, celiac disease, inflammatory bowel disease, chronic pancreatitis, appendicitis, carcinoid tumor, glucose-galactose malabsorption, intestinal enterokinase deficiency, intussusception, Crohn disease, hyperthyroidism, pediatric malabsorption syndromes (Nemeth & Pfleghaar, 2022).

In cases of infection and intake of harmful food or substances, diarrhea appears as a defense mechanism by which the body tries to eliminate harmful microbes or inanimate agents.

So, it is not always rational to block or limit this mechanism, clearly except in cases when diarrhea is excessive and/or represents a danger. It is important to emphasize that in 2020, among the world causes of death, diarrheal diseases ranged seventh with 1,519,229 deaths (World Cause of Death, 2020). Therefore, when diarrhea is excessive, in particular because of the dehydration it causes and especially in children and subjects already vulnerable, adequate treatment (rehydration mainly) is an indispensable priority.

When diarrhea is caused by an infection, diarrhea disperses the infectious agent in the environment and can be another case of host manipulation (s. above the subsections *1) Cough* and *2) Sneeze*). The most evident case is the infection from *Vibrio cholerae*, in which the bacterium, through a particular substance (cholera toxin), stimulates violent aqueous diarrhea, which is often deadly for the guest while being very useful for its dispersion (Hsueh & Waters, 2019).

On the contrary, a 1973 work showed that, for *Shigella* infection in volunteers, the treatment with a drug that blocked diarrhea (Lomotil) determined a doubling of the times with feverish subjects and in a state of malaise (DuPont & Hornick, 1973).

8. **Lacrimation**

The term lacrimation refers to tear secretion when it is increased.

In fact, in the absence of stimuli or events, there is a basal secretion of tears, which serves to keep the cornea moist, lubricated, and free from dust. Moreover, some of the substances in tears (such as lysozyme and lactoferrin) hinder cornea and conjunctiva bacterial infections. Besides, other substances suppress inflammation, maintain innervation, or participate in epithelial or stromal wound healing (Pflugfelder & Stern, 2020a). Furthermore, tears maintain a lacrimal film on the corneal surface that is essential for the refractive role of the cornea (Murube, 2009).

When the nerve endings on the corneal surface are stimulated in any way, even by slight potential threats to the integrity of the cornea and the tear film that covers it (foreign bodies, lesions of any kind, gaseous or liquid irritants), the reflex lacrimation is immediately activated (Bylsma et al., 2019).

Reflex lacrimation is a complex phenomenon regulated by the interaction of multiple nervous centers (Dartt, 2009). In any case, the abundant tears try to remove the foreign bodies or the irritating substances that have met the surface of the eye. Surely, the reflex lacrimation has an adaptive function and must not be inhibited.

When tears are triggered by sensations or emotions (physical pain, emotional stress, anger, suffering, mourning, but also pleasure, intense humor, and laughter), they are described as crying or weeping or as emotional tears (Bylsma et al., 2019). "Whereas producing emotional tears appears to be a uniquely human behavior, human infants share the production of distress calls with the young of most other mammals and birds, which are typically displayed when they are separated from their mothers (Gračanin et al., 2018).... in animals crying is limited to vocal forms of expression (i.e., distress vocalizations, separation calls), in humans the shedding of visible tears is an essential additional feature ..." (Bylsma et al., 2019).

The importance of tears is demonstrated by:

- familial dysautonomia, an autosomal recessive disease, in which there is impaired secretion of basal, reflex, and emotional tears resulting in a neurotrophic keratopathy and so corneal opacity that cause visual loss in 27% of patients (Mendoza-Santiesteban et al., 2017).
- a different condition characterized by insufficient lacrimation and alterations of tear composition and termed dry eye, with about 10 million cases in the US alone in 2017 (Farrand et al., 2017). It causes suffering of the corneal epithelium, accompanied by inflammation of the cornea, pain, and impaired visiotn, and is defined as keratoconjunctivitis sicca (Bron et al., 2017; Pflugfelder & Stern, 2020b).

6.2 Troubles Related to Pregnancy

Pregnancy involves some forms of ailments and troubles that are physiological and related to such a state. However, these ailments or troubles are often cured as diseases.

1. **Morning sickness**

Morning sickness, or nausea and vomiting during pregnancy, is present in most pregnant women between the fourth and 16th week of pregnancy (Festin, 2009). It is triggered by the smell or taste, or even the mere sight, of foods such as strong-tasting vegetables, meats, fish, poultry, eggs, alcohol, caffeine, and, in general, foods with peculiar taste or smell (Profet, 1992).

According to a non-evolutionary interpretation, morning sickness could be due to the rise in human chorionic gonadotrophin concentration during pregnancy, and therefore, appropriate medicines should be used to limit the symptoms (e.g., antihistamines, ginger, vitamin B_6, domperidone, metoclopramide, phenothiazines) (Festin, 2009).

A completely different interpretation was proposed, which must rightly be considered as evolutionary. Nausea and vomiting reduce the pregnant woman's feeding capacity, and this is a disadvantage in evolutionary terms. However, unpleasant smells and tastes in plant and animal foods lead to avoiding such foods or even inducing nausea and vomiting to expel them after they have been ingested. These actions clearly have a protective value that outweighs the possible nutritional value of the foods avoided. In fact, in pregnant women, in addition to the useful, necessary protection of the mother, there is also the need to protect the embryo from the possible teratogenic effects of the aforesaid foods. So, morning sickness is an amplification of a protective mechanism already present in non-pregnant subjects, activated especially in the period of maximum vulnerability of the embryo: "... pregnancy sickness represents a lowering of the usual human threshold of tolerance to toxins in order to compensate for the extreme vulnerability of the embryo to toxins during organogenesis ..." (Profet, 1992).

This adaptive interpretation of morning sickness was later confirmed: "(i) symptoms peak when embryonic organogenesis is most susceptible to chemical disruption (weeks 6-18), (ii) women who experience morning sickness are significantly less likely to miscarry than women who do not (9 of 9 studies), (iii) women who vomit suffer fewer miscarriages than those who experience nausea alone ... Animal products may be dangerous to pregnant women and their embryos because they often contain parasites and pathogens, especially when stored at room temperatures in warm climates. Avoiding foodborne microorganisms is particularly important to pregnant women because they are immunosuppressed, presumably to reduce the chances of rejecting tissues of their own offspring ... As a result, pregnant women are more vulnerable to serious, often deadly infections." (Flaxman & Sherman, 2000).

Regarding morning sickness, it is essential to avoid:

1. considering it as a disease rather than a useful adaptation;
2. drugs to counteract it because, if effective, they reduce or cancel an important defense;
3. drugs surely not tested on our species for the pregnancy period because it is not ethical to test a substance with an unknown effect on a pregnant woman, and tests on other species do not give results that are reliable for our species. In this regard, precisely concerning the "cure" of morning sickness, there is a terrible experience that is still vivid in the collective memory: "Thalidomide was released in the late 1950' s as a nonaddictive, nonbarbiturate sedative ... Thalidomide was very effective and quickly discovered to also be an effective antiemetic and used to treat morning sickness in pregnant women. Thalidomide was marketed and distributed in 46 countries around the world using different names.... Thalidomide became one of the world's largest selling drugs, and was marketed heavily and advertised as completely safe ... right up until it was eventually banned in November 1961. Indeed, sample packets of the drug were given out to physicians to distribute freely to patients suffering from morning sickness ... Precisely how many women were given the drug will never be known.... It was not until 1961 that thalidomide was confirmed ... to be the cause of the largest man-made medical disaster in history ... with huge numbers (over 10,000) of severe birth defects in children. In addition, there were reports of increased miscarriage rates during this period ..." (Vargesson, 2015)

Based on the criteria above, Thalidomide absolutely should not have been authorized or used, both because there was no proof of its safety and also because it is irrational to consider morning sickness a disease and, therefore, to treat it.

2. **Iron deficiency or withholding**

Iron deficiency is usually observed during pregnancy, resulting in mild anemia (Means, 2020) (Fig. 6.1).

Various studies have looked for a relationship between anemia/iron deficiency and damage to the child, but: "The literature remains inconclusive on the association of maternal anemia and neonatal mortality and morbidities. A meta-analysis estimated a decreased risk of perinatal mortality by 28% [RR = 0.72 (95% CI: 0.62-0.89)] with each 10-g/L increase in hemoglobin (Stoltzfus et al., 2004). However, a large prospective cohort study involving 163,313 live births found no relationship (Zhang et al., 2009). Multiple studies reported a U-shaped relationship between maternal hemoglobin and preterm birth and low birth weight, respectively (Rasmussen, 2001)." (Kozuki et al., 2012).

According to the same study, moderate or severe maternal anemia is associated with an increased risk of small-for-gestational-age outcomes, while there is no risk for mild anemia (Kozuki et al., 2012).

However, instead of considering iron deficiency and subsequent anemia in pregnancy as a risk factor that should be corrected, it is more rational to consider this phenomenon in an evolutionary context. In other words, it is necessary to consider whether this phenomenon has any advantages, mainly if iron deficiency represents an important defense of the organism against microbial infections, both outside of pregnancy and even more so in a vulnerable period such as pregnancy. The evidence in this regard is remarkable.

Lactoferrins are iron-binding glycoproteins with many important protective activities, antimicrobial activity included, which are present in many body fluids such as colostrum, milk, tears, nasal/bronchial secretions, saliva, bile/pancreatic secretions, urine, seminal/vaginal fluids (Gruden & Poklar Ulrih, 2021).

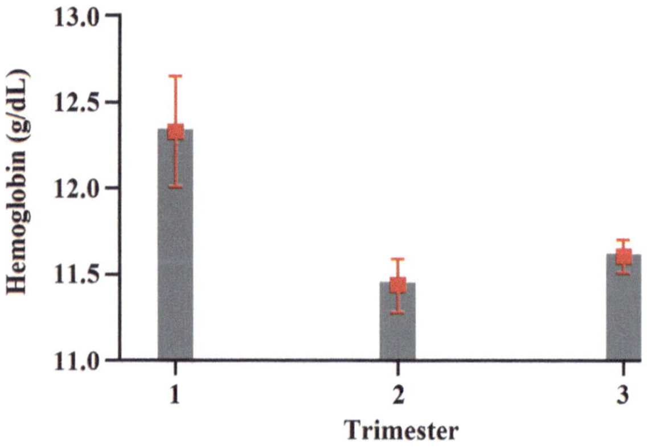

Fig. 6.1 Physiological anemia in pregnancy; data from (Milman et al., 2000; James et al., 2008; Shen et al., 2010), reported by (Means, 2020)

In tears, two substances, lactoferrin, and lipocalin, show antibacterial activity (Pflugfelder & Stern, 2020a): as iron is necessary for bacterial growth, they act by binding free iron or interfering with free iron uptake (Fluckinger et al., 2004; Flanagan & Willcox, 2009; Dartt, 2011).

Transferrin is another protein with meaningful antimicrobial action: "In recent years the critical role of iron in microbial growth and pathogenesis has garnered increasing attention. Virtually all microbes must obtain iron in order to survive and propagate …, including disease-causing pathogens that establish infections in mammalian hosts. Many microbes have thus evolved specialized mechanisms to acquire this limited resource. Conversely, an evolutionarily conserved, common host strategy to control microbial growth is to strictly regulate levels of available free iron.… Normal levels of unsaturated transferrin help to maintain the concentration of free iron in tissue fluids at levels that are restrictive for uncontrolled microbial growth.… Numerous studies have demonstrated the ability of transferrin to restrict microbial growth in vitro due to its iron sequestration capacity …" (Bruhn & Spellberg, 2015).

With these premises, an iron deficiency, or more likely a greater iron withholding by the host organisms, should be observed in case of infectious diseases. Treating this presumed defect with iron supply should increase the severity of infectious diseases, particularly in more vulnerable individuals. This would seem to be confirmed by various observations:

1. "There is convincing evidence that iron deficiency protects against many infectious diseases such as malaria, plague, and tuberculosis as shown by diverse medical, historical, and anthropologic studies." (Denic & Agarwal, 2007)
2. The availability of iron is essential for the development of a microorganism. There is complexity and redundancy in both host and microorganism mechanisms for acquiring iron and controlling its availability. "Unless the host immune response is impaired by severe iron deficiency, there is rarely an urgency to supplement iron and it is likely to contribute little to host iron status due to the block on absorption associated with inflammation. In the presence of intracellular infections such as tuberculosis or chronic inflammatory or immunosuppressive diseases (e.g., HIV), the decision to supplement iron must be considered on an individual basis, because the potential exists to benefit a pathogen rather than the host." (Doherty, 2007)
3. "In Bangladesh, in children aged <12 months, iron supplementation resulted in a 49% increase ($P = 0.03$) in the number of episodes of dysentery (Mitra et al., 1997). An earlier study in the Gambia that included only children who were anemic at baseline (hemoglobin < 3rd percentile of reference population) found that iron treatment was associated with an increase in fever-associated severe malaria (Smith et al., 1989).… a third study from Tanzania found, in children with severe anemia (hemoglobin $\leq$ 5.0 g/L), a significant increase in morbidity from infectious causes and a significantly higher incidence of pneumonia in the iron group (both: $P = 0.004$) (van den Hombergh et al., 1996). A large trial conducted in Zanzibar, Tanzania ($n = 24\,076$) found a 12% (95% CI: 2%, 23%) greater risk of severe illness leading to hospitalization or death and a 16% (95%

CI: 2%, 32%) greater risk of adverse events due to malaria associated with iron and folic acid supplementation (Sazawal et al., 2006)." (Iannotti et al., 2006)

4. In Polynesia, the incidence of sepsis by Escherichia coli was 20 times higher in Polynesian newborns compared with European newborns in a 5-year period in which Polynesian newborns were given intramuscular iron dextran "to cure" an iron deficiency. The impaired immunity of treated newborns ceased when the "cure" was stopped (Barry & Reeve, 1977).
5. "Our bodies have a related defense mechanism, of which most people are unaware and which physicians sometimes unwittingly attempt to frustrate. Here are some clues about how it works. A patient with chronic tuberculosis is found to have a low level of iron in his blood. A physician concludes that correcting the anemia may increase the patient's resistance, so she gives him an iron supplement. The patient's infection gets worse. Another clue: Zulu men often drink beer made in iron pots and often get serious liver infections caused by an amoeba. In contrast, less than 10% of Masai tribesmen have amoebic infections. They are herdsmen and drink large amounts of milk. When a group of Masai were given iron supplements, 88% soon got an amoebic infection. In another study, well-meaning investigators gave iron to supplement the low levels found in Somali nomads. At the end of one month, 38% had infections versus 8% of those who had not taken the supplements." (Nesse & Williams, 1994, p. 29)
6. A review observed, based on multiple studies, that malaria induces anemia due to "maldistribution of iron", but administration of iron to treat the anemia increases malaria incidence and malaria-related morbidity (Prentice et al., 2007).
7. "In northeastern Tanzania, where malaria and iron deficiency are common, we found that placental malaria was less prevalent (8.5% vs. 47.3% of women; $P < .0001$) and less severe (median parasite density, 4.2% vs. 6.3% of placental red blood cells; $P < .04$) among women with iron deficiency than among women with sufficient iron stores, especially during the first pregnancy. Multivariate analysis revealed that iron deficiency ($P < .0001$) and multigravidity ($P < .002$) significantly decreased the risk of placental malaria." (Kabyemela et al., 2008)
8. "Oral iron has been associated with increased rates of clinical malaria (5 of 9 studies) and increased morbidity from other infectious disease (4 of 8 studies).... No studies in malarial regions showed benefits ... Experimental studies in laboratory animals uniformly show reversible deleterious effects of iron administration on tests of functional immunity. These may occur even in mild deficiency.... Experimental and in vitro animal studies suggest that organisms that spend part of their life cycle intracellularly, such as plasmodia, mycobacteria and invasive salmonellae, may be enhanced by iron therapy." (Oppenheimer, 2001)
9. In the 1980s, on infants of Papua New Guinea, where iron deficiency and malarial transmission were common, a large prospective, randomized, and double-blind study was carried out by Oppenheimer and colleagues about the effects of iron supplementation. After 6 months, in the group treated with iron

compared with the control group, severe low respiratory infections, acute otitis media, and malaria-associated admissions to the hospital were more frequent (Oppenheimer, 2001).

10. A study carried out in sub-Saharan Africa showed that hematic iron levels were positively associated with increased pregnancy-associated malaria (Moya-Alvarez et al., 2015).
11. In 2021, a systematic review and meta-analysis concluded that "intravenous iron was associated with an increased risk of infection." (Shah et al., 2021)

Other studies do not appear to confirm a harmful effect of iron administration in case of infectious diseases, e.g., (Mantadakis et al., 2020). This apparent contradiction could be explained by the fact that, in non-vulnerable subjects, the body's defenses can be sufficient to overcome disease without damage, even without the host-determined limitation of iron use by pathogenic microorganisms.

6.3 Mental and Behavioral Disorders as an Adaptation to Particular Situations

The topics discussed in this section are extensively and much better debated in Randolph Nesse's excellent book (Nesse, 2019). This is no surprise as the author is indeed both one of the founders of Evolutionary Medicine and a psychiatrist.

Human behavior, i.e., instincts and predispositions to certain types of action that influence it, has been shaped by natural selection to deal with a wide range of situations that can often be faced in different and even opposite ways.

For example, in a dangerous condition such as a possible attack by a predator, we may run away as soon as possible, face the danger, or block any movement (i.e., freezing behavior), hoping that the predator will move away. There is no unique strategy, and human behavior is guided by conflicting impulses, each of which prevails depending on the case and also on how the genes that influence them are modulated in the single individual.

In this set of impulses that influence our behavior and often determine our destiny, there are characteristics that we often consider harmful, and indeed, if excessive, are rightly considered diseases. All this can be evaluated erroneously or distorted when not analyzed in the context of natural selection that has forged these characteristics. Moreover, it is necessary to consider that our way of life, that is, our ecological niche, has changed: the same characteristics, beneficial in the ecological niche to which we are adapted, can give rise to disturbances certainly classifiable as diseases.

Following in the footsteps of Nesse, let us now try to summarize these physiological characteristics of our behavior, which are not diseases but can become such in modern living conditions.

1. **Anxiety** and **fear**

Anxiety can be defined as the worry of a possible danger, while fear as the dread of a danger now evident. The two emotions are closely related and can also be considered as two gradations of the same emotion that occur in relation to a possible or real threat (Öhman, 2000).

Anxiety has a clear evolutionary motivation: "individual with a capacity for anxiety are more likely to escape from dangerous situations now and to avoid them in the future" (Nesse, 2019, p. 71), and the same is possible to say for fear.

Fear is accompanied by physiological reactions (e.g., increased heart and breath rate, sweating, mydriasis, constriction of blood vessels, liberation of metabolic energy sources for muscular action, etc., caused by the internal release of catecholamine hormones), which represent the body's predisposition to possible actions (flight, fight, resistance to wounds) and are collectively defined as fight-or-flight response (McCarty, 2016).

Both excessive anxiety and too little anxiety (hypophobia) are conditions that modify survival skills, but the second condition is often disregarded: "Hypophobia, a lack of adaptive aversive emotions in dangerous situations, is often maladaptive." (Kennair, 2007); "Hypophobia is serious and potentially fatal but underrecognized and rarely treated" (Nesse, 2019, p. 73).

Indeed, it does not seem likely that a person would go to the doctor complaining of a lack of anxiety. Furthermore, a person with reduced anxiety will be more capable of performing risky tasks and may be more successful in life, i.e., in evolutionary terms, having a greater ability to outrun rivals and have offspring, even if this carries a greater risk of death.

A condition closely related to anxiety is the phobia of particular animals or situations, e.g., snakes (ophidiophobia), spiders (arachnophobia), heights (acrophobia), and open spaces (agoraphobia). Some of these phobias are built-in automatic responses (Öhman, 1986). Others are not built-in: "Fear of snakes, for instance is not built-in, but the brain is prewired to learn it fast" (Nesse, 2019, p. 75). If a young monkey raised in a laboratory sees another monkey fearfully avoiding a snake toy, she immediately acquires a fear of snakes she did not previously have. This does not happen with a toy with the shape of a flower (Mineka et al., 1980). Other phobias can originate from unpleasant experiences (for example, being attacked by a dog may cause a phobia for dogs or cynophobia) or be transmitted by other people (Nesse, 2019, p. 76).

In general, built-in phobias or phobias for which there is a predisposition have an evolutionary motivation: snakes and spiders are dangerous, heights can lead to disastrous falls, and open spaces make one more vulnerable to attack by a predator (Nesse, 2019, p. 75).

In short, in evolutionarily "normal" conditions and specific situations, anxiety, fear, and some phobias, despite being unpleasant sensations, are physiological conditions with an adaptive meaning and should not be considered and treated as diseases.

The cases in which these conditions become accentuated and complicated, showing other displeasing manifestations, are different:

(i) Anxiety that grows to excessive levels and manifests itself under inappropriate conditions;
(ii) Acute crises of anxiety (panic attacks);
(iii) Generalized anxiety disorder;
(iv) Accentuated phobias or phobias for objects and situations that do not represent dangerous conditions;
(v) Multiform or generalized forms of anxiety about one's health condition (hypochondriasis);
(vi) Posttraumatic Stress Disorder because of acute or, more frequently, prolonged life-threatening stress.

These conditions may not be considered physiological characteristics, representing pathological alterations of the physiological events described above.

The detailed discussion about the mechanisms and conditions that give rise to these disorders is not the subject of this work. However, sharing Nesse's position (Nesse, 2019), it is possible to maintain that, to a large extent, these disturbances are due to living conditions that are very different from those to which we are adapted. This concept and the great prevalence and incidence of anxiety disorders have been discussed in *§3.2.7 Altered social interactions.*

2. **Low mood** and **depression**

Low mood and depression are practically the same mental state, in two different degrees of intensity, which manifests itself in relation to adverse events or situations. "Sadness can feel very similar to low mood, but it results from specific losses … Grief is the special kind of sadness caused by death of a loved one or other major loss" (Nesse, 2019, p. 88).

Low mood and depression are generally characterized or accompanied by pessimism, risk avoidance, low energy, social withdrawal, slow thinking, submissive behavior, tendency to low self-esteem, sensation of low energy, hopelessness, demoralization, loss of motivation, reduced pleasure or loss of interest for things that would be normally pleasant or interesting, altered sense of well-being (Gilbert, 2007; de Zwart et al., 2019; Nesse, 2019, p. 88). This set of manifestations in the mildest forms represents a normal reaction to adverse events of various kinds, such as the loss of a loved one, physical, mental, or sexual abuse, bullying, dismissal or unemployment, separation or divorce, medical diagnosis of a serious illness, catastrophic injury, social isolation or rejection, "situations … in which efforts to reach a goal are failing", "a loss in a status competition" (Nesse, 2019, p. 88).

Since these sensations, in their various manifestations, appear as a physiological response to events that also occur in normal conditions, it is necessary to search for their adaptive meaning in the evolutionarily "normal" conditions to which our species is adapted. In this regard, it is very useful to follow the careful discussion proposed by Nesse in *Chap. 6—Low mood and the art of giving up* of his book (Nesse, 2019).

In a group of chickens, a subject that loses a fight descends in the pecking order, behaves submissively, and tends to stay away from other chickens to reduce the chances of attacks by other chickens further up in the pecking order (Price, 1967). The vervet monkeys live in groups with a dominant male who is the only one to mate with females and, as a distinctive feature, displays bright blue testicles. If the dominant male is defeated by another male, he withdraws, acts depressed, and his testicles turn dark gray (Price & Sloman, 1987).

These phenomena have been interpreted as a loss in a status competition, and the subsequent behaviors are like what we would define as low mood and depression in a human being.

In one study, in events characterized by "humiliation and entrapment", an episode of depression followed in 75% of cases. In contrast, such episodes occurred in 20% of cases when there a family member was lost and 5% after a danger (Brown & Harris, 1978).

"The involuntary yielding hypothesis seems correct for many cases of depression I have treated. Myriads of spouses limit their achievements, and even their view of their own abilities, to preserve their marriages. The deceiving down social strategies prevents attacks by those with more power, at the price of depressive symptoms." (Nesse, 2019, p. 92).

Symptoms of depression also occur in cases of illness in which there is fever and inflammation (e.g., infections, traumatic injuries). Such events have been referred to as sickness behavior (Hart, 1988; Johnson, 2002) and are mediated by cytokines released when there is inflammation and fever (Raison et al., 2006; Dantzer & Kelley, 2007). Sickness behavior appears to have an important adaptive value as it induces the organism to rest, diverting the available energies to overcome the cause of illness and to avoid conditions such as running away from a predator or hunting for prey without being in the fullness of strength (Hart, 1988).

However, there are other explanations for low mood and depression that nevertheless appear farther from the original adaptive causes and more related to the complications of modern life:

1. "... while about half of my patients with depression seem to be trapped pursuing unreachable goals, many of those goals are not about social position. Is unrequited love a pursuit of a status goal? What about trying to find effective treatment for a child with cancer?" (Nesse, 2019, p. 93)
2. ... social withdrawal and thinking a lot can be useful when one encounters a dead end in life ... depressive withdrawal and intense cogitation can improve coping when major life problems require a major change but can also leave some people stuck in unproductive depression. Major life failures may motivate allocating enormous effort into finding new strategies." (Nesse, 2019, p. 94)

The overall impression is that low mood and depression are a behavioral, physiological module that represents a valuable adaptation to cope with certain conditions. However, in modern times, in cases where the manifestations characterizing low mood and depression are accentuated and prolonged and without signs of remission, we have the transformation of these behaviors into diseases such as pathological depression and major depressive episodes, which can be complicated by a suicide attempt and suicide. This has been discussed in *§3.2.7 Altered social interactions*, subsections *Pathologic depression* and *Suicide*, highlighting the great diffusion of these phenomena and their origin in alterations of the ecological niche.

3. **Jealousy**, **envy**, and **anger**

Jealousy involves the wish to keep the attention or fidelity of another person (e.g., those of parents for a child; those of a consort for the other consort).

Envy involves the wish to get what other individuals have.

Anger is helpful to stop the threatening behavior of a possible aggressor and when there is a threat to the individual exhibiting anger or to the family members of the same individual.

The adaptive significance of these emotions is clear. They become diseases or symptoms of a disease when they are accentuated, prolonged, or triggered in inappropriate situations.

References

Bányai, K., Estes, M. K., Martella, V., & Parashar, U. D. (2018). Viral gastroenteritis. *The Lancet, 392*(10142), 175–186. https://doi.org/10.1016/S0140-6736(18)31128-0

Barry, D. M., & Reeve, A. W. (1977). Increased incidence of gram-negative neonatal sepsis with intramuscula iron administration. *Pediatrics, 60*(6), 908–912.

Bobba, S., Spencer, S. K. R., Fox, O. J. K., Agar, A., Coroneo, M. T., & Francis, I. C. (2019). Management of the photic sneeze reflex utilising the philtral pressure technique. *Eye (London), 33*(7), 1186–1187. https://doi.org/10.1038/s41433-019-0368-4

Boulton, A. J. (2012). Diabetic foot–what can we learn from leprosy? Legacy of Dr Paul W. Brand. *Diabetes/Metabolism Research and Reviews, 28*(Suppl 1), 3–7. https://doi.org/10.1002/dmrr.2230

Breitenbach, R. A., Swisher, P. K., Kim, M. K., & Patel, B. S. (1993). The photic sneeze reflex as a risk factor to combat pilots. *Military Medicine, 158*(12), 806–909.

Bron, A. J., de Paiva, C. S., Chauhan, S. K., Bonini, S., Gabison, E. E., Jain, S., Knop, E., et al. (2017). TFOS DEWS II pathophysiology report. *The Ocular Surface, 15*(3), 438–510. https://doi.org/10.1016/j.jtos.2017.05.011. Erratum in: Ocul Surf, 2019, 17(4), 842.

Brown, G. W., & Harris, T. (Eds.). (1978). *Social origins of depression: A study of psychiatric disorder in women*. Tavistock Publications.

Bruhn, K. W., & Spellberg, B. (2015). Transferrin-mediated iron sequestration as a novel therapy for bacterial and fungal infections. *Current Opinion in Microbiology, 27*, 57–61. https://doi.org/10.1016/j.mib.2015.07.005

Bylsma, L. M., Gračanin, A., & Vingerhoets, A. J. J. M. (2019). The neurobiology of human crying. *Clinical Autonomic Research, 29*(1), 63–73. https://doi.org/10.1007/s10286-018-0526-y

Cevikbas, F., & Lerner, E. A. (2020). Physiology and pathophysiology of itch. *Physiological Reviews, 100*(3), 945–982. https://doi.org/10.1152/physrev.00017.2019

Dantzer, R., & Kelley, K. W. (2007). Twenty years of research on cytokine-induced sickness behavior. *Brain, Behavior, & Immunity, 21*(2), 153–160. https://doi.org/10.1016/j.bbi.2006.09.006

Dartt, D. A. (2011). Tear lipocalin: Structure and function. *The Ocular Surface, 9*(3), 126–138. https://doi.org/10.1016/s1542-0124(11)70022-2

Dartt, D. A. (2009). Neural regulation of lacrimal gland secretory processes: Relevance in dry eye diseases. *Progress in Retinal and Eye Research, 28*, 155–177. https://doi.org/10.1016/j.preteyeres.2009.04.003

Dean, L. (2012). ACHOO syndrome. In V. M. Pratt, S. A. Scott, M. Pirmohamed, et al. (Eds.), *Medical genetics summaries [Internet]*. National Center for Biotechnology Information (US). https://www.ncbi.nlm.nih.gov/books/NBK109193/

Denic, S., & Agarwal, M. M. (2007). Nutritional iron deficiency: An evolutionary perspective. *Nutrition, 23*(7–8), 603–614. https://doi.org/10.1016/j.nut.2007.05.002

de Truchis, P., & de Truchis, A. (2007). Diarrhées aiguës infectieuses [acute infectious diarrhea, in French]. *La Presse Médicale, 36*(4 Pt 2), 695–705. https://doi.org/10.1016/j.lpm.2006.11.023

de Zwart, P. L., Jeronimus, B. F., & de Jonge, P. (2019). Empirical evidence for definitions of episode, remission, recovery, relapse and recurrence in depression: A systematic review. *Epidemiology and Psychiatric Sciences, 28*(5), 544–562. https://doi.org/10.1017/S2045796018000227

Doherty, C. P. (2007). Host-pathogen interactions: The role of iron. *The Journal of Nutrition, 137*(5), 1341–1344. https://doi.org/10.1093/jn/137.5.1341

DuPont, H. L., & Hornick, R. B. (1973). Adverse effect of lomotil therapy in shigellosis. *JAMA, 226*(13), 1525–1528.

Earn, D. J., Andrews, P. W., & Bolker, B. M. (2014). Population-level effects of suppressing fever. *Proceedings. Biological sciences, 281*(1778), 20132570. https://doi.org/10.1098/rspb.2013.2570

Egloff, N., Cámara, R. J., von Känel, R., Klingler, N., Marti, E., & Ferrari, M. L. (2014). Hypersensitivity and hyperalgesia in somatoform pain disorders. *General Hospital Psychiatry, 36*(3), 284–290. https://doi.org/10.1016/j.genhosppsych.2014.01.011

Evans, S. S., Repasky, E. A., & Fisher, D. T. (2015). Fever and the thermal regulation of immunity: The immune system feels the heat. *Nature Reviews. Immunology, 15*(6), 335–349. https://doi.org/10.1038/nri3843

Farrand, K. F., Fridman, M., Stillman, I. Ö., & Schaumberg, D. A. (2017). Prevalence of diagnosed dry eye disease in the United States among adults aged 18 years and older. *American Journal of Ophthalmology, 182*, 90–98. https://doi.org/10.1016/j.ajo.2017.06.033

Festin, M. (2009). Nausea and vomiting in early pregnancy. *BioMed Central Clinical Evidence, 2009*, 1405.

Flanagan, J. L., & Willcox, M. D. (2009). Role of lactoferrin in the tear film. *Biochimie, 91*(1), 35–43. https://doi.org/10.1016/j.biochi.2008.07.007

Flaxman, S. M., & Sherman, P. W. (2000). Morning sickness: A mechanism for protecting mother and embryo. *The Quarterly Review of Biology, 75*(2), 113–148. https://doi.org/10.1086/393377

Fleckenstein, J. M., Matthew Kuhlmann, F., & Sheikh, A. (2021). Acute bacterial gastroenteritis. *Gastroenterology Clinics of North America, 50*(2), 283–304. https://doi.org/10.1016/j.gtc.2021.02.002

Fluckinger, M., Haas, H., Merschak, P., Glasgow, B. J., & Redl, B. (2004). Human tear lipocalin exhibits antimicrobial activity by scavenging microbial siderophores. *Antimicrobial Agents and Chemotherapy, 48*(9), 3367–3372. https://doi.org/10.1128/AAC.48.9.3367-3372.2004

Gaudio, F. G., & Grissom, C. K. (2016). Cooling methods in heat stroke. *The Journal of Emergency Medicine, 50*(4), 607–616. https://doi.org/10.1016/j.jemermed.2015.09.014

Gilbert, P. (2007). *Psychotherapy and counselling for depression* (3rd ed.). Sage.

Goldman, L., & Schafer, A. I. (2019). *Goldman-Cecil medicine* (26th ed.). Elsevier.

Gračanin, A., Bylsma, L. M., & Vingerhoets, A. J. J. M. (2018). Why only humans shed emotional tears: Evolutionary and cultural perspectives. *Human Nature, 29*(2), 104–133. https://doi.org/10.1007/s12110-018-9312-8

Graves, N. S. (2013). Acute gastroenteritis. *Primary Care, 40*(3), 727–741. https://doi.org/10.1016/j.pop.2013.05.006

Gruden, Š., & Poklar Ulrih, N. (2021). Diverse mechanisms of antimicrobial activities of Lactoferrins, Lactoferricins, and other Lactoferrin-derived peptides. *International Journal of Molecular Sciences, 22*(20), 11264. https://doi.org/10.3390/ijms222011264

Hanada, H. (2016). Specific problems with accidental deaths in Japan–cardiac arrest from hot baths and foreign body airway obstruction in elders. *Circulation Journal, 80*(7), 1523–1524. https://doi.org/10.1253/circj.CJ-16-0500

Harden, L. M., Kent, S., Pittman, Q. J., & Roth, J. (2015). Fever and sickness behavior: Friend or foe? *Brain, Behavior, and Immunity, 50*, 322–333. https://doi.org/10.1016/j.bbi.2015.07.012

Hart, B. L. (1988). Biological basis of the behavior of sick animals. *Neuroscience and Biobehavioral Reviews, 12*(2), 123–137. https://doi.org/10.1016/s0149-7634(88)80004-6

Hsueh, B. Y., & Waters, C. M. (2019). Combating cholera. *F1000Research, 8*, F1000 Faculty Rev-589. https://doi.org/10.12688/f1000research.18093.1

Iannotti, L. L., Tielsch, J. M., Black, M. M., & Black, R. E. (2006). Iron supplementation in early childhood: Health benefits and risks. *The American Journal of Clinicl Nutrition, 84*(6), 1261–1276. https://doi.org/10.1093/ajcn/84.6.1261

Ikoma, A., Steinhoff, M., Ständer, S., Yosipovitch, G., & Schmelz, M. (2006). The neurobiology of itch. *Nature Reviews. Neuroscience, 7*(7), 535–547. https://doi.org/10.1038/nrn1950

James, T. R., Reid, H. L., & Mullings, A. M. (2008). Are published standards for haematological indices in pregnancy applicable across populations: An evaluation in healthy pregnant Jamaican women. *BMC Pregnancy and Childbirth, 8*, 8. https://doi.org/10.1186/1471-2393-8-8

Johnson, R. W. (2002). The concept of sickness behavior: A brief chronological account of four key discoveries. *Veterinary Immunology and Immunopathology, 87*(3–4), 443–450. https://doi.org/10.1016/s0165-2427(02)00069-7

Kabyemela, E. R., Fried, M., Kurtis, J. D., Mutabingwa, T. K., & Duffy, P. E. (2008). Decreased susceptibility to *plasmodium falciparum* infection in pregnant women with iron deficiency. *The Journal of Infectious Diseases, 198*(2), 163–166. https://doi.org/10.1086/589512

Kelley, K. W., Bluthé, R. M., Dantzer, R., Zhou, J. H., Shen, W. H., Johnson, R. W., & Broussard, S. R. (2003). Cytokine-induced sickness behavior. *Brain, Behavior, and Immunity, 17*(Suppl 1), S112–S118. https://doi.org/10.1016/S0889-1591(02)00077-6

Kennair, L. E. O. (2007). Fear and fitness revisited. *Journal of Evolutionary Psychology, 5*(1–4), 105–117. https://doi.org/10.1556/JEP.2007.1020

Kluger, M. J. (1979). Phylogeny of fever. *Federation Proceedings, 38*(1), 30–34.

Kluger, M. J., Kozak, W., Conn, C. A., Leon, L. R., & Soszynski, D. (1996). The adaptive value of fever. *Infectious Disease Clinics of North America, 10*(1), 1–20. https://doi.org/10.1016/S0891-5520(05)70282-8

Kozuki, N., Lee, A. C., & Katz, J. (2012). Child health epidemiology reference group. Moderate to severe, but not mild, maternal anemia is associated with increased risk of small-for-gestational-age outcomes. *The Journal of Nutrition, 142*(2), 358–362. https://doi.org/10.3945/jn.111.149237

Landa, A., Peterson, B. S., & Fallon, B. A. (2012). Somatoform pain: A developmental theory and translational research review. *Psychosomatic Medicine, 74*(7), 717–727. https://doi.org/10.1097/PSY.0b013e3182688e8b

Leon, L. R., & Bouchama, A. (2015). Heat stroke. *Comprehensive Physiology, 5*(2), 611–647. https://doi.org/10.1002/cphy.c140017

Li, F., Jiang, H., Shen, X., Yang, W., Guo, C., Wang, Z., et al. (2021). Sneezing reflex is mediated by a peptidergic pathway from nose to brainstem. *Cell, 184*(14), P3762–P3773. https://doi.org/10.1016/j.cell.2021.05.017

Locht, C., & Antoine, R. (2021). The history of pertussis toxin. *Toxins (Basel), 13*(9), 623. https://doi.org/10.3390/toxins13090623

Loscalzo, J., Fauci, A., Kasper, D., et al. (2022). *Harrison's principles of internal medicine* (21st ed.). McGraw-Hill Education.

Mantadakis, E., Chatzimichael, E., & Zikidou, P. (2020). Iron deficiency anemia in children residing in high and low-income countries: Risk factors, prevention, diagnosis and therapy. *Mediterranean Journal of Hematology and Infectious Diseases, 12*(1), e2020041. https://doi.org/10.4084/MJHID.2020.041

McCarty, R. (2016). Chapter 4–The fight-or-flight response: A cornerstone of stress research. In G. Fink (Ed.), *Stress: Concepts, cognition, emotion, and behavior* (pp. 33–37). Academic. https://doi.org/10.1016/B978-0-12-800951-2.00004-2

Means, R. T. (2020). Iron deficiency and iron deficiency anemia: Implications and impact in pregnancy, fetal development, and early childhood parameters. *Nutrients, 12*(2), 447. https://doi.org/10.3390/nu12020447

Mendoza-Santiesteban, C. E., Palma, J. A., Norcliffe-Kaufmann, L., & Kaufmann, H. (2017). Familial dysautonomia: A disease with hidden tears. *Journal of Neurology, 264*(6), 1290–1291. https://doi.org/10.1007/s00415-017-8486-z

Milman, N., Byg, K. E., & Agger, A. O. (2000). Hemoglobin and erythrocyte indices during normal pregnancy and postpartum in 206 women with and without iron supplementation. *Acta Obstetricia et Gynecologica Scandinavica, 79*, 89–98. https://doi.org/10.1034/j.1600-0412.2000.079002089.x

Mineka, S., Keir, R., & Price, V. (1980). Fear of snakes in wild-and laboratory-reared rhesus monkeys (Macaca mulatta). *Animal Learning & Behavior, 8*(4), 653–663.

Mitra, A. K., Akramuzzaman, S. M., Fuchs, G. J., Rahman, M. M., & Mahalanabis, D. (1997). Long-term oral supplementation with iron is not harmful for young children in a poor community of Bangladesh. *The Journal of Nutrition, 127*(8), 1451–1455. https://doi.org/10.1093/jn/127.8.1451

Moya-Alvarez, V., Cottrell, G., Ouédraogo, S., Accrombessi, M., Massougbodgi, A., & Cot, M. (2015). Does iron increase the risk of malaria in pregnancy? *Open Forum Infectious Diseases, 2*(2), ofv038. https://doi.org/10.1093/ofid/ofv038

Murube, J. (2009). Basal, reflex, and psycho-emotional tears. *The Ocular Surface, 7*(2), 60–66. https://doi.org/10.1016/S1542-0124(12)70296-3

Nemeth, V., & Pfleghaar, N. (2022). Diarrhea. In *StatPearls [Internet]*. StatPearls Publishing. https://www.ncbi.nlm.nih.gov/books/NBK448082/

Nesse, R. M. (2019). *Good reasons for bad feelings. Insights from the frontier of evolutionary psychiatry*. Penguin.

Nesse, R. M., & Williams, G. C. (1994). *Why we get sick: The new science of Darwinian medicine*. Times Books.

Nijs, J., Lahousse, A., Kapreli, E., Bilika, P., Saraçoğlu, İ., Malfliet, A., et al. (2021). Nociplastic pain criteria or recognition of central sensitization? Pain phenotyping in the past, present and future. *Journal of Clinical Medicine, 10*(15), 3203. https://doi.org/10.3390/jcm10153203

Öhman, A. (1986). Face the beast and fear the face: Animal and social fears as prototypes for evolutionary analyses of emotion. *Psychophysiology, 23*(2), 123–145. https://doi.org/10.1111/j.1469-8986.1986.tb00608.x

Öhman, A. (2000). Fear and anxiety: Evolutionary, cognitive, and clinical perspectives. In M. Lewis, L. F. Barret, & J. M. Haviland-Jones (Eds.), *Handbook of emotions* (pp. 573–579). The Guilford Press.

Oppenheimer, S. J. (2001). Iron and its relation to immunity and infectious disease. *The Journal of Nutrition, 131*(2S-2), S616–S633. https://doi.org/10.1093/jn/131.2.616S

Pasikhova, Y., Ludlow, S., & Baluch, A. (2017). Fever in patients with cancer. *Cancer Control, 24*(2), 193–197. https://doi.org/10.1177/107327481702400212

Pflugfelder, S. C., & Stern, M. E. (2020a). Biological functions of tear film. *Experimental Eye Research, 197*, 108115. https://doi.org/10.1016/j.exer.2020.108115

Pflugfelder, S. C., & Stern, M. E. (2020b). The cornea in keratoconjunctivitis sicca. *Experimental Eye Research, 201*, 108295. https://doi.org/10.1016/j.exer.2020.108295

Prentice, A. M., Ghattas, H., Doherty, C., & Cox, S. E. (2007). Iron metabolism and malaria. *Food and Nutrition Bulletin, 28*(4 Suppl), S524–S539. https://doi.org/10.1177/15648265070284S406

Price, J. (1967). The dominance hierarchy and the evolution of mental illness. *The Lancet, 290*(7509), 243–246. https://doi.org/10.1016/S0140-6736(67)92306-9

Price, J., & Sloman, L. (1987). Depression as yielding behavior: An animal model based on Schjelderup-Ebbe's pecking order. *Ethology and Sociobiology, 8*(Supp 1), 85–98. https://doi.org/10.1016/0162-3095(87)90021-5

Profet, M. (1992). Pregnancy sickness as adaptation: A deterrent to maternal ingestion of teratogens. In J. Barkow, L. Cosmides, & J. Tooby (Eds.), *The adapted mind: Evolutionary psychology and the generation of culture* (pp. 327–366). Oxford University Press.

Raison, C. L., Capuron, L., & Miller, A. H. (2006). Cytokines sing the blues: Inflammation and the pathogenesis of depression. *Trends in Immunology, 27*(1), 24–31. https://doi.org/10.1016/j.it.2005.11.006

Rasmussen, K. (2001). Is there a causal relationship between iron deficiency or iron-deficiency anemia and weight at birth, length of gestation and perinatal mortality? *The Journal of Nutrition, 131*(2S-2), S590–S601. https://doi.org/10.1093/jn/131.2.590S

Ryan, M., & Levy, M. M. (2003). Clinical review: Fever in intensive care unit patients. *Critical Care, 7*(3), 221–225. https://doi.org/10.1186/cc1879

Roh, Y. S., Choi, J., Sutaria, N., & Kwatra, S. G. (2022). Itch: Epidemiology, clinical presentation, and diagnostic workup. *Journal of the American Academy of Dermatology, 86*(1), 1–14. https://doi.org/10.1016/j.jaad.2021.07.076

Salih, A. M., Alfaki, M., & Alam-Elhuda, D. M. (2016). Airway foreign bodies: A critical review for a common pediatric emergency. *World Journal of Emergency Medicine, 7*(1), 5–12. https://doi.org/10.5847/wjem.j.1920-8642.2016.01.001

Sazawal, S., Black, R. E., Ramsan, M., Chwaya, H. M., Stoltzfus, R. J., Dutta, A., et al. (2006). Effects of routine prophylactic supplementation with iron and folic acid on admission to hospital and mortality in preschool children in a high malaria transmission setting: Community-based, randomised, placebo-controlled trial. *The Lancet, 367*(9505), 133–143. https://doi.org/10.1016/S0140-6736(06)67962-2. Erratum in: The Lancet, 2006, 367(9507), 302.

Scarborough, B. M., & Smith, C. B. (2018). Optimal pain management for patients with cancer in the modern era. *CA: a Cancer Journal for Clinicians, 68*(3), 182–196. https://doi.org/10.3322/caac.21453

Schon, K. R., Parker, A. P. J., & Woods, C. G. (2018). Congenital insensitivity to pain overview. In M. P. Adam, D. B. Everman, G. M. Mirzaa, et al. (Eds.), *GeneReviews® [Internet]*. University of Washington. https://www.ncbi.nlm.nih.gov/books/NBK481553/

Schulman, C. I., Namias, N., Doherty, J., Manning, R. J., Li, P., Elhaddad, A., Lasko, D., et al. (2005). The effect of antipyretic therapy upon outcomes in critically ill patients: A randomized, prospective study. *Surgical Infections (Larchmont), 6*(4), 369–375. https://doi.org/10.1089/sur.2005.6.369

Shah, A. A., Donovan, K., Seeley, C., Dickson, E. A., Palmer, A. J. R., Doree, C., et al. (2021). Risk of infection associated with Administration of Intravenous Iron: A systematic review and meta-analysis. *JAMA Network Open, 4*(11), e2133935. https://doi.org/10.1001/jamanetworkopen.2021.33935. Erratum in: JAMA Network Open, 2022, 5(1), e2146637.

Sharma, S., Hashmi, M. F., & Alhajjaj, M. S. (2023). Cough. In *StatPearls [Internet]*. StatPearls Publishing. https://www.ncbi.nlm.nih.gov/books/NBK493221/

Shen, C., Jiang, Y. M., Shi, H., Liu, J. H., Zhou, W. J., Dai, Q. K., & Yang, H. (2010). A prospective, sequential and longitudinal study of haematological profile during normal pregnancy in Chinese women. *Journal of Obstetrics and Gynaecology, 30*, 357–361. https://doi.org/10.3109/01443611003681444

Small, P., Keith, P. K., & Kim, H. (2018). Allergic rhinitis. *Allergy, Asthma, and Clinical Immunology, 14*(Suppl 2), 51. https://doi.org/10.1186/s13223-018-0280-7

Smith, A. W., Hendrickse, R. G., Harrison, C., Hayes, R. J., & Greenwood, B. M. (1989). The effects on malaria of treatment of iron-deficiency anaemia with oral iron in Gambian children. *Annals of Tropical Paediatrics, 9*(1), 17–23. https://doi.org/10.1080/02724936.1989

Sneddon, L. U. (2018). Comparative physiology of nociception and pain. *Physiology (Bethesda, Md.), 33*(1), 63–73. https://doi.org/10.1152/physiol.00022.2017

Songu, M., & Önerci, T. M. (2023). Physiology and pathophysiology of sneezing and itching: Mechanisms of the symptoms. In Ö. Ö. Celebi & T. M. Önerci (Eds.), *Nasal physiology and pathophysiology of nasal disorders* (2nd ed., pp. 131–144). Springer. https://doi.org/10.1007/978-3-031-12386-3_12

Steeds, C. E. (2009). The anatomy and physiology of pain. *Surgery (Oxford), 27*(12), 507–511. https://doi.org/10.1016/j.mpsur.2009.10.013

Stoltzfus, R. J., Mullany, L. C., & Black, R. E. (2004). Iron deficiency anemia. In M. Ezzati, A. D. Lopez, A. Rodgers, & C. J. L. Murray (Eds.), *Comparative quantification of health risks: Global and regional burden of disease attributable to selected major risk factors* (Vol. 1, pp. 163–209). World Health Organization.

Treede, R. D., Jensen, T. S., Campbell, J. N., Cruccu, G., Dostrovsky, J. O., Griffin, J. W., et al. (2008). Neuropathic pain: Redefinition and a grading system for clinical and research purposes. *Neurology, 70*(18), 1630–1635. https://doi.org/10.1212/01.wnl.0000282763.29778.59

Trouvin, A. P., & Perrot, S. (2019). New concepts of pain. *Best Practice & Research. Clinical Rheumatology, 33*(3), 101415. https://doi.org/10.1016/j.berh.2019.04.007

van den Hombergh, J., Dalderop, E., & Smit, Y. (1996). Does iron therapy benefit children with severe malaria-associated anaemia? A clinical trial with 12 weeks supplementation of oral iron in young children from the Turiani division, Tanzania. *Journal of Tropical Pediatrics, 42*(4), 220–227. https://doi.org/10.1093/tropej/42.4.220

Vargesson, N. (2015). Thalidomide-induced teratogenesis: History and mechanisms. *Birth Defects Research. Part C, Embryo Today, 105*(2), 140–156. https://doi.org/10.1002/bdrc.21096

Weisman, A., Quintner, J., & Masharawi, Y. (2019). Congenital insensitivity to pain: A misnomer. *The Journal of Pain, 20*(9), 1011–1014. https://doi.org/10.1016/j.jpain.2019.01.331

World Cause of Death (2020) (Data from WHO). Accessed December 15, 2022, from www.worldlifeexpectancy.com/world-rankings-total-deaths

Yousaf, N., Monteiro, W., Matos, S., Birring, S. S., & Pavord, I. D. (2013). Cough frequency in health and disease. *The European Respiratory Journal, 41*(1), 241–243. https://doi.org/10.1183/09031936.00089312

Zhang, Q., Ananth, C. V., Rhoads, G. G., & Li, Z. (2009). The impact of maternal anemia on perinatal mortality: A population-based, prospective cohort study in China. *Annals of Epidemiology, 19*(11), 793–799. https://doi.org/10.1016/j.annepidem.2009.06.002

Zhong, W., Shahbaz, O., Teskey, G., Beever, A., Kachour, N., Venketaraman, V., & Darmani, N. A. (2021). Mechanisms of nausea and vomiting: Current knowledge and recent advances in intracellular emetic signaling systems. *International Journal of Molecular Sciences, 22*(11), 5797. https://doi.org/10.3390/ijms22115797

Chapter 7
Physiological Phenomena that Cause Troubles and Suffering or Even Death: Part II: Phenoptotic Phenomena Different from Slow Phenoptosis

There are numberless phenomena, collectively defined as phenoptosis, in which natural selection, instead of favoring mechanisms enhancing individual survival, on the contrary, favors mechanisms that kill the individual, increase the probability of death, or even determine the death of close blood relatives of individuals. These events would seem impossible in an evolutionary logic, yet they exist and are determined by natural selection on the basis of supra-individual selection. The consideration and understanding of these phenomena are indispensable for the interpretation of many biological phenomena and also of many phenomena that, in various ways, concern medicine.

Phenoptotic phenomena have been extensively discussed in (Libertini et al., 2021, *Chap. 2—Evolution and Phenoptosis*), to which reference is made for further information. This chapter will present a synthesis favoring the most interesting facts and concepts for evolutionary medicine.

7.1 Phenoptotic Phenomena

The concept of phenoptosis was proposed over 25 years ago to indicate the "programmed death of an organism" (Skulachev, 1997; Skulachev, 1999) and the term was inspired by the phenomenon apoptosis, i.e., the programmed death of a cell. In the following years, aging, precisely defined as "increasing mortality with increasing chronological age in populations in the wild" (Libertini, 1988), and as such manifested by individuals of our species, and also of many other species, was described as "slow phenoptosis" (Skulachev, 2002a).

Later, the definition of phenoptosis was expanded: "Phenoptosis is the death of an individual caused by its own actions or by actions of close relatives (siblicide; in particular, the parent-caused death of an offspring or filial infanticide) and not

G. Libertini et al., *The Essential Revolution of Evolutionary Medicine*, Advances in Studies of Aging and Health 4, https://doi.org/10.1007/978-3-031-84877-3_7

caused primarily by accidents or diseases or external factors, which is determined, regulated or influenced by genes favored by natural selection." (Libertini, 2012), and it was highlighted how phenoptotic phenomena are very varied and widespread among living species (Libertini, 2012; Libertini et al., 2021).

The concept of phenoptosis has also been proposed as "programmed organismal death (POD)" (Pepper et al., 2013) without however including the aging phenomenon.

7.1.1 Selective Mechanisms Favoring Phenoptotic Phenomena

Following the classic idea of evolution as a consequence of the "survival of the fittest" (Darwin, 1869)—an idea attributed to Darwin but actually proposed a few years earlier by Herbert Spencer (Spencer, 1864)—natural selection can be described as the different diffusion of a gene from one generation to the next as a consequence of the advantage or disadvantage that a gene entails:

$$\Delta_c \propto S\,P \tag{7.1}$$

That is, the frequency variation (Δ_c) between one generation and the next of a gene C that acts in the individual I is proportional to the advantage ($S > 0$) or disadvantage ($S < 0$) caused by the gene for I, and also as a function of the residual reproductive capacity (P) of I at the age in which the gene manifests its action.

As any gene determining a phenoptotic phenomenon is by definition disadvantageous for I, i.e., by definition $S < 0$, it follows that natural selection, as described by formula (7.1), cannot favor the diffusion of a gene causing phenoptosis.

However, many years after Darwin, the concepts of "inclusive fitness" and "kin selection" were proposed (Hamilton, 1964, 1970; Trivers, 1971; Wilson, 1975). They demonstrated how it was theoretically possible that natural selection could favor genes with $S < 0$, i.e., genes harmful to the survival of the individual.

To use the concept of "inclusive fitness" for the calculations of natural selection, it is necessary to consider both the advantage or disadvantage for the individual in which the gene acts (I_1) and also all the possible advantages or disadvantages for z other individuals genetically related to I_1 (I_2, I_3, ..., I_z). Therefore, the frequency variation of gene C from one generation to another is described by the equation:

$$\Delta_C \propto \Sigma_{x=1}^{z}\left(S_x \cdot P_x \cdot r_x\right) \tag{7.2}$$

where:

S_x = advantage or disadvantage for the individual I_x determined by C; r_x = coefficient of relationship between I_x and I_1; P_x = reproductive value of an individual I_x at the age when the gene C acts.

It is clear that C is favored by natural selection when the result of the summation is positive and that C is progressively eliminated by natural selection when the summation is negative. When C acts only on I_1, as by definition $r_1 = 1$, the formula (7.2) becomes:

$$\Delta_c \propto S_1 P_1 = S P \tag{7.3}$$

which is the same as (7.1).

Under opportune conditions, formula (7.2) can also describe forms of group selection [see (Libertini et al., 2021), *§2.2 Supra-Individual Selection*]. Also, particular population models can have effects similar to kin selection (Nowak et al., 2010).

However, these theoretical arguments, and their support by the evidence, showed that:

(i) The formulas regarding inclusive fitness and the aforementioned particular population models were not in contradiction with the central concept of natural selection expressed by (7.1);
(ii) Natural selection can favor genes that are harmful to the individual in which they act.

Moreover, Darwin himself mentioned that the behavior of an individual of a tribe, determining his sacrifice for the common good, would have been favored by natural selection (Darwin, 1871). Furthermore, parental care can only be explained by considering the inclusive fitness of both parents and children and not only the individual advantage of the parents (Libertini et al., 2021).

7.1.2 Classification of Phenoptotic Phenomena

Countless phenomena that fall under the definition of phenoptosis have long been known and described. An authoritative text (Finch, 1990) described many of them without, however, highlighting the common underlying evolutionary logic nor, therefore, using a unitary definition for them.

A classification of phenoptotic phenomena was proposed in 2012 (Libertini, 2012) (Fig. 7.1).

Below is this classification with some significant examples:

(A) Obligatory and rapid phenoptosis

When phenoptosis happens for all the individuals of a species in a specific period of the lifecycle and is quite rapid in its manifestation, it is defined as "Obligatory and rapid phenoptosis". This type of phenoptosis is described by Finch as "Rapid senescence and sudden death" (Finch, 1990, p. 43). However, the use of the term senescence should be avoided because it can lead to confusion with the "Obligatory and slow phenoptosis" (s. below). There are two general types of Obligatory and rapid phenoptosis:

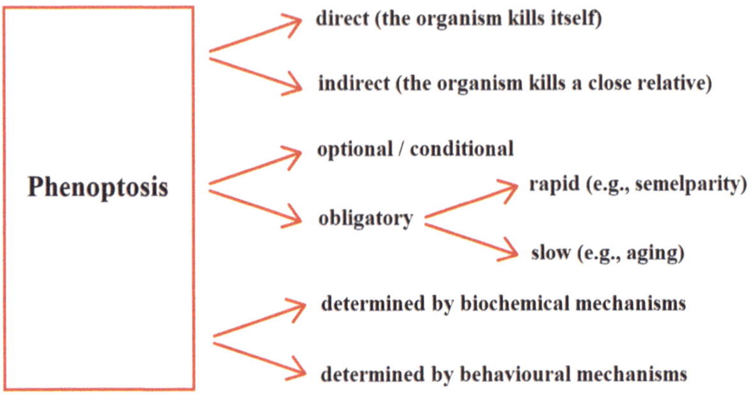

Fig. 7.1 Types of phenoptotic phenomena used for their classification (Libertini, 2012)

A-1) *Related to the reproductive cycle*

- It is a common type of phenoptosis and generally consists of a single phase of reproduction, soon followed by death (semelparity). This is a well-known phenomenon for many plant species, in particular for monocarpic angiosperms (Finch, 1990), and there is no doubt that it is a programmed and genetically determined process: "Many botanists emphasize that plant senescence is an orderly and active process …" (Finch, 1990, p. 98).

- Semelparity is also present in other groups of living beings. For example, it can be observed among fish (various species of *Anguilliformes* and *Salmoniformes*) and mammals (some rodents and some didelphid and dasyurid marsupials) (Finch, 1990).

- The definition of Obligatory and rapid phenoptosis does not mean that lifespan is necessarily short. "Various species of the thick-stemmed bamboos (*Phyllostachys*) have prolonged phases of vegetative growth that last for many years or decades (7, 30, 60, or 120 years) according to the species, before suddenly flowering and dying … Their death is often 'gregarious', such that local populations show synchrony … The 120-year flowering cycles of *P. bambusoides* and *P. henonis* are documented in Japan, going back 1000 years (Kawamura, 1927)." (Finch, 1990, p. 101)
  - Different cases of phenoptosis of this subtype are the "endotokic matricide" shown by some invertebrates (e.g., the nematode *Ascaris nigrovenosa*), where "the young kill their mother by boring through her body wall" (Finch, 1990, p. 102), or when the larvae feed on mother's body (e.g., the beetle *Micromalthus debilis*) (Finch, 1990, p. 103).

A-2) *Deriving in general from characteristics of the life cycle*

- In many insects, the larval stage feeds and accumulates energy, while adults serve for reproduction and use, mainly or exclusively, the reserves accumu-

Fig. 7.2 Honeybee queen and drone copulating in flight (left); ant queen and drone while mating (right)

lated in the previous stage. As an extreme but quite common case, the adults of many insects (e.g., mayflies, *Epheromoptera*) are entirely incapable of feeding, a characteristic defined as aphagy: "Aphagy from defective mouthparts or digestive organs is very common during the adult phases of insects … and is *the* limiting factor in the adult lifespan of many short-lived species. This phenomenon is, inarguably, programmed senescence …." (Finch, 1990, p. 49)

- Other insects in the adult phase can live only a few weeks as they do not have mechanisms for renewing their anatomical structures. The housefly (*Musca domestica*) and the fruit fly (*Drosophila melanogaster*) are examples of these insects (Finch, 1990, pp. 62–66).
- Male rotifers are much shorter-lived than females and survive only hours or a few days, as they, unlike females, "lack an intestine, anus, or other excretory and digestive organs …" (Finch, 1990, p. 57).
- In other cases, the male (drone) dies shortly after mating with the female, e.g., male bees (Winston, 1987) and ants (Holldobler & Wilson, 1990) (Fig. 7.2).

(B) Obligatory and slow phenoptosis

Phenoptosis is obligatory and slow when, in any individual of a species, there is a progressive slow decline of fitness, i.e., an increasing probability of death.

There are two subtypes:

B-1) *Duplications-related increasing probability of apoptosis in monocellular eukaryotes*

- In yeast, a monocellular eukaryotic species, the mother cell divides into two cells ("daughter" and "mother" cell) with different characteristics. Cells of the "daughter" lineage show all the physiological characteristics of healthy yeast cells. On the contrary, cells of the "mother" lineage show increasing metabolic alterations and an increasing probability of death by apoptosis in relation to the number of previous duplications, with a death rate increment following an exponential dynamic (Laun et al., 2007).

B-2) *Age-related increasing mortality in multicellular eukaryotes*

- Some species of multicellular eukaryotes (e.g., our species and most of the animals that are familiar to us) show an "increasing mortality with increasing chronological age in populations the wild" (Libertini, 1988), also described as "actuarial senescence" (Holmes & Austad, 1995) and "Gradual senescence with definite lifespan" (Finch, 1990). This type of phenoptosis, whose existence was denied or underestimated in the past (Kirkwood & Austad, 2000), is well documented in wild conditions (Ricklefs, 1998; Nussey et al., 2013) and will be discussed in *Chap. 8 -Physiological phenomena that cause troubles and suffering or even death—Part III: Slow phenoptosis or aging.*

(C) Optional phenoptosis

- Phenoptosis is optional (or conditional) when, under the determination or influence of genetically determined mechanisms, the death, or a behavior that determines a risk of death, happens only in particular conditions.

There are two subtypes:

C-1) *Determined by biochemical mechanisms*

- *Among unicellular eukaryotes*

In yeast, for cells of the mother lineage, there is an increasing vulnerability to death by apoptosis in relation to the number of previous duplications (Laun et al., 2001; Herker et al., 2004; Büttner et al., 2006; Fabrizio & Longo, 2008). In proportion to the increasing vulnerability, apoptosis is triggered by: (i) the shortage of nutrients (Granot et al., 2003); (ii) harmful chemical alterations (Madeo et al., 1999); and (iii) unsuccessful mating (Büttner et al., 2006).

As well as for apoptosis in multicellular eukaryotes, in yeast, the cellular parts of the individuals died by apoptosis are usefully absorbed by the other cells, which so "… are able to survive longer with substances released by dying cells." (Herker et al., 2004).

Therefore, in conditions of nutrient deficiency, the apoptosis of part of the population—which for yeast is also phenoptosis as the species is single-celled—appears an adaptive phenomenon because it releases cellular components then used by the other individuals of the population allowing the survival of the deme (Skulachev, 2002b, 2004; Fabrizio et al., 2004; Herker et al., 2004; Longo et al., 2005; Skulachev & Longo, 2005; Mitteldorf, 2006). As yeast species is divided into small demes, each with one or a few clones genetically identical, this interpretation appears plausible.

In proportion to the number of previous duplications, the increasing vulnerability of the individuals of the mother lineage appears to be a way to differentiate those destined for sacrifice in case of need among genetically uniform single-celled individuals.

A different case is apoptosis triggered by toxins secreted by enemy yeast tribes, as in this case, the competitors appear to exploit an adaptive mechanism to their advantage (Büttner et al., 2006).

– *Among prokaryotes*

It may seem strange or unlikely that single-celled prokaryotic organisms, with their simple structure, can express phenoptotic phenomena such as mass suicides in favor of other individuals. Yet, phenomena of this type are not a rare curiosity and are justified in adaptive terms (Lane, 2008). For example:

– Mass suicide of bacterial phytoplankton as a means of defense to block the spread of phages (bactericidal viruses) is well known (Lane, 2008);
– Phage infection triggers bacterial suicide "thereby curtailing viral multiplication and protecting nearby E. coli from infection" (Raff, 1998);
– "In E. coli, three suicide mechanisms that are activated by the appearance of a phage in the cell interior have been described" (Skulachev, 2004), and there is a "built-in suicide module" triggered by the presence of antibiotics as a defense against the propagation of other bacterial strains that produce them (Engelberg-Kulka et al., 2004).

The cases of "programmed cell death" in bacterial phytoplankton (Lane, 2008) and of "programmed death in bacteria" (Skulachev, 2004) need mechanisms of natural selection at a supra-individual level and adaptive explanations, which could be:

(i) defense against the propagation of phages (Raff, 1998; Lane, 2008);
(ii) elimination of damaged individuals that reduce the available resources for other individuals. "Most bacterial species actually do not live as planktonic suspensions in vivo but form complex biofilms, tightly knit communities of cells (Costerton et al., 1999). From this perspective, programmed death of damaged cells may be beneficial to a multicellular bacterial community." (Lewis, 2000)

C-2) *Determined by behavioral mechanisms*

In particular cases, behaviors that are risky for the individual who manifests them, while advantageous for the survival of genetically related individuals, can be favored by natural selection at the supra-individual level:

1. Among invertebrates, individual sacrifices are well known and widespread in the species of insects defined as eusocial (bees, ants, termites) (Wilson, 1975);
2. Among species of vertebrates with social organization, behaviors that are dangerous to one's own survival but which increase the chances of survival of other group members are common. For example, the dominant males of two baboon species—*Papio ursinus* (chacma baboons) (Hall, 1960) and *Papio cynocephalus* (yellow baboons) (Altmann & Altmann, 1970)—at considerable individual risk defend the other members of their group from predators (Fig. 7.3);
3. Some species of birds, facing severe risk, show behaviors to distract the attention of predators and save their offspring (Armstrong, 1947; Gramza, 1967), i.e., distraction behaviors (Brown, 1962);

Fig. 7.3 Some baboons who, risking their lives, attack a lioness to defend their group

4. For the males of many species, in search of a mate or to defend the exclusivity of their mates, the struggle to overcome rivals is dangerous and sometimes even deadly: e.g., for the males of the southern sea lion, *Otaria byronia* (Campagna et al., 2021). This type of behavior has the characteristics of optional phenoptosis;
5. The male of the *Mantis religiosa* (European mantis), in mating with the female, exposes itself to a serious risk, as the female, during copulation or just after, tries to kill him and then eat him, succeeding in about a third of the cases (Lawrence, 1992). Copulation cannibalism in mantis raises the male's investment in the offspring and significantly increases the number of eggs the female lays (Brown & Barry, 2016). At extremely low populational density, when a second meeting between a male and a female becomes unlikely, self-sacrifice by investing maximal resources by both partners in the only offspring generation could be beneficial. On the contrary, in case of a moderate population density and high recourse availability, the males prefer heavier, well-fed females with higher fecundity and lower risk of cannibalizing them (Lelito & Brown, 2008).

(D) Indirect phenoptosis

There is indirect phenoptosis when an individual, in particular conditions, determines or favors the death of a close relative.

There are two types of indirect phenoptosis.

D-1) *Determined by biochemical mechanisms*

- Among mice, when there is a new male partner, for a certain period, he kills the female's newborns generated by the previous partner as they do not share his genes. As an opposite strategy with an adaptive meaning, when there is a new partner, the pregnant female aborts through an internal biochemical mechanism. In this way, she saves the energy required to continue the gestation of an offspring that will be killed by the new partner (Bruce, 1959; Hattori et al., 2017).

- In vertebrates, a host's immune system must discriminate between its own antigens and those of a parasite. In contrast, the parasites try to avoid recognition by the immune system by using coat proteins with the same antigenic characteristics as the host (antigen mimicry) to counteract host defenses. In turn, the individuals of the host species try to limit the antigen mimicry of the parasites through the greatest possible variety of their own antigenic formulas. So, parasites cannot get an antigen mimicry that allows them to neutralize the antigenic recognition of all the hosts: the similarities between the antigenic formulas of parasites and host should determine a greater vulnerability to infections, while, conversely, more significant differences would offer greater resistance (Libertini, 1983, *Chap. IV—Antigenic polymorphism*).

In humans, the major histocompatibility complex (MHC) is one of the main groups of alleles that allow maximum antigenic variability for the host. There are known correlations between specific MHC alleles and resistance or susceptibility to infectious diseases (Lechler & Warrens, 2000; Shiina et al., 2004; Shiina et al., 2009).

As the best progeny is characterized by the greatest antigenic variability, there are two known mechanisms aimed at increasing the antigenic variability of the progeny:

(i) MHC-mediated mate choice. This mechanism is documented for various vertebrate taxa (Slev et al., 2006) and also for our species: (a) Women of college students rated the odors of men with different MCH alleles as being 'more pleasant' than those of men with similar MHC alleles (Wedekind et al., 1995; Wedekind & Füri, 1997); (b) In a community that was isolated and ethnically homogenous (Hutterites), statistically significantly fewer pairs were found to be matched according to a 16-locus MHC haplotype (Ober et al., 1997; Ober et al., 1999).

(ii) "post-copulatory selection (also called cryptic female choice)" (Loisel et al., 2008). In this case, which is a form of indirect phenoptosis, spontaneous miscarriages eliminate the fetuses with lesser antigen variability (Apanius et al., 1997). This phenomenon is well-documented for animals (Tregenza & Wedell, 2000). About our species, see the next *§7.2 Phenoptotic phenomena in our species different from slow phenoptosis.*

D-2) *Determined by behavioral mechanisms*

The killing of an offspring by a kin individual (kin infanticide) may be caused by a parent, through direct actions or abandonment (parental or filial infanticide), or by full- or half-sibs (siblicide). In the animal world, kin infanticide, both parental infanticide and siblicide, is widespread and often is followed by cannibalism (Hausfater & Hrdy, 1984, Introduction and Overview).

For many avian species, both parental infanticide and siblicide are well documented, and these behaviors appear well explainable in terms of inclusive fitness: "… ornithological field data demonstrate that [infanticide] is regularly the most important source of nestling mortality in certain avian species … Social behavior mediating intrabrood mortality is performed largely by the chicks themselves, although parents are also potentially important … siblicidal behavior reflects an evolutionary compromise between the direct and indirect components of the elder sib's fitness (Brown & Brown, 1981) …" (Mock, 1984).

Analogous actions occur in many species of fish (Dominey & Blumer, 1984) and invertebrates (Polis, 1984).

7.2 Phenoptotic Phenomena in our Species Different from Slow Phenoptosis

As an individual may die only once, a possible wrong deduction is that each species can present not more than a single phenoptotic phenomenon. However, a species comprises many individuals, and various types of phenoptosis show up only under certain circumstances. Therefore, it is possible that a species can exhibit various types of phenoptotic phenomena, as is the case with our species.

Before Birth

– *Elimination of fetuses with imperfections.*

In many cases, the fetus presents anomalies, and this, in various ways and times, leads to the termination of the pregnancy. This has been discussed in detail in *§2.2 Calculation of the frequencies expected for diseases caused by alterations of the genotype*. The mechanisms causing these phenomena are of fundamental importance to eliminate genetic alterations. In quantitative terms, more than two-thirds of the fetuses are suppressed: "Human reproduction is remarkably inefficient; nearly 70% of human conceptions do not survive to live birth … Overall, 50%-70% of specimens from sporadic spontaneous losses show some type of cytogenetic abnormality" (Hyde & Schust, 2015).

The comment "Human reproduction is remarkably inefficient …" (Hyde & Schust, 2015) does not consider the extreme importance of the mechanisms that determine this "inefficiency".

– *Elimination of fetuses with lesser antigen variability*

An excess of heterozygotes for MHC was found in newborns (Dorak et al., 2002), and spontaneous miscarriages eliminated the fetuses with lesser antigen variability, which are more vulnerable to infections (Apanius et al., 1997).

In ethnically homogenous and isolated communities, comparing couples with shared or not shared HLA-DR alleles, in the first case, there was a greater interval

Fig. 7.4 The elimination of a healthy twin (before the birth by spontaneous abortion, or after the birth by infanticide) is an adaptation in evolutionary terms

between pregnancies (Ober, 1992), a greater pregnancy loss rate (Ober et al., 1998), and significantly fewer children (Ober & van der Ven, 1997).

– *Reduction of twin to a singleton pregnancy*

In wild conditions, it is difficult or even death-causing for our species to rear two children simultaneously. This may be the evolutionary explanation of the phenomenon defined as the "vanishing twin" (Landy & Keith, 1998): "… sonograms of women in the first trimester of pregnancy reveal that twins are conceived two to four times more often than they are born; in the majority of cases, the smaller of the two foetuses disappears by the third trimester and is apparently reabsorbed by the mother (Robinson & Caines, 1977; Varma, 1979)." (Hausfater & Hrdy, 1984, Introduction and Overview).

In a recent review, the phenomenon is well described: "Vanishing twin syndrome … is a condition in which one of a set of twins or multiple embryos dies in utero, disappear, or gets resorbed partially or entirely, with an outcome of a spontaneous reduction of a multi-fetus pregnancy to a singleton pregnancy, portraying the image of a vanishing twin." (Zamani & Parekh, 2022). However, the authors do not consider the possible evolutionary cause of it nor the possible relationships with the infanticide observed in various primitive societies when twin pregnancies are carried to term (s. below) (Fig. 7.4).

After Birth, in Childhood

– The infanticide has been documented for various people, constituting a form of *D Indirect phenoptosis—D-2) Determined by behavioral mechanisms*. For the infanticides found in 60 different mostly primitive societies and reported by Daly and Wilson in "Table I. Circumstances of alleged infanticide in society" (Daly & Wilson, 1984), there were various events that appeared to motivate the killings:

Adulterous conception (15 societies),
Deformed or very ill (21),
Twins (14),
Birth too soon or too many (11),
No male support (6),
Mother dead (6),
Mother unwed (14),
Economic hardship (3).

Scrimshaw recalls that infanticide in human populations is described in both ancient and modern times (Scrimshaw, 1984). In Table I of her work (*Decision maker and proximate reasons for infanticide*), she reports that infanticide is decided (in order of importance) by the mother or father, family, society, and birth attendant. As reasons leading to this decision, it reports, inter alia: twins, deformed, physical birth defects, mental defects of infant, dead mother or father, bad omens at birth, wrong sex (usually a girl), infant unattractive or disliked, mother labeled "defective", hatred of spouse, illegitimacy, overpopulation, famine, or other disaster, too large a family, too close spacing, pathological behavior (Scrimshaw, 1984). The main types of motivation appear to be: (i) insufficient resources to support the infant; and (ii) elimination of infants who are unhealthy in order not to take away resources from any subsequent healthy infants.

By the standards of modern morality, the terrible cases of the elimination of defenseless infants certainly arouse shivers of indignation. However, in modern societies all over the world, the inconveniences deriving from unwanted births or births that one does not think can be sustained economically, apart from the cases in which infanticide is still performed, are faced either with contraceptive methods or through recourse to induced abortion.

In 2020, it has been estimated that 61% of all unintended pregnancies and 29% of all pregnancies end in induced abortions, with 73 million induced abortions each year (Bearak et al., 2020).

This extraordinary number of fetuses eliminated each year may be considered irrelevant for the calculation of mortality because regards fetuses, i.e., lives deemed to have not yet begun, or, on the contrary, should be compared with the total number of deaths from any cause, which was equal to 50,050,083 in 2020 (World Cause of Death, 2020). The choice between these two views is ethical or religious and not scientific.

In both pre-modern and modern societies, the elimination of offspring because (i) conceived in undesirable conditions; or (ii) they are physically or mentally abnormal; or (iii) there is a lack of economic capacity, or for other reasons, is a common event, and the difference, due to technological disparities, is between the suppression of already born subjects, by direct actions or abandonment, and the elimination of fetuses through induced abortion.

In Adulthood

- There are many circumstances in which an individual puts his or her life in danger, or even sacrifices it, to defend or save one or more genetically related or even non-genetically related individuals (Fig. 7.5) or even to win or keep a mate.
- For species that are capable of social organization—e.g., chimpanzees (Gilby et al., 2013)—the organization into groups for various purposes is typical, including more effective hunting, but mainly to defend or expand the territory by actions defined as wars. In our species, as social organization has become more extensive and effective, wars have increased extraordinarily in terms of killing capacity and number of deaths. Approximately 9–11 million soldiers died in World War I (out of a total of 15–22 million casualties) (Haythornthwaite, 1993) and, in World War II, 21–25 million soldiers (out of a total of 70–85 million casualties) (Doyle, 2013) (Fig. 7.6). Wars are cases in which individuals expose themselves to serious dangers of death and therefore must be considered as a type of optional phenoptosis.
- For all individuals who have not died of other causes, there is the common fate of death due to causes related to aging (obligatory slow phenoptosis), which will be discussed in the next chapter.

At this point, a brief discussion of what is defined as "behavior" may be helpful.

A "behavior" certainly implies that there is a nervous system that determines it through instincts, biochemical mechanisms, and, where there are, intelligence, awareness, and will (however defined or conceived).

For species such as eusocial insects, it is immediate to admit that behaviors are determined by instincts that guide them through sophisticated algorithms.

Fig. 7.5 Artistic representations of a dangerous rescue (left) and of a sacrifice to save other lives (right)

Fig. 7.6 Images from World War I (left) and World War II (right)

For vertebrates other than our species, there is surely an instinctive component, while the existence of specific genes that always determine a stereotyped behavior is doubtful. It is more likely that behaviors are guided in a varying way by genes that determine neuroendocrine development and a set of hormones that, in particular conditions, favor different strategies. For example, in the case of danger, it will be possible to flee or face the danger or freeze in immobility with different chances of success.

The choice of the strategy is also strongly influenced by the possible consequences of the strategy for related individuals. A mother with offspring will probably strenuously oppose a possible aggressor while, more likely, she will escape in the absence of children to protect.

In the case of our species, but to a certain extent also for other rather intelligent species, the instincts and the influence of the neuroendocrine system are modulated and modified by what we can define as consciousness, intelligence, and will, making the behavior less predictable and more articulated.

However, we must not forget that even these capacities have been shaped by natural selection to make behaviors more effective and valuable and do not constitute a complete alternative to instincts and neuroendocrine mechanisms.

References

Altmann, S. A., & Altmann, J. (1970). *Baboon ecology: African field research*. The University of Chicago Press.

Apanius, V., Penn, D., Slev, P. R., Ruff, L. R., & Potts, W. K. (1997). The nature of selection on the major histocompatibility complex. *Critical Reviews in Immunology, 17*(2), 179–224. https://doi.org/10.1615/critrevimmunol.v17.i2.40

Armstrong, E. A. (1947). *Bird display and behaviour: An introduction to the study of bird psychology*. Lindsay Drummond. Reprint (1965): New York: Dover.

Bearak, J., Popinchalk, A., Ganatra, B., Moller, A.-B., Tunçalp, Ö., Beavin, C., et al. (2020). Unintended pregnancy and abortion by income, region, and the legal status of abortion: Estimates from a comprehensive model for 1990–2019. *The Lancet. Global Health, 8*(9), e1152–e1161. https://doi.org/10.1016/S2214-109X(20)30315-6

Brown, J. L., & Brown, E. R. (1981). Kin selection and individual selection in babblers. In R. D. Alexander & D. D. Tinkle (Eds.), *Natural selection and social behavior* (pp. 244–256). Chiron Press.
Brown, R. G. (1962). The aggressive and distraction behaviour of the Western sandpiper, Ereunetes mauri. *Ibis, 104*(1), 1–12. https://doi.org/10.1111/j.1474-919X.1962.tb08624.x
Brown, W. D., & Barry, K. L. (2016). Sexual cannibalism increases male material investment in offspring: Quantifying terminal reproductive effort in a praying mantis. *Proceedings. Biological sciences, 283*(1833), 20160656. https://doi.org/10.1098/rspb.2016.0656
Bruce, H. M. (1959). An exteroceptive block to pregnancy in the mouse. *Nature, 184*(4680), 105. https://doi.org/10.1038/184105a0
Büttner, S., Eisenberg, T., Herker, E., Carmona-Gutierrez, D., Kroemer, G., & Madeo, F. (2006). Why yeast cells can undergo apoptosis: Death in times of peace, love, and war. *The Journal of Cell Biology, 175*(4), 521–525. https://doi.org/10.1083/jcb.200608098
Campagna, C., Le Boeuf, B., & Bisioli, C. (2021). Mating games and raiding parties in Southern Sea lions. In C. Campagna & R. Harcourt (Eds.), *Ethology and behavioral ecology of Otariids and the Odobenid. Ethology and behavioral ecology of marine mammals*. Springer. https://doi.org/10.1007/978-3-030-59184-7_9
Costerton, J. W., Stewart, P. S., & Greenberg, E. P. (1999). Bacterial biofilms: A common cause of persistent infections. *Science, 284*(5418), 1318–1322. https://doi.org/10.1126/science.284.5418.1318
Daly, M., & Wilson, M. (1984). A sociobiological analysis of human infanticide. In G. Hausfater & S. B. Hrdy (Eds.), *Infanticide: Comparative and evolutionary perspectives (pp. 487-502).* Aldine Transaction.
Darwin, C. R. (1869). *Origin of species* (5th ed.). John Murray.
Darwin, C. R. (1871). *The descent of man, and selection in relation to sex*. John Murray.
Dominey, W. J., & Blumer, L. S. (1984). Cannibalism of early life stages in fishes. In G. Hausfater & S. B. Hrdy (Eds.), *Infanticide: Comparative and evolutionary perspectives (pp. 43-64).* Aldine Transaction.
Dorak, M. T., Lawson, T., Machulla, H. K., Mills, K. I., & Burnett, A. K. (2002). Increased heterozygosity for MHC class II lineages in newborn males. *Genes and Immunity, 3*(5), 263–269. https://doi.org/10.1038/sj.gene.6363862
Doyle, P. (2013). *World war II in numbers*. Bloomsbury Publishing Plc.
Engelberg-Kulka, H., Sat, B., Reches, M., Amitai, S., & Hazan, R. (2004). Bacterial programmed cell death systems as targets for antibiotics. *Trends in Microbiology, 12*(2), 66–71. https://doi.org/10.1016/j.tim.2003.12.008
Fabrizio, P., Battistella, L., Vardavas, R., Gattazzo, C., Liou, L. L., Diaspro, A., et al. (2004). Superoxide is a mediator of an altruistic aging program in Saccharomyces cerevisiae. *The Journal of Cell Biology, 166*(7), 1055–1067. https://doi.org/10.1083/jcb.200404002
Fabrizio, P., & Longo, V. D. (2008). Chronological aging-induced apoptosis in yeast. *Biochimica et Biophysica Acta, 1783*(7), 1280–1285. https://doi.org/10.1016/j.bbamcr.2008.03.017
Finch, C. E. (1990). *Longevity, senescence, and the genome*. University of Chicago Press.
Gilby, I. C., Brent, L. J., Wroblewski, E. E., Rudicell, R. S., Hahn, B. H., Goodall, J., & Pusey, A. E. (2013). Fitness benefits of coalitionary aggression in male chimpanzees. *Behavioral Ecology and Sociobiology, 67*(3), 373–381. https://doi.org/10.1007/s00265-012-1457-6
Gramza, A. F. (1967). Responses of brooding nighthawks to a disturbance stimulus. *Auk, 84*(1), 72–86. www.jstor.org/stable/4083256
Granot, D., Levine, A., & Dor-Hefetz, E. (2003). Sugar-induced apoptosis in yeast cells. *FEMS Yeast Research, 4*(1), 7–13. https://doi.org/10.1016/S1567-1356(03)00154-5
Hall, K. R. L. (1960). Social vigilance behaviour of the chacma baboon, Papio ursinus. *Behaviour, 16*(3–4), 261–294. https://doi.org/10.1163/156853960X00188
Hamilton, W. D. (1964). The Genetical evolution of social behaviour. II. *Journal of Theoretical Biology, 7*(1), 1–52. https://doi.org/10.1016/0022-5193(64)90039-6

Hamilton, W. D. (1970). Selfish and spiteful behaviour in an evolutionary model. *Nature, 228*(5277), 1218–1220. https://doi.org/10.1038/2281218a0

Hattori, T., Osakada, T., Masaoka, T., Ooyama, R., Horio, N., Mogi, K., et al. (2017). Exocrine gland-secreting peptide 1 is a key chemosensory signal responsible for the Bruce effect in mice. *Current Biology, 27*(20), 3197–3201.e3. https://doi.org/10.1016/j.cub.2017.09.013

Hausfater, G., & Hrdy, S. B. (Eds.). (1984). *Infanticide: Comparative and evolutionary perspectives*. AldineTransaction.

Haythornthwaite, P. J. (1993). *The world war one source book*. Arms and Armour.

Herker, E., Jungwirth, H., Lehmann, K. A., Maldener, C., Fröhlich, K. U., Wissing, S., et al. (2004). Chronological aging leads to apoptosis in yeast. *The Journal of Cell Biology, 164*(4), 501–507. https://doi.org/10.1083/jcb.200310014

Holldobler, B., & Wilson, E. O. (1990). *The ants*. Springer.

Holmes, D. J., & Austad, S. N. (1995). Birds as animal models for the comparative biology of aging: A prospectus. *The Journals of Gerontology. Series A, Biological Sciences and Medical Sciences, 50*(2), B59–B66. https://doi.org/10.1093/gerona/50a.2.b59

Hyde, K. J., & Schust, D. J. (2015). Genetic considerations in recurrent pregnancy loss. *Cold Spring Harbor Perspectives in Medicine, 5*(3), a023119023119. https://doi.org/10.1101/cshperspect.a023119

Kawamura, S. (1927). On the periodical flowering of the bamboo. *Japanese Journal of Botany, 3*, 335–349.

Kirkwood, T. B., & Austad, S. N. (2000). Why do we age? *Nature, 408*(6809), 233–238. https://doi.org/10.1038/35041682

Landy, H. J., & Keith, L. G. (1998). The vanishing twin: A review. *Human Reproduction Update, 4*(2), 177–183. https://doi.org/10.1093/humupd/4.2.177

Lane, N. (2008). Marine microbiology: Origins of death. *Nature, 453*(7195), 583–585. https://doi.org/10.1038/453583a

Laun, P., Bruschi, C. V., Dickinson, J. R., Rinnerthaler, M., Heeren, G., Schwimbersky, R., et al. (2007). Yeast mother cell-specific ageing, genetic (in)stability, and the somatic mutation theory of ageing. *Nucleic Acids Research, 35*(22), 7514–7526. https://doi.org/10.1093/nar/gkm919

Laun, P., Pichova, A., Madeo, F., Fuchs, J., Ellinger, A., Kohlwein, S., et al. (2001). Aged mother cells of *Saccharomyces cerevisiae* show markers of oxidative stress and apoptosis. *Molecular Microbiology, 39*(5), 1166–1173. https://doi.org/10.1111/j.1365-2958.2001.02317.x

Lawrence, S. E. (1992). Sexual cannibalism in the praying mantid, Mantis religiosa: A field study. *Animal Behaviour, 43*(4), 569–583. https://doi.org/10.1016/S0003-3472(05)81017-6

Lechler, R., & Warrens, A. (2000). *HLA in health and disease*. Academic.

Lelito, J. P., & Brown, W. D. (2008). Mate attraction by females in a sexually cannibalistic praying mantis. *Behavioral Ecology and Sociobiology, 63*, 313–320. https://doi.org/10.1007/s00265-008-0663-8

Lewis, K. (2000). Programmed death in bacteria. *Microbiology and Molecular Biology Reviews, 64*(3), 503–514. https://doi.org/10.1128/mmbr.64.3.503-514.2000

Libertini, G. (1983). *Ragionamenti evoluzionistici [book in Italian]*. Società Editrice Napoletana. English edition (2011): *Evolutionary arguments on aging, disease, and other topics*. Azinet Press.

Libertini, G. (1988). An adaptive theory of the increasing mortality with increasing chronological age in populations in the wild. *Journal of Theoretical Biology, 132*(2), 145–162. https://doi.org/10.1016/s0022-5193(88)80153-x

Libertini, G. (2012). Classification of Phenoptotic Phenomena. *Biochemistry (Moscow), 77*(7), 707–715. https://doi.org/10.1134/S0006297912070024

Libertini, G., Corbi, G., Conti, V., Shubernetskaya, O., & Ferrara, N. (2021). *Evolutionary gerontology and geriatrics: Why and how we age*. In *Advances in studies of aging and health, vol. 2*. Springer Nature.

Loisel, D. A., Alberts, S. C., & Ober, C. (2008). Functional significance of MHC variation in mate choice, reproductive outcome, and disease risk. In S. C. Stearns & J. C. Koella (Eds.), *Evolution in health and disease* (2nd ed., pp. 95–108). Oxford University Press.

Longo, V. D., Mitteldorf, J., & Skulachev, V. P. (2005). Programmed and altruistic ageing. *Nature Reviews. Genetics, 6*(11), 866–872. https://doi.org/10.1038/nrg1706

Madeo, F., Fröhlich, E., Ligr, M., Grey, M., Sigrist, S. J., Wolf, D. H., & Fröhlich, K. U. (1999). Oxygen stress: A regulator of apoptosis in yeast. *The Journal of Cell Biology, 145*(4), 757–767. https://doi.org/10.1083/jcb.145.4.757

Mitteldorf, J. (2006). How evolutionary thinking affects people's ideas about aging interventions. *Rejuvenation Resarch, 9*(2), 346–350. https://doi.org/10.1089/rej.2006.9.346

Mock, D. W. (1984). Infanticide, siblicide, and avian nestling mortality. In G. Hausfater & S. B. Hrdy (Eds.), *Infanticide: Comparative and evolutionary perspectives* (pp. 3–30). AldineTransaction.

Nowak, A. M., Tarnita, C. E., & Wilson, E. O. (2010). The evolution of eusociality. *Nature, 466*(7310), 1057–1062. https://doi.org/10.1038/nature09205

Nussey, D. H., Froy, H., Lemaitre, J. F., Gaillard, J. M., & Austad, S. N. (2013). Senescence in natural populations of animals: Widespread evidence and its implications for bio-gerontology. *Ageing Research Reviews, 12*(1), 214–225. https://doi.org/10.1016/j.arr.2012.07.004

Ober, C. (1992). The maternal-fetal relationship in human pregnancy: An immunogenetic perspective. *Experimental and Clinical Immunogenetics, 9*(1), 1–14.

Ober, C., Hyslop, T., Elias, S., Weitkamp, L. R., & Hauck, W. W. (1998). Human leukocyte antigen matching and fetal loss: Results of a 10 year prospective study. *Human Reproduction (Oxford), 13*(1), 33–38. https://doi.org/10.1093/humrep/13.1.33

Ober, C., & van der Ven, K. (1997). Immunogenetics of reproduction: An overview. In L. B. Olding (Ed.), *Reproductive immunology* (pp. 1–23). Springer.

Ober, C., Weitkamp, L. R., Cox, N., Dytch, H., Kostyu, D., & Elias, S. (1997). HLA and mate choice in humans. *American Journal of Human Genetics, 61*(3), 497–504. https://doi.org/10.1086/515511

Ober, C., Weitkamp, L. R., & Cox, N. (1999). HLA and mate choice. In R. E. Johnston, D. Müller-Schwarze, & P. W. Sorenson (Eds.), *Advances in chemical signals in vertebrates* (pp. 189–199). Kluwer Academic Press.

Pepper, J. W., Shelton, D. E., Rashidi, A., & Durand, P. M. (2013). Are internal, death-promoting mechanisms ever adaptive? *Journal of Phylogenetics & Evolutionary Biology, 1*, 113. https://doi.org/10.4172/2329-9002.1000113

Polis, G. A. (1984). Intraspecific predation and "infant killing" among invertebrates. In G. Hausfater & S. B. Hrdy (Eds.), *Infanticide: Comparative and evolutionary perspectives* (pp. 87–104). AldineTransaction.

Raff, M. (1998). Cell suicide for beginners. *Nature, 396*(6707), 119–122. https://doi.org/10.1038/24055

Ricklefs, R. E. (1998). Evolutionary theories of aging: Confirmation of a fundamental prediction, with implications for the genetic basis and evolution of life span. *The American Naturalist, 152*(1), 24–44. https://doi.org/10.1086/286147

Robinson, H. P., & Caines, J. S. (1977). Sonar evidence of early pregnancy failure in patients with twin conceptions. *British Journal of Obstetrics and Gynaecology, 84*(1), 22–25. https://doi.org/10.1111/j.1471-0528.1977.tb12460.x

Scrimshaw, S. C. M. (1984). Infanticide in human population: Societal and human concerns. In G. Hausfater & S. B. Hrdy (Eds.), *Infanticide: Comparative and evolutionary perspectives* (pp. 439–462). AldineTransaction.

Shiina, T., Hosomichi, K., Inoko, H., & Kulski, J. K. (2009). The HLA genomic loci map: Expression, interaction, diversity and disease. *Journal of Human Genetics, 54*(1), 15–39. https://doi.org/10.1038/jhg.2008.5

Shiina, T., Inoko, H., & Kulski, J. K. (2004). An update of the HLA genomic region, locus information and disease associations. *Tissue Antigens, 64*(6), 631–649. https://doi.org/10.1111/j.1399-0039.2004.00327.x

Skulachev, V. P. (1997). Aging is a specific biological function rather than the result of a disorder in complex living systems: Biochemical evidence in support of Weismann's hypothesis. *Biochemistry (Moscow), 62*(11), 1191–1195. protein.bio.msu.ru/biokhimiya/contents/v62/full/62111394.html.

Skulachev, V. P. (1999). Phenoptosis: Programmed death of an organism. *Biochemistry (Moscow), 64*(12), 1418–1426. protein.bio.msu.ru/biokhimiya/contents/v64/full/64121679.html.

Skulachev, V. P. (2002a). Programmed death phenomena: From organelle to organism. *Annals of the New York Academy of Sciences, 959*, 214–237. https://doi.org/10.1111/j.1749-6632.2002.tb02095.x

Skulachev, V. P. (2002b). Programmed death in yeast as adaptation? *FEBS Letters, 528*(1–3), 23–26. https://doi.org/10.1016/s0014-5793(02)03319-7

Skulachev, V. P. (2004). Aging and the programmed death phenomena. In T. Nyström & H. D. Osiewacz (Eds.), *Topics in current genetics, Vol. 3, model systems in aging* (pp. 191–238). Springer.

Skulachev, V. P., & Longo, V. D. (2005). Aging as a mitochondria-mediated atavistic program: Can aging be switched off? *Annals of the New York Academy of Sciences, 1057*, 145–164. https://doi.org/10.1196/annals.1356.009

Slev, P. R., Nelson, A. C., & Potts, W. K. (2006). Sensory neurons with MHC-like peptide binding properties: Disease consequences. *Current Opinion in Immunology, 18*(5), 608–616. https://doi.org/10.1016/j.coi.2006.07.012

Spencer, H. (1864). *The principles of biology*. Williams and Norgate.

Tregenza, T., & Wedell, N. (2000). Genetic compatibility, mate choice and patterns of parentage: Invited review. *Molecular Ecology, 9*(8), 1013–1027. https://doi.org/10.1046/j.1365-294x.2000.00964.x

Trivers, R. L. (1971). The evolution of reciprocal altruism. *The Quarterly Review of Biology, 46*(1), 35–57. https://doi.org/10.1086/406755

Varma, T. R. (1979). Ultrasound evidence of early pregnancy failure in patients with multiple conceptions. *British Journal of Obstetrics and Gynaecology, 86*(4), 290–292. https://doi.org/10.1111/j.1471-0528.1979.tb11258.x

Wedekind, C., & Füri, S. (1997). Body odour preferences in men and women: Do they aim for specific MHC combinations or simply heterozygosity? *Proceedings. Biological sciences, 264*(1387), 1471–1479. https://doi.org/10.1098/rspb.1997.0204

Wedekind, C., Seebeck, T., Bettens, F., & Paepke, A. J. (1995). MHC-dependent mate preferences in humans. *Proceedings. Biological sciences, 260*(1359), 245–249. https://doi.org/10.1098/rspb.1995.0087

Wilson, E. O. (1975). *Sociobiology, the new synthesis*. Harvard University Press.

Winston, M. L. (1987). *The biology of the honey bee*. Harvard University Press.

World Cause of Death. (2020). (Data from WHO), Accessed December 15, 2022, from www.worldlifeexpectancy.com/world-rankings-total-deaths

Zamani, Z., & Parekh, U. (2022). Vanishing twin syndrome. In *StatPearls [Internet]*. StatPearls Publishing. https://www.ncbi.nlm.nih.gov/books/NBK563220/

Chapter 8
Physiological Phenomena that Cause Troubles and Suffering or Even Death: Part III: Slow Phenoptosis or Aging

8.1 Slow Phenoptosis or Aging: Why we Age

In this first section, the thesis of aging as a non-adaptive phenomenon will be shown as completely untenable. At the same time, it will be highlighted that various arguments strongly support the thesis of aging as an adaptive phenomenon.

In the second section, *§8.2 Slow phenoptosis or aging—How we age at the cellular level*, the cellular program that determines aging is described.

In the third and fourth sub-sections, *§8.3 Effects of cellular aging on tissues and organs* and *§8.4 Effects of cellular aging on the whole organism*, the effects of cellular aging on tissues, organs, the entire organism and our state of health are discussed.

8.1.1 Two Opposite Interpretations of Aging

Like many other species of multicellular eukaryotes familiar to us, our species shows, starting from a certain age, a progressive decline of physiological functions and, therefore, a reduction of fitness (i.e., survival capacity) commonly known as "aging". However, the term "aging" is used in many different contexts with different meanings. Hence, a more objective description, such as: "increasing mortality with increasing chronological age in populations the wild" (Libertini, 1988), or more briefly, "age-related progressive mortality increase (= fitness decline) in the wild" (Libertini, Corbi, et al., 2021a, p. 121), or "actuarial senescence" (Holmes & Austad, 1995) is preferable.

For this phenomenon, there are two completely opposite interpretations that deserve the definition of opposite paradigms (Libertini, Corbi, et al., 2021a, p. 121).

G. Libertini et al., *The Essential Revolution of Evolutionary Medicine*, Advances in Studies of Aging and Health 4, https://doi.org/10.1007/978-3-031-84877-3_8

The first, "non-adaptive (or non-programmed) aging paradigm", explains the phenomenon as something certainly harmful and for which it is therefore unlikely to propose an adaptive meaning or a genetically determined and regulated program that causes it. Consequently, aging is the consequence of the action of harmful factors that natural selection is unable to counteract sufficiently.

The second paradigm, the "adaptive (or programmed) aging paradigm", explains aging as something that, although harmful at the individual level, is beneficial at the supra-individual level under certain conditions. Therefore, aging has an adaptive value, is favored by natural selection under such conditions, and is necessarily determined by specific genetically determined and regulated mechanisms. In short, aging is a particular type of phenoptosis, i.e., "slow phenoptosis" (Skulachev, 2002).

8.1.2 Untenable Objections against Aging as an Adaptive Phenomenon

Regarding the hypothesis of aging as an adaptive phenomenon, three objections would seem to exclude this possibility *a priori* but are completely unfounded:

1. **Aging is certainly harmful to the individual and consequently cannot be favored by natural selection and have an adaptive significance**

This argument is clearly expressed in various works. For example: "The anomalous nature of aging as a putative adaptation is that it is bad for the individual in which the process is exhibited. An animal that grows to maturity and thereafter reproduces indefinitely has, other things being equal, a greater Darwinian fitness than one that grows to maturity and then survives and reproduces for only a fixed period of time." (Kirkwood & Melov, 2011); "... any hypothetical 'accelerated aging gene' would be disadvantageous to the individual. It is therefore difficult to see how genes for accelerated aging could be maintained in stable equilibrium, as individuals in whom the genes were inactivated by mutation would enjoy a selection advantage." (Kirkwood & Austad, 2000).

The fallacy of this argument is that it only considers selection at the individual level, whereby any trait harmful to the individual cannot be favored by natural selection. However, as we have seen in the case of the discussion regarding phenoptosis, considering selection at the supra-individual level, harmful phenomena at the individual level can be favored by natural selection if they involve an advantage in terms of kin selection or appropriate group selection models. If this were not the case, the innumerable forms of phenoptosis shown throughout the living world (Finch, 1990; Libertini, Corbi et al., 2021a, Chap. 2—Evolution and Phenoptosis) would be completely inexplicable.

2. **Aging is a phenomenon intrinsic to the nature of everything, including living beings, and therefore, being universal and inevitable, it cannot have an adaptive meaning**

Hayflick declared a few years ago: "Unlike any disease, age changes ... occur in every multicellular animal that reaches a fixed size at reproductive maturity ... occur in virtually all animate and inanimate matter." (Hayflick, 2000).

According to this belief, more or less rapid aging is possible, but it is excluded that there is no aging for a living being. Therefore, no adaptive justification is necessary for this phenomenon because it is shown by any species as it is universal for inanimate matter.

This belief is entirely contradicted by the evidence.

First, species showing a gradual age-related fitness decline are a small minority among all living species.

Limiting the discussion to multicellular species, many of these show life cycles very different from that of the gradual age-related fitness decline, as widely documented in an authoritative textbook (Finch, 1990).

As discussed with regard to phenoptosis, many species, after a period showing no age-related decline, reproduce and then die (semelparity). Other species have life cycles in which they pass through various stages of development and then reach an adult condition where there is death after a short time due to the inability to feed (aphagy) or inability to renew tissues and cells (see examples reported in *§7.1.2 Classification of phenoptotic phenomena*).

For all these species, the life cycle, including conditions leading to death, is genetically determined and has an adaptive significance (Finch, 1990).

Then there are species that, under natural conditions, show no age-related fitness decline and no detectable age-related mortality increase. These species have been defined as species with "Negligible Senescence" (Finch, 1990). Even for some of these species, since the size increases over the years and this reduces the probability of being preyed on, a reduction in age-related mortality is observed and defined as "negative senescence" (Vaupel et al., 2004). However, this definition is misleading as they are species showing no age-related functional decline with a mortality decrease due to the increase in size and not to any improvement of functions.

Therefore, aging, defined as age-related fitness decline, is shown only by a minority of species, and even some species show a constant (or also decreasing) mortality in relation to age. All of this contradicts the idea that age changes are universal in the living world as well as in the inanimate world.

3. **As old individuals are not present in the wild, aging cannot be influenced by natural selection and, therefore, cannot have any adaptive significance**

This argument is clearly expressed in various works maintaining that senescence is rare or nonexistent in the wild and therefore cannot be influenced by natural selection (Hayflick, 2000; Kirkwood & Austad, 2000). Moreover:

(i) "Data on age-related mortality patterns in wild animal populations reveal that, in many species, individuals rarely survive to ages when senescent deterioration becomes apparent (Medawar, 1952; Lack, 1954; Finch, 1990)." (Kirkwood, 2005);
(ii) "... senescence-associated increases in age-related mortality are far from ubiquitous, and ..., even where they are observed (Finch, 1990; Jones et al., 2008), they contribute only to a relatively small fraction of deaths within the population, ..." (Kirkwood & Melov, 2011).

However, this argument is disproved by the evidence.

There is extensive documentation, for many species in the wild, that there is an age-related increase in mortality and that this increase significantly contributes to the reduction of the average lifespan (Ricklefs, 1998; Nussey et al., 2013).

In particular, in the study by Nussey et al. (2013), where one of the authors is Austad, who subscribed to completely different opinions 13 years earlier—(Kirkwood & Austad, 2000), strong evidence is stated: "The recent emergence of long-term field studies presents irrefutable evidence that senescence is commonly detected in nature. We found such evidence in 175 different animal species from 340 separate studies." (Nussey et al., 2013).

Furthermore, as early as 1988, on the basis of some life tables of species studied in the wild, it was highlighted that considering all individuals, the mean lifespan was halved by the progressive increase of mortality. Moreover, excluding individuals who died before reaching maturity, the average duration of the subsequent life was reduced by two-thirds (Libertini, 1988).

Therefore, since an age-related mortality increase is well present in the wild and significantly changes aging species' mean duration of life, possible genes that determine aging or modify its rhythms are subjects of natural selection.

A completely different concept is the non-existence in natural conditions of subjects in an advanced aging state, such as a centenarian for our species or a subject of equivalent age for species with different lifespans. This does not contradict the well-documented fact of the existence in the wild of an age-related mortality increase (i.e., age-related decline of functions) up to a point where the decline of function makes survival impossible.

It is essential to avoid "the confusion of the process of senescence with the state of senility" (Williams, 1957), i.e., it is important to distinguish between age-related fitness decline (i.e., aging) and the condition, definable as "state of senility", in which this decline in fitness has so severely impaired survival skills that death cannot be avoided. While in the period in which there is a mortality increase, i.e., before the state of senility is reached, the causes of this mortality increase are surely subject to natural selection, on the contrary, the state of senility, being never or very rarely observable in the wild, cannot be the object of effective natural selection. Williams succinctly expressed these concepts: "No one would consider a man in his thirties senile, yet, according to athletic records and life tables, senescence is rampant during this decade. Surely this part of the human life-cycle concerns natural selection." (Williams, 1957).

8.1.3 Arguments against Non-adaptive Aging Theories

Since the idea of aging as an adaptive phenomenon appeared untenable due to the three arguments discussed above, which have been proven false and of no value, various theories or groups of theories were proposed to explain aging as a non-adaptive phenomenon. These theories have in common the concept that aging is a harmful phenomenon that the organism is unable to counteract. In some of these theories, some attempt is made to consider natural selection, while in others, natural selection is wholly neglected.

However, these theories appear to be flawed by various inconsistencies that make them untenable:

– **Mutation Accumulation theory** (Medawar, 1952; Mueller, 1987)

The formula (7.1):

$$\Delta_c \propto SP$$

shows that the action of natural selection is related to both the advantage or disadvantage of a gene (*S*) and the residual reproductive value of an individual (*P*), which decreases over the years due to the diminishing number of survivors.

On the basis of this simple concept, Medawar, in 1952, proposed that a harmful gene acting at a higher age (let us suppose at age t and call it a "t-gene") exerts less damage and is, therefore, less opposed by natural selection. Therefore, assuming the existence of a certain number of t-genes, their cumulative effect could cause the increasing damage that characterizes aging (Medawar, 1952). In 1988, a simple mathematical model was proposed demonstrating that even the effect of very many t-genes only slightly modified a life table of a non-aging species and absolutely did not lead to a curve of the type of an aging species (Libertini, 1988) (Fig. 8.1). The model has been reproposed in other works (Libertini et al., 2017; Libertini, Corbi, et al., 2021a) but has never been invalidated.

Apart from this argument, the Mutation Accumulation theory does not explain the different rates of aging among species and the fact that species with negligible senescence do not show a detectable increase in age-related mortality. It should be postulated that the number of t-genes varies according to the rhythms of aging and that these t-genes are not present in species with negligible senescence. Furthermore, it would also be necessary to postulate that natural selection is unable to develop mechanisms for correcting DNA mutations that are more effective than those already existing in each species. This statement is contradicted by observations on various species that demonstrated an inverse correlation between lifespan and somatic mutation rate per year (where clearly only mutations not corrected by specific mechanisms are considered) (Cagan et al., 2022). This greater efficacy of the mutation correction mechanisms in longer-lived species is justified by the fact that without it, greater longevity would cause a greater number of mutated cells and, therefore, a more significant possibility of cancer onset (Gorelick & Naxerova,

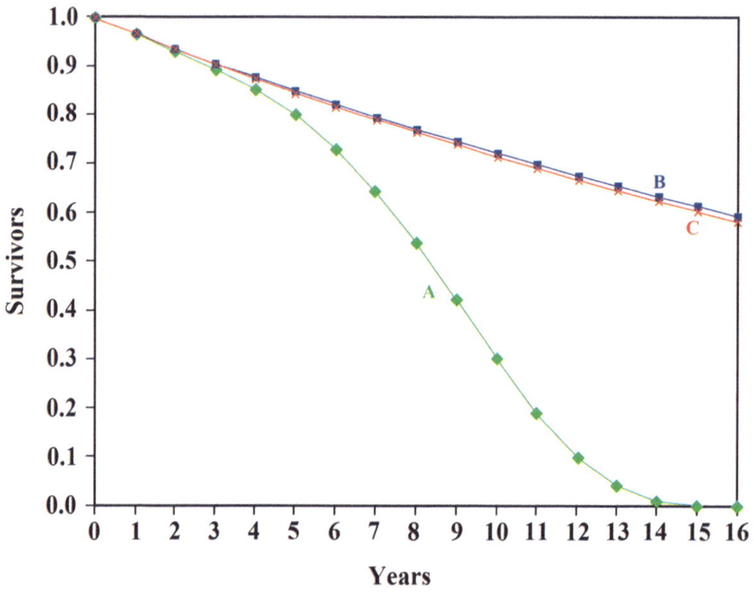

Fig. 8.1 Hypothetical effects of many t-genes on a life table. Curve A is the real life table in the wild of *Panthera leo*; mortality at time t described by Weibull's equation, $m_t = m_0 + \alpha\, t^\beta$, and the values $m_0 = 0.032$; $\alpha = 0.000252$; $\beta = 3$; data from (Ricklefs, 1998); Curve B = hypothetical life table showing the same Curve A without the age-related mortality increase, i.e. assuming a constant extrinsic mortality (the same value of m_0 in the Curve A; formula $m_t = m_0$); Curve C = hypothetical life table showing the effects on Curve B of 1000 t-genes for each year (with $v = .000001$). Figure, modified, from (Libertini, Corbi, et al., 2021a, *Chap. 4—The contrast between two paradigms*), where the formula used to obtain Curve C is discussed in detail

2022). However, it also demonstrates that the organisms, where necessary, can limit the number of uncorrected mutations. Furthermore, it has been documented that individuals with genetic alterations that cause a higher rate of mutations do not show signs of early or accelerated aging (Robinson et al., 2021). It has rightly been observed that this "argues against a straightforward connection between the accumulation of mutations and aging" (Gorelick & Naxerova, 2022).

In addition to these basic contradictions, for the Mutation Accumulation theory, evidence regarding the existence and effects of hypothetical t-genes is lacking or absent.

– **Antagonistic Pleiotropy theory** (Williams, 1957; Rose, 1991)

To overcome the insufficiencies of the Mutation Accumulation theory, a variant of this theory, the Antagonistic Pleiotropy theory, was proposed. According to this theory, there are genes with multiple actions (pleiotropic genes) that are beneficial in youth but harmful in older ages. Since the advantages at a young age are supposed to be greater than the disadvantages at an older age (also because at that age, there are fewer survivors), these pleiotropic genes are favored by selection, and the combined effect of many of these genes would cause aging.

Unfortunately, this theory, too, needs several postulates. First, among the various species, the number of such genes should vary according to the rapidity of aging, and for species with negligible senescence, such genes should not be present. Moreover, it is also necessary to postulate the non-existence of genes with analogous advantages in youth but without any disadvantage at older ages.

Apart from the various required postulates that, as for the previous theory, remove the theory from the scientific field and transform it into a mere hypothesis without foundation, it should be emphasized that there is no documentation of genes with the characteristics hypothesized for pleiotropic genes.

– **Disposable Soma theory** (Kirkwood, 1977, 1981; Kirkwood & Holliday, 1979)

A further variant of the Mutation Accumulation theory is the Disposable Soma theory. This hypothesis postulates that the organism has limited availability of one or more undefined resources. The organism must distribute this limited availability between the needs of reproduction and those of the "maintenance" of the functions, i.e., of the actions that contrast aging. The necessary reduction of resources intended for the second purpose determines the progressive decay of the organism, i.e., aging: "Evolutionary considerations suggest aging is caused not by active gene programming but by evolved limitations in somatic maintenance, resulting in a build-up of damage." (Kirkwood, 2005).

This theory, in addition to the postulated limit of undefined resources, necessarily postulates both the diversity of this limit as a function of the aging rate of a species and the absence of a limit for non-aging species. Furthermore, it is also implicitly postulated that there is an obligatory alternative between reproduction and maintenance.

Apart from these objections, which alone invalidate the scientific nature of the theory, there is no evidence of one or more limiting resources, nor has any general correlation been described between greater reproduction and more rapid aging.

Finally, suppose energy availability is the limiting resource. In that case, caloric restriction should limit lifespan, but the evidence appears to indicate the opposite (a more detailed discussion is in (Libertini, Corbi, et al., 2021a), *§4.4.5—Effects of caloric restriction on lifespan*).

The whole theory appears to be an unfounded generalization of some particular cases, which, while sometimes apparently observable, refer to specific adaptations of the organismal lifecycle to peculiar environmental conditions and not to the aging process. Thus, in some marsupials (i.e., from the genus *Antechinus*), reproduction does exhaust all the organismal resources for the males, and all of them die at the end of the short mating period (Naylor et al., 2008). Still, in that case, the *Antechinus* males refuse to eat or sleep during the mating month and subsequently die young with general exhaustion and do not face aging.

– **Damage Accumulation theories**

There is a large group of theories that explain aging as the effect of the progressive accumulation of various types of damage. The list of possible causes to which aging is attributed is long (Comfort, 1979; Libertini, Corbi, et al., 2021a, Chap.

1—Introduction) and seems limited only by human inventiveness. Among the most original causes proposed, there are: accumulation of heavy water (Hackh & Westling, 1934), the effect of an Aristotelian "entelechy" (Driesch, 1941; Bürger, 1954), toxic product of intestinal bacteria (Metchnikoff, 1904, 1907; Lorand, 1929; Metalnikov, 1937), and the inevitable increase in entropy (Hayflick, 2007a).

We could attribute a greater likelihood to the theories that explain aging as due to cellular "wear and tear" (Weismann, 1882; Pearl, 1928; Warthin, 1929) or to oxidative effects. Indeed, one of the hypothetical causes that most frequently occurs in Damage Accumulation theories is the deleterious effect of oxidation determined by free radicals on DNA (Bohr & Anson, 1995; Weinert & Timiras, 2003) or mitochondria (Harman, 1972; Miquel et al., 1980; Trifunovic et al., 2004; Balaban et al., 2005; Sanz & Stefanatos, 2008); or on the whole body (Harman, 1956; Croteau & Bohr, 1997; Beckman & Ames, 1998; Oliveira et al., 2010).

In some cases, the starting point of a theory is the known association between a certain factor and older ages. For example, since inflammatory phenomena and their consequences (e.g., immunological alterations) grow in relation to age, they have been proposed as factors determining aging (Franceschi et al., 2000; Fülöp et al., 2014; Franceschi et al., 2018), without excluding that the reverse could be more likely.

This objection could also be expressed for the theories in which aging is a consequence of oxidative phenomena. In fact, through appropriate cellular manipulations, it is possible to bring differentiated cells with an accumulation of oxidized substances back to the conditions of embryonic or stem cells in which such accumulations do not exist. For example, the cellular introduction of some factors reprograms adult fibroblasts into induced pluripotent stem cells, which show the chemical and growth properties of embryonic stem cells (Takahashi & Yamanaka, 2006).

In some cases, the association between a proposed cause and aging appears contradicted by the evidence. Thus, if aging is influenced by a faster metabolism, as flight requires an accelerated metabolism, smaller longevity for birds would be expected, but "The maximum life-span in birds … is materially longer than in mammals of comparable size and activity" (Comfort, 1979).

Regarding the possibility that aging is due to cell damage of various types (cellular wear and tear), on the basis of the evidence from many works, Fossel stated that: "Cells do not senesce because of wear and tear, but because they permit wear and tear to occur because of an altered pattern of gene expression." (Fossel, 2004, p. 53).

Apart from this, the Damage accumulation theories do not explain the differences in the aging rates between the various species on the basis of the proposed hypothetical cause of aging, nor the existence of species with "negligible senescence".

8.1.4 Adaptive Aging Theories

All the theories mentioned up to now have in common the concept that aging, being a harmful phenomenon for the individual, cannot in any way have an adaptive value. However, theories and simulation models were also proposed in which aging was favored by selection at the supra-individual level, and so had an adaptive value and was a programmed phenomenon.

Alfred Russel Wallace, one of the two proponents, together with Charles Darwin, of the theory of natural evolution, was also the first to suggest that aging could have an adaptive meaning, as he wrote in a letter to August Weismann reported by the same: "… for it is evident that when one or more individuals have provided a sufficient number of successors, they themselves, as consumers of nourishment in a constantly increasing degree, are an injury to those successors. Natural selection therefore weeds them out, and in many cases favors such races as die almost immediately after they have successors." (Wallace 1865–1870 in Weismann, 1889, vol. I).

The same idea was reiterated by Weismann (1889 and 1891, pp. 6–10 and 24–25) and commented on in modern times (Kirkwood & Cremer, 1982).

However, Weismann did not develop the idea of an adaptive meaning of aging in more detail and with more precise arguments, and some years later, he abandoned it (Weismann, 1892; Kirkwood & Cremer, 1982).

For many years, the idea remained ignored, but in 1961, Aldo Carl Leopold, a botanist, following the intuition of Wallace and Weismann, again suggested that aging increased the speed of evolutionary adaptability of a species because it accelerated the turnover of generations and explicitly proposed that aging was adaptive and programmed (Leopold, 1961).

In 1983, in a non-peer-reviewed book (Libertini, 1983) and then in a 1988 article (Libertini, 1988), for the first time, a theory was proposed that interpreted aging as an adaptive phenomenon favored by natural selection in terms of supra-individual selection by kin selection, in particular ecological conditions including spatially structured populations and K-selection. This proposal, later reaffirmed and deepened in other works (Libertini, 2006, 2008, 2009a, 2013), among other things, envisaged an inverse relationship between the proportion of senescent deaths and extrinsic mortality. This phenomenon will be briefly exposed in a subsequent sub-section.

Since 2004, other works have proposed theories in which, among other things, an evolutionary advantage for aging in spatially structured populations was shown (Travis, 2004; Martins, 2011; Yang, 2013; Mitteldorf & Martins, 2014).

The theoretical models of these works also predicted an inverse correlation between extrinsic mortality and precocious aging and there was an interesting new consideration: "I also predict that the molecular mechanism of ageing and age-related diseases should lie in mechanisms where it is difficult for selfish mutants to arise or spread …" (Yang, 2013).

In other papers, in accordance with the idea of aging as an adaptive phenomenon:

(i) It was highlighted that aging is favored by natural selection as it increases the speed of evolution or evolvability (Goldsmith, 2004, 2008);
(ii) In 1997, aging was defined as a form of "phenoptosis", i.e., a case of adaptive and programmed death determined by the same organism (Skulachev, 1997, 1999), and later was described as "slow phenoptosis" (Skulachev, 2002);
(iii) In 2009, aging was suggested to be useful for limiting the spread of diseases, with a theory analogous to the Red Queen hypothesis on the adaptive meaning of sex (Mitteldorf & Pepper, 2009).

8.1.5 Kin Selection Aging Theory

In 1988, for the first time, a theoretical model proposed that the "increasing mortality with increasing chronological age in populations in the wild" (IMICAW), already documented for various species studied in natural conditions and commonly defined as aging, could have an adaptive significance in certain ecological conditions (Libertini, 1988).

When this theory was proposed, the practically unanimous belief was that an adaptive meaning for aging was impossible. Furthermore, the concept of phenoptosis, which, together with many other different phenomena, includes aging in its meaning, had not yet been proposed. Finally, even accepting the theoretical arguments supporting the proposed theory, there was no evidence or documented hypothesis regarding the mechanisms by which aging could be determined.

In these pages, the theory above, defined as "kin selection aging theory", will be presented in a synthetic way.

In 1988, some studies on wild populations already demonstrated the existence of age-related increasing mortality in the wild (Deevey Jr., 1947; Beverton & Holt, 1959; Laws, 1966, 1968; Laws & Parker, 1968; Spinage, 1970, 1972). The number of species exhibiting the IMICAW phenomenon was significantly expanded in a 1998 work (Ricklefs, 1998). As already mentioned, the existence of the phenomenon was underestimated or denied 2 years later—"... there is scant evidence that senescence contributes significantly to mortality in the wild ..." (Kirkwood & Austad, 2000). However, afterward, an authoritative review demonstrated with "irrefutable evidence" that the phenomenon was well detectable in the wild (Nussey et al., 2013).

It was also known that there are species not showing any age-related increasing mortality in natural conditions (Deevey Jr., 1947; Beverton & Holt, 1959; Bourlière, 1957; Comfort, 1979). They were defined as "non-IMICAW" (Libertini, 1988) and subsequently described as species with "negligible senescence" (Finch, 1990).

Disregarding the early stages of life, which for various reasons have a high mortality rate, the age-related mortality increase is described by the Gompertz-Makeham equation—see (Comfort, 1979)—and by similar formulas (Ricklefs, 1998):

$$m_t = m_0 e^{\alpha t} + m_e \tag{8.1}$$

where m_t and m_0= mortality rate at times t and 0, respectively; e = Euler's number, the base of natural logarithms; α = slope constant.

The term $m_0\, e^{\alpha t}$ is the age-specific mortality, or "intrinsic mortality" (m_i).

The term m_e is the non-age-specific mortality or "extrinsic mortality".

The "mean duration of life in the wild" (ML) of the individuals of a population is in function of both m_i and m_e.

In a non-IMICAW species, by definition, ML is in function only of extrinsic mortality, i.e.:

$$m_t = m_e \tag{8.2}$$

The ML of an IMICAW species, i.e., a species with aging, is not necessarily smaller than the ML of a non-IMICAW species. For example, the value reported for the ML of the Impala (*Aepyceros melampus*), an IMICAW species, is 5.8 years (Spinage, 1972), while that of the Robin (*Turdus migratorius*), a non-IMICAW species, is 1.01 years (Deevey Jr., 1947).

However, for two IMICAW species with equal extrinsic mortality, the species with lower intrinsic mortality has a longer ML.

The time, after the early stages of life, when the mortality rate is at its lowest value (m_{min} or m_0), is defined as "τ". When the increment of the mortality rate exceeds an arbitrary threshold value, this time (τ^*) is defined as the beginning of the aging period. Figure 8.2 illustrates these definitions.

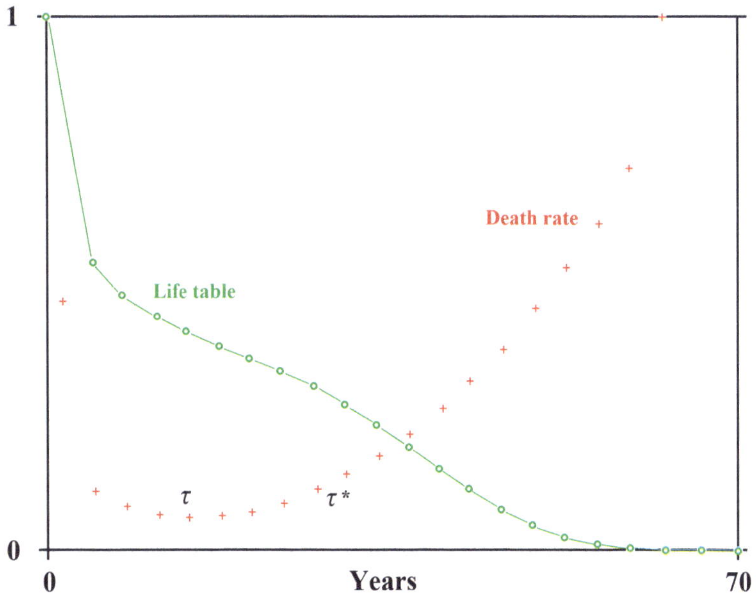

Fig. 8.2 Life table and death rates in the wild of elephant (*Loxodonta africana*); data from (Laws, 1966)

- **Effects of the variation of an allele advantage and ML variation**

Evolution can be described as the continuous increase, from one generation to the next, in the frequency of alleles that, for whatever reason, have a selective advantage and, conversely, the continuous decrease in the frequency of alleles with a selective disadvantage.

Considering two species with different ML, the species with lower ML will have a more rapid succession of generations. Consequently, all else being equal, the species with lower ML will evolve more rapidly than a species with higher ML.

In mathematical terms, let us consider an allele C, which carries an advantage *s* over an allele C′.

The frequency of C at generation n + 1 (C_{n+1}) will vary with respect to its frequency at generation n (C_n) according to the formula:

$$C_{n+1} = \frac{C_n(1+s)}{C_n(1+s)+C'_n} = \frac{C_n(1+s)}{1+C_n s} \tag{8.3}$$

For details about this and other related formulas, see Appendix—Part A in (Libertini, Corbi, et al., 2021a).

The iterative application of the formula (8.3) gives the frequency curves of Fig. 8.3, where the value of the ML is the same for all the curves ($ML = 1 = ML_k$), while *s* is different for each curve. It is clearly shown that the diffusion rate of gene C is a function of the value of *s*.

The same formula (8.3) is used to obtain the curves of Fig. 8.4 where *s* has the same value for all the curves ($s = 0.01 = s_k$), while the ML value of curve z (ML_z) is given by the formula (8.4).

$$ML_z = ML_k / (s_z / s_k) = 1/(s_z / 0.01) \tag{8.4}$$

where ML_k = the constant value of ML in the previous figure = 1; s_z = value of *s* in the z curve of the previous figure.

The curves clearly show that a decrease or an increase in the value of *s*, has the same effect as an increase or a decrease in the ML value, respectively. For a mathematical demonstration of this inverse relationship, see (Libertini, 1988, Appendix).

- **Effects of IMICAW on ML**

Estimating the effects of age-related increasing mortality (IMICAW) on mean duration of life (ML) is essential.

In the aforementioned 1988 work (Libertini, 1988), the data obtained from studies in the wild concerning only 8 species of mammals were considered—Zebra (Spinage, 1972); Hippopotamus (Laws, 1968); Elephant (Laws, 1966); Waterbuck (Spinage, 1970); Warthog (Spinage, 1972); Impala (Spinage, 1972); Buffalo (Spinage, 1972); and Dall mountain sheep (Deevey Jr., 1947) -, the only ones that could be found at the time.

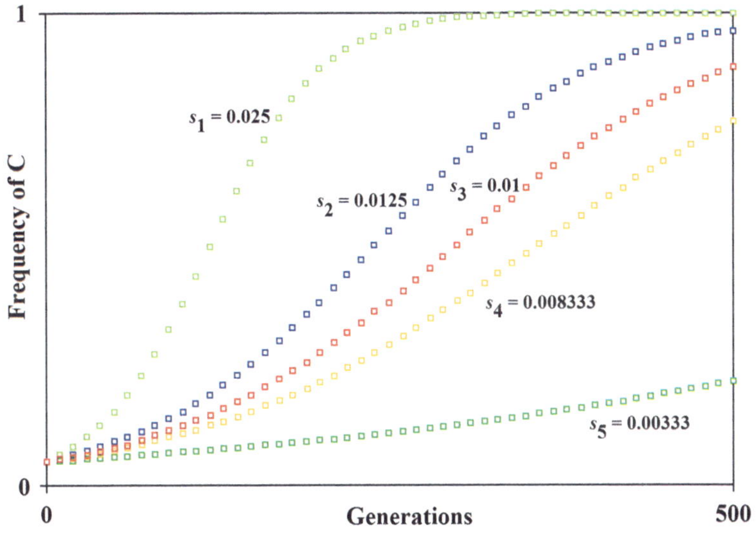

Fig. 8.3 Spreading of a gene (C) according to the variation of s (values indicated near each curve). The value of ML (= 1 = ML_k) is the same for all curves. Curves 1–5 are calculated using formula (8.3)

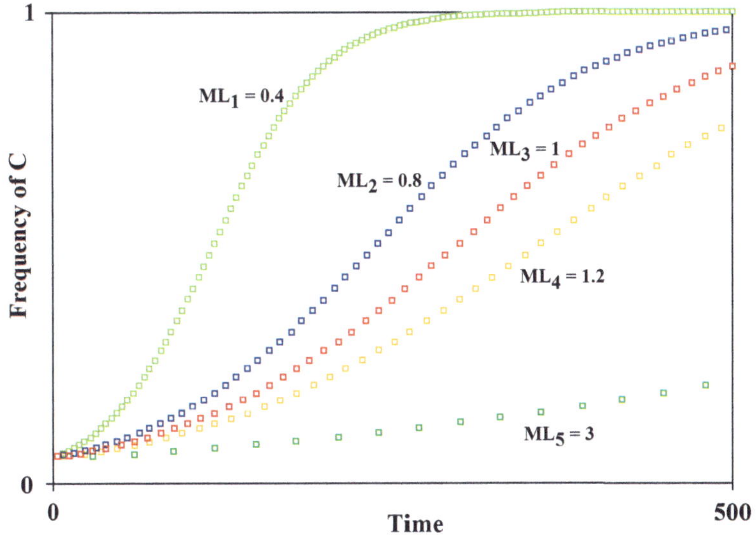

Fig. 8.4 Spreading of a gene (C) according to the variation of ML (values indicated near each curve). The value of s (= 0.01 = s_k) is the same for all curves. Curves 1–5 are calculated using formula 8.3

It was deduced from these data that the ratio between a hypothetical ML in the absence of intrinsic mortality (HML) and the ML observed in the wild ranged from 1.55 to 3.21, with an average value equal to 2.34.

Considering then only the individuals survived at the time (τ) in which mortality had its minimum value, the ratio between the hypothetical average residual life of the individuals who survived the time τ without the increase due to intrinsic mortality (HMLτ—τ) and the average residual life, in the wild, of the individuals surviving at time τ (MLτ τ), i.e., the ratio (HMLτ—τ)/ (MLτ τ), was in the range from 2.42 to 5.09, with an average value equal to 3.67.

This indicated that the intrinsic mortality, or aging, roughly halved the ML under natural conditions ($1/2.34 = 0.427 \approx 0.5$). Furthermore, considering only the residual life of the surviving individuals at time τ, aging appeared to reduce residual life to less than a third ($1/3.67 = 0.272 < 1/3$).

An example of these data is offered in Fig. 8.5, which illustrates the data available in 1988 for the zebra (*Equus burchelli boehmi*) and the life table with and without the effects of IMICAW (real and hypothetical life table, respectively) (Libertini, 1988).

More extensive and reliable data were offered by Ricklefs' study (1998): "Another way to assess the impact of senescence is to compare the average life span (T) of adult individuals in natural populations to that in hypothetical nonaging populations ($T = 1/m_o$)…. Senescence reduces average life span by only 2% when $m = 1.0$ yr-[1] but by almost 80% when $m = 0.01$ yr-[1]" (Ricklefs, 1998).

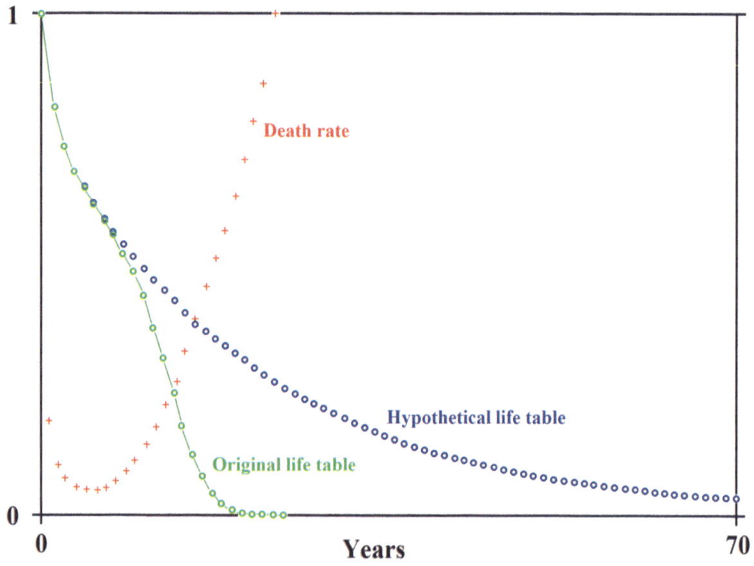

Fig. 8.5 Life table and death rate in the wild of zebra (*Equus burchelli boehmi*); data from (Spinage, 1972). The hypothetical life table without intrinsic mortality has been added. For zebra: $\tau = 6$ years; $m_{min} = 4.63\%/y$; ML = 8.48 y; HML = 17.23 y; ratio HML/ML = 2.03; $ML_\tau—\tau = 6.73$ y; $HML_\tau—\tau = 21.55$ y; ratio $(HML_\tau—\tau) / (ML_\tau—\tau) = 3.20$

In short, although in previous works this was denied by various authors (Hayflick, 2000; Kirkwood & Austad, 2000), in natural conditions, age-related increasing mortality greatly reduces the ML for many species, and so it is surely under the action of natural selection.

– **Possible advantage of a gene causing IMICAW**

Comparing two species, or two groups of individuals of the same species, with different ML, the species or group with smaller ML, all other conditions being equal, will have the advantage of a more rapid succession of generations and, therefore, of more rapid evolution.

However, this does not exclude that within the species or group with smaller ML, individuals with a longer lifespan have more opportunities to reproduce and so are advantaged by natural selection. This means that any genes that shorten life span would be eliminated by natural selection.

Now, let us consider the effects of selection at the supra-individual level, as we did to explain the existence of phenoptotic phenomena (s. *§7.1.1 Selective mechanisms favoring phenoptotic phenomena*).

Suppose we want to calculate the inclusive fitness of a gene C that somehow reduces the average life span. In that case, it is necessary to pose two preliminary conditions, which will then be discussed in the following subsection:

1. The population must be numerically stable, for example, because of limited ecological resources, so that only when an individual dies, there is space for another individual;
2. The preferential replacement of individuals who died prematurely by individuals genetically related to them is necessary. An individual (I_1) who dies prematurely as a consequence of the action of C is replaced by another individual who has, on average, a fraction *r* (coefficient of relationship) of genes equal to those of I_1.

Given these conditions, we have:

(i) The inactive allele C′ causes an ML ($ML_{C'}$) equal to 1 arbitrary unity of time, while the active allele C caused an ML (ML_C) that is smaller than $ML_{C'}$;
(ii) The smaller ML causes for I_1 a disadvantage *s*', due to a smaller reproductive capacity;
(iii) For any gene that is spreading (or decaying) within the species, a reduction of ML is equivalent to a proportional increase of the advantage (or disadvantage) *s* caused by the gene (see the previous subsection *Effects of the variation of an allele advantage and ML variation* and Figs. 8.3 and 8.4).

Therefore, for individuals with C′, as $ML_{C'} = 1$:

$$s_{C'} = s / ML_{C'} = s \tag{8.5}$$

while for individuals with C, as $ML_C < ML_{C'}$:

$$s_C = s / ML_C > s \tag{8.6}$$

Now, if we consider *m* genes that are spreading within the species (for simplicity, for all these genes $s_x = s$), the overall advantage ($S_{C'}$) for each individual with the gene C will be:

$$S_{C'} = \sum_{x=1}^{m} (rs_x / ML_{C'}) = rms \tag{8.7}$$

while for the individuals with the gene C, it will be:

$$S_C = \sum_{x=1}^{m} (rs_x / ML_C) = rms / ML_C > S_{C'} \tag{8.8}$$

Therefore, the spreading, or decay, within a population of a gene C that causes a reduced ML, may be expressed by the recursion formula:

$$C_{n+1} = \frac{C_n\left[1+(Sc - Sc') - s'\right]}{1 + C_n\left[(Sc - Sc') - s'\right]} \tag{8.9}$$

where C will be favored by kin selection and will spread within the species if:

$$Sc - Sc' = rm\frac{s}{MLc} - rm\frac{s}{MLc'} = rms\left(\frac{1}{MLc} - 1\right) > s'(8.10) \tag{8.10}$$

According to this theory, gene C would be a particular unselfish gene that is favored by a "hitchhiking" mechanism (Hill & Robertson, 1966; Felsenstein, 1974) or the "hitchhiking effect" (Strobeck et al., 1976).

Figure 8.6 has been obtained from formula (8.10). In the simulations, it has been hypothesized $S_C \gg s'$, since S_C is the sum of the advantages of *m* genes spreading within the species: "$S \gg S'$, since *S* sums up the advantages of the *m* genes that are spreading within the species" (Libertini, 1988).

The fact that S_C is the sum of many advantages was disregarded in a criticism of this model, which judged it as unlikely: "The equation depends on the selection advantage, *S*, of the beneficial gene on which C hitchhikes, ... If more realistic values are used, the aging gene C disappears" (Kowald & Kirkwood, 2016).

- **The two preliminary conditions and possible objections**

As before said, kin selection aging theory requires two preliminary conditions for the possible advantage of a shorter ML:

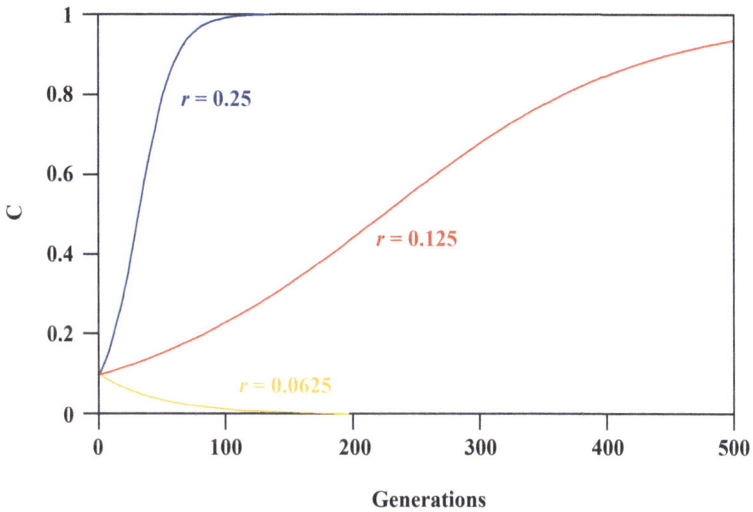

Fig. 8.6 Spreading (or decay) of a gene causing IMICAW if the dead individual (I_1) is substituted by a kin individual, i.e., with $r > 0$. The value of r for each curve is indicated in the figure. For all curves: $S_C = 0.5$, $s' = 0.05$, $ML_C = 0.5$; $C_0 = 0.1$

1. Numerically stable population

In 1970, Pianka proposed the division of populations into two main types: K-selected and r-selected (Pianka, 1970). In the first type of population, the condition of a stable number of individuals is generally true (Leopold, 1961). For r-selected species, on the contrary, this condition is unlikely, and insufficient living space for new individuals is of secondary importance compared to the reproductive capacity, which is the key factor for the evolutionary success of such species.

2. Preferential replacement of predeceased individuals.

This condition is possible in the case of species divided into small populations (demes) with a limited interdemic genetic flow. It is probable that among individuals of a deme, the coefficient of relationship (r) is greater than the mean r of the entire population and that a predeceased individual frees up living space for individuals of the same deme. This second condition occurs with territorial species, plants, and sessile animals because for these populations, which are likely K-selected species, a predeceased individual is more probably replaced by genetically related individuals.

As these conditions are essential for the model, it was predicted that the species showing age-related increasing mortality should be K-selected, divided into small demes, and territorial or non-mobile. On the contrary, for r-selected species, age-related increasing mortality should be unlikely. "I believe that such previsions are in accordance with the data of natural observation. That is, only for K-selected species we will have survival curves of type I [i.e., that characteristic of aging species]; see

(Pianka, 1970). Moreover, a certain parallelism must be observed between IMICAW and unselfish and social behaviors (Wilson, 1975), in particular Pianka's table as modified in chap. IV. This parallelism is not casual since in our theory a gene causing IMICAW is a kind of unselfish gene." (Libertini, 1988).

However, other factors surely cause an earlier or later IMICAW for species showing age-related increasing mortality. In fact, individuals with a longer lifespan have: (i) a lesser incidence of the total length of life of the more vulnerable life period, i.e., the early stages of life (development and growth); (ii) a better development and exploitation of learning abilities. Due to these factors, both if IMICAW is adaptive or if the contrary is true, two positive correlations are expected and found: (1) between ML and body mass (Bourlière, 1957, 1960; Sacher, 1959), as well as (2) between ML and brain weight in mammals, which is likely related to the ability of learning (Sacher, 1959; Comfort, 1979).

About kin selection aging theory, i.e., the hypothesis that aging is caused by a type of unselfish genes favored by kin selection in particular conditions, as those present in K-selected species, the aforesaid preliminary conditions give rise to two possible objections:

- In a deme, for the first individual (or few individuals) having an IMICAW-causing gene C, there are no kin individuals (or there are few of them) with copies of C that could enjoy the suggested advantage of its action. This objection already had its answer before the theory was proposed. In fact, a similar problem exists for the early stages of spreading of any "unselfish" gene. The proposed solution was that random non-selective mechanisms are important in the early stage and up to a critical frequency (Boorman & Levitt, 1973). For simple statistical considerations, the random increase in the frequency of an "unselfish" gene or of gene C is possible in a small population but unlikely for a large population. For this reason, it is essential that the population is divided into demes or that there is territoriality of the species for the initial diffusion of a C gene.
- The second objection is that the individuals of many IMICAW species showing age-related increasing mortality (e.g., peregrine birds, cetaceans, many species of herbivorous, etc.) emigrate each year from one area to another distant area and afterward come back to the first area. Both in the phase of the migration and at least in one of the two areas where they live, there is no territoriality, i.e., in these phases, the conditions that might favor an IMICAW-causing gene C are inexistent. However, if the individuals of these species, at least in a single particular period, e.g., in the mating and reproduction period, live in well-defined areas with the characteristics of territoriality, i.e., areas that are distinct and stable for each group, and if the habitat is saturated in this territorial phase, then the species are K-selected in these periods, and this can favor the spreading of IMICAW-causing genes (Libertini, 1988).

8.1.6 *Inverse Relation Between Extrinsic Mortality and Proportion of Deaths Due to Intrinsic Mortality*

Non-adaptive aging theories maintain that the main factor determining the evolution of longevity is the intensity of environmental or extrinsic mortality (m_e). If m_e is high, the force of natural selection rapidly weakens in relation to age and contrasts aging with ever less effectiveness. The opposite occurs when m_e is low. Therefore, the prediction of these theories is a direct correlation between m_e and the precocity of aging (Kirkwood & Cremer, 1982; Kirkwood & Austad, 2000).

Non-adaptive aging theories always see aging as something inevitably harmful contrasted by natural selection. So, if m_e (i.e., mortality different from age-related increased mortality) is high, natural selection against aging acts less effectively, and aging is predicted to come earlier.

According to the opposite thesis, from the time of the first theory interpreting aging as an adaptive phenomenon, it was observed that higher mortality would reduce the hypothetical advantage caused by a shorter lifespan because of aging. Thus, when comparing aging among the various species, higher extrinsic mortality was predicted to be associated with lower intrinsic mortality (i.e., due to aging). There was no clear evidence about this paradoxical phenomenon, which was defined with a suggestive name such as the "Methuselah effect" (Libertini, 1983, 1988). However, it was predicted that aging would not be observed in the case of high extrinsic mortality, i.e., species would have been non-IMICAW. It was underlined that Comfort so wrote about this possibility: "... populations of many species of fish, studied in the wild, show an age structure and a pattern of death similar to that found in birds, i.e. a high constant mortality unrelated to age and a virtually constant expectation of life ..." (Comfort, 1979, p. 92).

Furthermore, examples of non-IMICAW species with high extrinsic mortality were mentioned: "*Callionymus lyra, Leuresthes tenuis, Leucichthys kiyi, Cottus gobio, Clupea sprattus, Clupea pallasi* (Beverton & Holt, 1959), the blackbird, the song thrush, the robin, the starling and the lapwing (Deevey Jr., 1947)" (Libertini, 1988).

Some years later, in 1998, Ricklefs, while trying to document the direct relationship between extrinsic mortality and the "proportion of senescent deaths" (P_s) (i.e., intrinsic mortality), which was the common prediction of non-adaptive aging theories (see above), contradicting the expectations of such theories, found the inverse relation predicted by the adaptive aging theory: "Senescence reduces average life span by only 2% when $m_0 = 1.0$ yr^{-1} but by almost 80% when $m_0 = 0.01$ yr^{-1}". (Ricklefs, 1998).

There was a clear contradiction between what Ricklefs was trying to document (testified by the title of the article: "Evolutionary theories of aging: confirmation of a fundamental prediction, with implications for the genetic basis and evolution of life span") and what resulted from the data, as highlighted by the same author (Ricklefs, 1998).

The data found by Ricklefs regarding the inverse relation between extrinsic mortality (m_e) and the proportion of deaths due to the age-related mortality increase—"proportion of senescent deaths", P_s (Ricklefs, 1998)—are illustrated in Fig. 8.7, where Ricklefs data are reported together with those obtained from the study of a human population in the wild (Hill & Hurtado, 1996).

In 2004, in a paper that proposed an adaptive meaning for aging, it was observed that: "Evolution selects for earlier programmed death when the probability of death is lower." (Travis, 2004).

In a subsequent paper, it was hinted that: "... adaptive hypothesis ... appears indispensable to explain the observed inverse correlation between extrinsic mortality and the proportion of deaths due to intrinsic mortality" (Libertini, 2008).

Later, in the discussion about a model suggesting that aging was advantageous in spatially structured populations and that predicted the same inverse relationship between extrinsic mortality and P_s, it was observed that this inverse relationship was necessarily a common prediction of all adaptive aging theories: "We found that evolved life span varies directly with mortality m [i.e., m_e], so that senescent mortality tends to complement background mortality. Both contribute to the population turnover rate, and thus to evolvability ... this complementary relationship between background death [i.e., m_e] and evolved senescence is characteristic of adaptive theories of aging. A high background death rate leads to a *longer* evolved life span. This contrasts with classical theories, in which a high background death rate leads to a *shorter* evolved life span." (Mitteldorf & Martins, 2014).

For the observed inverse relationship between extrinsic and intrinsic mortality, predicted by adaptive aging theories, while the opposite is predicted by

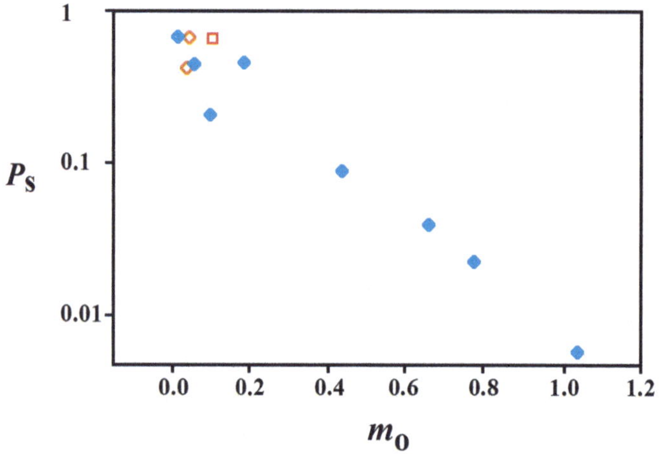

Fig. 8.7 Inverse relationship between extrinsic mortality (m_e) and the proportion of senescent deaths (P_s), i.e., due to intrinsic mortality (m_i) or aging. Ordinates are in logarithmic scale; open rhombs = mammal species; solid rhombs = bird species; open square = Ache people in wild conditions (data of mammal and bird species from (Ricklefs, 1998), data for Ache people from (Hill & Hurtado, 1996); figure from (Libertini, 2013), redrawn

non-adaptive aging theories, no explanation has been proposed that could be compatible with the second type of theories.

8.2 Slow Phenoptosis or Aging: How we Age at the Cellular Level

In the previous section, arguments and empirical data have been presented to support the idea that aging is an adaptive programmed phenomenon and against the opposing paradigm of aging as a non-adaptive phenomenon.

However, as modern science is always, in all its parts, firmly based on empirical evidence, a critic could rightly object that the above can also be persuasive and captivating but does not represent definitive proof in support of the thesis of aging as an adaptive programmed phenomenon, so excluding indisputably the opposite view.

To give a valid answer to this criticism, it is first necessary to consider that there is a fundamental distinction between the two paradigms:

- if aging is an adaptive and genetically programmed and modulated phenomenon, there is necessarily a special program for aging that can be described in its parts with the support of multiple and valid scientific works;
- if aging is not an adaptive phenomenon, there is not and cannot exist any genetically defined and modulated program that determines aging.

Suppose it is possible to describe the aging program required by the adaptive aging paradigm in a detailed and scientifically valid way. In that case, this will be a strong confirmation of this paradigm and an equally strong demonstration that the opposite paradigm is unlikely and hardly tenable. This description would permanently eliminate rooted beliefs, which can be summarized as follows:

1. "Evolutionary considerations suggest aging is caused not by active gene programming but by evolved limitations in somatic maintenance, resulting in a build-up of damage.... there is scant evidence for the existence of such a program, and there are powerful arguments why it should not exist." (Kirkwood, 2005)
2. Many authors present non-adaptive aging theories as the only ones that explain or could explain correctly aging (Olshansky et al., 2002; De Grey, 2005; Hayflick, 2007b; Kirkwood & Melov, 2011; Gladyshev, 2016). It is also maintained that three theories (Mutation Accumulation theory, Antagonistic Pleiotropy theory, and Disposable Soma theory) provide complementary explanations for aging (Kirkwood & Austad, 2000).

However, a complex and sophisticated mechanism that appears to explain how aging is determined and modulated has been precisely described in some recent works and defined as the "subtelomere-telomere theory of aging" (Libertini et al., 2020; Libertini, Shubernetskaya, et al., 2021b; Libertini et al., 2022).

This mechanism reworks a previous theory that can be defined as the "telomere theory of aging". Furthermore, the subtelomere-telomere theory of aging is not the result of the ideas of one or few researchers but represents the fruit of decades of work by many valid researchers who, in general, did not have the aim of demonstrating the nature of aging as an adaptive phenomenon.

Consequently, their results are not undermined or misdirected by the possible belief that aging is an adaptive phenomenon, and indeed, some of these researchers are convinced supporters of the opposite thesis. An evident case is that of Leonard Hayflick, who was the first to demonstrate the limits of cellular duplication (Hayflick & Moorhead, 1961), which before were completely excluded (Carrel, 1912, 1913; Carrel & Ebeling, 1921; Parker, 1961). Nevertheless, after elucidating what will result as a fundamental step to understanding the mechanisms of aging, this author has long been a firm supporter of the idea of aging as a non-adaptive phenomenon (Hayflick, 2000).

In this section, the fundamental points of the subtelomere-telomere theory will be illustrated, with the main works in support of each of them. In particular, the features of the proposed mechanism that appear incompatible with the non-adaptive aging thesis will be highlighted.

8.2.1 Telomere Theory of Aging

At the end of the nineteenth century, August Weismann hypothesized that the main mechanism at the origin of aging was the slowing down or blocking of tissue and organ renewal: "... death takes place because a worn-out tissue cannot persistently renew itself, and because a capacity for increase by means of cell-division is not everlasting, but finite ... Functional disturbances will appear as soon as the rate at which the worn-out cells are renewed becomes slow and insufficient." (Weismann, 1892, pp. 21–22); "... the organism did not finally cease to renew the worn-out cell material because the nature of the cells did not permit them to multiply indefinitely, but because the power of multiplying indefinitely was lost when it ceased to be of use" (Weismann, 1892, p. 25).

Later, Weismann's hypothesis appeared falsified by the experiments of Alexis Carrel, winner of the Nobel Prize in Physiology or Medicine in 1912, which seemed to demonstrate an unlimited capacity of cellular reproduction (Carrel, 1912, 1913; Ebeling, 1913; Carrel & Ebeling, 1921). Even in later years, the idea of unlimited capacity of normal cells was the rule (Parker, 1961).

However, Weismann's hypothesis became tenable again when Hayflick's experiments showed that those of Carrel were flawed by errors and that *in vitro* duplication capabilities of human diploid fibroblast-like cells (HDFCs) were limited (Hayflick & Moorhead, 1961; Hayflick, 1965).

Furthermore, when comparing fetal and adult HDFCs, there was a marked difference in the number of potential duplications (about 50 and 20–30, respectively) (Hayflick & Moorhead, 1961), and skin HDFCs obtained from donors of various ages showed a reduction in potential duplications of approximately 0.2 duplications/year of donor life (Martin et al., 1970).

Later, the limit in cell duplication capability (the Hayflick limit) was demonstrated *in vivo* (Schneider & Mitsui, 1976) and for other types of human cells: epidermal keratinocytes (Rheinwald & Green, 1975), arterial smooth muscle cells (Bierman, 1978), lens epithelial cells (Tassin et al., 1979). It was also demonstrated that Hayflick limit was earlier in syndromes that partly reproduce the characteristics of aging, i.e., in progeria (Hutchinson-Gilford syndrome) (Goldstein, 1969; Martin et al., 1970; Hayflick, 1977) and Werner's syndrome (Martin et al., 1970).

It was also shown that: (i) in HDFCs, telomeres shorten in relation to aging in proportion to the number of previous duplications (Harley et al., 1990); (ii) in immortal human cell lines there was unrestricted telomerase activity (Morin, 1989); and (iii) if telomerase was activated, cell acquired the capacity of unlimited duplications (Bodnar et al., 1998; Counter et al., 1998; Vaziri, 1998; Vaziri & Benchimol, 1998; de Lange & Jacks, 1999).

Moreover, cell cultures of mutated *Tetrahymena* with inactivated telomerase showed a reduction in duplication capacity (Yu et al., 1990), and specific proteins appeared to regulate telomerase activity (van Steensel & de Lange, 1997).

Finally, the Hayflick limit appeared to be correlated, at least approximately, with the longevity of the species (Röhme, 1981), and "Mice have a lifespan of three years and a Hayflick Limit of fifteen divisions [for fibroblasts], while the Galapagos turtle, which lives for 200 years, has a Hayflick Limit of around 110 divisions. Human fibroblasts have a Hayflick Limit of between forty and sixty divisions" (Fossel, 2004, p. 22).

The important implications of the limits in duplication capabilities of cells other than germ line cells were soon appropriately underlined by Hayflick: "The importance of a correct interpretation of this phenomenon cannot be overemphasized since the apparent indefinite multiplication of isolated normal vertebrate cells in culture, as purportedly demonstrated by Carrel, has often been cited as evidence for the thesis that senescence in higher animals is a phenomenon resulting from the effects of events at the supracellular level (Pearl, 1922; Bidder, 1925; Medawar, 1940; Cowdry, 1952; Comfort, 1964; Maynard Smith, 1962). It follows, therefore, that if normal animal cells do indeed have only a limited capacity for division in cell culture, the manifestations of aging might very well have an intracellular basis…." (Hayflick, 1977, p. 163).

As an explanation for the Hayflick limit, it was observed, first by Olovnikov and then by Watson, that DNA duplication by the enzyme DNA polymerase was incomplete and left out a small end of the molecule. Therefore, at each duplication, DNA is shortened, and this could explain the limits in cell duplication capacities (Olovnikov, 1971; Watson, 1972).

A little later, Alexey Olovnikov pointed out that, to justify both the absence of the limit in germline cells and the limit with a higher number of duplications in stem cells, an enzyme was needed to remedy the lack of duplication of the terminal part of DNA molecule (Olovnikov, 1973). This enzyme (telomerase) was identified 12 years later (Greider & Blackburn, 1985).

Meanwhile, the structure of the two terminal parts of the DNA molecule, called telomeres, was clarified. In a protozoan species, it was demonstrated that the telomere was a short sequence of nucleotides (TTGGGG), defined as a motif, repeated many times (Blackburn & Gall, 1978). Almost the same sequence (TTAGGG) was present in mammalian telomeres (Moyzis et al., 1988), trypanosomes, molds, non-mammal vertebrates, and many organisms of other phyla (Blackburn, 1991; Telomerase Database, 2020).

Another nearly identical motif (TTAGG) is present in many species (e.g., *Stegobium paniceum*, drugstore beetle; *Bombyx mori*, domestic silkworm; *Apis mellifera*, honeybee; *Manica yessensis*, an ant species; and *Giardia lamblia*, a flagellated protozoan) (Telomerase Databases, 2020).

The fact that the motif was extraordinarily conserved in species even phylogenetically distant, showed that the telomere had a key importance.

These facts led to the proposal of a hypothesis that seemed plausible to explain the basic mechanism of aging. As the succession of cellular duplications shortened the telomeres, this could justify the progressive alterations of the cells and the subsequent impairment of the body's functions, i.e., the phenomena that characterize aging.

However, this hypothesis, which could be called the "telomere theory of aging", was clearly in serious contrast with the results of various empirical data.

In fact, as a logical consequence of this theory, in the comparison among species, there should be a direct relationship between the life span and the initial length of the telomere (i.e., the length in germ line cells) and an analogous relationship with the activity of telomerase. However, there were evident discrepancies between these predictions and the evidence:

1. among the rodent species, life span appeared to have no relationship with telomere length (Seluanov et al., 2007) and telomerase activity (Seluanov et al., 2007; Gorbunova et al., 2008). Conversely, body mass and telomerase activity showed a direct relationship (Seluanov et al., 2007);
2. two *Mus* strains with different telomere lengths (10 kb and 20 kb) show equal life span and rhythms of increments of age-related mortality (Fossel, 2004, p. 60);
3. in comparison with our species, hamsters and mice have much longer telomeres and quite shorter life spans (Slijepcevic & Hande, 1999);
4. in mice, despite the short life span, most somatic cells show a baseline activity of telomerase (Prowse & Greider, 1995);
5. the first cell of cloned animals obtained from somatic cells of a donor have shorter telomeres than the first cell of the donor (Shiels et al., 1999). Nevertheless, the life span and rhythms of aging appeared the same for a cloned and a donor animal (Kubota et al., 2000; Lanza et al., 2001; Fossel, 2004, p. 60);
6. telomeres shorten at each generation in telomerase knockout (mTR−/−) mice, which have inactivated telomerase. For these mice reared in protected laboratory conditions, viability and fertility are compromised only when telomeres are quite shortened after four generations (Herrera et al., 1999) or up to six generations (Blasco et al., 1997). In the previous generations, despite the reduction in telo-

mere length compared to wild strains, the life span appeared unchanged, although dysfunctions were found in organs with highly proliferative cells (e.g., abnormal spermatogenesis and hematopoiesis (Lee et al., 1998; Herrera et al., 1999);
7. there are cells, defined as perennial cells (e.g., most neurons), which show no duplication activity and turnover and so cannot have telomere shortening due to duplications but age as non-perennial cells.

Because of these manifest contradictions between evidence and its predictions, the telomere theory of aging became untenable. So, telomere shortening in relation to the number of previous duplications—imprecisely described as telomere attrition also in recent works, e.g., (Turner et al., 2019)—seemed to be another type of damage that accumulated in the cells over time, in accordance with the paradigm of aging as a phenomenon non-adaptive, and not dependent on a program.

8.2.2 *Subtelomere-Telomere Theory of Aging*

The facts illustrated in the previous sub-section forced either the abandonment of the telomere theory of aging or its radical reformulation. However, in pursuing the goal of achieving a sound radical reformulation of the telomere theory of aging or suggesting its abandonment, other phenomena had to be considered:

1. **Cell senescence**

In the famous experiments that brought to light the existence of a limit in the duplications of normal human fetal fibroblasts in culture (Hayflick & Moorhead, 1961; Hayflick, 1965), it was found that the cells no longer able to duplicate had peculiar characteristics, becoming what are defined as senescent cells (Smith & Pereira-Smith, 1989).

The state of cell senescence is an "active process" (Smith & Pereira-Smith, 1989), and it has been observed that a senescent cell is not simply an old cell but a cell in which a specific cellular program has been triggered (Ben-Porath & Weinberg, 2005). This program can be activated by various damaging factors, including oxidative stress, DNA damage, and altered culture conditions (Collado et al., 2007; Acosta et al., 2008) but also by telomere shortening (d'Adda di Fagagna et al., 2003).

The alterations of cellular senescence are precisely defined:

(i) block of duplication capacity or replicative senescence (Cristofalo & Pignolo, 1993; Kwon et al., 2019);
(ii) specific alterations of cellular functions (Hayflick, 1977, pp. 169–171; Shelton et al., 1999; Kirkland & Tchkonia, 2017; Kwon et al., 2019) and "profound transcriptional changes" (van Deursen, 2014);
(iii) specific alterations of extracellular secretions, which are defined as senescence-associated secretory phenotype (SASP) (Coppé et al., 2006; Coppé et al., 2008; Rodier et al., 2009; Xu, Palmer, et al., 2015; Xu, Tchkonia, et al., 2015b);
(iv) resistance to apoptosis (Wang, 1995; Kirkland & Tchkonia, 2017).

Cells in the state of cell senescence cells may be identified through the expression of p16^{Ink4a}. Their number, both as an absolute number and as a fraction of all cells, shows an age-related increase (Krishnamurthy et al., 2004; Baker et al., 2016), which is related to the manifestations of aging and age-related diseases (Baker et al., 2004; Baker et al., 2008).

2. **Probabilistic activation of cell senescence in relation to telomere shortening**

For some time, there was the belief that the mechanism leading to cell senescence was activated when the telomere reached a critical length. Blackburn observed that the activation of this mechanism was not dependent on a critical length of the telomere, while the probability of activation of the mechanism grew in relation to telomere shortening. Moreover, there could be a transition to cell senescence with little shortened telomeres or even with telomeres not shortened at all: "Cellular senescence is commonly described as cells being able to divide a finite number of times, with senescence occurring only when telomeres reach a critical short length. Usually implicit is the idea that the 'young' cells, which are all dividing at the beginning of the passaging, are free of 'aged' phenotypes until, following a period of telomere shortening during the passaging, critically short telomeres appear in the 'old' cells. Implicit is the idea of an inbuilt delay before senescence is reached. But a significant (although infrequently cited) literature dealing with the behaviour of individual cells reveals a different picture (for example (Pontèn et al., 1983; Jones et al., 1985); reviewed in (Holliday, 1996)." (Blackburn, 2000).

In fact, as underlined by Blackburn, evidence showed that the decline of the duplication capacities of the cells in culture did not begin after a certain number of duplications, while there was a progressive increase in the reduction of growth potential (Pontèn et al., 1983; Jones et al., 1985).

Blackburn proposed that the telomere was covered by a cap made up of distinctive molecules ("sequence-specific DNA-binding proteins"), forming a telomeric "DNA-protein complex" (Blackburn, 2000). The bond between telomere and cap was not rigid and, therefore, the complex oscillated between two states:

(i) "capped" telomere, in which proteins and telomere were bound, and thus telomere was protected, and the cell resulted resistant to the transition to a "non-cycling state" or cell senescence;
(ii) "uncapped" telomere, in which the bond between proteins and telomere was at least partly loosened, the telomere lost its protection, and the cell became vulnerable to the transition to "non-cycling" state.

According to this hypothesis, which gave a brilliant explanation to the experimental data, the strength of the bond between the proteins and the telomere of the DNA-protein complex was somehow related to a telomere length-dependent regulation (Fig. 8.8).

3. **Telomeric position effect and particular phenomena observed in yeast**

In 1990, it was observed in yeast that a gene inserted very close to the telomere was reversibly repressed ("telomeric position effect") (Gottschling et al., 1990).

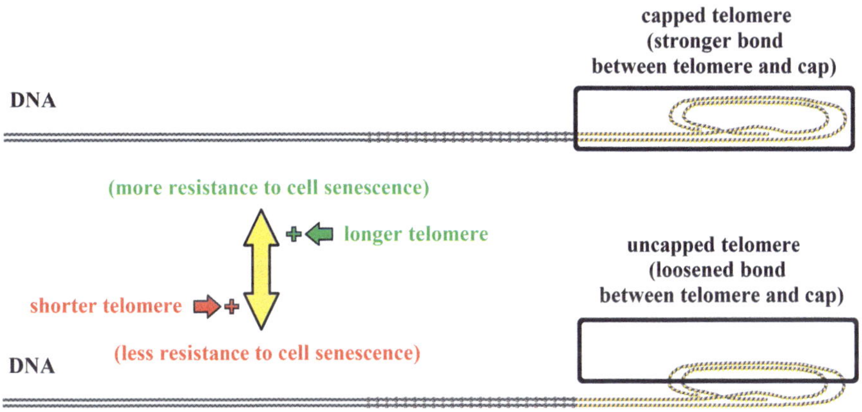

Fig. 8.8 Oscillation of telomere between capped and uncapped conditions. The probability of the uncapped state is in relation to telomere shortening, and in this condition, the probability of triggering cell senescence increases. Figure from (Libertini, Corbi, et al., 2021a), Fig. 5.4, modified

Some years later the phenomenon was shown for our species (Baur et al., 2001) and other mammal species (Baur et al., 2004).

In "ring 17 syndrome", a very rare genetic disease, there is the loss of part of the two terminal arms of chromosome 17, and the expression levels of genes located very close to the chromosome ends are significantly downregulated in comparison with controls. This is explained as a type of telomeric position effect due to the shortening of the two telomeres of chromosome 17 (Surace et al., 2014).

In yeast, some phenomena are important to consider for the study of aging and the telomeric position effect:

– **Yeast particular apoptosis**

Yeast monocellular individuals show a phenomenon very similar to the apoptosis of multicellular eukaryotes and, therefore, also defined as apoptosis (Madeo et al., 1997).

Indeed, the overexpression of a particular factor (human Bcl-2) inhibited apoptosis in multicellular eukaryotes and delayed the process leading to apoptosis in yeast (Longo et al., 1997), and the overexpression of another factor (mammalian BAX) triggered apoptosis in multicellular eukaryotes and stimulated apoptosis in yeast (Ligr et al., 1998).

Furthermore, yeast apoptosis, as the apoptosis of the multicellular eukaryotes, results in the orderly cleavage of the cell into parts and substances that are not harmful to other monocellular individuals, which phagocytizing or absorbing such parts or substances can survive longer (Herker et al., 2004).

These facts, and others, among which conserved proteasomal, mitochondrial, and histone-regulated apoptotic pathways and several orthologues of apoptotic mammalian proteins (Büttner et al., 2006), fully justified the hypothesis of a common phylogenetic origin for yeast and multicellular eukaryote apoptosis and the definition of apoptosis for the phenomenon above in yeast (Madeo et al., 1999; Longo et al., 2005; Kaeberlein et al., 2007).

– **Differences between yeast cells of "daughter" and "mother" lineages**

In wild-strain yeast, telomerase is always active, so telomere length does not shorten at each duplication (D'Mello & Jazwinski, 1991; Maringele & Lydall, 2004).

Each yeast cell reproduces by division into two different cells, the "mother" cell and the "daughter" cell. While cells of the "daughter" lineage have the capacity of unlimited divisions (Maringele & Lydall, 2004), cells of the "mother" lineage may reproduce only a limited number of times, i.e., 25–35 duplications in 3 days (Jazwinski, 1993). Cells of this second lineage show, in relation to the number of previous duplications: (i) progressive accumulation of particular substances (extra-chromosomal ribosomal DNA circles, or ERCs) on the part of DNA molecule that is adjacent to the telomere (here defined as subtelomere) (Sinclair & Guarente, 1997); (ii) increasing functional alterations; (iii) decreasing resistance to sustain stress; and (iv) increasing probability of the block of replicative capacity, immediately followed by apoptosis (Jazwinski, 1993; Laun et al., 2001; Lesur & Campbell, 2004; Herker et al., 2004; Büttner et al., 2006; Fabrizio & Longo, 2008).

For cells of the mother lineage, when the greatest number of duplications is reached, death by apoptosis occurs (Laun et al., 2001). However, in the wild, it is unlikely that this number of duplications is reached due to the progressive increase in functional alterations and decrease in the ability to withstand stress without activating the apoptosis mechanism. These alterations follow an exponential dynamic (Laun et al., 2007), analogous to the age-related mortality increase of aging species (Ricklefs, 1998; Nussey et al., 2013).

– **Yeast *tlc1Δ* mutants**

A strain of yeast mutant (*tlc1Δ* mutants) shows deficient telomerase activity and, therefore, telomeres shorten at each duplication in both mother and daughter lineage cells. In cells of the daughter lineage, which show no ERC accumulation as for wild strains, there is, in proportion to the number of previous duplications, decreasing resistance to stress and a transcriptome that appears the same as wild strain mother cells with the same number of duplications (Lesur & Campbell, 2004).

Regarding the mechanisms that cause the decline of cellular functions and ability to withstand stress without activating apoptosis, for the cells of the mother lineage of wild-type strains and for the cells of the daughter lineage of *tlc1Δ* mutant strains, the interpretations are as follows:

(i) for cells of the mother lineage of wild strains, accumulation of ERC molecules on the subtelomere determines repression of the same (Sinclair & Guarente, 1997), which in turn causes alterations in other parts of the DNA;
(ii) for the cells of the daughter lineage of *tlc1Δ* mutant strains, in which there is no accumulation of ERCs, it has been hypothesized that, as a consequence of the shortening of the telomere due to insufficient telomerase activity, there would be the sliding of the cap (or hood) covering the telomere on the adjacent telo-

mere with progressive repression of the subtelomere (Fossel, 2004; Libertini, 2009a) (Fig. 8.9), which in turn, as in the previous case, causes alterations in other parts of the DNA: "One model of telomere-gene expression linkage is an altered chromosomal structure (Ferguson et al., 1991), such as a heterochromatin 'hood' that covers the telomere and a variable length of the subtelomeric chromosome (Fossel, 1996; Villeponteau, 1997; Wright et al., 1999). As the telomere shortens, the hood slides further down the chromosome (the heterochromatin hood remains invariant in size and simply moves with the shortening terminus) … the result is an alteration of transcription from portions of the chromosome immediately adjacent to the telomeric complex, usually causing transcriptional silencing, although the control is doubtless more complex than merely telomere effect through propinquity (Aparicio & Gottschling, 1994; Singer et al., 1998; Stevenson & Gottschling, 1999). These silenced genes may in turn modulate other, more distant genes (or sets of genes). There is some direct evidence for such modulation in the subtelomere …" (Fossel, 2004, p. 50).

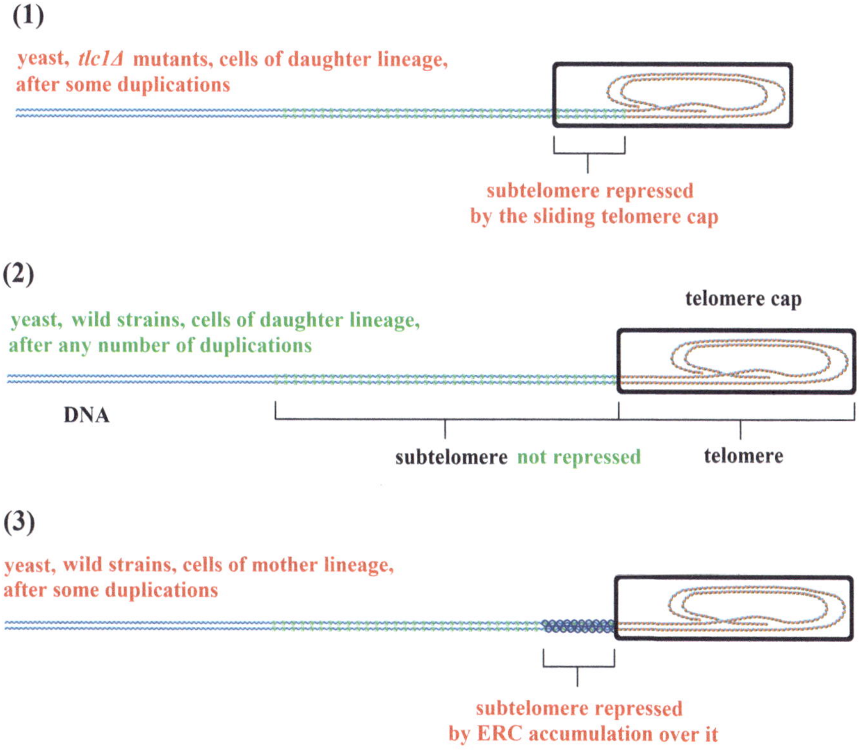

Fig. 8.9 Some phenomena shown by yeast. (1): in *tlc1Δ* mutants, in cells of daughter lineage, no ERC accumulation, but telomere shortening, telomere repression, and cell alterations; (2): in wild strains, in cells of daughter lineage, no ERC accumulation, telomere repression, and cell alterations; (3): in wild strains, in cells of mother lineage, ERC accumulation, telomere repression, and cell alterations; (Fig. 5.8 from (Libertini, Corbi, et al., 2021a), modified)

In both cases, the mechanism is a form of telomeric position effect, which, in one case, is determined by the accumulation of ERCs, while in the other, it is caused by the sliding of the telomeric cap on the subtelomere.

4. **Gradual Cell Senescence**

In cultures of yeast, where cell senescence causes immediate apoptosis (Laun et al., 2001; Herker et al., 2004), we have seen above that:

(A) in cells of the mother lineage of wild strains, in proportion to the number of previous duplications, there are (i) subtelomere repression caused by ERC accumulation and (ii) increasing functional alterations and vulnerability to cell senescence (Laun et al., 2001; Herker et al., 2004);
(B) in cells of the daughter lineage of mutant *tlc1Δ* strains (where telomerase is inactive and so telomeres shorten at each duplication, likely with progressive subtelomere inhibition in absence of ERC accumulation), in proportion to the number of previous duplications, there are (i) functional alterations with increasing vulnerability to cell senescence; and (ii) a transcriptome equal to that of mother cells of wild strains with the same number of previous duplications (Lesur & Campbell, 2004).

Now, it is necessary to clarify whether this phenomenon, defined as "gradual cell senescence" (Libertini, Corbi, et al., 2021a; Libertini, 2022; Libertini et al., 2022), is also somehow present in cells of multicellular eukaryotes.

In cultures of cells of this type, it is possible to observe functional alterations of the whole of the cells, but it is not immediate to distinguish between those due to cells in the state of cell senescence and those due to cells in a state of partial senescence: "There is substantial variability in the degree of senescence and few if any fully senescent cells, but a significant degree of altered gene expression within a percentage of partially senescent cells." (Fossel, 2004, p. 148).

This results in little knowledge or unawareness of gradual cell senescence as an entity distinct from cell senescence. However, there is evidence regarding the sure distinction between gradual cell senescence and cell senescence:

(i) In a study about the consequences of telomere shortening, it was highlighted that there are widespread changes in gene expression (up- or down-regulation) in relation to telomere shortening before such shortening activates the cell senescence program which blocks cell division (Robin et al., 2014);
(ii) In vitro, in a culture of duplicating mesenchymal stem cells (MSCs), in relationship with the number of previous duplications, it is possible to observe gradual changes in mRNA expression showing "a consistent pattern of alterations in the global gene expression … These changes are not restricted to later passages, but are continuously acquired with increasing passages." (Wagner et al., 2008). Furthermore, in relationship with the number of previous duplications, MSCs show gradual changes in DNA methylation, whose measurement can be used to calculate the number of duplications (Koch, 2012; Schellenberg, 2014; Fernandez-Rebollo, 2020).

It is difficult to support the hypothesis that the progressive alterations manifested by the gradual cell senescence are due to the progressive accumulation of harmful substances or random actions of other factors since there are experimental data that show the complete reversibility of such alterations by appropriate cellular reprogramming:

(A) The above-reported increasing functional alterations of MSCs in relationship to the number of previous duplications may be canceled by reprogramming these cells in induced pluripotent stem cells (iPSCs) (Spitzhorn, 2019). These iPSCs, whatever the age of the donor and the source of the cell, have the profile of a young cell (Spitzhorn, 2019).
(B) From iPSCs, it is possible to obtain induced MSCs (iMSCs) showing better cellular functions and fewer epigenetic changes (Hynes, 2013).
(C) For iMSCs, "DNA methylation, related to age, was completely erased, and iMSCs reacquired senescence-associated DNA methylation during culture in vitro." (Zhou et al., 2020)

5. **Telomere Length**

Telomeric DNA, or simply telomere, is the terminal part of both ends of a DNA molecule and is composed of a nucleotide sequence (motif) repeated several times. Above, it has already been reported that the motif shows great stability during evolution, with phylogenetically distant species having the same motif or some minor variation.

For the length of the telomere, it is different.

First, it should be emphasized that there are many telomeres in every cell: two for each DNA molecule, of which there are two copies in each chromosome. So, to get the number of telomeres (n), it is necessary to multiply the number of chromosomes by 4 (for example, for our species, $n = 23 \bullet 4 = 92$). Now, considering the initial length of each telomere before any duplication, i.e., their length in the germ cell, it is not the same for all telomeres, and indeed: "… telomere lengths within the same cell are heterogeneous and certain chromosome arms typically have either short or long telomeres." (Londoño-Vallejo et al., 2001). However, their initial length is hereditary, as demonstrated by their equal length in monozygotic twins but not in dizygotic twins (Graakjaer et al., 2003; Hjelmborg et al., 2015).

Apart from these differences in initial telomere length, even in a single individual, there is also significant variation in mean telomere length when comparing species. Indeed, while in yeast the length of the telomere is about 450 base pairs (Runge & Zakian, 1989), among rodents (Seluanov et al., 2007) and more generally among mammals (Gomes et al., 2011), the length ranges from 5 to 50 kb without correlation with the life span.

These data indicate that any reformulation of the telomere theory of aging cannot have the initial length of the telomere as a central parameter.

– **The proposal of the subtelomere-telomere theory of aging**

To overcome the difficulties that made the telomere theory of aging untenable and also to explain the phenomena described in this section, a radical reformulation of the theory, the "subtelomere-telomere theory of aging", was proposed (Fossel, 2004; Libertini & Ferrara, 2016a; Libertini, 2017).

The main point of this reformulation was the pivotal importance attributed to the subtelomere, which can be deduced from some phenomena among the aforementioned ones and was hinted at by Fossel in the quotation already reported in the previous sub-section *Yeast tlc1Δ mutants* (Fossel, 2004, p. 50).

The fundamental points of the subtelomere-telomere theory of aging are the following:

1. Each telomere is covered by a cap (or hood), with a size modeled on the basis of the length of the telomere in the first cell of the organism and not changed in all subsequent duplications. As telomeres have various lengths, the caps have sizes that are in proportion to these lengths;
2. When telomeres shorten at each duplication (if telomerase is inactive), the caps, having fixed lengths, gradually cover a greater extent of the part of the DNA molecule adjacent to the telomere (i.e., the subtelomere). The progressive covering of the subtelomere can also be described, less faithfully, as the sliding of the hood on the subtelomere;
3. The progressive covering of the subtelomere represses an increasing part of the subtelomere, which is hypothesized to have sequences, defined as "r-sequences", with regulatory actions on other parts of the DNA;
4. The increasing repression of the subtelomere would be the cause of the increasing alterations of the cellular functions defined as gradual cell senescence. In the daughter cells of the yeast *tlc1Δ* mutant strains, but also in the cells of multicellular eukaryotes (when telomerase is inactive), the progressive subtelomeric repression would be due to the progressive telomeric shortening at each duplication and to the consequential "sliding" of the cap. In yeast mother cells of wild strains, where there is no shortening as telomerase is always active, the repression would be caused by the progressive accumulation of ERCs;
5. Among the repressed regulatory actions, there would also be the regulation of the stability of the telomere-cap complex, whereby as the subtelomere is progressively repressed, the complex becomes more loosened, and the probability of activation of the cell senescence program increases.

The subtelomere-telomere theory of aging thus formulated made it possible to overcome some fundamental objections regarding the telomere theory of aging:

(A) If the size of the cap is fixed for each telomere in the first cell of the organism on the basis of telomere length, and the progressive subtelomeric repression depends on the shortening of the telomere and not on its initial length, this explains very well (i) the lack of correlation between life span and average initial telomere length; and (ii) the lack of importance of the variety of telomere lengths already from the first cell of the organism;

(B) As the subtelomeric repression is progressive, this is compatible (i) with the progressiveness of the alterations of gradual cell senescence and (ii) with the progressive increasing vulnerability to cell senescence, not linked to a critical shortening of the telomere;

(C) As regards the aging of perennial cells, i.e., although they do not duplicate and so cannot have telomere shortening due to duplications, their aging could well be explained by the dependence of these perennial cells on other trophic cells, which are non-perennial and so have cell duplications and shorten their telomeres as all non-perennial cells (Libertini & Ferrara, 2016b; Libertini, 2018).

The thesis of the subtelomere-telomere theory of aging can be well summarized graphically (Fig. 8.10) considering:

(i) the condition of cells of multicellular organisms with active telomerase as analogous to that of daughter lineage cells in wild yeast strains, where telomerase is always active (s. Figure 8.9);
(ii) the condition of cells of multicellular organisms with telomerase repressed as analogous to that of daughter lineage cells in *tlc1Δ* mutant strains of yeast, where telomerase is inactive (s. Figure 8.9).

However, although the subtelomere-telomere theory appeared rationally consistent and its predictions agreed with the empirical data, for the hypothetical existence of the subtelomeric "r-sequences", there was no direct experimental proof. Moreover,

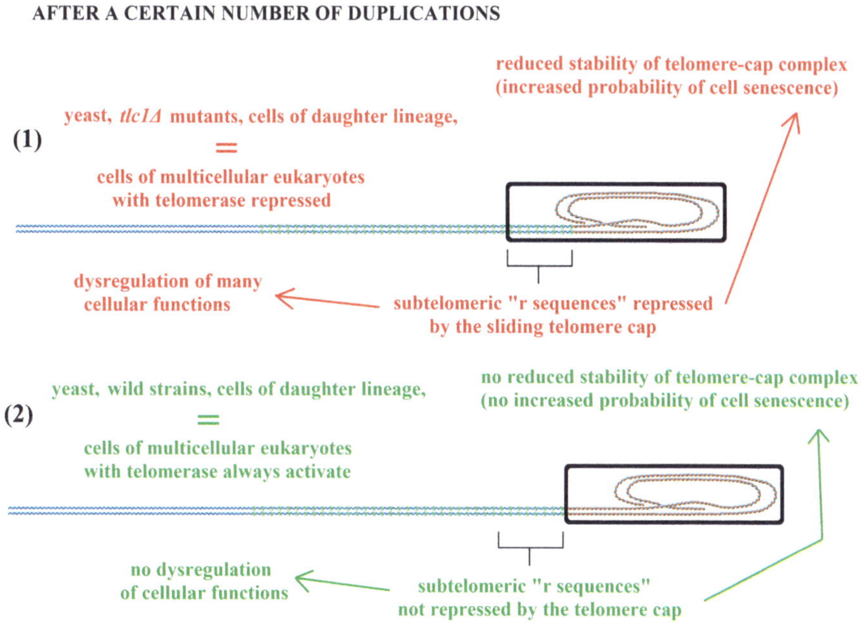

Fig. 8.10 The condition of two types of yeast cells (1: cells of daughter lineage of *tlc1Δ* mutant strains; 2: cells of daughter lineage of wild strains) is used as a faithful model for the condition of two types of multicellular eukaryote cells (1: cells with telomerase repressed; 2: cells with telomerase active)

for other characteristics suggested by the theory (e.g., fixed size of the telomere cap, cellular regulations dependent on r-sequences—including the stability of the telomere-cap complex -, and ways in which these regulations were implemented), there were no specific empirical data.

Therefore, the theory could be considered plausible but not yet confirmed by the evidence.

8.2.3 Subtelomere-Telomere Theory of Aging Integrated with TERRA Sequences

In the immense edifice of science, different specific searches take place in separate rooms, and those who arrive at a result in one room are often unaware of what has been discovered in other rooms, which may be essential for their research.

While the existence of fundamental regulatory sequences in the subtelomere (the aforesaid r-sequences) was hypothesized by the subtelomere-telomere theory of aging based on indirect clues, at the same time, very valid researchers were studying and describing sequences with the characteristics of these hypothetical sequences, without having in any way the aim of demonstrating aging mechanisms. The following is a brief review of their results that constitute a fundamental contribution to the subtelomere-telomere theory of aging, transforming the same from a plausible hypothesis to a theory with ample confirmations in the evidence.

– **TERRA sequences**

Over 30 years ago, two noncoding subtelomeric sequences (TelBam3.4 and TelSau2.0) were identified (Brown et al., 1990). Later, it was shown that they had two evolutionarily conserved regions with peculiar structures, one 1.6 kb long and the other 1.3 kb long, which were precisely described (Nergadze et al., 2009). These sequences were called TElomeric Repeat-containing RNA (TERRA; here, "T-sequences" for brevity).

About these T-sequences:

1. They are not protein-coding but are transcribed producing RNA sequences (here "T-transcripts" for brevity) (Rudenko & Van der Ploeg, 1989; Solovei et al., 1994), and have been described in our species (Azzalin et al., 2007; Schoeftner & Blasco, 2008), Zebrafish and mice (Schoeftner & Blasco, 2008), yeast (Luke et al., 2008; Bah et al., 2012; Greenwood & Cooper, 2012) and plants (Vrbsky et al., 2010);
2. T-sequences "are emerging as new key players in several important biological processes" (Diman & Decottignies, 2018) and are "... evolutionarily conserved in vertebrates." (Azzalin & Lingner, 2008). Because such sequences are a general characteristic of eukaryotic cells and appear evolutionarily conserved, they should have a pivotal function from ancestral times;

3. In mammals, the transcription of T-sequences starts from a promoter that is present in the subtelomeric DNA and proceeds towards the repeated motif of the telomere using some of them in the transcription (Azzalin et al., 2007; Schoeftner & Blasco, 2008; Feuerhahn et al., 2010). This transcription occurs in at least two-thirds of the telomeres (Nergadze et al., 2009; Porro et al., 2014; Diman et al., 2016);
4. In humans, the subtelomeric promoters of T-sequences transcription are constituted by DNA islands rich in CpG dinucleotides. These islands show a particular structure with repeated sequences, at about 1 kb from the telomeric sequence (Diman & Decottignies, 2018);
5. The T-sequences are observed in the subtelomeric sections of almost all chromosomes, e.g., chromosomes 2, 9, 13, 18, and sex chromosomes (Chu, Cifuentes-Rojas, et al., 2017a);
6. T-transcripts have DNA targets (i) in every part of the genome, including sites far from the telomere; (ii) also within genes, where introns are favored; (iii) in trans near or within genes; and (iii) in cis at telomeres (Chu, Cifuentes-Rojas, et al., 2017a);
7. The vast majority of the binding sites of T-transcripts are observed in distal intergenic and intronic sections of chromosomes where they regulate gene expression (Diman and Decottignies);
8. T-transcripts bind at many sites in non-subtelomeric or telomeric parts of DNA, expressing a likely regulatory action on other noncoding DNA sequences that appear to have relevant regulatory functions for the expression of numberless genes (Chu et al., 2017a, 2017b);
9. It has been shown that physical exercise increases T-transcripts levels in skeletal muscle biopsies in healthy subjects (Diman et al., 2016), which is compatible with the idea that physical exercise has a beneficial protective action against aging.

- **T-sequences and telomere protection**

Regarding the relationship between the effects of T-sequences and telomere protection against the possible activation of the cell senescence program:

(A) in embryonic stem cells of mice, reduced levels of T-transcripts appear to correlate with lower telomere protection (Chu, Cifuentes-Rojas, et al., 2017a; Chu, Froberg, et al., 2017b);
(B) blockade of T-sequences is correlated with alterations of the function of the telomere cap, and a loss of telomere integrity has been demonstrated after the decline of the levels of T-transcripts (Chu, Cifuentes-Rojas, et al., 2017a);
(C) the deletion of the 20q locus causes a substantial decline in the levels of T-transcripts and then a massive DNA damage response, which is a likely "demonstration in any organism of the essential role of TERRA in the maintenance of telomeres" (Montero et al., 2016);
(D) the inhibition of the transcription of T-sequences activates the mechanisms for telomeric DNA damage response (Bettin et al., 2019);

(E) ATRX, a protein studied for alpha thalassemia mental retardation X-related syndrome, and T-transcripts share hundreds of target genes for which they are functionally antagonistic. In general, ATRX represses a target gene while T-transcripts activate it. T-transcripts contribute to telomere stability binding to telomeric DNA and are antagonized by ATRX (Chu, Cifuentes-Rojas, et al., 2017a).

– **Age-related epigenetic modifications**

Age-related epigenetic modifications of cellular DNA are well documented:

1. Various forms of epigenetic modifications show a relationship with increasing age and with cell and tissue types (Illingworth et al., 2008; Bernstein et al., 2010);
2. The most known and studied type of age-related epigenetic modification is the cytosine-5 methylation within CpG dinucleotides, or DNA methylation (Horvath, 2013; Mammalian Methylation Consortium, 2021);
3. In embryonic cells and induced pluripotent stem cells (iPSCs), there are virtually no points modified by DNA methylation, while the number of these sites grows in relation to the number of previous cellular duplications (Horvath, 2013; Mammalian Methylation Consortium, 2021);
4. These epigenetic changes are reversible. In fact, after the transformation of adult somatic cells into iPSCs, no points appear modified by DNA methylation as for embryonic cells (Horvath, 2013);
5. CpG sequences modified by DNA methylation with age (Rakyan et al., 2010; Teschendorff et al., 2010; Horvath et al., 2012; Bell et al., 2012) are restricted to specific parts of DNA molecule, in particular CpG islands (CGIs), i.e., DNA stretches where CpG nucleotides are about 1 per 10 bp and constituting only 2% of the entire DNA (Illingworth et al., 2008). CGIs are often the transcription start sites of genes (Bird, 2002), and their methylation appears correlated with the silencing of the gene promoters (Stein et al., 1982). Vice versa, the demethylation restores promoter expression (Hansen & Gartler, 1990);
6. Age-related epigenetic modifications of CGIs in some cases are hypermethylation, and in others hypomethylation (Bollati et al., 2009; Christensen et al., 2009; Bell et al., 2012; Horvath et al., 2012);
7. These types of age-related DNA methylation have been proposed as reliable indices for assessing age in individuals of our species. The best index proposed shows a correlation with age equal to 0.96 with an error of 3.6 years (Horvath, 2013);
8. The same epigenetic modifications are documented for mammals in general. For 128 mammal species, it was proposed a Horvath's-like index that showed a correlation greater than 0.96 and a mean error of less than 3.5% (Mammalian Methylation Consortium, 2021). In general, CGIs appear so evolutionarily conserved to allow the definition of this reliable index (Mammalian Methylation Consortium, 2021);
9. In addition to DNA methylation, other age-related epigenetic changes are known (Greer & Shi, 2012; Booth & Brunet, 2016; Pal & Tyler, 2016). However, no

better or more reliable indicators like the two mentioned above have been proposed regarding these epigenetic changes.

– **Relations between T-sequences and age-related epigenetic modifications**

There are documented relations between T-sequences and age-related epigenetic modifications:

1. In Terc (– / -) mice, the levels of methylation of subtelomeric DNA and telomere shortening are related (Benetti et al., 2007);
2. It is documented that subtelomeric CGIs "promote transcription of TERRA molecules." (Nergadze et al., 2009);
3. In mice, there is a correlation between subtelomere methylation and telomere shortening (Blasco, 2007);
4. The block of some master epigenetic regulators for DNA methylation is related to the loss of telomere-length control, and telomere shortening to a critical length compromises the epigenetic condition of subtelomeres and telomeres (Blasco, 2007);
5. In healthy subjects, telomere length shows an age-related decrease. Furthermore, there is an association between longer telomeres and hypermethylation of subtelomeric sequence and between shorter telomeres and hypomethylation. The same associations were observed for sarcoidosis patients (Maeda et al., 2009);
6. In human leukocytes, it was observed the association between shorter telomeres and decreased methylation level of CpG sites in subtelomeric loci (i.e., within 4 Mb of the telomere) (Buxton et al., 2014);
7. Telomere shortening modifies gene expression and increases the gravity and the risk of age-related diseases (Buxton et al., 2014).

So, longer telomeres are related to hypermethylation of T-sequences, and shorter telomeres are related to hypomethylation of T-sequences, which should be related to less and greater repression of T-sequences, respectively.

– **Effects of T-sequences on other regulatory sequences**

Regarding the regulatory functions of T-sequences on other regulatory sequences, important indications can be obtained by considering: (i) the variation of their regulatory activity in relation to the number of duplications; and (ii) their condition in cell senescence when there should be great alterations of cellular function in relation to maximal repression of T-sequences:

1. In mesenchymal stem cells (MSCs), for the cellular DNA: "expansion of MSC has a very consistent impact on DNA-methylation profiles"; "517 CpG sites were consistently differentially methylated in early versus late passages" (Schellenberg et al., 2011);
2. For MSCs in cell senescence, there is an association with some histone marks of aging (e.g., trimethylation at specific targets) and DNA methylation in specific CGIs (Schellenberg et al., 2011);

3. In MSCs, in relation to age, almost one third of the CpG sites shows changes in DNA methylation (60% hypomethylation and 40% hypermethylation) (Zhou et al., 2020).

These data indicate a relation between T-sequence repression and epigenetic modifications in many parts of the DNA molecule. As T-sequences appear to have regulatory actions on many other sequences that, in turn, regulate other cellular functions, they may be defined as "first-level regulators" acting on "second-level regulators".

– **Identity of T-sequences and r-sequences**

Considering the experimental data reported above, the characteristics of the T-sequences correspond exactly to those hypothesized in the subtelomere-telomere theory of aging for the r-sequences.

In fact, T-sequences:

(A) are positioned in the subtelomere;
(B) are repressed with telomere shortening;
(C) originate transcripts that show regulatory actions (i) over innumerable other sequences with regulatory functions positioned in near or distant parts of the DNA molecule in which T-sequences are present; and (ii) over other regulatory sequences in other chromosomes of the same cell;
(D) variously influence numberless cellular functions by these actions;
(E) regulate the stability of the cell telomere-cap complex, which is inversely related to the probability of activation of the cell senescence program;
(F) in relation to their progressive repression, determine the alterations that can explain the gradual cell senescence;
(G) are widespread and evolutionarily stable, so their function is certainly of great importance.

Based on these characteristics, in the subtelomere-telomere theory of aging, it is possible to replace the hypothetical r-sequences with the real T-sequences without modifying in anything the theory (Fig. 8.11).

– **The telomeric heterochromatin cap**

About the telomeric heterochromatin cap (or hood), the subtelomere-telomere theory of aging requires that the cap must have a size determined in the first cell based on the length of each telomere and not modified in any subsequent cell duplication even if the telomere shortens.

It is known that the telomere is covered by a chain of shelterin protein complexes consisting of the proteins RAP1, TIN2, TRF1, TRF2, POT1, and TPP1 (Stewart et al., 2012; Jones et al., 2016). The reciprocal arrangement of these proteins was proposed some years ago (Stewart et al., 2012), and based on this suggestion, the possible link between two contiguous shelterin complexes was hinted (Libertini, Corbi, et al., 2021a, Chap. 5).

Suppose the hypothesis of the fixed size of the telomeric cap is true. In that case, a likely prediction is that the amount of shelterin proteins in a cell should be

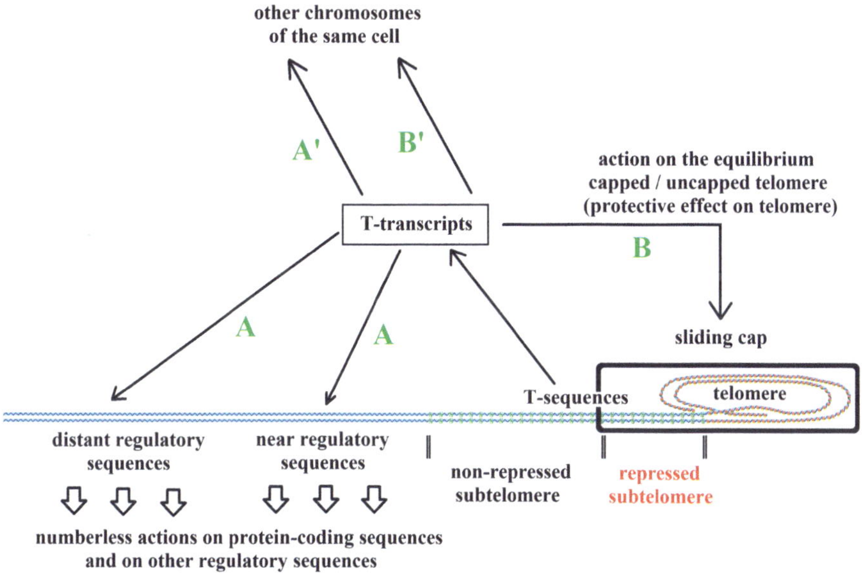

Fig. 8.11 Scheme of the subtelomere-telomere theory of aging (Fig. 1 from (Libertini, Shubernetskaya, et al., 2021b), simplified and redrawn). T-sequences, progressively repressed by telomere shortening and the consequent cap "sliding" over the subtelomere, act by their T-transcripts: (**a**) on near and distant regulatory sequences on the same chromosome; (**a'**) on similar regulatory sequences in other chromosomes; (**b**) on the equilibrium between capped and uncapped conditions of the two telomeres of the same chromosome; (**b′**) the same of B for other chromosomes

constant and not related to the total length of the telomeres, which varies with the number of duplications if telomerase is restricted. On the contrary, if the hypothesis is false, i.e., the size of the cap changes with telomere shortening, a reduction in the amount of shelterin proteins, proportional to telomere shortening, should be found.

However, there is evidence regarding the constancy of the amount of shelterin proteins even if the length of the telomeres is shortened. In fact, a key experiment showed that the amounts of shelterin complex proteins were equal in primary and transformed cells and were not related to telomere length (Takai et al., 2010). This appears to confirm the subtelomere-telomere theory of aging about the invariance of telomeric cap size after its formation in the first cell of the organism.

– **Aging as an epigenetic phenomenon**

Considering the same experimental data reported before, aging can rightly be described as a physiological phenomenon regulated by epigenetic modifications.

This statement does not mean that aging is an exception among the functions of the organism.

In fact, when it was discovered that protein sequences were encoded by specific sections of DNA, defined as genes, knowing the innumerable capabilities of proteins, it was easy to believe that all functions, both cellular and related to the overall organization of the organism, depended on protein-coding parts of the DNA. However, it was later observed that:

(i) Only a small part of the DNA encodes protein sequences. So, it was even thought that the remaining parts of the DNA, as they did not originate proteins, were functionless or "junk" DNA (Ohno, 1972);
(ii) The ENCODE and the Human Genome Projects demonstrated that only 2% of the DNA is protein-coding, while >90% is transcribed producing RNA sequences with functions that are mostly unknown (Chu, Cifuentes-Rojas, et al., 2017a);
(iii) Species that are very different in the complexity of the body's organization instead of differing in equal proportion in the number of genes showed that this number was practically equivalent. It was unexpected to observe that a rather unsophisticated nematode and our species had an equal number of protein-coding sequences (Moraes & Góes, 2016).

The logical deduction is that the "program" that defines every function of the organism, including the development of a complex organism starting from a single cell, the functional organization of every organ or apparatus, any simple or complex function, and therefore also aging described as a function, did not derive solely from the innumerable activities of proteins. On the contrary, the main part of the "program" consists of the regulatory functions carried out by DNA sections that do not encode proteins but regulate every characteristic of the organism. This, at the cellular level and within the overall morphological and functional organization of the organism, by mechanisms of up-regulation/down-regulation (activation/repression) both of protein-coding and non-protein-coding sections of the DNA.

A splendid confirmation of this theoretical argument is the recent demonstration of the differences in epigenetic marks between the vast diversity of the normal cell types of an organism (Loyfer et al., 2023). These epigenetic modifications are so specific and precise that it is possible to indicate the developmental affinities of various cell types ("... DNA methylation atlas of normal human cell types ... recapitulates key elements of tissue ontogeny ..." (Loyfer et al., 2023).

Consequently, the development and the functioning of the organism in all its parts are continuously regulated by the non-protein-coding parts of the DNA through continuous precise epigenetic modifications. So, it is no wonder that aging also follows the same general rule.

8.2.4 Some Characteristics of Aging Mechanisms that Are Incompatible with the Non-programmed Paradigm of Aging

Regarding the cellular mechanisms of aging discussed above, some of their characteristics appear difficult to explain for the paradigm of aging conceived as a non-adaptive phenomenon and so not determined and modulated by a specific program:

– **Cell senescence**

The increase of the number of cells in the state of cell senescence, both in absolute terms and as a percentage of cells, is undoubtedly harmful to the functionality of the tissues and the overall fitness of the organism, contributing to the manifestations of aging and age-related diseases. This is also demonstrated by the fact that the selective elimination of senescent cells reduces and improves these manifestations (Baker et al., 2008; Chang, Wang, et al., 2016a). Therefore, active research is underway for the selective elimination of these cells by drugs called senolytics (Chang, Wang, et al., 2016a; Fuhrmann-Stroissnigg et al., 2017), a subject deepened in (Libertini et al., 2018).

The increase in the number of harmful cells, such as senescent cells, is perfectly compatible with the mechanisms of aging within the paradigm of aging as an adaptive and programmed phenomenon, according to what was previously exposed. On the contrary, an explanation for their existence is necessary for the opposite thesis.

In this regard, it has been proposed as an explanation that cell senescence, since it blocks the replicative capacity of cells and therefore slows down a possible pathological cellular proliferation, would be a primary general defense against cancer (Campisi, 2000, 2003; Rodier & Campisi, 2011). Furthermore, as cell senescence, in addition to blocking the duplication capacity, causes alterations both in senescent cells and in cells that in any way depend on their correct functionality, this contradiction has been explained as an example of an evolutionary trade-off between the harmful effects of senescent cells and their advantageous action against the possible proliferation of cancer cells, i.e., as an example of antagonistic pleiotropy (Campisi, 2013).

This explanation is untenable for several reasons:

1. A senescent cell, because of the senescence-associated secretory phenotype (SASP), secretes "myriad factors associated with inflammation and malignancy" (Coppé et al., 2008), so increasing the risk of cancer (Parrinello et al., 2005; Coppé et al., 2008);
2. In a paper, the authors maintain that cellular senescence is indispensable to counter tumorigenesis because halts the proliferation of damaged cells. Then they observe that senescent cells secrete various substances that promote carcinogenesis and that the eliminations of senescent cells delays the onset of cancer and prolongs lifespan in vivo (Ghosh & Capell, 2016);
3. In another paper, the authors, while maintaining that cell senescence counteracts cancer by blocking the proliferation of malignant cells, then observe that cell senescence is induced by many anti-cancer chemotherapeutic drugs, but that the elimination of such cells reduces various side effects of such drugs, in the short and long term, including cancer recurrence (Demaria et al., 2017);
4. In a third work, always taking as certain the action of cell senescence as a powerful antitumor mechanism, there is then the observation that the same phenomenon favors degenerative and proliferative pathologies probably by stimulating a chronic inflammation. To justify this contradiction, it is therefore hypothesized

that cell senescence is justified by a mechanism of antagonistic pleiotropy (Campisi, 2013);

5. "Senescent cells are present in premalignant lesions and sites of tissue damage and accumulate in tissues with age." (Biran et al., 2017);
6. In mice, the selective elimination of senescent cells, in addition to reducing the age-related alterations and increasing the lifespan, delayed the progression of cancer (Baker et al., 2016);
7. For many animal species studied in the wild, a progressive age-related increase in mortality that significantly reduces life span is well documented (Nussey et al., 2013). However, for none of these species, the mortality due to cancer appears to influence the age-related increase in mortality significantly. As for our species, in a hunter-gatherer population studied in the wild before contact with modern populations, no cancer was observed before the age of 70, to which about 15–20% of individuals survived (Hill & Hurtado, 1996). For older ages, even if no cases of cancer were reported, it is possible that there were a few cases. However, since in any case, both for our species and for other animal species studied in the wild, mortality due to cancer did not appear to be a significant cause of death, "This completely disproves the hypothesis that the reduction of cell duplication capacities would be a defense against cancer: it would be like arguing that a defense against a deadly disease has the effect of mass killing before the disease begins to kill." (Libertini, 2013). This argument will also be proposed in the next subsection regarding the limits in telomerase enzyme activity;
8. Cell senescence, in addition to blocking the cell duplication capacity, with a possible useful action against possible carcinogenic proliferations, also shows a series of negative effects. So, it is not explained why senescent cells are resistant to apoptosis, causing chronic damage that is inexplicable if the role of cell senescence in the aging mechanism is not admitted (Mitteldorf, 2013).

– **Limits in telomerase activity**

Restrictions in the activity of telomerase enzyme have also been hypothesized as a defense against possible cancerous proliferation (Wright & Shay, 2005), i.e., another example of a trade-off between negative effects of telomerase restrictions and positive anticancer effects (Stone et al., 2016; Young, 2018). However, this hypothesis is contradicted by several facts:

1. Some species show no age-related mortality increase and are defined as species with "negligible senescence" (Finch, 1990). Young and old individuals of animals of this type, such as rainbow trout and lobster (Klapper, Heidorn, et al., 1998a; Klapper, Kühne, et al., 1998b) and rockfish species (Black, 2002), show the same levels of telomerase activity. For these species, an increased oncogenic risk is unlikely, as demonstrated by their constant mortality rate, and this proves that telomerase activity is not itself oncogenic (Libertini, 2008);

2. In normal mice, artificially induced telomerase activity delays aging and increases lifespan without increasing oncogenic risk (Bernardes de Jesus et al., 2012);
3. The absence of a relationship between telomerase activity and oncogenic risk is also demonstrated in other works (Tomás-Loba et al., 2008; Jaskelioff et al., 2011; Muñoz-Lorente et al., 2018);
4. If telomerase is inactive, telomeres shorten at each duplication, but when telomere length reaches a critically low value, this determines a telomere dysfunction and instability of chromosomal DNA, increasing the vulnerability to cancer (DePinho, 2000; Artandi, 2002; Wu et al., 2003; Artandi & DePinho, 2010; Ma et al., 2011);
5. In subjects suffering from *Dyskeratosis congenita*, a genetic disease in which telomerase activity is altered, a high incidence of cancer is observed (Dokal, 2000), so demonstrating that telomerase is not an oncogenic factor and, on the contrary, protects from cancer.
6. In mice, the treatment with AAV9-*Tert* induces telomerase reactivation and does not cause carcinogenesis (Whittemore et al., 2019);
7. In the sequence of oncogenesis, unrestrained telomerase activity is often observed but comes after the onset of cancer and not before: "The role of the telomere in chromosomal stability (Blagosklonny, 2001; Campisi et al., 2001; Hackett et al., 2001) argues that telomerase protects against carcinogenesis (Chang et al., 2001; Gisselsson et al., 2002), especially early in carcinogenesis when genetic stability is critical (Elmore & Holt, 2000; Kim & Hruszkewycz, 2001; Rudolph et al., 2001), as well as protecting against aneuploidy and secondary speciation (Pathak et al., 2002). The role of telomerase depends on the stage of malignancy as well as cofactors (Ohmura et al., 2000); expression is late and permissive, not causal (Seger et al., 2002)." (Fossel, 2004, p. 78);
8. The penultimate argument for cell senescence proposed in the previous subsection is also likely valid for the limits in telomerase activity.

– **Gradual cell senescence**

The subsection dedicated to gradual cell senescence shows that this phenomenon is distinct from cell senescence.

As previously discussed, cell senescence, because it causes replicative senescence, has been justified as a general defense against cancerous proliferation, although such an explanation is untenable for various reasons mentioned above.

For gradual cell senescence, as this phenomenon is characterized by alterations of cellular functions in relationship with the number of previous cell duplications and telomere shortening (when telomerase activity is restricted), but not by replicative senescence, the proposed explanation for cell senescence cannot be suggested.

The progressive decay of cellular functions is perfectly compatible with the paradigm of aging as an adaptive and programmed phenomenon. On the contrary, for the opposite paradigm, which interprets aging as the progressive accumulation of alterations insufficiently opposed by natural selection, it is necessary to explain whether gradual cell senescence has in some way an adaptive meaning or why natural selection is not able to counteract the progressive decay of cellular functions at each duplication, for example by allowing higher activity of the telomerase enzyme.

– **Position of T-sequences**

As widely indicated by the works reported in the previous section, T-sequences, "... key players in several important biological processes" (Diman & Decottignies, 2018), perform important general regulatory functions in the cell. It is also true that these sequences are in the immediate vicinity of the telomere, i.e., in the most vulnerable position for possible repression caused by the "telomeric position effect" (Gottschling et al., 1990). Indeed, we have seen that this distinctive position allows their progressive repression in relationship with telomere shortening, which is a key element of the mechanisms proposed to explain aging. Not wanting to accept this explanation, which is perfectly consistent with the thesis of aging as a programmed phenomenon, anyone wishing to support the opposite thesis would have to justify why pivotal sequences are in such a vulnerable position.

– **Reversibility of the alterations of cellular functions related to aging**

The interpretation of aging as a non-adaptive phenomenon caused by the progressive accumulation of random alterations, outside of any hypothetical program and determined by factors that natural selection fails to contrast, implies that these alterations are without order and not easily reversible.

However, many experimental data, some of which already mentioned, show that, at the cellular level, the aforesaid alterations are totally and precisely reversible by means of appropriate specific actions:

1. Senescence is a precise multi-step process involving the $p16^{Ink4a}$ and/or p53-p21 pathways (van Deursen, 2014). When cell senescence is fully established—"full senescence" (van Deursen, 2014) -, it is not reversible by telomerase activation, but *in vitro*, it is completely reversible by the inactivation of both p53 and $p16^{Ink4a}$ (Beauséjour et al., 2003);
2. A characteristic of senescent cells is the strong resistance to apoptosis. As a rule, a cell with impaired functions, such as a senescent cell, should trigger its elimination through apoptosis. However, a part of the cell senescence program is the activation of Senescent Cell Anti-apoptotic Pathways (SCAPs), which block the activation of the apoptosis machinery. These SCAPs comprise various pathways (e.g., PI3K/AKT, BCL-2/BCL-XL, tyrosine kinase, p53/p21/serpines), and precisely the unblocking of these apoptosis-activating pathways constitutes the main objective of the senolytics, i.e., drugs aiming to eliminate senescent cells selectively (Kirkland & Tchkonia, 2017);
3. In 2006, it was shown, by introducing in the cell four factors (Oct3/4, Sox2, c-Myc, and Klf4), the reprogramming of differentiated cells into induced pluripotent stem cells (iPSCs), which show the characteristics of embryonic stem cells (Takahashi & Yamanaka, 2006). The epigenetic modifications of these iPSCs are practically inexistent as for embryonic cells (Horvath, 2013);

4. Mesenchymal stem cells (MSCs) show increasing alterations in relationship with the number of previous duplications that can be erased entirely by reprogramming MSCs into iPSCs. These iPSCs, whatever the number of previous duplications, the origin of the cells, and the age of the donor, show the profile of young cells (Spitzhorn, 2019);
5. It is possible to obtain from iPSCs-induced MSCs (iMSCs) for which "DNA methylation, related to age, was completely erased …" (Zhou et al., 2020). Such "… iMSCs reacquired senescence-associated DNA methylation during culture in vitro." (Zhou et al., 2020).

These experimental data show that the cellular alterations related to duplications are not random alterations but expressions of complex pathways or program modules, which opportune manipulations can modify and even cancel. This contrasts with the interpretation of progressive cellular alterations due to factors determining chaotic irreversible damage.

8.3 Effects of Cellular Aging on Tissues and Organs

Until now, the focus has been on changes at the cellular level that could be the likely substrate of aging. It is now necessary to investigate how these cellular modifications determine the supracellular manifestations of aging.

8.3.1 Cell Turnover

Apart from cellular deaths due to accidental events (e.g., traumas, infections), most of the cells in our body continuously die spontaneously and are replaced based on genetically regulated events defined as "programmed cell death" (PCD). This continuous cellular replacement, defined as cell turnover, is complete and faster in young and healthy subjects, while it tends to slow down and be incomplete in elderly individuals. Some types of PCD and the related cell turnovers have been well-known for some time (Hall & Hall, 2020; Fox & Rompolski, 2021):

(i) The cells of the mucous membranes continuously detach and are replaced by new cells;
(ii) The keratinized epidermal cells detach and are replaced by new cells originating in the basal parts of the epidermis;
(iii) The keratinized cells that form nails and hair, highly differentiated cells of the epidermis, are eliminated by detachment or wear;
(iv) The erythroblasts are transformed into erythrocytes, cells without a nucleus, and after about 3 months, are eliminated by the phagocytes and replaced with new erythrocytes;

(v) The cells of the intestinal villi detach at the ends of the villi and are replaced by cells that originate from stem cells present in the intestinal crypts.

In addition to these types of PCD, the phenomenon of apoptosis was not known until 1972. It affects many cell types previously thought to have no turnover. Apoptosis is an ordered form of cellular self-destruction that does not constitute a pathological phenomenon, is not caused by accidental events, and does not cause inflammation. Furthermore, the cell parts that are released by apoptotic cells are usefully utilized by other cells (Kerr et al., 1972).

Phylogenetically, apoptosis is a very ancient phenomenon and is even present in unicellular eukaryote species, e.g., *Saccharomyces cerevisiae* (Laun et al., 2007; Kaeberlein et al., 2007), where, as this species is unicellular, it has the meaning of a particular form of phenoptosis.

In multicellular eukaryotic organisms, apoptosis allows cell turnover for many cell types in healthy organisms (Wyllie et al., 1980; Lynch et al., 1986; Medh & Thompson, 2000), but also performs other functions such as embryonic development, the correct development and functioning of the immune system, atrophy caused by hormones (Elmore, 2007).

Regarding the function related to cell turnover, apoptosis, since its discovery, has been documented for most cell types, including: adipocytes (Prins & O'Rahilly, 1997); alveolar epithelial cells of type II (Sutherland et al., 2001); biliary epithelial cells (Harada et al., 2000); cardiac myocytes (Anversa et al., 2006); chondrocytes (Héraud et al., 2000); hepatocytes (Benedetti et al., 1988); gliocytes (Pontèn et al., 1983); kidney cells (Cardani & Zavanella, 2000); osteocytes (Spelsberg et al., 1999); pancreatic β-cells (Finegood et al., 1995); prostatic cells (Xia et al., 2001); skeletal myocytes (Migheli et al., 1997; Pollack & Leeuwenburgh, 2001); thyrocytes (Dremier et al., 1994).

Apoptosis is so widespread that the term is imprecisely used as a synonym of PCD, e.g., (Jacobson et al., 1997).

For the turnover of cells eliminated by the various forms of PCD, replacing the lost cells with others obtained by duplicating specific stem cells is necessary. Cell turnover rates are very different depending on cell type (Richardson et al., 2014), ranging from cells in which turnover occurs in a few days—e.g., 3–6 days for the cells of intestinal epithelium (Alberts et al., 2013)—to other cells in which turnover rates are measured in years—e.g., about 4.5 years for heart myocytes (Anversa et al., 2006) and 10 years for osteocytes (Alberts et al., 2013).

Cellular turnover is a massive and continuous phenomenon. In our species, it was estimated that every day, an average of 50–70 billion cells (i.e., roughly 700,000 cells per second) die and are replaced, and in a year, the mass of these cells is roughly equal to that of our body weight (Reed, 1999).

According to a more recent and precise estimate (Sender & Milo, 2021), a person weighing 70 kg has about 30 trillion cells ($=30*10^{12}$). Such a number of cells is replaced in 80 days ($= 30*10^{12}/80 = 3.75*10^{11}$ every day, which means, dividing by 86,400, the number of seconds in a day, approximately 4.35 million cells per second), while an equivalent mass of cells is replaced in about a year and a half, with marked disparities both in turnover times and cellular masses among the various cell types.

Cellular renewal is vigorous and continuous in healthy and young organisms. However, with aging, as we will see in the next sections, the renewal slows down more and more with a progressive impoverishment of the number of cells and a decay of their functions.

8.3.2 Replicative Senescence and Decline of Cell Turnover

For the subtelomere-telomere theory of aging, a simple interpretation could be that:

(i) cells with rapid turnover, and therefore with rapid duplication of specific stem cells, should show a reduction in telomere length correlated with cell turnover rate, and aged individuals should have critically shortened telomeres;
(ii) consequently, cells with higher turnover and shorter telomeres should show more signs of cellular aging, and the reverse should be true for cell types with slower turnover.

However, these predictions are contradicted by experimental results, and furthermore, this would not explain aging for perennial cells in which there is no cell turnover.

In a review where the reduction in telomere length from neonates to centenarians was studied (Takubo et al., 2010), it was observed, in relation to age, a uniform reduction in the length of telomeres, except the cells with very low or no cell turnover (cardiac myocytes and most neurons), where no sensible reduction was pointed out. For a few cellular types, the annual reduction of telomere lengths was critical: e.g., for liver cells, telomere length was 13.7 ± 2.5 kbp in neonates and 8.7 ± 1.4 in centenarians with a reduction rate of 120 bp/year. For most cell types, the reduction rates were lower, in the range of 20–60 bp/year (Takubo et al., 2010). In another study on some types of cells with quite different cell turnover rates (skeletal muscle, leukocytes, subcutaneous adipocytes, and epithelial cells), the rates of telomere shortening were shown to be similar (Daniali et al., 2013).

These observations did not explain the apparent aging temporal homogeneity in the various parts of the organism and the aging in perennial cells, i.e., in cells without duplications or renewal.

However, the same study highlighted that, while telomere lengths were practically the same in the different organs and tissues of a fetus—as already known from previous studies, e.g. (Okuda et al., 2002)—in stem cells of cellular types with high turnover the length of telomeres was shorter than in stem cells of cells with slow turnover.

From these data, it was deduced that the progenitor stem cells of the various cell types actively duplicate themselves in the growth period proportionally to the subsequent cell turnover rates. So, there were multiple duplications (i.e., notable expansion) for cells with high turnover (e.g., hematopoietic cells). On the contrary, there was reduced expansion for cells with low turnover (e.g., myocytes). In this phase, the length of the telomeres decreased with the number of duplications (i.e., the extent of expansion), while in the following phases, the telomeres decreased at a constant rate (Daniali et al., 2013) (Fig. 8.12).

It was necessary to underline that, with the reduction of telomere lengths, even in stem cells, there is always the possibility of passing to the condition of cell senescence in which there is the block of replicative capacities (Blackburn, 2000), and this gradually reduces the number of the stem cells with consequent slowing down of the speed and capacity of cell turnover. In short, the decline of cell turnover is explained not by the shortening of telomere length at a critical level but by the gradual transition of a growing fraction of stem cells to a cell senescence state in which replication is blocked.

To this, it is necessary to add:

(i) the progressive increase of cells in the condition of cell senescence, which determines not only the block of replicative capacity but also alterations of cellular functions and abnormal secretions, defined as SASP. Indeed, the selective elimination of these cells is indicated as a goal to reduce the manifestations of aging (Libertini et al., 2018);
(ii) in relation to the shortening of telomeres, there are gradual alterations of cellular functions, a condition defined as gradual cell senescence and discussed in the previous subsection *§8.2.2 Subtelomere-telomere theory of aging* in this chapter.

In the case of the perennial cells, i.e., in cells without apparent turnover (most of the neurons of the central nervous system, among which retina photoreceptor cells and lens fiber cells of the eyes), the phenomena reported above, such as the decline of cell turnover, increasing gradual cell senescence, increasing number of the fraction of cells in cell senescence, cannot happen. However, all these processes apply to cells with turnover that are indispensable for the perennial cells and play the role of trophic or satellite cells (gliocytes of various types, retina pigmented cells included, and lens epithelium cells). The decline of these satellite cells in number and functionality determines the aging of perennial cells and the tissues of which they are the main components (Libertini & Ferrara, 2016b; Libertini, 2018).

8.3.3 The "Atrophic Syndrome"

The phenomena previously discussed (cell senescence; gradual cell senescence; related epigenetic modifications; related alterations of cellular secretions, or SASP; the increasing number of cells in replicative senescence, in particular, stem cells) lead progressively, for each tissue or organ, to a condition of functional and morphological alterations defined as "atrophic syndrome" (Libertini, 2009a, 2014; Libertini, Corbi, et al., 2021a), characterized by:

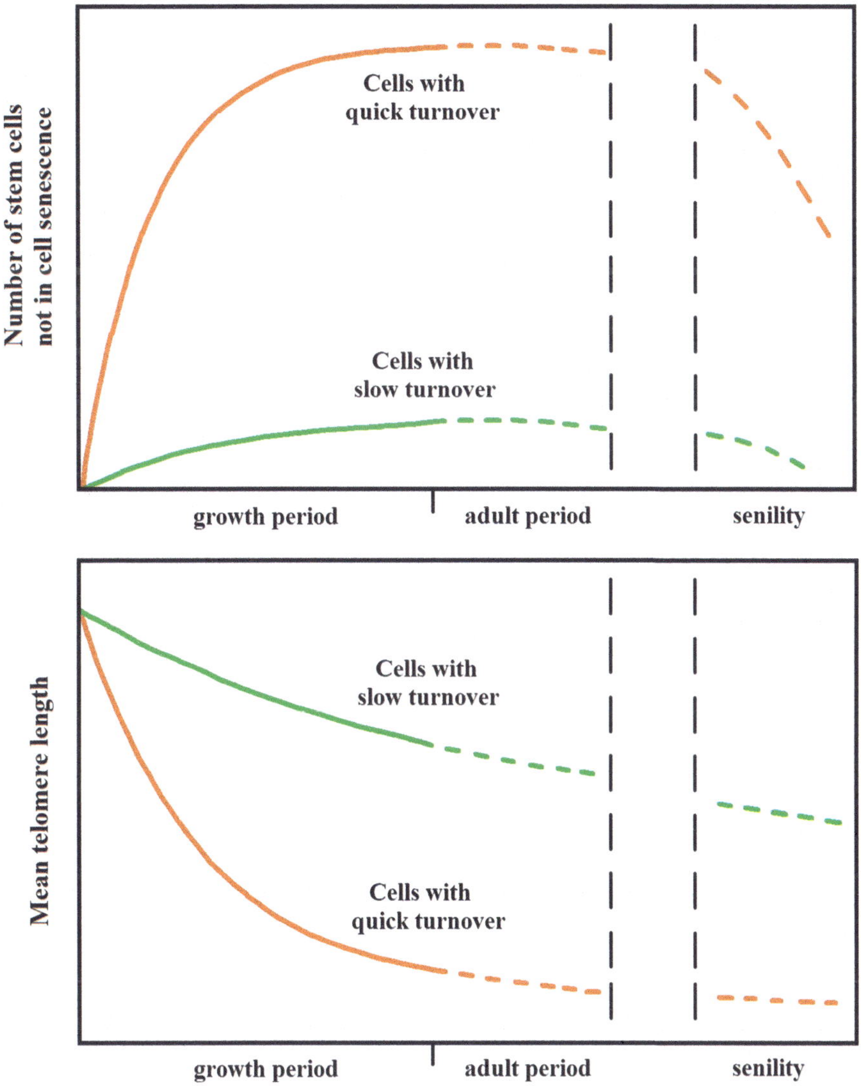

Fig. 8.12 Top: Progenitor cells for cell types with high turnover (e.g., hematopoietic cells) in the growth period show massive expansion, while for cell types with low turnover (e.g., skeletal muscle cells), they show modest expansion. Bottom: In the growth period, telomere shortening is proportional to the expansion, while subsequently, the rate of telomere shortening is equal for both types of cells. Afterward (senility, right side of the images), the number of stem cells declines for their progressive transition to cell senescence. Both images' left sides are inspired by Fig. 4 from (Daniali et al., 2013)

1. increasing number of senescent cells, which show altered cellular functions, including the block of replicative capacity and changed cellular secretions (SASP);
2. increasing number of cells in gradual cell senescence, which show altered cellular functions in proportion to telomere shortening;
3. slowdown and insufficiency of cell turnover due to the progressive reduction of overall replicative capacity, in particular for cell senescence of an increasing fraction of stem cells;
4. because of the progressive insufficiency of cell turnover, decline of the specific cells of a tissue or organ, hypertrophy of the remaining specific cells, and replacement of the missing cells with non-functional connective cells;
5. because of (i) numerical decline of functional cells, (ii) altered functions of cells in cell senescence or in gradual cell senescence, and (iii) altered cellular secretions (SASP), dysfunctions of the cells depending on missing cells and/or on the normality of the intercellular fluids;
6. because of the alterations above, in the case of cells that are indispensable for the vitality of perennial cells, death and functional alterations for these cells;
7. because of the alterations referred to in the previous points, progressive functional and anatomical alterations of all tissues and organs and of the overall functionality of the organism.

A scheme of some of the characteristics of tissue aging is illustrated in Fig. 8.13.

In cases with a marked reduction of telomere length, there is also the possibility of telomere instability and increased risk of cancer (DePinho, 2000; Artandi, 2002). For our species, this occurrence does not appear to be a frequent condition in the wild, where cancer is rare or unobserved (Hill & Hurtado, 1996; Libertini, 2013). On the contrary, this happens in particular pathological conditions, such as chronic liver damage from alcoholism or chronic active hepatitis, which leads to cirrhosis and, hence, hepatocarcinoma (In der Stroth et al., 2020).

8.3.4 "Direct" and "Indirect" Aging

The manifestations and effects of atrophic syndrome concern all tissues, organs, and systems and, as a whole, constitute the description of the body's aging. The division of these effects into two main categories may be helpful (Libertini, 2017, 2022; Libertini, Corbi, et al., 2021a):

(i) **"Direct" aging**—For tissues or organs in which all cell types are subject to turnover, the effects of atrophic syndrome directly concern all cells. In this case, it may be useful to define this phenomenon as "direct" aging (Fig. 8.14);

(ii) **"Indirect" aging**—In some cases, even if most of the cells of a tissue or organ are subject to turnover, one or more principal cell types are perennial, i.e., are

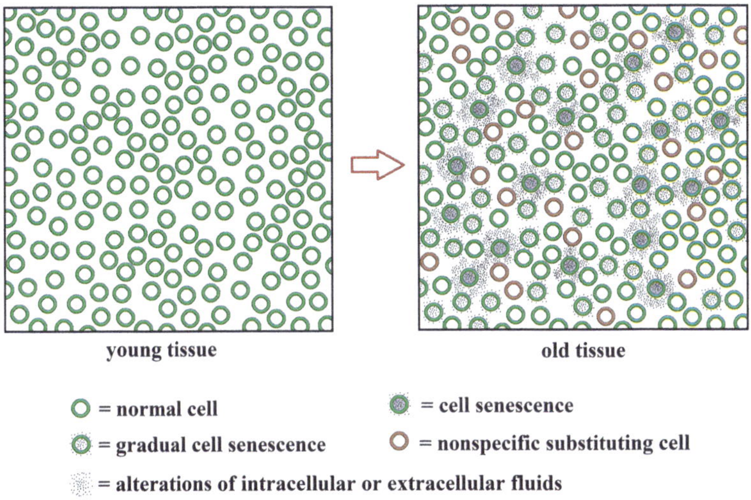

Fig. 8.13 Scheme of the differences between a young and an old tissue

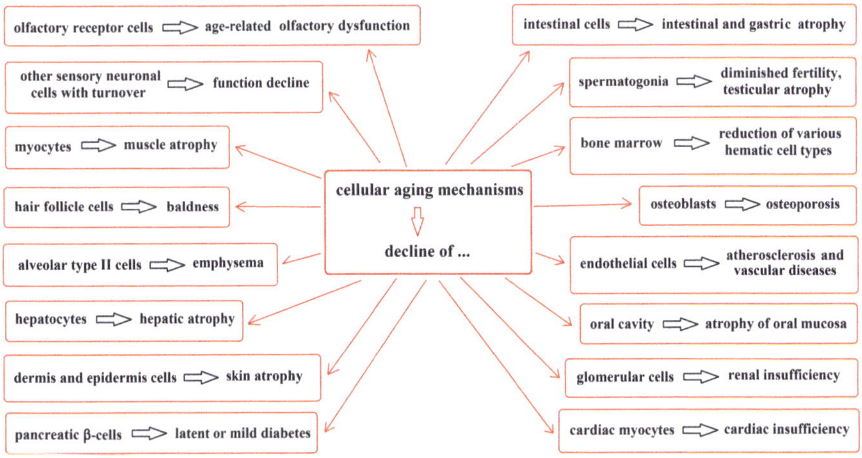

Fig. 8.14 "Direct" aging: most cells show turnover and are directly affected by cellular aging mechanisms

not subject to turnover but depend for their functionality and vitality on other cells that are not permanent. For these cells, the atrophic syndrome has no direct action but acts indirectly on the cells that are trophic for the perennial cells. These cases, in relation to aging, will be defined as "indirect" aging (Fig. 8.15).

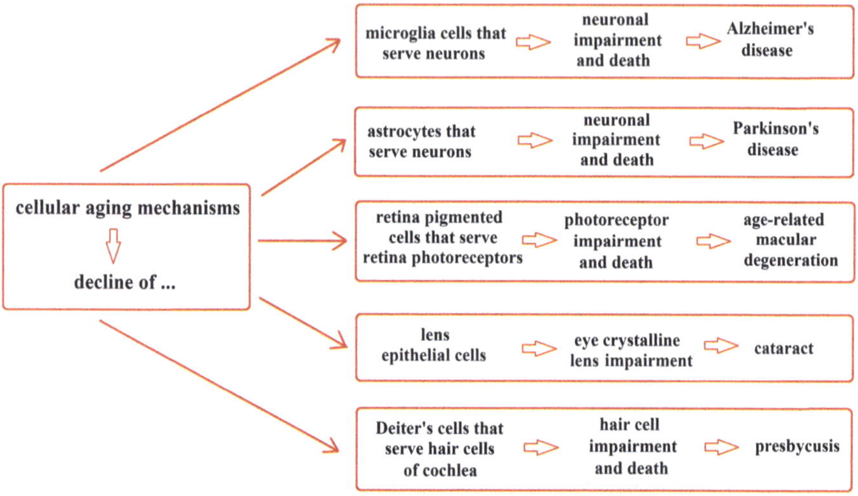

Fig. 8.15 "Indirect" aging: some cell types, defined as perennial, do not show turnover and are not directly affected by the mechanisms of cellular aging. However, these cells depend for their vitality and functionality on other cells that show turnover and undergo the effects of cellular aging mechanisms. As a consequence, perennial cells indirectly suffer the effects of cellular aging mechanisms

8.3.5 Why some Types of Cells Are Not Subject to Turnover

Most cell types are subject to turnover, while a minority, the so-called perennial cells, are without turnover. Since most perennial cells are neurons, one might believe that their peculiar condition is somehow due to the fact that they are nerve cells.

However, this idea is contradicted by the fact that there are nerve cells not subject to turnover (e.g., olfactory receptor cells and other sensory neuronal cells). Furthermore, the eye lens fiber cells are not neurons and are not subject to turnover.

The presence or absence of cell turnover can be easily explained on the basis of structural reasons related to the single cell types.

– **Eye lens fiber cells (a non-nerve cell without turnover)**

The lens fiber cells constitute the bulk of the crystalline lens, which focuses the light rays on the retina in the best possible way, so it must be transparent and have a regular shape (Fig. 8.16).

The outermost part of the lens is the lens capsule, a smooth, elastic, and transparent membrane rich in collagen synthesized by lens epithelium cells (Forrester et al., 2020), which are only on the anterior side of the lens and are essential for its functionality (Candia, 2004).

As main functions, lens epithelium cells: (i) are progenitor cells for lens fiber cells in all periods of life, from embryo to adulthood. The new lens fiber cells are added as outer layers in the equatorial area of the lens epithelium, the so-called germinative zone, then they lengthen, detach from lens epithelium cells, synthesize

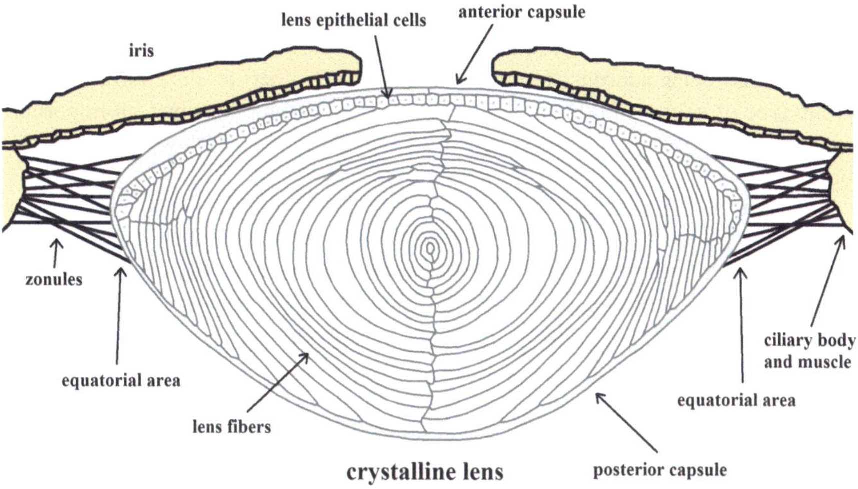

Fig. 8.16 Scheme of the eye lens and its anatomical connections

crystallin protein, and lose their nucleus; and (ii) are trophic cells for the same lens fiber cells, which cannot be autonomous for the lack of the nucleus and so are dependent for their vitality and also for their transparency on lens epithelium cells (Forrester et al., 2020).

The lens continues to grow throughout life, and lens fiber cells are never replaced, but due to their dependence on lens epithelium cells, which are subject to turnover (Tassin et al., 1979), when the turnover of these cells slows down and becomes insufficient, the functionality of the lens fiber cells, in particular the transparency, is compromised and cataracts occur (Tassin et al., 1979).

However, it has long been argued that senile cataract is due to alterations in crystalline proteins and that cataract in Werner syndrome is caused by alterations in the lens epithelial cells (Martin & Oshima, 2000).

As the stability of the conformation and arrangement of lens fiber cells is essential for their function, the possibility of their periodic replacement appears difficult, and this justifies their condition as perennial cells, enucleated and therefore incapable of functional autonomy and necessarily dependent on trophic cells.

– **Neurons without turnover**

According to one estimate, in the central nervous system (CNS), there are 10^{11} neurons interconnected by 10^{14} synapses (Williams & Herrup, 1988). This estimation means that, on average, there are about 1000 synapses for each neuron. The enormous number of synapses and the variety of connections depending on the neuron exclude that such interconnections are genetically determined. On the contrary, brain functions and their variability from individual to individual likely depend on the progressive formation and the modeling of these interconnections on the basis of experience. Given this complexity, it is unlikely that neurons can be replaced without any alterations or losses in the functions performed. This appears confirmed by the absence of cell turnover for CNS neurons, with some exceptions for some particular neurons (Horner & Gage, 2000; Zhao et al., 2008), e.g., for certain type of neurons in the olfactory bulb and in the hippocampus, an important organ for some types of memory and learning (Zhao et al., 2008), where it is probable that new connections and new neurons are required.

Two important types of CNS neurons without a turnover but dependent on other cells subject to turnover (photoreceptor cells, neurons of the cerebral cortex, and basal nuclei) will be briefly discussed in the next section, *§8.3.7 Effects on some cell types without cell turnover*.

- **Neurons with turnover**

The connections with other nerve cells are simple for some types of nerve cells, e.g., olfactory receptor cells (ORCs) and other sensory neuronal cells. Each ORC receives the olfactory signal from a single branching dendrite and, through the single axon, transmits it to another neuron that processes the signal (Bermingham-McDonogh & Reh, 2011).

As these connections are simple and standardized, the replacement of ORCs by turnover is possible and common to all vertebrates (Graziadei & Monti Graziadei, 1978). In humans, ORCs are replaced every few months (Bermingham-McDonogh & Reh, 2011) through the activity of primary stem cells (globose basal cells) and intermediate amplifying progenitor cells (horizontal basal cells) (Caggiano et al., 1994; Huard et al., 1998; Chen et al., 2004; Leung et al., 2007; Iwai et al., 2008).

8.3.6 Effects on some Cell Types with Turnover

The effects of cellular mechanisms of aging on different tissues or organs are detailed in *Chap. 6—Aging in the Human Species* of (Libertini, Corbi, et al., 2021a). The description of these effects will not be repeated here, both for the sake of brevity and because they are not the specific subject of this book. However, some general considerations will be discussed.

- **Some cell types formerly considered as perennial**

Heart myocytes were thought to be perennial cells; therefore, the heart as a whole was believed to be an organ that could not regenerate itself. This paradigm has been overcome in recent times: "A Paradigm Shift in Cardiac Biology … The Heart Is a Self-Renewing Organ … In the normal human heart, there are $\approx 5\times10^9$ parenchymal cells in the left ventricle inclusive of the interventricular septum (Olivetti et al., 1991; Olivetti et al., 1995) and $\approx 3\times10^6$ terminally differentiated myocytes are generated per day … by commitment and growth of CSCs [cardiac stem cells] (Kajstura et al., 1998; Beltrami et al., 2001; Urbanek et al., 2003, 2005). This degree of myocyte formation ensures that the entire cell population of the heart is replaced approximatively every 4.5 years. Thus, parenchymal cells cannot live, as is generally believed, as long as the organism, ≈80 years (Sanderson & Scherbov, 2005). The human heart replaces completely its myocyte population ≈18 times during the course of life, independently from cardiac diseases." (Anversa et al., 2006).

In short, in a normal heart, about 3 million cardiac myocytes die by apoptosis every day (about 35 cells every second) and are replaced by new myocytes originating from the duplication and differentiation of specific stem cells. In relation to age (i.e., in relation to the number of duplications), these stem cells show telomeric shortening and an increasing number of senescent cells (Leri et al., 2001; Chimenti et al., 2003; Urbanek et al., 2003).

In the elderly, the heart shows a reduced number of myocytes with a progressive increase in myocyte cell volume per nucleus (Olivetti et al., 1991), which is caused by the decline of duplication capacity and the consequent progressive decline and insufficiency of myocyte turnover (Anversa et al., 2006).

The erroneous belief that the heart muscle is without renewal was probably derived from the slowness of cell turnover, which, however, allows a complete replacement of the muscle in about 5 years. Moreover, an increased renewal of the heart muscle cell can be stimulated in a regulated way by external factors, even at an advanced age, leading to cardiac regeneration (Weinberger & Eschenhagen, 2021; Zhao et al., 2023). Several adaptations, including enhanced myocardial regeneration capacity, are evolutionarily favored in populations constantly living at high altitudes, decreasing heart-disease-associated morbidity and death rate (Mallet et al., 2021). A similar adaptive response can be observed even in individuals constantly living at low altitudes as a response to transient high-altitude exposure and potentially can be used in therapeutic approaches. Thus, hypoxia can directly stimulate cardiomyocyte proliferation, which potentially can be used in the therapy of cardiac diseases, including regeneration after infarction (Kimura et al., 2017; Mallet et al., 2021; Zhao et al., 2023).

- **The atrophic syndrome determines a decrease in the size of the active parenchyma of an organ**

In general, the atrophic syndrome causes a progressive slowing down of cell turnover with the reduction of the number of specific cells of a tissue or organ partially replaced by non-specific cells. This generally leads to a reduction in the volume of the active parenchyma of an organ. For example, about:

- **Skin**—The atrophy of the various parts of the skin (epidermis, dermis with all its components and structures) is carefully and in detail described by Griffiths (1998) and is correlated with age (Katzberg, 1958; Gilchrest et al., 1969; Gilchrest et al., 1982; Grove, 1989; Branchet et al., 1990; Calleja-Agius et al., 2013).
- **Myocytes of skeletal muscle**—The myocytes of skeletal muscle are cells with turnover (Richardson et al., 2014), and with the decline of their turnover, age-related muscle atrophy is observed (Young et al., 1984; Davies et al., 1986; Vandervoort & McComas, 1986; Klitgaard et al., 1990; Rutherford & Jones, 1992; Cumming, 1998; Richardson et al., 2014; Lim, 2016), both as reduction of

the cross-sectional area of the muscles and as maximum voluntary isometric force (Grimby et al., 1982; Lexell et al., 1988).

- **Kidneys**—Age-related renal changes include a reduction in weight of the kidney corresponding to the general decrease of other organs. There is a reduction in the number and size of nephrons and of glomeruli, and widening of the interstitial spaces with an increase in the interstitial connective tissue (Jassal et al., 1998)
- **Liver**—Hepatic atrophy is related to age (Wakabayashi et al., 2002; Grizzi et al., 2013; Kim et al., 2015). Liver volume shows an age-related decline (Marchesini et al., 1988), both in proportion to body weight and as absolute value (Wynne et al., 1989). Between the ages of 24 and 91, there is a reduction in volume of about 37% and in blood flow through the liver of approximately 53% (Marchesini et al., 1988). During this period, liver volume (and likely the number of hepatocytes) declines, but there is an increase in the mean size of hepatocytes, whereas, in liver atrophy caused by starvation, hepatocyte size decreases (Watanabe & Tanaka, 1982; David & Reinke, 1988).
- Other organs would appear to violate the rule of atrophy accompanied by volume reduction. However, in any case, even if the overall volume is not reduced, there is always a reduction in the volume of the functional parenchyma:
- For the **heart**, with aging, the number of myocytes is reduced, accompanied by an increase in the cellular volume of myocytes (Olivetti et al., 1991). So, there is, at the same time, an atrophy as regards the number of myocytes and a hypertrophy as regards the volume of the muscle mass if there are conditions, such as hypertension, requiring a greater chronic effort of the heart. However, in congestive heart failure due to left ventricular systolic dysfunction, there is dilatation of the left ventricle (Aronow, 1998). In atrial fibrillation, due, among other things, to the substantial decline of the cells of the sinus atrial node—the number of pacemaker cells of the node at the age of 75 is reduced to 10% compared to those present at 20 years (Davies, 1988) -, there is atrial dilatation.
- These phenomena, even if the term hypertrophy is often used to describe them, originate from the atrophic syndrome of the myocardial tissue, and the term hypertrophy should be substituted with other terms to avoid misunderstandings.
- In the elderly, the atrophy of the lungs, due to the constraint of the size of the rib cage, does not involve a decrease in external size but rarefaction and dilatation of alveoli, i.e., in emphysema: ".... the area of the alveoli falls and the alveoli and alveoli ducts enlarge.... residual capacity, residual volume, and compliance increase...." (Connolly, 1998)
- The severity of emphysema and related alterations is age-related (Verbeken et al., 1992; Enright et al., 1993; Martinez et al., 2017). However, parameters related to lung functionality, such as Forced Expiratory Volume in the first second (FEV1) and Forced Vital Capacity (FVC), show an age-related reduction (Enright et al., 1993).
- The bone is not reduced in its external dimensions, but there is a reduction in its density and in the thickness of the cortical bone and a widening of its internal cavities, as is well shown by the dilatation of the spaces between the bony trabeculae in the vertebral bodies. Regarding these atrophic manifestations, described as senile osteoporosis, starting from the age of 35–40, women lose

35–50% of trabecular bone and 25–30% of cortical bone, whereas men 15–45% and 5–15%, respectively (Francis, 1998). In a subsequent phase, when the trabecular network in vertebral bodies becomes too thin and with reduced resistance, there are phenomena of structural failure of the vertebral bodies with consequent crushing and reduction of the volume of these bodies (Dieppe & Tobias, 1998).

– **The effects of atrophic syndrome are not synonymous with wear**

The idea that aging is due to wear and tear phenomena is ancient and still widespread today. For example, it is easy and common for a doctor the explanation to an elderly person of why the joints hurt and the height decreases with age, attributing these phenomena to the wear and tear of bones and joints, either because of the need to give an easy and understandable explanation or because there is a belief about it.

Indeed, both osteocytes (Spelsberg et al., 1999; Richardson et al., 2014) and chondrocytes (Héraud et al., 2000; Chang, Liu, et al., 2016b) are subject to cell turnover. So, the correct answer is that, in the elderly, the renewal of bone and joint cells is increasingly slow and incomplete with age, and this causes the progressive atrophy of bones and joints.

In the elderly, the vertebral bodies do not fail because their external parts are worn out, but as a result of bone atrophy, especially for the internal trabeculae, which have no contact with other anatomical parts as a possible cause of wear. Similarly, the elderly's femur is at risk of breaking following a fall, not because it is worn out but because of internal bone atrophy.

The joints then do not age because they are worn out but because the progressive atrophy of both the cartilage and the underlying bone, caused by the mechanisms of cellular aging, allows their progressive wear. In short, it is true to say that aging causes wear of the joints, while it is false to say that wear of the joints causes their aging. In accordance with this idea, a modern orientation for the non-surgical treatment of joint alterations is the search for a restoration of the renewal capacity of cartilage cells by means of mesenchymal stem cells (Chang, Liu, et al., 2016b).

As another example, the muscles of an athlete who has used them maximally throughout a lifetime do not age because they are worn out. Conversely, with or without continuous and demanding use of the muscles, i.e., with or without hypothetical wearing factors, there is a progressive age-related decline in muscle mass and in the maximum strength the muscles can express (Young et al., 1984, 1985; Davies et al., 1986; Vandervoort & McComas, 1986; Klitgaard et al., 1990; Rutherford & Jones, 1992; Phillips et al., 1992).

8.3.7 Effects on some Cell Types without Turnover

The age-related decline of perennial cells is explained by the decline of their trophic cells, which are cells with turnover, as suggested and discussed in detail in other works (Libertini, 2009a, 2009b; Libertini & Ferrara, 2016b; Libertini, 2018; Libertini, Corbi, et al., 2021a, Chap. 5).

– **Retina photoreceptors (cones and rods)**

The retina and optic nerve are integral parts of CNS and anatomically constitute an extension of it. Retina photoreceptor sensory cells (cones and rods), highly differentiated neurons, detect the light rays and perform, together with other retinal neurons, the first processing of the information given by these rays. Photoreceptors are perennial cells and are never replaced due to the complexity of the connections with other retinal neurons. However, their parts are continuously renewed with the help of retina pigmented cells (RPCs), which are highly differentiated gliocytes and form a unicellular layer (retina pigmented epithelium). The apex of each photoreceptor leans on this layer and is surrounded by part of the cytoplasm of an RPC (Fig. 8.17).

Photoreceptors contain particular membranes or discs on which there are photopsin molecules sensitive to light that progressively oxidize. These discs are continuously produced by the photoreceptor cell while the oxidized discs are removed by the satellite RPC. In fact, every day, RPCs actively eliminate 10% of the discs by phagocytosis and subsequent metabolization. Since each RPC serves approximately 50 photoreceptors, high metabolic activity is required for its function.

As RPCs are subject to cell turnover and photoreceptors cannot remain viable without the trophic activity of RPCs (Fine et al., 2000), the functionality and vitality of perennial cells such as cones and rods depend on cells with turnover. With the progressive decline and insufficiency of RPC turnover, these satellite cells enlarge, each serving a more significant number of photoreceptors. However, for their increasing decline, there is a progressive accumulation of particular substances: A2E, derived from vitamin A catabolism (Crouch et al., 2015), lipofuscin (Crouch et al., 2015), and also β-amyloid protein (Ermilov & Nesterova, 2016).

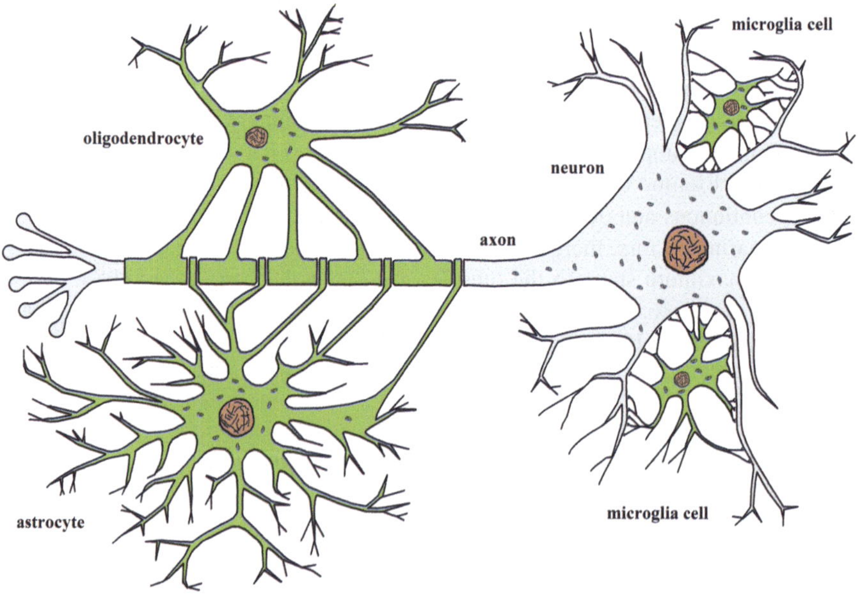

Fig. 8.17 Scheme of a neuron with some trophic cells. Image from (Libertini, 2018), modified and redrawn

In a further phase of RPC turnover decline, gaps form in the retina-pigmented epithelium, and photoreceptors die because of the absence of their trophic cells (Berger et al., 1999). This process is functionally and clinically more evident where there is a greater density of photoreceptors, i.e., the macula, with a greater accumulation of A2E (Sparrow, 2003). The disorder is called "age-related macular degeneration" (AMD), as the initial symptoms are caused by macular damage (Fine et al., 2000), but it affects the entire retina.

– **-Neurons of the central nervous system**

In analogy to what happens for photoreceptors, it was proposed that, for other types of perennial neurons of the CNS (e.g., those of cerebral cortex and basal nuclei), there was a similar dependence on the trophic action of specialized types of gliocytes subject to turnover. The decline of this turnover causes suffering and death of the served neurons with consequent pathological manifestations (Libertini, 2009a, 2009b; Libertini & Ferrara, 2016b; Libertini, 2018). An analogous proposal was suggested by other authors without using the photoreceptors as a term of comparison (Fossel, 1996, 2004; Flanary, 2009).

A) **Alzheimer's disease**

CNS neurons are served by specific gliocytes with cell turnover, which degrade β-amyloid protein (Qiu et al., 1998; Vekrellis et al., 2000). Fossel suggested that Alzheimer's disease (AD) was caused by the decline of neuronal trophic gliocytes, microglia cells, determined by telomere shortening (Fossel, 1996; Fossel, 2004). In particular:

(i) "One function of the microglia (Vekrellis et al., 2000) is degradation of β-amyloid through insulin-degrading enzyme (IDE), a function known to falter in Alzheimer disease (Bertram et al., 2000)" (Fossel, 2004, p. 233);
(ii) "telomere lengths of circulating monocytes can serve as an independent predictor in at least vascular dementia (von Zglinicki et al., 2000)" (Fossel, 2004, p. 235).

A different widespread hypothesis explains AD as the consequence of β-amyloid and tau protein accumulation. However, drugs or vaccines proposed by pharmaceutical societies against β-amyloid accumulation or for eliminating β-amyloid plaques, though effective against β-amyloid accumulation and plaques, have proved a clinical failure (Gorelick, 2004; Abbott, 2008). In an experimental trial, post-mortem analyses showed that amyloid plaques were reduced in nearly all patients although severe dementia had occurred before death (Abbott, 2008). This observation is likely due to the fact that β-amyloid and tau protein accumulation is not the cause of AD but one of the consequences of the suggested real cause of AD, i.e., the decline of neuronal trophic cells.

B) **Parkinson's disease and related disorders**

AD is defined as a tauopathy because there is an accumulation of tau protein, which forms neurofibrillary tangles (Galpern & Lang, 2006). There are three diseases defined as synucleinopathies because due to the accumulation of alpha-synuclein (AS), which forms inclusions known as Lewy bodies (Davie, 2008; Schulz-Schaeffer, 2010): (i) Parkinson's disease (PD), a relatively common degenerative disorder of the CNS; (ii) Dementia with Lewy bodies, primary parkinsonism with some additional features (Samii et al., 2004); and (iii) Multiple System Atrophy, a rare genetic disease with AS accumulation in oligodendrocytes (Sturm & Stefanova, 2014).

There is overlapping of clinical manifestations between these three neuropathies and between them and AD (Aarsland et al., 2009). E.g.:

1. PD is arbitrarily considered to affect mainly the motor system with typical movement disorders, but it also shows various non-motor symptoms such as sensory deficits (Barnett-Cowan et al., 2010) and mood and behavior alterations (Jankovic, 2008);
2. A typical manifestation of AD is cognitive decline, which is also observed in advanced stages of PD (Galpern & Lang, 2006), with a risk of dementia 2–6 times that of the whole population (Caballol et al., 2007; Jankovic, 2008) and increasing in relation to the duration of the disease (Caballol et al., 2007);
3. Neurofibrillary tangles, considered a typical manifestation of AD, are also shown in the brains of PD subjects (Galpern & Lang, 2006).

The symptoms of PD and related diseases are explained as a consequence of neuron death in specific parts of the brain, but the mechanism causing their death is controversial (Obeso et al., 2008). However:

1. The accumulation of AS in astroglial cells may be a crucial factor for PD pathogenesis (Halliday & Stevens, 2011);
2. Astroglial cells may be important for neuronal survival by secreting and producing neurotrophic factors and antioxidants against damage induced by dead neurons (Brück et al., 2016);
3. Microglial cells are important for the clearance of debris and dead cells, so supporting neuronal survival and synaptic functions. On the other hand, these cells can be over-activated in the course of the disease so contributing to its progression through inflammatory processes (Brück et al., 2016);
4. The age-related decline of neuronal remyelination might be explained by a decreasing number of progenitor cells of oligodendrocytes, which are trophic cells for the axons (Sim et al., 2002; Segel et al., 2019).

These observations suggest that AS accumulation is due to the decline of specific gliocytes with trophic functions, such as astrocytes (Morales et al., 2015) and oligodendrocytes (Sim et al., 2002; Segel et al., 2019), in particular those dedicated to axon trophism, and this, in turn, determines the activation of microglial and astroglial cells. The decline of astrocytes and oligodendrocytes, both in function and

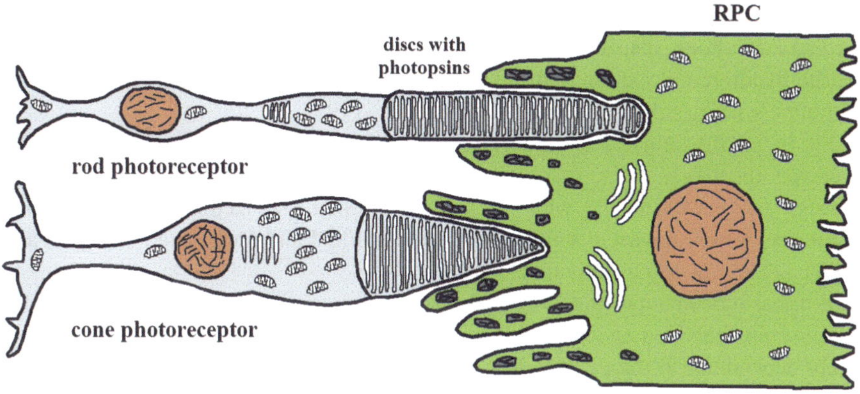

Fig. 8.18 Scheme of two photoreceptors (a cone and a rod) served by a retinal pigmented cell. Image from (Libertini, 2018), modified and redrawn

number, would be caused by the decline of their turnover and functionality, as for the decline of RPCs and the consequent AMD.

The concepts expressed about AD and synucleinopathies are illustrated in Fig. 8.18.

As in the same individual there can be, to varying degrees, the decline of various types of neuronal trophic gliocytes, this would explain the overlapping of symptoms between AD and PD and related diseases. In fact, these various diseases would have a single origin, with different manifestations depending on the various types of cells more or less affected.

8.3.8 Endothelial Cell Turnover: A Useful Indicator for Aging Process

The turnover of endothelial cells, which constitute the layer of blood vessels closest to the lumen, depends on the duplication of particular progenitor cells (endothelial progenitor cells, EPCs), which originate from the bone marrow (Hill et al., 2003).

The decline in the number of EPCs is related to age but also to some known risk factors for cardiovascular disease (diabetes, overweight and obesity, smoking, hypertension). The Framingham risk score considered these factors to calculate the risk of developing cardiovascular diseases (Wilson et al., 1987). It was later found that the decline of EPCs had a predictive value for cardiovascular risk equal to this important risk score (Hill et al., 2003; Werner et al., 2005). The fact was explained: "… continuous endothelial damage or dysfunction leads to an eventual depletion or exhaustion of a presumed finite supply of endothelial progenitor cell … continuous risk-factor-induced injury may lead to eventual depletion of circulating endothelial cells" (Hill et al., 2003).

Diseases associated with compromised endothelial function (cerebral ischemia, cardiac infarctions, insufficiencies of the peripheral circulation, and other diseases caused by altered blood circulation) increase exponentially after a certain age, also without cardiovascular risk factors (Tallis et al., 1998).

So, in healthy subjects without risk factors, the age-related decline of EPC turnover was the likely cause of cardiovascular diseases, i.e., was part of the aging process, while, in subjects with the aforementioned risk factors, probability and occurrence of cardiovascular diseases were amplified and anticipated (Tallis et al., 1998), i.e., aging was accelerated.

It has also been underlined, on the basis of many works, that there are analogous relationships between endothelial dysfunction and a series of age-related disorders likely dependent on the decline in cell turnover, both for cell types with turnover (direct aging) and for cell types without turnover (indirect aging). Furthermore, some categories of drugs with "protective" effects on endothelial dysfunction and its consequences—statins (Su, 2015; Davidson, 2007), ACE inhibitors (ACE-i), and sartans (ARBs) (Chłopicki & Gryglewski, 2005; Su, 2015)—showed similar actions on other age-related disorders. These relationships are summarized in Table 8.1, a synthesis of Tables 7.8 and 8.2 from (Libertini, Corbi, et al., 2021a), where the references supporting each relationship indicated in the table are reported (here are omitted for the sake of brevity).

In general, a large parallelism is observed between risk factors or protective effects for endothelial dysfunction and the same actions for other age-related dysfunctions. There are some exceptions, highlighted in gray in the Table, which are discussed in (Libertini, Corbi, et al., 2021a).

These relationships support the hypothesis of a common mechanism underlying age-related dysfunctions, namely the decline in cell turnover, which in turn originates from the previously discussed mechanisms of aging at the cellular level.

8.4 Effects of Cellular Aging on the Whole Organism

The progressive alterations of the functions in the various organs and systems, in addition to their specific manifestations, reduce the overall efficiency of the organism and, therefore, its ability to survive, i.e., fitness. For our species, this reduction must first of all be evaluated in evolutionarily "normal" conditions, and then it is necessary to consider what happens in a different ecological niche, such as that of today's living conditions.

Table 8.1 Relations between some age-related dysfunctions and (i) "risk factors"; or (ii) "protective drugs" (ACE-I and ARBs)

Age-related dysfunction (cell turnover of the specific cells: Y/N)	Effect on the risk by:						Protective effect by:	
	Hypert.	Diabetes	Smoking	Obesity / dyslipidemia	Moderate alcohol use	Alcohol abuse	Statins	ACE-I / ARBs
Endothelial dysfunction (Y)	+	+	+	+	−	+	+	+
Olfactory dysfunction (Y)	+	+	+	+	.	+	+	.
Renal insufficiency (Y)	+	+	+	+	−	+	+	+
Atrophy of oral mucosa and salivary glands (Y)	+?	+	+	.	.	+	.	.
Intestinal and gastric atrophy (Y)	.	.	+	.	.	.	.	.
Alopecia (Y)	+	+	+	+	+?	+	.	.
Emphysema and related diseases (Y)	+	+	+	−	−	+	+	+
Skin atrophy (Y)	.	+	.	.	.	.	+	+
Osteoporosis (Y)	+	+	+	+	−	+	+	+
Atrophy of other sensory neuronal cells with turnover (Y)	.	+	+	.	.	+	.	.
Cataract (Y)	+	+	+	+	−	+	+?	+
Testicular atrophy (Y)	.	+	+	+	/	+	.	/
Muscle atrophy (Y)	.	+	+	+	.	+	−	+

(continued)

Table 8.1 (continued)

Age-related dysfunction (cell turnover of the specific cells: Y/N)	Effect on the risk by:						Protective effect by:	
	Hypert.	Diabetes	Smoking	Obesity / dyslipidemia	Moderate alcohol use	Alcohol abuse	Statins	ACE-I / ARBs
Cardiac insufficiency and related diseases (Y)	+	+	+	+	–	+	+	+
Reduction of various hematic cell types (Y)	–	–	–	–	?	+	–	.
Diabetes and impairment of glucose tolerance (Y)	+		+	+	–	+	–/	+
Hepatic atrophy and related disease (Y)s	.	+	+	+	.	+	–/	+
Age-related macular degeneration (N)	+	+	+	+	–/	+	?	-?
Alzheimer's disease (N)	+	+	+	+	–	+	+	+
Parkinson's disease (N)	+/	+	–	+	–	+	+	+
Hearing impairment (N)	+	+	+	+	–	+	+	+

Notes: +, –, / = increased, decreased, or unaltered risk or protective effect;? = doubtful results;. = no specific study

8.4.1 Age-Related Fitness Reduction under Evolutionarily "Normal" Conditions

A rough but certainly indicative evaluation of how and to what extent fitness is reduced in living conditions as close as possible to those to which our species is adapted was provided some years ago by the study of a population, the Ache of Paraguay, living in primitive conditions analogous to those to which our species is adapted (Hill & Hurtado, 1996).

Figure 8.19 shows the percentage of survivors and the mortality rate in the wild observed in this study for each age.

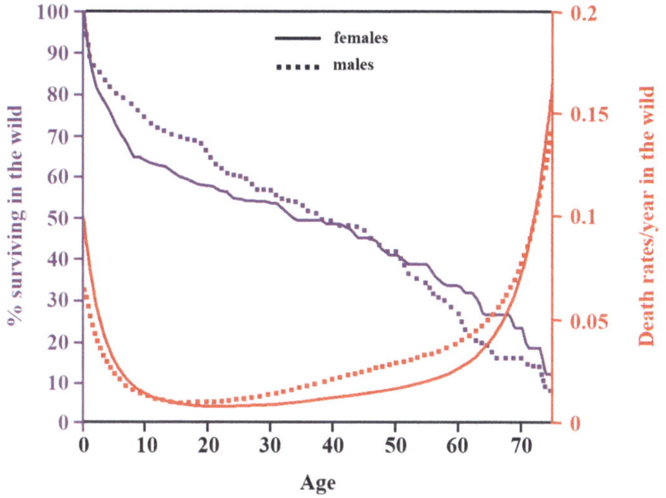

Fig. 8.19 Survivors (in %) and mortality rate/year observed for Ache people in the wild before contact with modern individuals. The figure has been obtained from the synthesis of Figs. 6.2 and 6.3 in [Hill & Hurtado, 1996]

Figure 8.20 is a reworking of the previous one in which the first stages of life (when the mortality is very high) and the subsequent period (up to the age of 30 when the mortality has the lowest value) have been excluded. From about this age, mortality begins to grow, at first slowly, and then, from about 60 years of age, with a progressive acceleration. The survivors at age 75 are about 10% of the original population and about 20% of those who reached the age of 30.

As numerical values, for the Ache people studied in the wild, the minimum mortality was at 20 years of age and equal to about under 0.9%/year, and the mean life span (ML) was 38.8 years (Hill & Hurtado, 1996). In the hypothetical case of the absence of an age-related mortality increase (i.e., in the hypothetical case of no aging), the hypothetical mean life span (HML) would have been 87.75 years, with an HML/ML ratio equal to 2.26. This showed that for our species, aging reduced the duration of life under natural conditions by less than half (Libertini, 2013).

Furthermore, considering only the survivors at the age of 20, their residual life (ML_{20}) was estimated to be equal to 38.1 years, for a total life span equal to 20 + 38.1 = 58.1 years. If the absence of a subsequent increase in mortality was assumed for this fraction of the population (i.e., in the hypothetical case of no aging), the hypothetical mean life span after 20 years (HML_{20}) would have been 116.04 years, for a total life span of 20 + 116.04 = 136.04 years, with an HML_{20}/ML_{20} ratio equal to 3.044. Therefore, excluding individuals who died before the age of 20, in practice, the part of the population with minimal reproductive activity, aging reduced the mean length of life in natural conditions by about two-thirds (Libertini, 2013). These data were in good agreement with those previously observed for 8 mammalian species (Libertini, 1988).

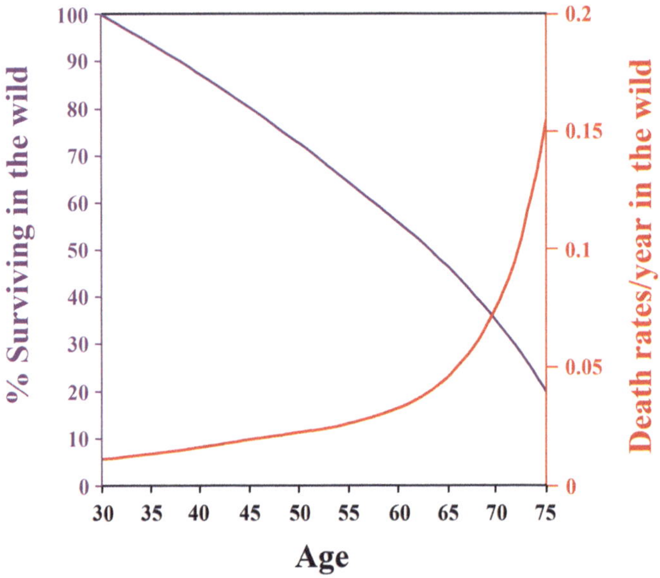

Fig. 8.20 Figure obtained from the previous one excluding the early stages of life in which mortality is very high and the subsequent period, up to the age of 30, in which—after a minimum at the age of 20—mortality is still slightly above this minimum. Furthermore, the data for males and females were considered together, and the curves were smoothed

8.4.2 Estimation of Fitness Variation under Modern Conditions

It is useful to consider if estimating the ability to survive (fitness) and its age-related decline in modern conditions is possible.

In fact, this is quite simple, and the data are already available.

First of all, it is necessary to remember that the word fitness has two meanings:

(i) in biology, it indicates the ability to survive and to propagate one's genes;
(ii) in sports and athletic activities, it indicates the state of health, the ability to perform a sporting activity and to win a competition.

Records for various sports activities are of two general types:

- absolute ones, i.e., without age limits;
- limited to certain age groups, generally from 30 years onwards, divided into 5-year groups.

Figures 8.21 and 8.22, relating to various types of sport, show both the records for ages under 30 (essentially the absolute records) and those for the age groups 30–35, 35–40, ... 95–100. The first figure concerns the records of running competitions over various distances, while the second concerns a more heterogeneous set of

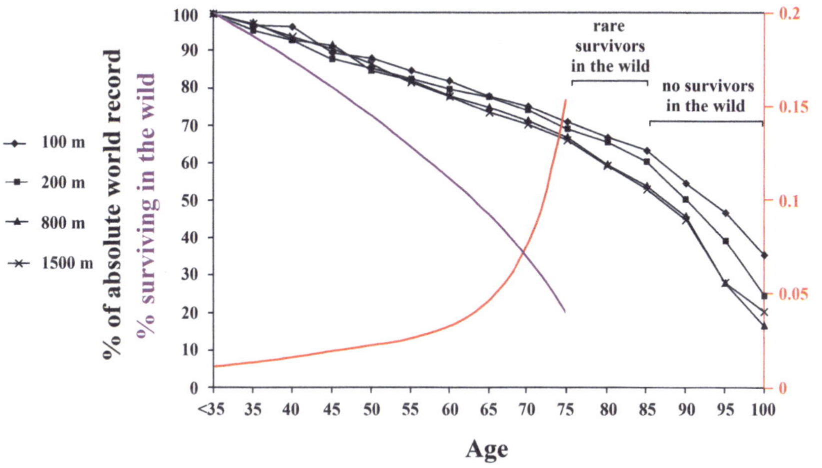

Fig. 8.21 Records for running competitions for ages under 30 (the absolute records) and age groups 30–35, 35–40, … 95–100, superimposed on the curves of mortality rates and survivors of a population living in natural conditions. Data sources for world records from (Wikipedia, 2023)

races. For both figures, the values are shown as percentages with respect to the absolute records, which have a reference value of 100%.

These records are significant because they are the best that a healthy human organism, with the utmost will and commitment, can express in each age group.

These values are superimposed on the curves of mortality rates and survivors of a population living in natural conditions.

The figures show that when, in natural conditions, the mortality rate significantly increases (around 20%/year) and the survivors are only 20% of those who have exceeded the age of 20 (about 10% of individuals at birth), there is a decline in records of about 30%, which, however, in natural conditions, means the inability to survive, or rather corresponds to fitness, in its biological meaning, equal to zero.

This indicates that the reduction of athletic ability (fitness in the sporting meaning) leads to an amplified reduction in the ability to survive, i.e., fitness in the biological meaning, but also of fitness in the meaning of being able to win a race.

These reports also indicate that for aging to reach levels where it is lethal, it is not necessary that an individual's physical abilities be reduced to zero, but it is sufficient that it even partially reduces these abilities.

8.4.3 Variation of Survival Capacity under Modern Life Conditions

Modern living conditions are very different from evolutionarily normal ones, and this certainly has many consequences for age-related mortality rates and survival.

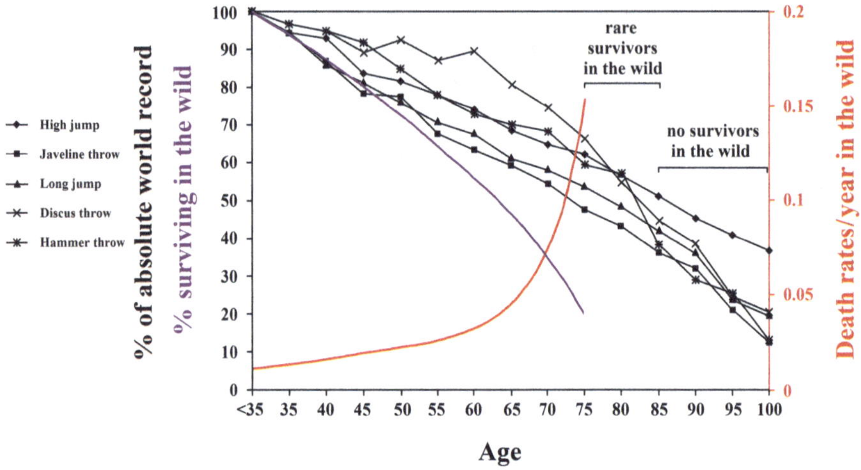

Fig. 8.22 Records for various types of non-running competitions for ages under 30 (the absolute records) and age groups 30–35, 35–40, ... 95–100, superimposed on the curves of mortality rates and survivors of a population living in natural conditions. Data sources for world records from (Wikipedia, 2023)

There are three main factors (s. Figure 8.23:

1. Reduction of the extrinsic (or environmental) mortality rate (i.e., that not due to aging). This shifts the death rate curve down and to the right and the survival curve to the right;
2. Multiple variations of the ecological niche with the diseases resulting from mismatches between species adaptation and new conditions. This increases the morbidity and mortality rates, i.e., shifts the death rate curve up and to the left and the survival curve to the left;
3. Medical and welfare care (absent in the wild) reduce morbidity and mortality rates, i.e., they shift the mortality rate curve down and to the right, and the survival curve to the right. This does not mean an improvement in living conditions, i.e., what is now called "the quality of life". Today, very many survive in conditions of complete dependence on care and assistance, that is to say, in conditions that would be completely incompatible with life in natural conditions.

8.4.4 Importance of Aging as a Cause of Death

We have seen above that under natural conditions, aging roughly halves the life span for all individuals and reduces it to a third for individuals who survive the age of 20.

Despite such enormous importance for life span, official statistics do not report aging as a cause of death (World Cause of Death, 2020).

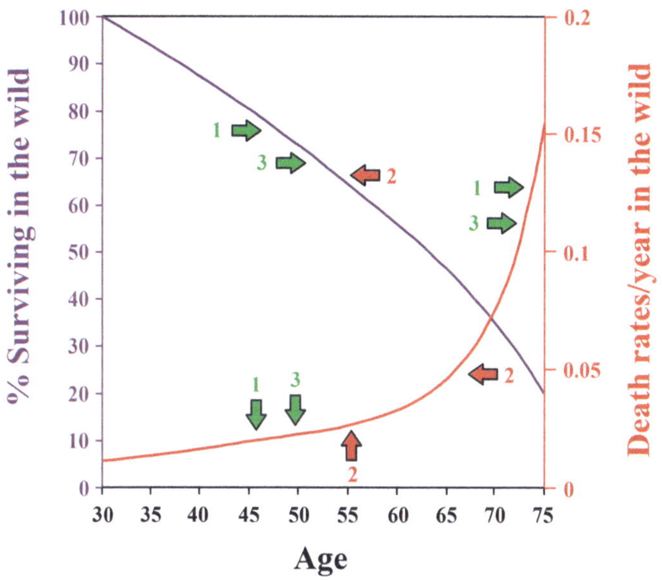

Fig. 8.23 Various factors modify survival capacity in modern life conditions (s. the text; the numbers indicate the factors that modify mortality rates and survival)

In part, this happens because aging is not recognized as a distinct phenomenon with its own physiology and also because of overlapping diseases that are high in frequency and worsen with age.

However, the main factor that determines this invisibility of aging as a cause of death is that it manifests itself as a growing vulnerability to any harmful or potentially lethal factor.

Let us see what happens in natural conditions, returning to the data obtained from the study of the Ache people. For this population, individuals die largely from violent events caused by other individuals. Then, there are events related to poisonings, infections, predation by wild animals, snakebites, and other causes. Only in some cases (6 cases) "old age" is reported as the cause of death in the absence of a defined pathology (Hill & Hurtado, 1996). This rarity of aging as a cause of death reported in that study contrasts with the solid age-related mortality increase observed.

In fact, aging, in most cases, does not kill directly but makes the individual more and more vulnerable to any damaging event, so explaining the increase in mortality in relation to age. Only in a few cases, in the older age group, in the absence of a clear harmful event to which death can be attributed, and with the progression of aging manifestations, does the opportunity arise to attribute the cause of death to aging itself.

In modern societies, with the remarkable progress of diagnostic skills and medical knowledge, even in very elderly subjects, an organ damaged by aging is identified as the cause of death, and the cause of death is attributed to the disease caused by this failure.

It is an anomalous situation that must somehow be analyzed and defined in a rational way, overcoming the absurdity of the official non-existence of aging as a cause of death.

References

Aarsland, D., Londos, E., & Ballard, C. (2009). Parkinson's disease dementia and dementia with Lewy bodies: Different aspects of one entity. *International Psychogeriatrics, 21*(2), 216–219. https://doi.org/10.1017/S1041610208008612

Abbott, A. (2008). The plaque plan. *Nature, 456*(7219), 161–164. https://doi.org/10.1038/456161a

Acosta, J. C., O'Loghlen, A., Banito, A., Guijarro, M. V., Augert, A., Raguz, S., et al. (2008). Chemokine signaling via the CXCR2 receptor reinforces senescence. *Cell, 133*(6), 1006–1018. https://doi.org/10.1016/j.cell.2008.03.038

Alberts, B., Bray, D., Hopkin, K., Johnson, A., Lewis, J., Raff, M., et al. (Eds.). (2013). *Essential cell biology* (4th ed.). Garland Science.

Anversa, P., Kajstura, J., Leri, A., & Bolli, R. (2006). Life and death of cardiac stem cells. *Circulation, 113*(11), 1451–1463. https://doi.org/10.1161/CIRCULATIONAHA.105.595181

Aparicio, O. M., & Gottschling, D. E. (1994). Overcoming telomeric silencing: A trans-activator competes to establish gene expression in a cell cycle-dependent way. *Genes & Development, 8*(10), 1133–1146. https://doi.org/10.1101/gad.8.10.1133

Aronow, W. S. (1998). Effects of aging on the heart. In R. C. Tallis, H. M. Fillit, & J. C. Brocklehurst (Eds.), *Brocklehurst's textbook of geriatric medicine and gerontology* (5th ed., pp. 255–262). Churchill Livingstone.

Artandi, S. E. (2002). Telomere shortening and cell fates in mouse models of neoplasia. *Trends in Molecular Medicine, 8*(1), 44–47. https://doi.org/10.1016/s1471-4914(01)02222-5

Artandi, S. E., & DePinho, R. A. (2010). Telomeres and telomerase in cancer. *Carcinogenesis, 31*(1), 9–18. https://doi.org/10.1093/carcin/bgp268

Azzalin, C. M., & Lingner, J. (2008). Telomeres: The silence is broken. *Cell Cycle, 7*(9), 1161–1165. https://doi.org/10.4161/cc.7.9.5836

Azzalin, C. M., Reichenbach, P., Khoriauli, L., Giulotto, E., & Lingner, J. (2007). Telomeric repeat containing RNA and RNA surveillance factors at mammalian chromosome ends. *Science, 318*(5851), 798–801. https://doi.org/10.1126/science.1147182

Bah, A., Wischnewski, H., Shchepachev, V., & Azzalin, C. M. (2012). The telomeric transcriptome of Schizosaccharomyces pombe. *Nucleic Acids Research, 40*(7), 2995–3005. https://doi.org/10.1093/nar/gkr1153

Baker, D. J., Childs, B. G., Durik, M., Wijers, M. E., Sieben, C. J., Zhong, J., et al. (2016). Naturally occurring p16(Ink4a)-positive cells shorten healthy lifespan. *Nature, 530*(7589), 184–189. https://doi.org/10.1038/nature16932

Baker, D. J., Jeganathan, K. B., Cameron, J. D., Thompson, M., Juneja, S., Kopecka, A., et al. (2004). BubR1 insufficiency causes early onset of aging-associated phenotypes and infertility in mice. *Nature Genetics, 36*(7), 744–749. https://doi.org/10.1038/ng1382

Baker, D. J., Perez-Terzic, C., Jin, F., Pitel, K. S., Niederländer, N. J., Jeganathan, K., et al. (2008). Opposing roles for p16Ink4a and p19Arf in senescence and ageing caused by BubR1 insufficiency. *Nature Cell Biology, 10*(7), 825–836. https://doi.org/10.1038/ncb1744

Balaban, R. S., Nemoto, S., & Finkel, T. (2005). Mitochondria, oxidants, and aging. *Cell, 120*(4), 483–495. https://doi.org/10.1016/j.cell.2005.02.001

Barnett-Cowan, M., Dyde, R. T., Foxe, S. H., Moro, E., Hutchison, W. D., & Harris, L. R. (2010). Multisensory determinants of orientation perception in Parkinson's disease. *Neuroscience, 167*(4), 1138–1150. https://doi.org/10.1016/j.neuroscience.2010.02.065

Baur, J. A., Wright, W. E., & Shay, J. W. (2004). Analysis of mammalian telomere position effect. *Methods in Molecular Biology, 287*, 121–136. https://doi.org/10.1385/1-59259-828-5:121

Baur, J. A., Zou, Y., Shay, J. W., & Wright, W. E. (2001). Telomere position effect in human cells. *Science, 292*(5524), 2075–2077. https://doi.org/10.1126/science.1062329

Beauséjour, C. M., Krtolica, A., Galimi, F., Narita, M., Lowe, S. W., Yaswen, P., & Campisi, J. (2003). Reversal of human cellular senescence: Roles of the p53 and p16 pathways. *The EMBO Journal, 22*(16), 4212–4222. https://doi.org/10.1093/emboj/cdg417

Beckman, K. B., & Ames, B. N. (1998). The free radical theory of aging matures. *Physiological Reviews, 78*(2), 547–581. https://doi.org/10.1152/physrev.1998.78.2.547

Bell, J. T., Tsai, P. C., Yang, T. P., Pidsley, R., Nisbet, J., Glass, D., et al. (2012). Epigenome-wide scans identify differentially methylated regions for age and age-related phenotypes in a healthy ageing population. *PLoS Genetics, 8*(4), e1002629. https://doi.org/10.1371/journal.gen.1002629

Beltrami, A. P., Urbanek, K., Kajstura, J., Yan, S. M., Finato, N., Bussani, R., et al. (2001). Evidence that human cardiac myocytes divide after myocardial infarction. *The New England Journal of Medicine, 334*(23), 1750–1757. https://doi.org/10.1056/NEJM200106073442303

Benedetti, A., Jezequel, A. M., & Orlandi, F. (1988). A quantitative evaluation of apoptotic bodies in rat liver. *Liver, 8*(3), 172–177. https://doi.org/10.1111/j.1600-0676.1988.tb00987.x

Benetti, R., García-Cao, M., & Blasco, M. A. (2007). Telomere length regulates the epigenetic status of mammalian telomeres and subtelomeres. *Nature Genetics, 39*(2), 243–250. https://doi.org/10.1038/ng1952

Ben-Porath, I., & Weinberg, R. (2005). The signals and pathways activating cellular senescence. *The International Journal of Biochemistry & Cell Biology, 37*(5), 961–976. https://doi.org/10.1016/j.biocel.2004.10.013

Berger, J. M., Fine, S. L., & Maguire, M. G. (1999). *Age-related macular degeneration*. Mosby.

Bermingham-McDonogh, O., & Reh, T. A. (2011). Regulated reprogramming in the regeneration of sensory receptor cells. *Neuron, 71*(3), 389–405. https://doi.org/10.1016/j.neuron.2011.07.015

Bernardes de Jesus, B., Vera, E., Schneeberger, K., Tejera, A. M., Ayuso, E., Bosch, F., & Blasco, M. A. (2012). Telomerase gene therapy in adult and old mice delays aging and increases longevity without increasing cancer. *EMBO Molecular Medicine, 4*(8), 691–704. https://doi.org/10.1002/emmm.201200245

Bernstein, B. E., Stamatoyannopoulos, J. A., Costello, J. F., Ren, B., Milosavljevic, A., Meissner, A., et al. (2010). The NIH roadmap epigenomics mapping consortium. *Nature Biotechnology, 28*, 1045–1048. https://doi.org/10.1038/nbt1010-1045

Bertram, L., Blacker, D., Mullin, K., Keeney, D., Jones, J., Basu, S., et al. (2000). Evidence for genetic linkage of Alzheimer's disease to chromosome 10q. *Science, 290*(5500), 2302–2303. https://doi.org/10.1126/science.290.5500.2302

Bettin, N., Oss Pegorar, C., & Cusanelli, E. (2019). The emerging roles of TERRA in telomere maintenance and genome stability. *Cells, 8*(3), 246. https://doi.org/10.3390/cells8030246

Beverton, R. J. H., & Holt, S. J. (1959). A review of the lifespans and mortality rates of fish in nature, and their relation to growth and other physiological characteristics. In G. E. W. Wolstenholme & M. O'Connor (Eds.), *The Lifespan of animals, CIBA Foundation Colloquia on Ageing* (Vol. 5, pp. 142–180). https://doi.org/10.1002/9780470715253.ch10

Bidder, G. P. (1925). The mortality of plaice. *Nature, 115*, 495–496. www.nature.com/articles/115495a0.

Bierman, E. L. (1978). The effect of donor age on the in vitro life span of cultured human arterial smooth-muscle cells. *In Vitro, 14*(11), 951–955. https://doi.org/10.1007/bf02616126

Biran, A., Zada, L., Abou Karam, P., Vadai, E., Roitman, L., Ovadya, Y., et al. (2017). Quantitative identification of senescent cells in aging and disease. *Aging Cell, 16*(4), 661–671. https://doi.org/10.1111/acel.12592

Bird, A. (2002). DNA methylation patterns and epigenetic memory. *Genes & Development, 16*(1), 6–21. https://doi.org/10.1101/gad.947102

Blackburn, E. H. (1991). Structure and function of telomeres. *Nature, 350*(6319), 569–573. https://doi.org/10.1038/350569a0

Blackburn, E. H. (2000). Telomere states and cell fates. *Nature, 408*(6808), 53–56. https://doi.org/10.1038/35040500

Blackburn, E. H., & Gall, J. G. (1978). A tandemly repeated sequence at the termini of the extrachromosomal ribosomal RNA genes in Tetrahymena. *Journal of Molecular Biology, 120*(1), 33–53. https://doi.org/10.1016/0022-2836(78)90294-2

Black, H. (2002). Fishing for answers to questions about the aging process. *Bioscience, 52*(1), 15–18. https://doi.org/10.1641/0006-3568(2002)052[0015:FFATQA]2.0.CO;2

Blagosklonny, M. V. (2001). How carcinogens (or telomere dysfunction) induce genetic instability: Associated-selection model. *FEBS Letters, 506*(3), 169–172. https://doi.org/10.1016/s0014-5793(01)02894-0

Blasco, M. A. (2007). The epigenetic regulation of mammalian telomeres. *Nature Reviews. Genetics, 8*(4), 299–309. https://doi.org/10.1038/nrg2047

Blasco, M. A., Lee, H. W., Hande, M. P., Samper, E., Lansdorp, P. M., DePinho, R. A., & Greider, C. W. (1997). Telomere shortening and tumor formation by mouse cells lacking telomerase RNA. *Cell, 91*(1), 25–34. https://doi.org/10.1016/s0092-8674(01)80006-4

Bodnar, A. G., Ouellette, M., Frolkis, M., Holt, S. E., Chiu, C. P., Morin, G. B., et al. (1998). Extension of life-span by introduction of telomerase into Normal human cells. *Science, 279*(5349), 349–352. https://doi.org/10.1126/science.279.5349.349

Bohr, V. A., & Anson, R. M. (1995). DNA damage, mutation and fine structure DNA repair in aging. *Mutation Research, 338*(1–6), 25–34. https://doi.org/10.1016/0921-8734(95)00008-t

Bollati, V., Schwartz, J., Wright, R., Litonjua, A., Tarantini, L., Suh, H., et al. (2009). Decline in genomic DNA methylation through aging in a cohort of elderly subjects. *Mechanisms of Ageing and Development, 130*(4), 234–239. https://doi.org/10.1016/j.mad.2008.12.003

Boorman, S. A., & Levitt, P. R. (1973). A frequency-dependent natural selection model for the evolution of social cooperation networks. *Proceedings of the National Academy of Sciences of the United States of America, 70*(1), 187–189. https://doi.org/10.1073/pnas.70.1.187

Booth, L. N., & Brunet, A. (2016). The aging epigenome. *Molecular Cell, 62*(5), 728–744. https://doi.org/10.1016/j.molcel.2016.05.013

Bourlière, F. (1957). The comparative biology of ageing: A physiological approach. In G. E. W. Wolstenholme & M. O'Connor (Eds.), *Methodology of the study of ageing. CIBA foundation colloquia on ageing* (Vol. 3, pp. 20–38). Little, Brown and Co.. https://doi.org/10.1002/9780470719039.ch3

Bourlière, F. (1960). Species differences in potential longevity of vertebrates and their physiological implications. In B. Strehler (Ed.), *The biology of aging* (pp. 128–131). American Institute of Biological Sciences.

Branchet, M. C., Boisnic, S., Frances, C., & Robert, A. M. (1990). Skin thickness changes in normal aging skin. *Gerontology, 36*(1), 28–35. https://doi.org/10.1159/000213172

Brown, W. R., MacKinnon, P. J., Villasanté, A., Spurr, N., Buckle, V. J., & Dobson, M. J. (1990). Structure and polymorphism of human telomere-associated DNA. *Cell, 63*(1), 119–132. https://doi.org/10.1016/0092-8674(90)90293-n

Brück, D., Wenning, G. K., Stefanova, N., & Fellner, L. (2016). Glia and alpha-synuclein in neurodegeneration: A complex interaction. *Neurobiology of Disease, 85*, 262–274. https://doi.org/10.1016/j.nbd.2015.03.003

Bürger, M. (1954). *Altern und Krankheit*. G. Thieme.

Büttner, S., Eisenberg, T., Herker, E., Carmona-Gutierrez, D., Kroemer, G., & Madeo, F. (2006). Why yeast cells can undergo apoptosis: Death in times of peace, love, and war. *The Journal of Cell Biology, 175*(4), 521–525. https://doi.org/10.1083/jcb.200608098

Buxton, J. L., Suderman, M., Pappas, J. J., Borghol, N., McArdle, W., Blakemore, A. I., et al. (2014). Human leukocyte telomere length is associated with DNA methylation levels in mul-

tiple subtelomeric and imprinted loci. *Scientific Reports, 4*, 4954. https://doi.org/10.1038/srep04954
Caballol, N., Martí, M. J., & Tolosa, E. (2007). Cognitive dysfunction and dementia in Parkinson disease. *Movement Disorders, 22*(Suppl 17), S358–S366. https://doi.org/10.1002/mds.21677
Cagan, A., Baez-Ortega, A., Brzozowska, N., Abascal, F., Coorens, T. H. H., Sanders, M. A., et al. (2022). Somatic mutation rates scale with lifespan across mammals. *Nature, 604*(7906), 517–524. https://doi.org/10.1038/s41586-022-04618-z
Caggiano, M., Kauer, J. S., & Hunter, D. D. (1994). Globose basal cells are neuronal progenitors in the olfactory epithelium: A lineage analysis using a replication-incompetent retrovirus. *Neuron, 13*(2), 339–352. https://doi.org/10.1016/0896-6273(94)90351-4
Calleja-Agius, J., Brincat, M., & Borg, M. (2013). Skin connective tissue and ageing. *Best Practice & Research. Clinical Obstetrics & Gynaecology, 27*(5), 727–740. https://doi.org/10.1016/j.bpobgyn.2013.06.004
Campisi, J. (2000). Cancer, aging and cellular senescence. *In Vivo, 14*(1), 183–188.
Campisi, J. (2003). Cancer and ageing: Rival demons? *Nature Reviews Cancer, 3*(5), 339–349. https://doi.org/10.1038/nrc1073
Campisi, J. (2013). Aging, cellular senescence, and cancer. *Annual Review of Physiology, 75*, 685–705. https://doi.org/10.1146/annurev-physiol-030212-183653
Campisi, J., Kim, S. H., Lim, C. S., & Rubio, M. (2001). Cellular senescence, cancer and aging: The telomere connection. *Experimental Gerontology, 36*(10), 1619–1637. https://doi.org/10.1016/s0531-5565(01)00160-7
Candia, O. A. (2004). Electrolyte and fluid transport across corneal, conjunctival and lens epithelia. *Experimental Eye Research, 78*(3), 527–535. https://doi.org/10.1016/j.exer.2003.08.015
Cardani, R., & Zavanella, T. (2000). Age-related cell proliferation and apoptosis in the kidney of male Fischer 344 rats with observations on a spontaneous tubular cell adenoma. *Toxicologic Pathology, 28*(6), 802–806. https://doi.org/10.1177/019262330002800606
Carrel, A. (1912). On the permanent life of tissue outside of the organism. *The Journal of Experimental Medicine, 15*(5), 516–528. https://doi.org/10.1084/jem.15.5.516
Carrel, A. (1913). Contributions to the study of the mechanism of the growth of connective tissue. *The Journal of Experimental Medicine, 18*(3), 287–298. https://doi.org/10.1084/jem.18.3.287
Carrel, A., & Ebeling, A. H. (1921). Age and multiplication of fibroblasts. *The Journal of Experimental Medicine, 34*(6), 599–623. https://doi.org/10.1084/jem.34.6.599
Chang, J., Wang, Y., Shao, L., Laberge, R. M., Demaria, M., Campisi, J., et al. (2016a). Clearance of senescent cells by ABT263 rejuvenates aged hematopoietic stem cells in mice. *Nature Medicine, 22*(1), 78–83. https://doi.org/10.1038/nm.4010
Chang, S., Khoo, C., & DePinho, R. A. (2001). Modeling chromosomal instability and epithelial carcinogenesis in the telomerase-deficient mouse. *Seminars in Cancer Biology, 11*(3), 227–239. https://doi.org/10.1006/scbi.2000.0374
Chang, Y. H., Liu, H. W., Wu, K. C., & Ding, D. C. (2016b). Mesenchymal stem cells and their clinical applications in osteoarthritis. *Cell Transplantation, 25*(5), 937–950. https://doi.org/10.3727/096368915X690288
Chen, X., Fang, H., & Schwob, J. E. (2004). Multipotency of purified, transplanted globose basal cells in olfactory epithelium. *The Journal of Comparative Neurology, 469*(4), 457–474. https://doi.org/10.1002/cne.11031
Chimenti, C., Kajstura, J., Torella, D., Urbanek, K., Heleniak, H., Colussi, C., et al. (2003). Senescence and death of primitive cells and myocytes lead to premature cardiac aging and heart failure. *Circulation Research, 93*(7), 604–613. https://doi.org/10.1161/01.RES.0000093985.76901.AF
Chłopicki, S., & Gryglewski, R. J. (2005). Angiotensin converting enzyme (ACE) and HydroxyMethylGlutaryl-CoA (HMG-CoA) reductase inhibitors in the forefront of pharmacology of endothelium. *Pharmacological Reports, 57S*, 86–96.
Christensen, B. C., Houseman, E. A., Marsit, C. J., Zheng, S., Wrensch, M. R., Wiemels, J. L., et al. (2009). Aging and environmental exposures alter tissue specific DNA methylation depen-

dent upon CpG Island context. *PLoS Genetics, 5*(8), e1000602. https://doi.org/10.1371/journal.pgen.1000602

Chu, H.-P., Cifuentes-Rojas, C., Kesner, B., Aeby, E., Lee, H.-G., Wei, C., et al. (2017a). TERRA RNA antagonizes ATRX and protects telomeres. *Cell, 170*(1), 86–101. https://doi.org/10.1016/j.cell.2017.06.017

Chu, H.-P., Froberg, J. E., Kesner, B., Oh, H. J., Ji, F., Sadreyev, R., et al. (2017b). PAR-TERRA directs homologous sex chromosome pairing. *Nature Structural & Molecular Biology, 24*(8), 620–631. https://doi.org/10.1038/nsmb.3432

Collado, M., Blasco, M. A., & Serrano, M. (2007). Cellular senescence in cancer and aging. *Cell, 130*(2), 223–233. https://doi.org/10.1016/j.cell.2007.07.003

Comfort, A. (1964). *Ageing: The biology of senescence*. Routledge & Kegan Paul Ltd..

Comfort, A. (1979). *The biology of senescence*. Livingstone.

Connolly, M. J. (1998). Age-related changes in the respiratory system. In R. C. Tallis, H. M. Fillit, & J. C. Brocklehurst (Eds.), *Brocklehurst's textbook of geriatric medicine and gerontology* (5th ed., pp. 1073–1078). Churchill Livingstone.

Coppé, J.-P., Kauser, K., Campisi, J., & Beauséjour, C. M. (2006). Secretion of vascular endothelial growth factor by primary human fibroblasts at senescence. *The Journal of Biological Chemistry, 281*(40), 29568–29574. https://doi.org/10.1074/jbc.M603307200

Coppé, J.-P., Patil, C. K., Rodier, F., Sun, Y., Muñoz, D. P., Goldstein, J., et al. (2008). Senescence-associated secretory phenotypes reveal cell-nonautonomous functions of oncogenic RAS and the p53 tumor suppressor. *PLoS Biology, 6*(12), 2853–2868. https://doi.org/10.1371/journal.pbio.0060301

Counter, C. M., Hahn, W. C., Wei, W., Caddle, S. D., Beijersbergen, R. L., Lansdorp, P. M., et al. (1998). Dissociation among in vitro telomerase activity, telomere maintenance, and cellular immortalization. *Proceedings of the National Academy of Sciences of the United States of America, 95*(25), 14723–14728. https://doi.org/10.1073/pnas.95.25.14723

Cowdry, E. V. (1952). Problems of aging. In A. I. Lansing (Ed.), *Cowdry's problems of ageing: Biological and medical aspects* (3rd ed.). Williams & Wilkins.

Cristofalo, V. J., & Pignolo, R. J. (1993). Replicative senescence of human fibroblast-like cells in culture. *Physiological Reviews, 73*(3), 617–638. https://doi.org/10.1152/physrev.1993.73.3.617

Croteau, D. L., & Bohr, V. A. (1997). Repair of oxidative damage to nuclear and mitochondrial DNA in mammalian cells. *The Journal of Biological Chemistry, 272*(41), 25409–25412. https://doi.org/10.1074/jbc.272.41.25409

Crouch, R. K., Koutalos, Y., Kono, M., Schey, K., & Ablonczy, Z. (2015). A2E and lipofuscin. *Progress in Molecular Biology and Translational Science, 134*, 449–463. https://doi.org/10.1016/bs.pmbts.2015.06.005

Cumming, W. J. K. (1998). Aging and neuromuscolar disease. In R. C. Tallis, H. M. Fillit, & J. C. Brocklehurst (Eds.), *Brocklehurst's textbook of geriatric medicine and gerontology* (5th ed., pp. 1115–1130). Churchill Livingstone.

d'Adda di Fagagna, F., Reaper, P. M., Clay-Farrace, L., Fiegler, H., Carr, P., von Zglinicki, T., et al. (2003). A DNA damage checkpoint response in telomere-initiated senescence. *Nature, 426*(6963), 194–198. https://doi.org/10.1038/nature02118

Daniali, L., Benetos, A., Susser, E., Kark, J. D., Labat, C., Kimura, M., et al. (2013). Telomeres shorten at equivalent rates in somatic tissues of adults. *Nature Communications, 4*, 1597. https://doi.org/10.1038/ncomms2602

David, H., & Reinke, P. (1988). Liver morphology with aging. In L. Bianchi, P. Holt, & O. F. W. James (Eds.), *Aging in liver and gastro-intestinal tracts* (pp. 143–159). MTP Press.

Davidson, M. H. (2007). Overview of prevention and treatment of atherosclerosis with lipid-altering therapy for pharmacy directors. *The American Journal of Managed Care, 13*(Suppl 10), S260–S269.

Davie, C. A. (2008). A review of Parkinson's disease. *British Medical Bulletin, 86*, 109–127. https://doi.org/10.1093/bmb/ldn013

Davies, C. T. M., Thomas, D. O., & White, M. J. (1986). Mechanical properties of young and elderly human muscle. *Acta Medica Scandinavica, 711*, 219–226. https://doi.org/10.1111/j.0954-6820.1986.tb08954.x

Davies, M. J. (1988). The pathological basis of arrhythmias. *Geriatric Cardiovascular Medicine, 1*(3), 181–183.

Deevey, E. S., Jr. (1947). Life tables for natural populations of animals. *The Quarterly Review of Biology, 22*(4), 283–314. https://doi.org/10.1086/395888

De Grey, A. (2005). The SENS challenge: $20,000 says the foreseeable defeat of aging is not laughable. *Rejuvenation Research, 8*(4), 207–210. https://doi.org/10.1089/rej.2005.8.207

de Lange, T., & Jacks, T. (1999). For better or worse? Telomerase inhibition and cancer. *Cell, 98*(3), 273–275. https://doi.org/10.1016/s0092-8674(00)81955-8

Demaria, M., O'Leary, M. N., Chang, J., Shao, L., Liu, S., Alimirah, F., et al. (2017). Cellular senescence promotes adverse effects of chemotherapy and cancer relapse. *Cancer Discovery, 7*(2), 165–176. https://doi.org/10.1158/2159-8290.CD-16-0241

DePinho, R. A. (2000). The age of cancer. *Nature, 408*(6809), 248–254. https://doi.org/10.1038/35041694

Dieppe, P., & Tobias, J. (1998). Bone and joint aging. In R. C. Tallis, H. M. Fillit, & J. C. Brocklehurst (Eds.), *Brocklehurst's textbook of geriatric medicine and gerontology* (5th ed., pp. 1131–1135). Churchill Livingstone.

Diman, A., Boros, J., Poulain, F., Rodriguez, J., Purnelle, M., Episkopou, H., et al. (2016). Nuclear respiratory factor 1 and endurance exercise promote human telomere transcription. *Science Advances, 2*(7), e1600031. https://doi.org/10.1126/sciadv.1600031

Diman, A., & Decottignies, A. (2018). Genomic origin and nuclear localization of TERRA telomeric repeat-containing RNA: From darkness to Dawn. *The FEBS Journal, 285*(8), 1389–1398. https://doi.org/10.1111/febs.14363

D'Mello, N. P., & Jazwinski, S. M. (1991). Telomere length constancy during aging of Saccharomyces cerevisiae. *Journal of Bacteriology, 173*(21), 6709–6713. https://doi.org/10.1128/jb.173.21.6709-6713.1991

Dokal, I. (2000). Dyskeratosis congenita in all its forms. *British Journal of Haematology, 110*(4), 768–779. https://doi.org/10.1046/j.1365-2141.2000.02109.x

Dremier, S., Golstein, J., Mosselmans, R., Dumont, J. E., Galand, P., & Robaye, B. (1994). Apoptosis in dog thyroid cells. *Biochemical and Biophysical Research Communications, 200*(1), 52–58. https://doi.org/10.1006/bbrc.1994.1412

Driesch, H. (1941). Zur Problematik des Alterns. *Zeitschrift für Alternforschung, 3*, 26–43.

Ebeling, A. (1913). The permanent life of connective tissue outside of the organism. *The Journal of Experimental Medicine, 17*(3), 273–285. https://doi.org/10.1084/jem.17.3.273

Elmore, L. W., & Holt, S. E. (2000). Telomerase and telomere stability: A new class of tumor suppressor? *Molecular Carcinogenesis, 28*(1), 1–4. https://doi.org/10.1002/(SICI)1098-2744(200005)28:1<1::AID-MC1>3.0.CO;2-N

Elmore, S. (2007). Apoptosis: A review of programmed cell death. *Toxicologic Pathology, 35*(4), 495–516. journals.sagepub.com/doi/full/10.1080/01926230701320337

Enright, P. L., Kronmal, R. A., Higgins, M., Schenker, M., & Haponik, E. F. (1993). Spirometry reference values for women and men 65 to 85 years of age. Cardiovascular health study. *The American Review of Respiratory Diseases, 147*(1), 125–133. https://doi.org/10.1164/ajrccm/147.1.125

Ermilov, V. V., & Nesterova, A. A. (2016). β-Amyloidopathy in the pathogenesis of age-related macular degeneration in correlation with neurodegenerative diseases. *Advances in Experimental Medicine and Biology, 854*, 119–125. https://doi.org/10.1007/978-3-319-17121-0_17

Fabrizio, P., & Longo, V. D. (2008). Chronological aging-induced apoptosis in yeast. *Biochimica et Biophysica Acta, 1783*(7), 1280–1285. https://doi.org/10.1016/j.bbamcr.2008.03.017

Felsenstein, J. (1974). The evolutionary advantage of recombination. *Genetics, 78*(2), 737–756. https://doi.org/10.1093/genetics/78.2.737

Ferguson, B. M., Brewer, B. J., & Fangman, W. L. (1991). Temporal control of DNA replication in yeast. *Cold Spring Harbor Symposia on Quantitative Biology, 56*, 293–302. https://doi.org/10.1101/sqb.1991.056.01.036

Fernandez-Rebollo, E. (2020). Senescence-associated metabolomic phenotype in primary and iPSC-derived mesenchymal stromal cells. *Stem Cell Reports, 14*, 201–209. https://doi.org/10.1016/j.stemcr.2019.12.012

Feuerhahn, S., Iglesias, N., Panza, A., Porro, A., & Lingner, J. (2010). TERRA biogenesis, turnover and implications for function. *FEBS Letters, 584*(17), 3812–3818. https://doi.org/10.1016/j.febslet.2010.07.032

Finch, C. E. (1990). *Longevity, senescence, and the genome*. University of Chicago Press.

Finegood, D. T., Scaglia, L., & Bonner-Weir, S. (1995). Dynamics of beta-cell mass in the growing rat pancreas. Estimation with a simple mathematical model. *Diabetes, 44*(3), 249–256. https://doi.org/10.2337/diab.44.3.249

Fine, S. L., Berger, J. W., Maguire, M. G., & Ho, A. C. (2000). Age-related macular degeneration. *The New England Journal of Medicine, 342*(7), 483–492. https://doi.org/10.1056/NEJM200002173420707

Flanary, B. (2009). Telomeres: Function, shortening, and lengthening. In L. Mancini (Ed.), *Telomeres: Function, shortening and lengthening* (pp. 379–386). Nova Science Publ.

Forrester, J. V., Dick, A. D., McMenamin, P., & Roberts, F. (2020). *The eye: Basic sciences in practice* (5th ed.). W. B. Saunders Company Ltd..

Fossel, M. B. (1996). *Reversing Human Aging*. William Morrow and Company.

Fossel, M. B. (2004). *Cells, aging and human disease*. Oxford University Press.

Fox, S. I., & Rompolski, K. (2021). *Human physiology* (16th ed.). McGraw-Hill Education.

Franceschi, C., Bonafè, M., Valensin, S., Olivieri, F., De Luca, M., Ottaviani, E., & De Benedictis, G. (2000). Inflamm-aging. An evolutionary perspective on immunosenescence. *Annals of the New York Academy of Sciences, 908*, 244–254. https://doi.org/10.1111/j.1749-6632.2000.tb06651.x

Franceschi, C., Garagnani, P., Morsiani, C., Conte, M., Santoro, A., Grignolio, A., et al. (2018). The continuum of aging and age-related diseases: Common mechanisms but different rates. *Frontiers in Medicine (Lausanne), 5*, 61. https://doi.org/10.3389/fmed.2018.00061

Francis, R. M. (1998). Metabolic bone disease. In R. C. Tallis, H. M. Fillit, & J. C. Brocklehurst (Eds.), *Brocklehurst's textbook of geriatric medicine and gerontology* (5th ed., pp. 1137–1154). Churchill Livingstone.

Fuhrmann-Stroissnigg, H., Ling, Y. Y., Zhao, J., McGowan, S. J., Zhu, Y., Brooks, R. W., et al. (2017). Identification of HSP90 inhibitors as a novel class of senolytics. *Nature Communications, 8*(1), 422. https://doi.org/10.1038/s41467-017-00314-z

Fülöp, T., Witkowski, J. M., Pawelec, G., Alan, C., & Larbi, A. (2014). On the immunological theory of aging. In *Interdisciplanary topics in gerontology* (Vol. 39, pp. 163–176). S. Karger AG. https://doi.org/10.1159/000358904

Galpern, W. R., & Lang, A. E. (2006). Interface between tauopathies and synucleinopathies: A tale of two proteins. *Annals of Neurology, 59*(3), 449–458. https://doi.org/10.1002/ana.20819

Ghosh, K., & Capell, B. C. (2016). The senescence-associated secretory phenotype: Critical effector in skin cancer and aging. *The Journal of Investigative Dermatology, 136*(11), 2133–2139. https://doi.org/10.1016/j.jid.2016.06.621

Gilchrest, B. A., Blog, F. B., & Szabo, G. (1969). Effects of aging and chronic Sun exposure on melanocytes in human skin. *The Journal of Investigative Dermatology, 73*(2), 141–143. https://doi.org/10.1111/1523-1747.ep12581580

Gilchrest, B. A., Stoff, J. S., & Soter, N. A. (1982). Chronologic aging alters the response to ultraviolet-induced inflammation in human skin. *The Journal of Investigative Dermatology, 79*(1), 11–15. https://doi.org/10.1111/1523-1747.ep12510417

Gisselsson, D., Jonson, T., Yu, C., Martins, C., Mandahl, N., Wiegant, J., et al. (2002). Centrosomal abnormalities, multipolar mitoses, and chromosomal instability in head and neck tumours with dysfunctional telomeres. *British Journal of Cancer, 87*(2), 202–207. https://doi.org/10.1038/sj.bjc.6600438

Gladyshev, V. N. (2016). Aging: Progressive decline in fitness due to the rising deleteriome adjusted by genetic, environmental, and stochastic processes. *Aging Cell, 15*, 594–602. https://doi.org/10.1111/acel.12480

Goldsmith, T. C. (2004). Aging as an evolved characteristic—Weismann's theory reconsidered. *Medical Hypotheses, 62*(2), 304–308. https://doi.org/10.1016/S0306-9877(03)00337-2

Goldsmith, T. C. (2008). Aging, evolvability, and the individual benefit requirement; medical implications of aging theory controversies. *Journal of Theoretical Biology, 252*(4), 764–768. https://doi.org/10.1016/j.jtbi.2008.02.035

Goldstein, S. (1969). Lifespan of cultured cells in progeria. *Lancet, 1*(7591), 424. https://doi.org/10.1016/s0140-6736(69)91404-4

Gomes, N. M., Ryder, O. A., Houck, M. L., Charter, S. J., Walker, W., Forsyth, N. R., et al. (2011). Comparative biology of mammalian telomeres: Hypotheses on ancestral states and the roles of telomeres in longevity determination. *Aging Cell, 10*(5), 761–768. https://doi.org/10.1111/j.1474-9726.2011.00718.x

Gorbunova, V., Bozzella, M. J., & Seluanov, A. (2008). Rodents for comparative aging studies: From mice to beavers. *Age, 30*(2–3), 111–119. https://doi.org/10.1007/s11357-008-9053-4

Gorelick, A. N., & Naxerova, K. (2022). Mutational clocks tick differently across species. *Nature, 604*(7906), 435–436. https://doi.org/10.1038/d41586-022-00976-w

Gorelick, P. B. (2004). Risk factors for vascular dementia and Alzheimer disease. *Stroke, 35*(11 Suppl 1), 2620–2622. https://doi.org/10.1161/01.STR.0000143318.70292.47

Gottschling, D. E., Aparicio, O. M., Billington, B. L., & Zakian, V. A. (1990). Position effect at S. Cerevisiae telomeres: Reversible repression of pol II transcription. *Cell, 63*(4), 751–762. https://doi.org/10.1016/0092-8674(90)90141-z

Graakjaer, J., Bischoff, C., Korsholm, L., Holstebroe, S., Vach, W., Bohr, V. A., et al. (2003). The pattern of chromosome-specific variations in telomere length in humans is determined by inherited, telomere-near factors and is maintained throughout life. *Mechanisms of Ageing and Development, 124*(5), 629–640. https://doi.org/10.1016/s0047-6374(03)00081-2

Graziadei, P. P. C., & Monti Graziadei, G. A. (1978). Continuous nerve cell renewal in the olfactory system. In M. Jacobson (Ed.), *Handbook of sensory physiology* (Development of sensory systems) (Vol. IX, pp. 55–83). Springer.

Greenwood, J., & Cooper, J. P. (2012). Non-coding telomeric and subtelomeric transcripts are differentially regulated by telomeric and heterochromatin assembly factors in fission yeast. *Nucleic Acids Research, 40*(7), 2956–2963. https://doi.org/10.1093/nar/gkr1155

Greer, E. L., & Shi, Y. (2012). Histone methylation: A dynamic mark in health, disease and inheritance. *Nature Reviews Genetics, 13*(5), 343–357. https://doi.org/10.1038/nrg31730

Greider, C. W., & Blackburn, E. H. (1985). Identification of a specific telomere terminal transferase activity in Tetrahymena extracts. *Cell, 43*(2 Pt 1), 405–413. https://doi.org/10.1016/0092-8674(85)90170-9

Griffiths, C. E. M. (1998). Aging of the skin. In R. C. Tallis, H. M. Fillit, & J. C. Brocklehurst (Eds.), *Brocklehurst's textbook of geriatric medicine and gerontology* (5th ed., pp. 1293–1298). Churchill Livingstone.

Grimby, G., Danneskiold-Samsøe, B., Hvid, K., & Saltin, B. (1982). Morphology and enzymatic capacity in arm and leg muscles in 78-81 year old men and women. *Acta Physiologica Scandinavica, 115*(1), 125–134. https://doi.org/10.1111/j.1748-1716.1982.tb07054.x

Grizzi, F., Di Caro, G., Laghi, L., Hermonat, P., Mazzola, P., Nguyen, D. D., et al. (2013). Mast cells and the liver aging process. *Immunity & Ageing, 10*(1), 9. https://doi.org/10.1186/1742-4933-10-9

Grove, G. L. (1989). Age-associated changes in human epidermal cell renewal and repair. In A. K. Balin & A. M. Kligman (Eds.), *Aging and the skin* (pp. 193–204). Raven Press.

Hackett, J. A., Feldser, D. M., & Greider, C. W. (2001). Telomere dysfunction increases mutation rate and genomic instability. *Cell, 106*(3), 275–286. https://doi.org/10.1016/s0092-8674(01)00457-3

Hackh, I. W. D., & Westling, E. H. (1934). A possible cause of old age. *Science, 79*(2045), 231. https://doi.org/10.1126/science.79.2045.231.a

Halliday, G. M., & Stevens, C. H. (2011). Glia: Initiators and progressors of pathology in Parkinson's disease. *Movement Disorders, 26*(1), 6–17. https://doi.org/10.1002/mds.23455

Hall, J. E., & Hall, M. E. (2020). *Guyton and Hall textbook of medical physiology* (14th ed.). Elsevier.

Hansen, R. S., & Gartler, S. M. (1990). 5-Azacytidine-induced reactivation of the human X chromosome-linked PGK1 gene is associated with a large region of cytosine demethylation in the 5' CpG Island. *Proceedings of the National Academy of Sciences of the United States of America, 87*(11), 4174–4178. https://doi.org/10.1073/pnas.87.11.4174

Harada, K., Iwata, M., Kono, N., Koda, W., Shimonishi, T., & Nakanuma, Y. (2000). Distribution of apoptotic cells and expression of apoptosis-related proteins along the intrahepatic biliary tree in normal and non-biliary diseased liver. *Histopathology, 37*(4), 347–354. https://doi.org/10.1046/j.1365-2559.2000.00982.x

Harley, C. B., Futcher, A. B., & Greider, C. W. (1990). Telomeres shorten during ageing of human fibroblasts. *Nature, 345*(6274), 458–460. https://doi.org/10.1038/345458a0

Harman, D. (1956). Aging: A theory based on free radical and radiation chemistry. *Journal of Gerontology, 11*(3), 298–300. https://doi.org/10.1093/geronj/11.3.298

Harman, D. (1972). The biologic clock: The mitochondria? *Journal of the American Geriatrics Society, 20*(4), 145–147. https://doi.org/10.1111/j.1532-5415.1972.tb00787.x

Hayflick, L. (1965). The limited *in vitro* lifetime of human diploid cell strains. *Experimental Cell Research, 37*, 614–636. https://doi.org/10.1016/0014-4827(65)90211-9

Hayflick, L. (1977). The cellular basis for biological aging. In C. E. Finch & L. Hayflick (Eds.), *Handbook of the biology of aging* (pp. 159–186). Van Nostrand Reinhold.

Hayflick, L. (2000). The future of ageing. *Nature, 408*(6809), 267–269. https://doi.org/10.1038/35041709

Hayflick, L. (2007a). Entropy explains aging, genetic determinism explains longevity, and undefined terminology explains misunderstanding both. *PLoS Genetics, 3*(12), e220. https://doi.org/10.1371/journal.pgen.0030220

Hayflick, L. (2007b). Biological aging is no longer an unsolved problem. *Annals of the New York Academy of Sciences, 1100*, 1–13. https://doi.org/10.1196/annals.1395.001

Hayflick, L., & Moorhead, P. S. (1961). The serial cultivation of human diploid cell strains. *Experimental Cell Research, 25*, 585–621. https://doi.org/10.1016/0014-4827(61)90192-6

Héraud, F., Héraud, A., & Harmand, M. F. (2000). Apoptosis in normal and osteoarthritic human articular cartilage. *Annals of the Rheumatic Diseases, 59*(12), 959–965. https://doi.org/10.1136/ard.59.12.959

Herker, E., Jungwirth, H., Lehmann, K. A., Maldener, C., Fröhlich, K. U., Wissing, S., et al. (2004). Chronological aging leads to apoptosis in yeast. *The Journal of Cell Biology, 164*(4), 501–507. https://doi.org/10.1083/jcb.200310014

Herrera, E., Samper, E., Martín-Caballero, J., Flores, J. M., Lee, H. W., & Blasco, M. A. (1999). Disease states associated with telomerase deficiency appear earlier in mice with short telomeres. *The EMBO Journal, 18*(11), 2950–2960. https://doi.org/10.1093/emboj/18.11.2950

Hill, J. M., Zalos, G., Halcox, J. P., Schenke, W. H., Waclawiw, M. A., Quyyumi, A. A., & Finkel, T. (2003). Circulating endothelial progenitor cells, vascular function, and cardiovascular risk. *The New England Journal of Medicine, 348*(7), 593–600. https://doi.org/10.1056/NEJMoa022287

Hill, K., & Hurtado, A. M. (1996). *Ache life history*. Aldine De Gruyter.

Hill, W. G., & Robertson, A. (1966). The effect of linkage on limits to artificial selection. *Genetic Research, 8*(3), 269–294.

Hjelmborg, J. B., Dalgård, C., Möller, S., Steenstrup, T., Kimura, M., Christensen, K., et al. (2015). The heritability of leucocyte telomere length dynamics. *Journal of Medical Genetics, 52*(5), 297–302. https://doi.org/10.1136/jmedgenet-2014-102736

Holliday, R. (1996). Endless quest. *BioEssays, 18*(1), 3–5. https://doi.org/10.1002/bies.950180103

Holmes, D. J., & Austad, S. N. (1995). Birds as animal models for the comparative biology of aging: A prospectus. *The Journals of Gerontology. Series A, Biological Sciences and Medical Sciences, 50*(2), B59–B66. https://doi.org/10.1093/gerona/50a.2.b59

Horner, P. J., & Gage, F. H. (2000). Regenerating the damaged central nervous system. *Nature, 407*, 963–970. https://doi.org/10.1038/35039559

Horvath, S. (2013). DNA methylation age of human tissues and cell types. *Genome Biology, 14*(10), R115. https://doi.org/10.1186/gb-2013-14-10-r115. Erratum in: Genome Biology, 2015, 16(1), 96. doi:10.1186/s13059-015-0649-6.

Horvath, S., Zhang, Y., Langfelder, P., Kahn, R., Boks, M., van Eijk, K., et al. (2012). Aging effects on DNA methylation modules in human brain and blood tissue. *Genome Biology, 13*(10), R97. https://doi.org/10.1186/gb-2012-13-10-r97

Huard, J. M., Youngentob, S. L., Goldstein, B. J., Luskin, M. B., & Schwob, J. E. (1998). Adult olfactory epithelium contains multipotent progenitors that give rise to neurons and non-neural cells. *The Journal of Comparative Neurology, 400*(4), 469–486. https://doi.org/10.1002/(SICI)1096-9861(19981102)400:4<469::AID-CNE3>3.0.CO;2-8

Hynes, K. (2013). Mesenchymal stem cells from iPS cells facilitate periodontal regeneration. *Journal of Dental Research, 92*, 833–839. https://doi.org/10.1177/0022034513498258

Illingworth, R., Kerr, A., Desousa, D., Jørgensen, H., Ellis, P., Stalker, J., et al. (2008). A novel CpG Island set identifies tissue-specific methylation at developmental gene loci. *PLoS Biology, 6*, e22. https://doi.org/10.1371/journal.pbio.0060022

In der Stroth, L., Tharehalli, U., Günes, C., & Lechel, A. (2020). Telomeres and telomerase in the development of liver cancer. *Cancers (Basel), 12*(8), 2048. https://doi.org/10.3390/cancers12082048

Iwai, N., Zhou, Z., Roop, D. R., & Behringer, R. R. (2008). Horizontal basal cells are multipotent progenitors in normal and injured adult olfactory epithelium. *Stem Cells, 26*(5), 1298–1306. https://doi.org/10.1634/stemcells.2007-0891

Jacobson, M. D., Weil, M., & Raff, M. C. (1997). Programmed cell death in animal development. *Cell, 88*(3), 347–354. https://doi.org/10.1016/s0092-8674(00)81873-5

Jankovic, J. (2008). Parkinson's disease: Clinical features and diagnosis. *Journal of Neurology, Neurosurgey, and Psychiatry, 79*(4), 368–376. https://doi.org/10.1136/jnnp.2007.131045

Jaskelioff, M., Muller, F. L., Paik, J. H., Thomas, E., Jiang, S., Adams, A. C., et al. (2011). Telomerase reactivation reverses tissue degeneration in aged telomerase-deficient mice. *Nature, 469*(7328), 102–106. https://doi.org/10.1038/nature09603

Jassal, V., Fillit, H., & Oreopoulos, D. G. (1998). Aging of the urinary tract. In R. C. Tallis, H. M. Fillit, & J. C. Brocklehurst (Eds.), *Brocklehurst's textbook of geriatric medicine and gerontology* (5th ed., pp. 919–924). Churchill Livingstone.

Jazwinski, S. M. (1993). The genetics of aging in the yeast *Saccharomyces cerevisiae*. *Genetica, 91*(1–3), 35–51. https://doi.org/10.1007/BF01435986

Jones, M., Bisht, K., Savage, S. A., Nandakumar, J., Keegan, C. E., & Maillard, I. (2016). The shelterin complex and hematopoiesis. *The Journal of Clinical Investigation, 126*(5), 1621–1629. https://doi.org/10.1172/JCI84547

Jones, O. R., Gaillard, J. M., Tuljapurkar, S., Alho, J. S., Armitage, K. B., Becker, P. H., et al. (2008). Senescence rates are determined by ranking on the fast-slow life-history continuum. *Ecology Letters, 11*(7), 664–673. https://doi.org/10.1111/j.1461-0248.2008.01187.x

Jones, R. B., Whitney, R. G., & Smith, J. R. (1985). Intramitotic variation in proliferative potential: Stochastic events in cellular aging. *Mechanisms of Ageing and Development, 29*(2), 143–149. https://doi.org/10.1016/0047-6374(85)90014-4

Kaeberlein, M., Burtner, C. R., & Kennedy, B. K. (2007). Recent developments in yeast aging. *PLoS Genetics, 3*(5), e84. https://doi.org/10.1371/journal.pgen.0030084

Kajstura, J., Leri, A., Finato, N., Di Loreto, C., Beltrami, C. A., & Anversa, P. (1998). Myocyte proliferation in end-stage cardiac failure in humans. *Proceedings of the National Academy of Sciences of the United States of America, 95*(15), 8801–8805. https://doi.org/10.1073/pnas.95.15.8801

Katzberg, A. A. (1958). The area of the dermo-epidermal junction in human skin. *The Anatomical Record, 131*(4), 717–725. https://doi.org/10.1002/ar.1091310408

Kerr, J. F., Wyllie, A. H., & Currie, A. R. (1972). Apoptosis: A basic biological phenomenon with wide-ranging implications in tissue kinetics. *British Journal of Cancer, 26*(4), 239–257. https://doi.org/10.1038/bjc.1972.33

Kim, I. H., Kisseleva, T., & Brenner, D. A. (2015). Aging and liver disease. *Current Opinion in Gastroenterology, 31*(3), 184–191. https://doi.org/10.1097/MOG.0000000000000176

Kim, N. W., & Hruszkewycz, A. M. (2001). Telomerase activity modulation in the prevention of prostate cancer. *Urology, 57*(4 Suppl 1), 148–153. https://doi.org/10.1016/s0090-4295(00)00962-6

Kimura, W., Nakada, Y., & Sadek, H. A. (2017). Hypoxia-induced myocardial regeneration. *Journal of Applied Physiology (1985), 123*(6), 1676–1681. https://doi.org/10.1152/japplphysiol.00328.2017

Kirkland, J. L., & Tchkonia, T. (2017). Cellular senescence: A translational perspective. *eBioMedicine, 21*, 21–28. https://doi.org/10.1016/j.ebiom.2017.04.013

Kirkwood, T. B. (1977). Evolution of ageing. *Nature, 270*(5635), 301–304. https://doi.org/10.1038/270301a0

Kirkwood, T. B. (1981). Repair and its evolution: Survival versus reproduction. In C. R. Townsend & P. Calow (Eds.), *Physiological ecology: An evolutionary approach to resource use* (pp. 165–189). Blackwell Scientific Publications.

Kirkwood, T. B. (2005). Understanding the odd science of aging. *Cell, 120*(4), 437–447. https://doi.org/10.1016/j.cell.2005.01.027

Kirkwood, T. B., & Austad, S. N. (2000). Why do we age? *Nature, 408*(6809), 233–238. https://doi.org/10.1038/35041682

Kirkwood, T. B., & Cremer, T. (1982). Cytogerontology since 1881: A reappraisal of august Weissmann and a review of modern Progress. *Human Genetics, 60*(2), 101–121. https://doi.org/10.1007/bf00569695

Kirkwood, T. B., & Holliday, R. (1979). The evolution of ageing and longevity. *Proceedings of the Royal Society of London. Series B, Biological sciences, 205*(1161), 531–546. https://doi.org/10.1098/rspb.1979.0083

Kirkwood, T. B., & Melov, S. (2011). On the programmed/non-programmed nature of ageing within the life history. *Current Biology, 21*(18), R701–R707. https://doi.org/10.1016/j.cub.2011.07.020

Klapper, W., Heidorn, H., Kühne, K., Parwaresch, R., & Krupp, G. (1998a). Telomerase in 'immortal fish. *FEBS Letters, 434*(3), 409–412. https://doi.org/10.1016/s0014-5793(98)01020-5

Klapper, W., Kühne, K., Singh, K. K., Heidorn, K., Parwaresch, R., & Krupp, G. (1998b). Longevity of lobsters is linked to ubiquitous telomerase expression. *FEBS Letters, 439*(1–2), 143–146. https://doi.org/10.1016/s0014-5793(98)01357-x

Klitgaard, H., Mantoni, M., Schiaffino, S., Ausoni, S., Gorza, L., Laurent-Winter, C., et al. (1990). Function. Morphology and protein expression of ageing skeletal muscle: A cross-sectional study of elderly men with different training backgrounds. *Acta Physiologica Scandinavica, 140*(1), 41–54. https://doi.org/10.1111/j.1748-1716.1990.tb08974.x

Koch, C. M. (2012). Monitoring of cellular senescence by DNA-methylation at specific CpG sites. *Aging Cell, 11*, 366–369. https://doi.org/10.1111/j.1474-9726.2011.00784.x

Kowald, A., & Kirkwood, T. B. (2016). Can aging be programmed? A critical literature review. *Aging Cell, 15*(6), 986–998. https://doi.org/10.1111/acel.12510

Krishnamurthy, J., Torrice, C., Ramsey, M. R., Kovalev, G. I., Al-Regaiey, K., Su, L., & Sharpless, N. E. (2004). Ink4a/Arf expression is a biomarker of aging. *The Journal of Clinical Investigation, 114*(9), 1299–1307. https://doi.org/10.1172/JCI22475

Kubota, C., Yamakuchi, H., Todoroki, J., Mizoshita, K., Tabara, N., Barber, M., & Yang, X. (2000). Six cloned calves produced from adult fibroblast cells after long-term culture. *Proceedings of the National Academy of Sciences of the United States of America, 97*(3), 990–995. https://doi.org/10.1073/pnas.97.3.990

Kwon, S. M., Hong, S. M., Lee, Y. K., Min, S., & Yoon, G. (2019). Metabolic features and regulation in cell senescence. *BMB Reports, 52*(1), 5–12. https://doi.org/10.5483/BMBRep.2019.52.1.291

Lack, D. (1954). *The natural regulation of animal numbers*. Clarendon Press.

Lanza, R. P., Cibelli, J. B., Faber, D., Sweeney, R. W., Henderson, B., Nevala, W., et al. (2001). Cloned cattle can be healthy and normal. *Science, 294*(5548), 1893–1894. https://doi.org/10.1126/science.1063440

Laun, P., Bruschi, C. V., Dickinson, J. R., Rinnerthaler, M., Heeren, G., Schwimbersky, R., et al. (2007). Yeast mother cell-specific ageing, genetic (in)stability, and the somatic mutation theory of ageing. *Nucleic Acids Research, 35*(22), 7514–7526. https://doi.org/10.1093/nar/gkm919

Laun, P., Pichova, A., Madeo, F., Fuchs, J., Ellinger, A., Kohlwein, S., et al. (2001). Aged mother cells of *Saccharomyces cerevisiae* show markers of oxidative stress and apoptosis. *Molecular Microbiology, 39*(5), 1166–1173. https://doi.org/10.1111/j.1365-2958.2001.02317.x

Laws, R. M. (1966). Age criteria for the African elephant, *Loxodonta a. africana*. *East African Wildlife Journal, 4*(1), 1–37. https://doi.org/10.1111/j.1365-2028.1966.tb00878.x

Laws, R. M. (1968). Dentition and ageing of the hippopotamus. *East African Wildlife Journal, 6*(1), 19–52. https://doi.org/10.1111/j.1365-2028.1968.tb00899.x

Laws, R. M., & Parker, I. S. C. (1968). Recent studies on elephant populations in East Africa. *Symposia of the Zoological Society of London, 21*, 319–359.

Lee, H. W., Blasco, M. A., Gottlieb, G. J., Horner, J. W., 2nd, Greider, C. W., & DePinho, R. A. (1998). Essential role of mouse telomerase in highly proliferative organs. *Nature, 392*(6676), 569–574. https://doi.org/10.1038/33345

Leopold, A. C. (1961). Senescence in plant development. *Science, 134*(3492), 1727–1732. https://doi.org/10.1126/science.134.3492.1727

Leri, A., Barlucchi, L., Limana, F., Deptala, A., Darzynkiewicz, Z., Hintze, T. H., et al. (2001). Telomerase expression and activity are coupled with myocyte proliferation and preservation of telomeric length in the failing heart. *Proceedings of the National Academy of Sciences of the United States of America, 98*(15), 8626–8631. https://doi.org/10.1073/pnas.151013298

Lesur, I., & Campbell, J. L. (2004). The transcriptome of prematurely aging yeast cells is similar to that of telomerase-deficient cells. *Molecular Biology of the Cell, 15*(3), 1297–1312. https://doi.org/10.1091/mbc.e03-10-0742

Leung, C. T., Coulombe, P. A., & Reed, R. R. (2007). Contribution of olfactory neural stem cells to tissue maintenance and regeneration. *Nature Neuroscience, 10*(6), 720–726. https://doi.org/10.1038/nn1882

Lexell, J., Taylor, C. C., & Sjöström, M. (1988). What is the cause of the ageing atrophy? Total number, size and proportion of different fiber types studied in whole vastus lateralis muscle from 15- to 83-year-old men. *Journal of the Neurological Sciences, 84*(2–3), 275–294. https://doi.org/10.1016/0022-510x(88)90132-3

Libertini, G. (1983). *Ragionamenti evoluzionistici [book in Italian]*. Società Editrice Napoletana. English edition (2011): *Evolutionary arguments on aging, disease, and other topics*. Crownsville (MD): Azinet Press.

Libertini, G. (1988). An adaptive theory of the increasing mortality with increasing chronological age in populations in the wild. *Journal of Theoretical Biology, 132*(2), 145–162. https://doi.org/10.1016/s0022-5193(88)80153-x

Libertini, G. (2006). Evolutionary explanations of the "actuarial senescence in the wild" and of the "state of senility". *The Scientific WorldJournal, 6*, 1086–1108. https://doi.org/10.1100/tsw.2006.209

Libertini, G. (2008). Empirical evidence for various evolutionary hypotheses on species demonstrating increasing mortality with increasing chronological age in the wild. *The Scientific World Journal, 8*, 183–193. https://doi.org/10.1100/tsw.2008.36

Libertini, G. (2009a). The role of telomere-telomerase system in age-related fitness decline, a tameable process. In L. Mancini (Ed.), *Telomeres: Function, shortening and lengthening* (pp. 77–132). Nova Science Publ.

Libertini, G. (2009b). Prospects of a longer life span beyond the beneficial effects of a healthy lifestyle. In J. V. Bentely & M. A. Keller (Eds.), *Handbook on longevity: Genetics, diet and disease* (pp. 35–95). Nova Science Publ.

Libertini, G. (2013). Evidence for aging theories from the study of a hunter-gatherer people (ache of Paraguay). *Biochemistry (Moscow), 78*(9), 1023–1032. https://doi.org/10.1134/S0006297913090083

Libertini, G. (2014). Programmed aging paradigm: How we get old. *Biochemistry (Moscow), 79*(10), 1004–1016. https://doi.org/10.1134/S0006297914100034

Libertini, G. (2017). The feasibility and necessity of a revolution in geriatric medicine. *OBM Geriatrics, 1*(2), 2. https://doi.org/10.21926/obm.geriat.1702002

Libertini, G. (2018). Programmed aging paradigm and aging of perennial neurons. In S. I. Ahmad (Ed.), *Aging. Exploring a complex phenomenon* (pp. 91–116). CRC Press–Taylor & Francis Group.

Libertini, G. (2022). Perspectives to modify and counter aging in the frame of subtelomere–telomere theory of aging. *OBM Geriatrics, 6*(4), 208. https://doi.org/10.21926/obm.geriatr.2204208

Libertini, G., Corbi, G., Conti, V., Shubernetskaya, O., & Ferrara, N. (2021a). *Evolutionary gerontology and geriatrics: Why and how we age. Advances in studies of aging and health* (Vol. 2). Springer Nature.

Libertini, G., Corbi, G., & Ferrara, N. (2020). Importance and meaning of TERRA sequences for aging mechanisms. *Biochemistry (Moscow), 85*(12), 1505–1517. https://doi.org/10.1134/S0006297920120044

Libertini, G., Corbi, G., Shubernetskaya, O., & Ferrara, N. (2022). Is human aging a form of Phenoptosis? *Biochemistry (Moscow), 87*(12), 1446–1464. https://doi.org/10.1134/S0006297922120033

Libertini, G., & Ferrara, N. (2016a). Possible interventions to modify aging. *Biochemistry (Moscow), 81*(12), 1413–1428. https://doi.org/10.1134/S0006297916120038

Libertini, G., & Ferrara, N. (2016b). Aging of perennial cells and organ parts according to the programmed aging paradigm. *Age (Dordrecht, Netherlands), 38*(2), 35. https://doi.org/10.1007/s11357-016-9895-0

Libertini, G., Ferrara, N., Rengo, G., & Corbi, G. (2018). Elimination of senescent cells: Prospects according to the subtelomere-telomere theory. *Biochemistry (Moscow), 83*(12), 1477–1488. https://doi.org/10.1134/S0006297918120064

Libertini, G., Rengo, G., & Ferrara, N. (2017). Aging and aging theories. *Journal of Gerontology and Geriatrics, 65*(1), 59–77. http://www.jgerontology-geriatrics.com/article/view/158

Libertini, G., Shubernetskaya, O., Corbi, G., & Ferrara, N. (2021b). Is evidence supporting the subtelomere-telomere theory of aging? *Biochemistry (Moscow), 86*(12), 1766–1781. https://doi.org/10.1134/S0006297921120026

Ligr, M., Madeo, F., Fröhlich, E., Hilt, W., Fröhlich, K. U., & Wolf, D. H. (1998). Mammalian Bax triggers apoptotic changes in yeast. *FEBS Letters, 438*(1), 61–65. https://doi.org/10.1016/s0014-5793(98)01227-7

Lim, J. Y. (2016). Therapeutic potential of eccentric exercises for age-related muscle atrophy. *Integrative Medicine Research, 5*(3), 176–181. https://doi.org/10.1016/j.imr.2016.06.003

Londoño-Vallejo, J. A., DerSarkissian, H., Cazes, L., & Thomas, G. (2001). Differences in telomere length between homologous chromosomes in humans. *Nucleic Acids Research, 29*(15), 3164–3171. https://doi.org/10.1093/nar/29.15.3164

Longo, V. D., Ellerby, L. M., Bredesen, D. E., Valentine, J. S., & Gralla, E. B. (1997). Human Bcl-2 reverses survival defects in yeast lacking superoxide dismutase and delays death of wild-type yeast. *The Journal of Cell Biology, 137*(7), 1581–1588. https://doi.org/10.1083/jcb.137.7.1581

Longo, V. D., Mitteldorf, J., & Skulachev, V. P. (2005). Programmed and altruistic ageing. *Nature Reviews. Genetics, 6*(11), 866–872. https://doi.org/10.1038/nrg1706

Lorand, A. (1929). Le rôle de l'auto-intoxication intestinale dans la production de la vieillesse. *Clinique (Paris, France), 24*, 205–207.
Loyfer, N., Magenheim, J., Peretz, A., Cann, G., Bredno, J., Klochendler, A., et al. (2023). A DNA methylation atlas of normal human cell types. *Nature, 613*(7943), 355–364. https://doi.org/10.1038/s41586-022-05580-6
Luke, B., Panza, A., Redon, S., Iglesias, N., Li, Z., & Lingner, J. (2008). The Rat1p 5′ to 3′ exonuclease degrades telomeric repeat-containing RNA and promotes telomere elongation in Saccharomyces cerevisiae. *Molecular Cell, 32*(4), 465–477. https://doi.org/10.1016/j.molcel.2008.10.019
Lynch, M. P., Nawaz, S., & Gerschenson, L. E. (1986). Evidence for soluble factors regulating cell death and cell proliferation in primary cultures of rabbit endometrial cells grown on collagen. *Proceedings of the National Academy of Sciences of the United States of America, 83*(13), 4784–4788. https://doi.org/10.1073/pnas.83.13.4784
Madeo, F., Fröhlich, E., & Fröhlich, K. U. (1997). A yeast mutant showing diagnostic markers of early and late apoptosis. *The Journal of Cell Biology, 139*(3), 729–734. https://doi.org/10.1083/jcb.139.3.729
Madeo, F., Fröhlich, E., Ligr, M., Grey, M., Sigrist, S. J., Wolf, D. H., & Fröhlich, K. U. (1999). Oxygen stress: A regulator of apoptosis in yeast. *The Journal of Cell Biology, 145*(4), 757–767. https://doi.org/10.1083/jcb.145.4.757
Maeda, T., Guan, J. Z., Higuchi, Y., Oyama, J., & Makino, N. (2009). Aging-related alterations of subtelomeric methylation in sarcoidosis patients. *The Journal of Gerontology. Series A, Biological Sciences and Medical Sciences, 64*(7), 752–760. https://doi.org/10.1093/gerona/glp049
Ma, H., Zhou, Z., Wei, S., Liu, Z., Pooley, K. A., Dunning, A. M., et al. (2011). Shortened telomere length is associated with increased risk of cancer: A meta-analysis. *PLoS One, 6*(6), e20466. https://doi.org/10.1371/journal.pone.0020466
Mallet, R. T., Burtscher, J., Richalet, J. P., Millet, G. P., & Burtscher, M. (2021). Impact of high altitude on cardiovascular health: Current perspectives. *Vascular Health and Risk Management, 17*, 317–335. https://doi.org/10.2147/VHRM.S294121
Mammalian Methylation Consortium. (2021). Universal DNA methylation age across mammalian tissues. *bioRxiv*, 2021.01.18.426733. https://doi.org/10.1101/2021.01.18.426733.
Marchesini, G., Bua, V., Brunori, A., Bianchi, G., Pisi, P., Fabbri, A., et al. (1988). Galactose elimination capacity and liver volume in aging man. *Hepatology, 8*(5), 1079–1083. https://doi.org/10.1002/hep.1840080516
Maringele, L., & Lydall, D. (2004). Telomerase- and recombination-independent immortalization of budding yeast. *Genes & Development, 18*(21), 2663–2675. https://doi.org/10.1101/gad.316504
Martinez, C. H., Diaz, A. A., Meldrum, C., Curtis, J. L., Cooper, C. B., Pirozzi, C., et al. . (2017). Age and small airway imaging abnormalities in subjects with and without airflow obstruction in SPIROMICS. *American Journal of Respiratory and Critical Care Medicine, 195*(4), 464–472. https://doi.org/10.1164/rccm.201604-0871OC.SPIROMICS Investigators
Martin, G. M., & Oshima, J. (2000). Lessons from human progeroid syndromes. *Nature, 408*(6809), 263–266. https://doi.org/10.1038/35041705
Martin, G. M., Sprague, C. A., & Epstein, C. J. (1970). Replicative life-span of cultivated human cells. Effects of Donor's age, tissue, and genotype. *Laboratory Investigation, 23*(1), 86–92.
Martins, A. C. (2011). Change and aging senescence as an adaptation. *PLoS One, 6*(9), e24328. https://doi.org/10.1371/journal.pone.0024328
Maynard Smith, J. (1962). Review lectures on senescence. I. The causes of ageing. *Proceedings of the Royal Society of London. Series B, Biological sciences, 157*(966), 115–127. https://doi.org/10.1098/rspb.1962.0065
Medawar, P. B. (1940). The growth, growth energy and ageing of the chicken's heart. *Proceedings of the Royal Society of London Series B, Biological sciences, 129*(856), 332–355. https://doi.org/10.1098/rspb.1940.0042

Medawar, P. B. (1952). *An unsolved problem in biology*. H. K. Lewis. Reprinted in: Medawar, P. B. (1957). The uniqueness of the individual. Methuen, London.
Medh, R. D., & Thompson, E. B. (2000). Hormonal regulation of physiological cell turnover and apoptosis. *Cell and Tissue Research, 301*(1), 101–124. https://doi.org/10.1007/s004419900159
Metalnikov, S. (1937). *La lutte contre la mort*. Gallimard.
Metchnikoff, E. (1904). *The nature of man*. Heinemann.
Metchnikoff, E. (1907). *The prolongation of life—Optimistic studies*. Heinemann.
Migheli, A., Mongini, T., Doriguzzi, C., Chiadò-Piat, L., Piva, R., Ugo, I., & Palmucci, L. (1997). Muscle apoptosis in humans occurs in normal and denervated muscle, but not in myotonic dystrophy, dystrophinopathies or inflammatory disease. *Neurogenetics, 1*(2), 81–87. https://doi.org/10.1007/s100480050012
Miquel, J., Economos, A. C., Fleming, J., & Johnson, J. E., Jr. (1980). Mitochondrial role in cell aging. *Experimental Gerontology, 15*(6), 575–591. https://doi.org/10.1016/0531-5565(80)90010-8
Mitteldorf, J. (2013). Telomere biology: Cancer firewall or aging clock? *Biochemistry (Moscow), 78*(9), 1054–1060. https://doi.org/10.1134/S0006297913090125
Mitteldorf, J., & Martins, A. C. (2014). Programmed life span in the context of evolvability. *The American Naturalist, 184*(3), 289–302. https://doi.org/10.1086/677387
Mitteldorf, J., & Pepper, J. (2009). Senescence as an adaptation to limit the spread of disease. *Journal of Theoretical Biology, 260*(2), 186–195. https://doi.org/10.1016/j.jtbi.2009.05.013
Montero, J. J., Lopez de Silanes, I., Grana, O., & Blasco, M. A. (2016). Telomeric RNAs are essential to maintain telomeres. *Nature Communications, 7*, 12534. https://doi.org/10.1038/ncomms12534
Moraes, F., & Góes, A. (2016). A decade of human genome project conclusion: Scientific diffusion about our genome knowledge. *Biochemistry and Molecular Biology Education, 44*(3), 215–223. https://doi.org/10.1002/bmb.20952
Morales, I., Sanchez, A., Rodriguez-Sabate, C., & Rodriguez, M. (2015). The degeneration of dopaminergic synapses in Parkinson's disease: A selective animal model. *Behavioural Brain Research, 289*, 19–28. https://doi.org/10.1016/j.bbr.2015.04.019
Morin, G. B. (1989). The human telomere terminal transferase enzyme is a ribonucleoprotein that synthesizes TTAGGG repeats. *Cell, 59*(3), 521–529. https://doi.org/10.1016/0092-8674(89)90035-4
Moyzis, R. K., Buckingham, J. M., Cram, L. S., Dani, M., Deaven, L. L., Jones, M. D., et al. (1988). A highly conserved repetitive DNA sequence (TTAGGG)n, present at the telomeres of human chromosomes. *Proceedings of the National Academy of Sciences of the United States of America, 85*(18), 6622–6626. https://doi.org/10.1073/pnas.85.18.6622
Mueller, L. D. (1987). Evolution of accelerated senescence in laboratory populations of drosophila. *Proceedings of the National Academy of Sciences of the United States of America, 84*(7), 1974–1977. https://doi.org/10.1073/pnas.84.7.1974
Muñoz-Lorente, M. A., Martínez, P., Tejera, Á., Whittemore, K., Moisés-Silva, A. C., Bosch, F., & Blasco, M. A. (2018). AAV9-mediated telomerase activation does not accelerate tumorigenesis in the context of oncogenic K-Ras-induced lung cancer. *PLoS Genetics, 14*(8), e1007562. https://doi.org/10.1371/journal.pgen.1007562
Naylor, R., Richardson, S. J., & McAllan, B. M. (2008). Boom and bust: A review of the physiology of the marsupial genus *Antechinus*. *Journal of Comparative Physiology. B, 178*, 545–562. https://doi.org/10.1007/s00360-007-0250-8
Nergadze, S. G., Farnung, B. O., Wischnewski, H., Khoriauli, L., Vitelli, V., Chawla, R., et al. (2009). CpG-Island promoters drive transcription of human telomeres. *RNA, 15*(12), 2186–2194. https://doi.org/10.1261/rna.1748309
Nussey, D. H., Froy, H., Lemaitre, J. F., Gaillard, J. M., & Austad, S. N. (2013). Senescence in natural populations of animals: Widespread evidence and its implications for bio-gerontology. *Ageing Research Reviews, 12*(1), 214–225. https://doi.org/10.1016/j.arr.2012.07.004
Obeso, J. A., Rodríguez-Oroz, M. C., Benitez-Temino, B., Blesa, F. J., Guridi, J., Marin, C., & Rodriguez, M. (2008). Functional organization of the basal ganglia: Therapeutic implica-

tions for Parkinson's disease. *Movement Disorders, 23*(Suppl 3), S548–S559. https://doi.org/10.1002/mds.22062

Ohmura, Y., Aoe, M., Andou, A., & Shimizu, N. (2000). Telomerase activity and Bcl-2 expression in non-small cell lung cancer. *Clinical Cancer Research, 6*(8), 2980–2987.

Ohno, S. (1972). So much "junk" DNA in our genome. In H. H. Smith (Ed.), *Evolution of genetic systems* (pp. 366–370). Gordon and Breach.

Okuda, K., Bardeguez, A., Gardner, J. P., Rodriguez, P., Ganesh, V., Kimura, M., et al. (2002). Telomere length in the newborn. *Pediatric Research, 52*(3), 377–381. https://doi.org/10.1203/00006450-200209000-00012

Oliveira, B. F., Nogueira-Machado, J.-A., & Chaves, M. M. (2010). The role of oxidative stress in the aging process. *The Scientific World Journal, 10*, 1121–1128. https://doi.org/10.1100/tsw.2010.94

Olivetti, G., Giordano, G., Corradi, D., Melissari, M., Lagrasta, C., Gambert, S. R., & Anversa, P. (1995). Gender differences and aging: Effects on the human heart. *Journal of the American College of Cardiology, 26*(4), 1068–1079. https://doi.org/10.1016/0735-1097(95)00282-8

Olivetti, G., Melissari, M., Capasso, J. M., & Anversa, P. (1991). Cardiomyopathy of the aging human heart. Myocyte loss and reactive cellular hypertrophy. *Circulation Research, 68*(6), 1560–1568. https://doi.org/10.1161/01.res.68.6.1560

Olovnikov, A. M. (1971). [Principle of marginotomy in template synthesis of polynucleotides] [in Russian]. *Doklady Akademii Nauk SSSR*, 201(6), 1496-1499. English version: Olovnikov, A. M. (1971)). Principle of marginotomy in template synthesis of polynucleotides. *Doklady Biochemistry, 201*, 394–397.

Olovnikov, A. M. (1973). A theory of marginotomy: The incomplete copying of template margin in enzyme synthesis of polynucleotides and biological significance of the problem. *Journal of Theoretical Biology, 41*(1), 181–190. https://doi.org/10.1016/0022-5193(73)90198-7

Olshansky, S. J., Hayflick, L., & Carnes, B. A. (2002). Position statement on human aging. *The Journals of Gerontology. Series A, Biological Sciences and Medical Sciences, 57*(8), B292–B297. https://doi.org/10.1093/gerona/57.8.b292

Pal, S., & Tyler, J. K. (2016). Epigenetics and aging. *Science. Advances, 2*(7), e1600584. https://doi.org/10.1126/sciadv.1600584

Parker, R. C. (1961). *Methods of tissue culture*. Harper & Row.

Parrinello, S., Coppé, J.-P., Krtolica, A., & Campisi, J. (2005). Stromal-epithelial interactions in aging and cancer: Senescent fibroblasts alter epithelial cell differentiation. *Journal of Cell Science, 118*(Pt 3), 485–496. https://doi.org/10.1242/jcs.01635

Pathak, S., Multani, A. S., Furlong, C. L., & Sohn, S. H. (2002). Telomere dynamics, aneuploidy, stem cells, and cancer (review). *International Journal of Oncology, 20*(3), 637–641. www.spandidos-publications.com/ijo/20/3/637

Pearl, R. (1922). *The biology of death*. J. B. Lippincott Co..

Pearl, R. (1928). *The rate of living*. Knopf.

Phillips, S. K., Bruce, S. A., Newton, D., & Woledge, R. C. (1992). The weakness of old age is not due to failure of muscle activation. *Journal of Gerontology, 47*(2), M45–M49. https://doi.org/10.1093/geronj/47.2.m45

Pianka, E. R. (1970). On r- and K-selection. *The American Naturalist, 104*(940), 592–597. https://doi.org/10.1086/282697

Pollack, M., & Leeuwenburgh, C. (2001). Apoptosis and aging: Role of the mitochondria. *The Journals of Gerontology. Series A, Biological Sciences and Medical Sciences, 56*(11), B475–B482. https://doi.org/10.1093/gerona/56.11.b475

Pontèn, J., Stein, W. D., & Shall, S. (1983). A quantitative analysis of the aging of human glial cells in culture. *Journal of Cellular Physiology, 117*(3), 342–352. https://doi.org/10.1002/jcp.1041170309

Porro, A., Feuerhahn, S., Delafontaine, J., Riethman, H., Rougemont, J., & Lingner, J. (2014). Functional characterization of the TERRA transcriptome at damaged telomeres. *Nature Communications, 5*, 5379. https://doi.org/10.1038/ncomms6379

Prins, J. B., & O'Rahilly, S. (1997). Regulation of adipose cell number in man. *Clinical Science (London), 92*(1), 3–11. https://doi.org/10.1042/cs0920003

Prowse, K. R., & Greider, C. W. (1995). Developmental and tissue-specific regulation of mouse telomerase and telomere length. *Proceedings of the National Academy of Sciences of the United States of America, 92*(11), 4818–4822. https://doi.org/10.1073/pnas.92.11.4818

Qiu, W. Q., Walsh, D. M., Ye, Z., Vekrellis, K., Zhang, J., Podlisny, M. B., et al. (1998). Insulin-degrading enzyme regulates extracellular levels of amyloid beta-protein by degradation. *The Journal of Biological Chemistry, 273*(49), 32730–32738. https://doi.org/10.1074/jbc.273.49.32730

Rakyan, V. K., Down, T. A., Maslau, S., Andrew, T., Yang, T. P., Beyan, H., et al. (2010). Human aging-associated DNA hypermethylation occurs preferentially at bivalent chromatin domains. *Genome Research, 20*(4), 434–439. https://doi.org/10.1101/gr.103101.109

Reed, J. C. (1999). Dysregulation of apoptosis in cancer. *Journal of Clinical Oncology, 17*(9), 2941–2953. https://doi.org/10.1200/JCO.1999.17.9.2941

Rheinwald, J. G., & Green, H. (1975). Serial cultivation of strains of human epidermal keratinocytes: The formation of keratinizing colonies from single cells. *Cell, 6*(3), 331–344. https://doi.org/10.1016/s0092-8674(75)80001-8

Richardson, B. R., Allan, D. S., & Le, Y. (2014). Greater organ involution in highly proliferative tissues associated with the early onset and acceleration of ageing in humans. *Experimental Gerontology, 55*, 80–91. https://doi.org/10.1016/j.exger.2014.03.015

Ricklefs, R. E. (1998). Evolutionary theories of aging: Confirmation of a fundamental prediction, with implications for the genetic basis and evolution of life span. *The American Naturalist, 152*(1), 24–44. https://doi.org/10.1086/286147

Robin, J. D., Ludlow, A. T., Batten, K., Magdinier, F., Stadler, G., Wagner, K. R., et al. (2014). Telomere position effect: Regulation of gene expression with progressive telomere shortening over long distances. *Genes & Development, 28*(22), 2464–2476. https://doi.org/10.1101/gad.251041.114

Robinson, P. S., Coorens, T. H. H., Palles, C., Mitchell, E., Abascal, F., Olafsson, S., et al. (2021). Increased somatic mutation burdens in normal human cells due to defective DNA polymerases. *Nature Genetics, 53*(10), 1434–1442. https://doi.org/10.1038/s41588-021-00930-y

Rodier, F., & Campisi, J. (2011). Four faces of cellular senescence. *The Journal of Cell Biology, 192*(4), 547–556. https://doi.org/10.1083/jcb.201009094

Rodier, F., Coppé, J. P., Patil, C. K., Hoeijmakers, W. A., Muñoz, D. P., Raza, S. R., et al. (2009). Persistent DNA damage signalling triggers senescence-associated inflammatory cytokine secretion. *Nature Cell Biology, 11*(8), 973–979. https://doi.org/10.1038/ncb1909. Erratum in: Nature Cell Biology, 2009, 11, 1272.

Röhme, D. (1981). Evidence for a relationship between longevity of mammalian species and life spans of normal fibroblasts *in vitro* and erythrocytes *in vivo*. *Proceedings of the National Academy of Sciences of the United States of America, 78*(8), 5009–5013. https://doi.org/10.1073/pnas.78.8.5009

Rose, M. R. (1991). *Evolutionary biology of aging*. Oxford University Press.

Rudenko, G., & Van der Ploeg, L. H. (1989). Transcription of telomere repeats in protozoa. *The EMBO Journal, 8*(9), 2633–2638. https://doi.org/10.1002/j.1460-2075.1989.tb08403.x

Rudolph, K. L., Millard, M., Bosenberg, M. W., & DePinho, R. A. (2001). Telomere dysfunction and evolution of intestinal carcinoma in mice and humans. *Nature Genetics, 28*(2), 155–159. https://doi.org/10.1038/88871

Runge, K. W., & Zakian, V. A. (1989). Introduction of extra telomeric DNA sequences into Saccharomyces cerevisiae results in telomere elongation. *Molecular and Cellular Biology, 9*(4), 1488–1497. https://doi.org/10.1128/mcb.9.4.1488

Rutherford, O. M., & Jones, D. A. (1992). The relationship of muscle and bone loss and activity levels with age in women. *Age and Ageing, 21*(4), 286–293. https://doi.org/10.1093/ageing/21.4.286

Sacher, G. A. (1959). Relation of lifespan to brain weight and body weight in mammals. In G. E. W. Wolstenholme & M. O'Connor (Eds.), *The lifespan of animals (pp. 115-133).*, CIBA Foundation Colloquia on Ageing (Vol. 5). Little, Brown and Co. https://doi.org/10.1002/9780470715253.ch9

Samii, A., Nutt, J. G., & Ransom, B. R. (2004). Parkinson's disease. *Lancet, 363*(9423), 1783–1793. https://doi.org/10.1016/S0140-6736(04)16305-8

Sanderson, W. C., & Scherbov, S. (2005). Average remaining lifetimes can increase as human populations age. *Nature, 435*(7043), 811–813. https://doi.org/10.1038/nature03593

Sanz, A., & Stefanatos, R. K. (2008). The mitochondrial free radical theory of aging: A critical view. *Current Aging Science, 1*(1), 10–21. https://doi.org/10.2174/1874609810801010010

Schellenberg, A. (2014). Proof of principle: Quality control of therapeutic cell preparations using senescence-associated DNA-methylation changes. *BioMed Central Research Notes, 7*, 254. https://doi.org/10.1186/1756-0500-7-254

Schellenberg, A., Lin, Q., Schüler, H., Koch, C. M., Joussen, S., Denecke, B., et al. (2011). Replicative senescence of mesenchymal stem cells causes DNA-methylation changes which correlate with repressive histone marks. *Aging (Albany NY), 3*(9), 873–888. https://doi.org/10.18632/aging.100391

Schneider, E. L., & Mitsui, Y. (1976). The relationship between in vitro cellular aging and in vivo human age. *Proceedings of the National Academy of Sciences of the United States of America, 73*(10), 3584–3588. https://doi.org/10.1073/pnas.73.10.3584

Schoeftner, S., & Blasco, M. A. (2008). Developmentally regulated transcription of mammalian telomeres by DNA-dependent RNA polymerase II. *Nature Cell Biology, 10*(2), 228–236. https://doi.org/10.1038/ncb1685

Schulz-Schaeffer, W. J. (2010). The synaptic pathology of alpha-synuclein aggregation in dementia with Lewy bodies, Parkinson's disease and Parkinson's disease dementia. *Acta Neuropathologica, 120*(2), 131–143. https://doi.org/10.1007/s00401-010-0711-0

Segel, M., Neumann, B., Hill, M. F. E., Weber, I. P., Viscomi, C., Zhao, C., et al. (2019). Niche stiffness underlies the ageing of central nervous system progenitor cells. *Nature, 573*(7772), 130–134. https://doi.org/10.1038/s41586-019-1484-9

Seger, Y. R., García-Cao, M., Piccinin, S., Cunsolo, C. L., Doglioni, C., Blasco, M. A., et al. (2002). Transformation of normal human cells in the absence of telomerase activation. *Cancer Cell, 2*(5), 401–413. https://doi.org/10.1016/s1535-6108(02)00183-6

Seluanov, A., Chen, Z., Hine, C., Sasahara, T. H., Ribeiro, A. A., Catania, K. C., et al. (2007). Telomerase activity coevolves with body mass not lifespan. *Aging Cell, 6*(1), 45–52. https://doi.org/10.1111/j.1474-9726.2006.00262.x

Sender, R., & Milo, R. (2021). The distribution of cellular turnover in the human body. *Nature Medicine, 27*(1), 45–48. https://doi.org/10.1038/s41591-020-01182-9

Shelton, D. N., Chang, E., Whittier, P. S., Choi, D., & Funk, W. D. (1999). Microarray analysis of replicative senescence. *Current Biology, 9*(17), 939–945. https://doi.org/10.1016/s0960-9822(99)80420-5

Shiels, P. G., Kind, A. J., Campbell, K. H., Waddington, D., Wilmut, I., Colman, A., & Schnieke, A. E. (1999). Analysis of telomere lengths in cloned sheep. *Nature, 399*(6734), 316–317. https://doi.org/10.1038/20580

Sim, F. J., Zhao, C., Penderis, J., & Franklin, R. J. M. (2002). The age-related decrease in CNS Remyelination efficiency is attributable to an impairment of both oligodendrocyte progenitor recruitment and differentiation. *The Journal of Neuroscience, 22*(7), 2451–2459. https://doi.org/10.1523/JNEUROSCI.22-07-02451.2002

Sinclair, D. A., & Guarente, L. (1997). Extrachromosomal rDNA circles–a cause of aging in yeast. *Cell, 91*(7), 1033–1042. https://doi.org/10.1016/s0092-8674(00)80493-6

Singer, M. S., Kahana, A., Wolf, A. J., Meisinger, L. L., Peterson, S. E., Goggin, C., et al. (1998). Identification of high-copy disruptors of telomeric silencing in Saccharomyces cerevisiae. *Genetics, 150*(2), 613–632. https://doi.org/10.1093/genetics/150.2.613

Skulachev, V. P. (1997). Aging is a specific biological function rather than the result of a disorder in complex living systems: Biochemical evidence in support of Weismann's hypothesis. *Biochemistry (Moscow), 62*(11), 1191–1195. protein.bio.msu.ru/biokhimiya/contents/v62/full/62111394.html.

Skulachev, V. P. (1999). Phenoptosis: Programmed death of an organism. *Biochemistry (Moscow), 64*(12), 1418–1426. protein.bio.msu.ru/biokhimiya/contents/v64/full/64121679.html.

Skulachev, V. P. (2002). Programmed death phenomena: From organelle to organism. *Annals of the New York Academy of Sciences, 959*, 214–237. https://doi.org/10.1111/j.1749-6632.2002.tb02095.x

Slijepcevic, P., & Hande, M. P. (1999). Chinese hamster telomeres are comparable in size to mouse telomeres. *Cytogenetic and Genome Research, 85*, 196–199. https://doi.org/10.1159/000015292

Smith, J. R., & Pereira-Smith, O. M. (1989). Altered gene expression during cellular aging. *Genome, 31*(1), 386–389. https://doi.org/10.1139/g89-058

Solovei, I., Gaginskaya, E. R., & Macgregor, H. C. (1994). The arrangement and transcription of telomere DNA sequences at the ends of lampbrush chromosomes of birds. *Chromosome Research, 2*(6), 460–470. https://doi.org/10.1007/bf01552869

Sparrow, J. R. (2003). Therapy for macular degeneration: Insights from acne. *Proceedings of the National Academy of Sciences of the United States of America, 100*(8), 4353–4354. https://doi.org/10.1073/pnas.1031478100

Spelsberg, T. C., Subramaniam, M., Riggs, B. L., & Khosla, S. (1999). The actions and interactions of sex steroids and growth factors/cytokines on the skeleton. *Molecular Endocrinology (Baltimore), 13*(6), 819–828. https://doi.org/10.1210/mend.13.6.0299

Spinage, C. A. (1970). Population dynamics of the Uganda Defassa waterbuck (*Kobus defassa Ugandae* Neumann) in the queen Elizabeth park, Uganda. *The Journal of Animal Ecology, 39*(1), 51–78. https://doi.org/10.2307/2889

Spinage, C. A. (1972). African ungulate life tables. *Ecology, 53*(4), 645–652. https://doi.org/10.2307/1934778

Spitzhorn, L. S. (2019). Human iPSC-derived MSCs (iMSCs) from aged individuals acquire a rejuvenation signature. *Stem Cell Research & Therapy, 10*, 100. https://doi.org/10.1186/s13287-019-1209-x

Stein, R., Razin, A., & Cedar, H. (1982). In vitro methylation of the hamster adenine phosphoribosyltransferase gene inhibits its expression in mouse L cells. *Proceedings of the National Academy of Sciences of the United States of America, 79*(11), 3418–3422. https://doi.org/10.1073/pnas.79.11.3418

Stevenson, J. B., & Gottschling, D. E. (1999). Telomeric chromatin modulates replication timing near chromosome ends. *Genes & Development, 13*(2), 146–151. https://doi.org/10.1101/gad.13.2.146

Stewart, J. A., Chaiken, M. F., Wang, F., & Price, C. M. (2012). Maintaining the end: Roles of telomere proteins in end-protection, telomere replication and length regulation. *Mutation Research, 730*(1–2), 12–19. https://doi.org/10.1016/j.mrfmmm.2011.08.011

Stone, R. C., Horvath, K., Kark, J. D., Susser, E., Tishkoff, S. A., & Aviv, A. (2016). Telomere length and the cancer–atherosclerosis trade-off. *PLoS Genetics, 12*(7), e1006144. https://doi.org/10.1371/journal.pgen.1006144

Strobeck, C., Maynard Smith, J., & Charlesworth, B. (1976). The effects of hitchhiking on a gene for recombination. *Genetics, 82*(3), 547–558. https://doi.org/10.1093/genetics/82.3.547

Sturm, E., & Stefanova, N. (2014). Multiple system atrophy: Genetic or epigenetic? *Experimental Neurobiology, 23*(4), 277–291. https://doi.org/10.5607/en.2014.23.4.277

Su, J. B. (2015). Vascular endothelial dysfunction and pharmacological treatment. *World Journal of Cardiology, 7*(11), 719–741. https://doi.org/10.4330/wjc.v7.i11.719

Surace, C., Berardinelli, F., Masotti, A., Roberti, M. C., Da Sacco, L., D'Elia, G., et al. (2014). Telomere shortening and telomere position effect in mild ring 17 syndrome. *Epigenetics & Chromatin, 7*(1), 1. https://doi.org/10.1186/1756-8935-7-1

Sutherland, L. M., Edwards, Y. S., & Murray, A. W. (2001). Alveolar type II cell apoptosis. Comparative biochemistry and physiology. *Part A, Molecular & integrative physiology, 129*(1), 267–285. https://doi.org/10.1016/s1095-6433(01)00323-3

Takahashi, K., & Yamanaka, S. (2006). Induction of pluripotent stem cells from mouse embryonic and adult fibroblast cultures by defined factors. *Cell, 126*(4), 663–676. https://doi.org/10.1016/j.cell.2006.07.024

Takai, K. K., Hooper, S., Blackwood, S., Gandhi, R., & de Lange, T. (2010). In vivo stoichiometry of shelterin components. *The Journal of Biological Chemistry, 285*(2), 1457–1467. https://doi.org/10.1074/jbc.M109.038026

Takubo, K., Aida, J., Izumiyama-Shimomura, N., Ishikawa, N., Sawabe, M., Kurabayashi, R., et al. (2010). Changes of telomere length with aging. *Geriatrics & Gerontology International, 10*(Suppl 1), S197–S206. https://doi.org/10.1111/j.1447-0594.2010.00605.x

Tallis, R. C., Fillit, H. M., & Brocklehurst, J. C. (Eds.). (1998). *Brocklehurst's textbook of geriatric medicine and gerontology* (5th ed.). Churchill Livingstone.

Tassin, J., Malaise, E., & Courtois, Y. (1979). Human lens cells have an in vitro proliferative capacity inversely proportional to the donor age. *Experimental Cell Research, 123*(2), 388–392. https://doi.org/10.1016/0014-4827(79)90483-x

Telomerase Database. (Updated August 2020). Accessed Mar 28, 2023, from telomerase.asu.edu/sequences_telomere.html

Teschendorff, A. E., Menon, U., Gentry-Maharaj, A., Ramus, S. J., Weisenberger, D. J., Shen, H., et al. (2010). Age-dependent DNA methylation of genes that are suppressed in stem cells is a hallmark of cancer. *Genome Research, 20*(4), 440–446. https://doi.org/10.1101/gr.103606.109

Tomás-Loba, A., Flores, I., Fernández-Marcos, P. J., Cayuela, M. L., Maraver, A., Tejera, A., et al. (2008). Telomerase reverse transcriptase delays aging in cancer-resistant mice. *Cell, 135*(4), 609–622. https://doi.org/10.1016/j.cell.2008.09.034

Travis, J. M. (2004). The evolution of programmed death in a spatially structured population. *The Journals of Gerontology. Series A, Biological Sciences and Medical Sciences, 59*(4), 301–305. https://doi.org/10.1093/gerona/59.4.b301

Trifunovic, A., Wredenberg, A., Falkenberg, M., Spelbrink, J. N., Rovio, A. T., Bruder, C. E., et al. (2004). Premature ageing in mice expressing defective mitochondrial DNA polymerase. *Nature, 429*(6990), 417–423. https://doi.org/10.1038/nature02517

Turner, K. J., Vasu, V., & Griffin, D. K. (2019). Telomere biology and human phenotype. *Cells, 8*(1), 73. https://doi.org/10.3390/cells8010073

Urbanek, K., Quaini, F., Tasca, G., Torella, D., Castaldo, C., Nadal-Ginard, B., et al. (2003). Intense myocyte formation from cardiac stem cells in human cardiac hypertrophy. *Proceedings of the National Academy of Sciences of the United States of America, 100*(18), 10440–10445. https://doi.org/10.1073/pnas.1832855100

Urbanek, K., Torella, D., Sheikh, F., De Angelis, A., Nurzynska, D., Silvestri, F., et al. (2005). Myocardial regeneration by activation of multipotent cardiac stem cells in ischemic heart failure. *Proceedings of the National Academy of Sciences of the United States of America, 102*(24), 8692–8697. https://doi.org/10.1073/pnas.0500169102

Vandervoort, A. A., & McComas, A. J. (1986). Contractile changes in opposing muscles of the human ankle joint with aging. *Journal of Applied Physiology, 61*(1), 361–367. https://doi.org/10.1152/jappl.1986.61.1.361

van Deursen, J. M. (2014). The role of senescent cells in ageing. *Nature, 509*(7501), 439–446. https://doi.org/10.1038/nature13193

van Steensel, B., & de Lange, T. (1997). Control of telomere length by the human telomeric protein TRF1. *Nature, 385*(6618), 740–743. https://doi.org/10.1038/385740a0

Vaupel, J. W., Baudisch, A., Dölling, M., Roach, D. A., & Gampe, J. (2004). The case for negative senescence. *Theoretical Population Biology, 65*(4), 339–351. https://doi.org/10.1016/j.tpb.2003.12.003

Vaziri, H. (1998). Extension of life span in normal human cells by telomerase activation: A revolution in cultural senescence. *Journal of Anti-Aging Medicine, 1*, 125–130. https://doi.org/10.1089/rej.1.1998.1.125

Vaziri, H., & Benchimol, S. (1998). Reconstitution of telomerase activity in normal cells leads to elongation of telomeres and extended replicative life span. *Current Biology, 8*(5), 279–282. https://doi.org/10.1016/s0960-9822(98)70109-5

Vekrellis, K., Ye, Z., Qiu, W. Q., Walsh, D., Hartley, D., Chesneau, V., et al. (2000). Neurons regulate extracellular levels of amyloid beta-protein via proteolysis by insulin-degrading enzyme. *The Journal of Neuroscience, 20*(5), 1657–1765.

Verbeken, E. K., Cauberghs, M., Mertens, I., Clement, J., Lauweryns, J. M., & Van de Woestijne, K. P. (1992). The senile lung. Comparison with normal and emphysematous lungs. 1. Structural aspects. *Chest, 101*(3), 793–799. https://doi.org/10.1378/chest.101.3.793

Villeponteau, B. (1997). The heterochromatin loss model of aging. *Experimental Gerontology, 32*(4–5), 383–394. https://doi.org/10.1016/s0531-5565(96)00155-6

von Zglinicki, T., Serra, V., Lorenz, M., Saretzki, G., Lenzen-Grossimlighaus, R., Gessner, R., et al. (2000). Short telomeres in patients with vascular dementia: An indicator of low antioxidative capacity and a possible risk factor? *Laboratory Investigation, 80*(11), 1739–1747. https://doi.org/10.1038/labinvest.3780184

Vrbsky, J., Akimcheva, S., Watson, J. M., Turner, T. L., Daxinger, L., Vyskot, B., et al. (2010). siRNA-mediated methylation of Arabidopsis telomeres. *PLoS Genetics, 6*(6), e1000986. https://doi.org/10.1371/journal.pgen.1000986

Wagner, W., Horn, P., Castoldi, M., Diehlmann, A., Bork, S., Saffrich, R., et al. (2008). Replicative senescence of mesenchymal stem cells: A continuous and organized process. *PLoS One, 3*(5), e2213. https://doi.org/10.1371/journal.pone.0002213

Wakabayashi, H., Nishiyama, Y., Ushiyama, T., Maeba, T., & Maeta, H. (2002). Evaluation of the effect of age on functioning hepatocyte mass and liver blood flow using liver scintigraphy in preoperative estimations for surgical patients: Comparison with CT volumetry. *The Journal of Surgery Research, 106*(2), 246–253. https://doi.org/10.1006/jsre.2002.6462

Wang, E. (1995). Senescent human fibroblasts resist programmed cell death, and failure to suppress bcl2 is involved. *Cancer Research, 55*(11), 2284–2292.

Warthin, A. S. (1929). *Old age, the major involution; the physiology and pathology of the ageing process*. Hoeber.

Watanabe, T., & Tanaka, Y. (1982). Age-related alterations in the size of human epatocytes: Study of mononuclear and binucleate cells. *Virchow Archive B Cell Pathology, 39*(1), 9–20. https://doi.org/10.1007/bf02892832

Watson, J. D. (1972). Origin of concatemeric T7 DNA. *Nature: New Biology, 239*(94), 197–201. https://doi.org/10.1038/newbio239197a0

Weinberger, F., & Eschenhagen, T. (2021). Cardiac regeneration: New hope for an old dream. *Annual Review of Physiology, 83*(1), 59–81. https://doi.org/10.1146/annurev-physiol-031120-103629

Weinert, B. T., & Timiras, P. S. (2003). Invited review: Theories of aging. *Journal of Applied Physiology, 95*(4), 1706–1716. https://doi.org/10.1152/japplphysiol.00288.2003

Weismann, A. (1882). Über die Dauer des Lebens. .

Weismann, A. (1889.); (1st ed. 1889, 2nd ed. 1891)). English translation. In E. B. Poulton, S. Schonland, & A. E. Shipley (Eds.), *Essays upon heredity and kindred biological Problems* (Vol. I). Clarendon Press.

Weismann, A. (1892). *Essays upon heredity and kindred biological problems* (Vol. II). Clarendon Press.

Werner, N., Kosiol, S., Schiegl, T., Ahlers, P., Walenta, K., Link, A., et al. (2005). Circulating endothelial progenitor cells and cardiovascular outcomes. *The New England Journal of Medicine, 353*(10), 999–1007. https://doi.org/10.1056/NEJMoa043814

Whittemore, K., Derevyanko, A., Martinez, P., Serrano, R., Pumarola, M., Bosch, F., & Blasco, M. A. (2019). Telomerase gene therapy ameliorates the effects of neurodegeneration associated

to short telomeres in mice. *Aging (Albany NY), 11*(10), 2916–2948. https://doi.org/10.18632/aging.101982
Wikipedia. (2023). *List of world records in masters athletics*. Accessed April 19, 2023, from en.wikipedia.org/wiki/List_of_world_records_in_masters_athletics
Williams, G. C. (1957). Pleiotropy, natural selection and the evolution of senescence. *Evolution, 11*, 398–411. https://doi.org/10.2307/2406060
Williams, R. W., & Herrup, K. (1988). The control of neuron number. *Annual Review of Neuroscience, 11*(1), 423–453.
Wilson, E. O. (1975). *Sociobiology, the new synthesis*. Harvard University Press.
Wilson, P. W., Castelli, W. P., & Kannel, W. B. (1987). Coronary risk prediction in adults (the Framingham heart study). *The American Journal of Cardiology, 59*(14), 91G–94G. https://doi.org/10.1016/0002-9149(87)90165-2. Erratum in: The American Journal of Cardiology, 1987, 60(13), A11.
World Cause of Death (2020). (Data from WHO). Accessed Dec 15, 2022, from www.worldlifeexpectancy.com/world-rankings-total-deaths
Wright, W. E., & Shay, J. W. (2005). Telomere biology in aging and cancer. *Journal of the American Geriatrics Society, 53*(9 Suppl), S292–S294. https://doi.org/10.1111/j.1532-5415.2005.53492.x
Wright, W. E., Tesmer, V. M., Liao, M. L., & Shay, J. W. (1999). Normal human telomeres are not late replicating. *Experimental Cell Research, 251*(2), 492–429. https://doi.org/10.1006/excr.1999.4602
Wu, X., Amos, C. I., Zhu, Y., Zhao, H., Grossman, B. H., Shay, J. W., et al. (2003). Telomere dysfunction: A potential cancer predisposition factor. *Journal of the National Cancer Institute, 95*(16), 1211–1218. https://doi.org/10.1093/jnci/djg011
Wyllie, A. H., Kerr, J. F. R., & Currie, A. R. (1980). Cell death: The significance of apoptosis. *International Review of Cytology, 68*, 251–306. https://doi.org/10.1016/s0074-7696(08)62312-8
Wynne, H. A., Cope, L. H., Mutch, E., Rawlins, M. D., Woodhouse, K. W., & James, O. F. (1989). The effect of age upon liver volume and apparent liver blood flow in healthy man. *Hepatology, 9*(2), 297–301. https://doi.org/10.1002/hep.1840090222
Xia, S. J., Xu, C. X., Tang, X. D., Wang, W. Z., & Du, D. L. (2001). Apoptosis and hormonal milieu in ductal system of normal prostate and benign prostatic hyperplasia. *Asian Journal of Andrology, 3*(2), 131–134.
Xu, M., Palmer, A. K., Ding, H., Weivoda, M. M., Pirtskhalava, T., White, T. A., et al. (2015a). Targeting senescent cells enhances adipogenesis and metabolic function in old age. *eLife, 2015*, e12997. https://doi.org/10.7554/eLife.12997
Xu, M., Tchkonia, T., Ding, H., Ogrodnik, M., Lubbers, E. R., Pirtskhalava, T., et al. (2015b). JAK inhibition alleviates the cellular senescence-associated secretory phenotype and frailty in old age. *Proceedings of the National Academy of Sciences of the United States of America, 112*(46), E6301–E6310. https://doi.org/10.1073/pnas.1515386112
Yang, J. N. (2013). Viscous populations evolve altruistic programmed ageing in ability conflict in a changing environment. *Evolutionary Ecology Research, 15*, 527–543.
Young, A. J. (2018). The role of telomeres in the mechanisms and evolution of life-history trade-offs and ageing. *Philosophical Transactions of the Royal Society of London. Series B, Biological Sciences, 373*(1741), 20160452. https://doi.org/10.1098/rstb.2016.0452
Young, A., Stokes, M., & Crowe, M. (1984). Size and strength of the quadriceps muscles of old and young women. *European Journal of Clinical Investigation, 14*(4), 282–287. https://doi.org/10.1111/j.1365-2362.1984.tb01182.x
Young, A., Stokes, M., & Crowe, M. (1985). The size and strength of the quadriceps muscles of old and young men. *Clinical Physiology, 5*(2), 145–154. https://doi.org/10.1111/j.1475-097x.1985.tb00590.x
Yu, G. L., Bradley, J. D., Attardi, L. D., & Blackburn, E. H. (1990). *In vivo* alteration of telomere sequences and senescence caused by mutated *Tetrahymena* telomerase RNAs. *Nature, 344*(6262), 126–132. https://doi.org/10.1038/344126a0

Zhao, C., Deng, W., & Gage, F. H. (2008). Mechanisms and functional implications of adult neurogenesis. *Cell, 132*(4), 645–660.

Zhao, Y., Xiong, W., Li, C., Zhao, R., Lu, H., Song, S., et al. (2023). Hypoxia-induced signaling in the cardiovascular system: Pathogenesis and therapeutic targets. *Signal Transduction and Targeted Therapy, 8*, 431. https://doi.org/10.1038/s41392-023-01652-9

Zhou, X., Hong, Y., Zhang, H., & Li, X. (2020). Mesenchymal stem cell senescence and rejuvenation: Current status and challenges. *Frontiers in Cell and Developmental Biology, 8*, 364. https://doi.org/10.3389/fcell.2020.00364

Chapter 9
Other Topics Studied by Evolutionary Medicine

In this chapter, heterogeneous topics specifically investigated by evolutionary medicine are briefly discussed.

9.1 Constraints Determined by Our Phylogenetic History

Some inconsistencies exist between some of our organism's anatomical or functional characteristics and a hypothetical, optimal design. This is due to the constraints imposed by the species' previous evolutionary history or phylogeny.

In fact, "evolution does no sensible planning. It always proceeds by just slightly modifying what it already has" (Nesse & Williams, 1994, p. 124). Alternatively, by recalling a famous statement by Gottfried Leibniz (1705): *Natura non facit saltus* (Nature does not make jumps): a species does not pass from an anatomical or functional condition to another condition, which from a rational point of view appears more logical, if this implies a complex and so improbable modification of the existing one.

9.1.1 Crossing Between Respiratory and Digestive Tracts

It is well known that, for vertebrates, the respiratory and digestive tracts cross. For our species, this occurs at the level of the epiglottis (Fig. 9.1). When a solid or liquid food passes through this point, the epiglottis must be lowered to prevent food from entering the trachea. In this event, the cough reflex is immediately activated to eliminate this abnormal passage of food in the respiratory tract, which can cause suffocation or dangerous lung infections (the so-called aspiration pneumonia).

G. Libertini et al., *The Essential Revolution of Evolutionary Medicine*, Advances in Studies of Aging and Health 4, https://doi.org/10.1007/978-3-031-84877-3_9

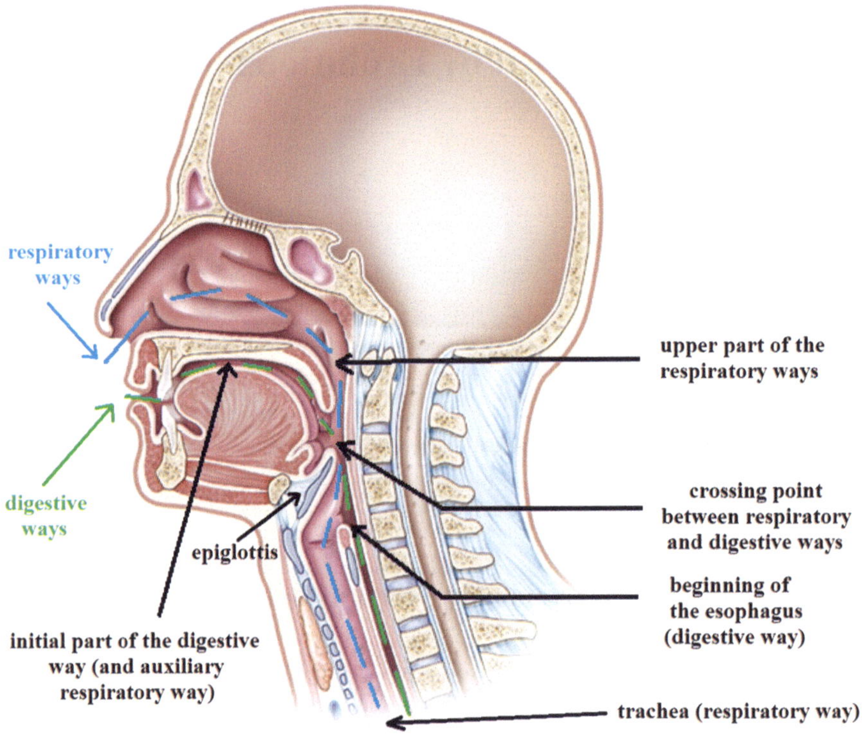

Fig. 9.1 Anatomical diagram of the upper parts of the respiratory and digestive tracts in which the crossing point between these ways is highlighted

These mechanisms generally work well, but serious exceptions exist in two main periods of life.

As already underlined, speaking of the cough reflex, the accidental introduction of a foreign body into the airways is a dangerous event quite frequent before the age of 3. It is the leading cause of accidental infant death and the fourth one among all causes in children ≤5 years old (Doody, 2004), particularly in the group 1–2 years of age (Salih et al., 2016).

Furthermore, concerning accidental deaths, in a 2014 study, the leading cause of death in Japan for all ages was airway obstruction by foreign bodies in elderly subjects (9806 deaths out of a total of 39,029 accidental deaths) (Hanada, 2016).

Both for very young subjects and for elderly subjects, the main causes that triggered the dangerous or even fatal episodes, in addition to the existence of the aforementioned dangerous crossing, were the insufficiency or inability to evaluate the danger and the insufficiency of neuromotor coordination (due to immaturity of development for very young subjects and degenerative phenomena for elderly subjects).

The existence of a cross between the respiratory and digestive tracts in vertebrates derives from the phylogenetic origin of the vertebrates based on anatomical conditions summarized in Fig. 9.2.

For a more precise analysis from an evolutionary point of view:

(i) For children, in 40% of all cases, there is often the introduction into the mouth and the subsequent passage in the airways of small artificial items, plastic bags, food particles such as peanuts (Park et al., 1996), i.e., objects not present in a natural environment.
(ii) For cases in elderly subjects, in Japan in 2014, the third leading cause of accidental death for all ages were deaths associated with drowning or immersion in the bathtub (5362 deaths), mainly in the elderly population (about 5000 cases) where this was the second causes of death (Hanada, 2016). Therefore, in the elderly, due to the deterioration of neuromotor and cognitive functions, the most frequent accidental deaths are airway occlusion by a foreign body and bath-related accidents, which consequently, as a primary cause, should be both attributed to aging, while the conditions that cause the fatal event are contingent, i.e., non-fatal in non-old age.

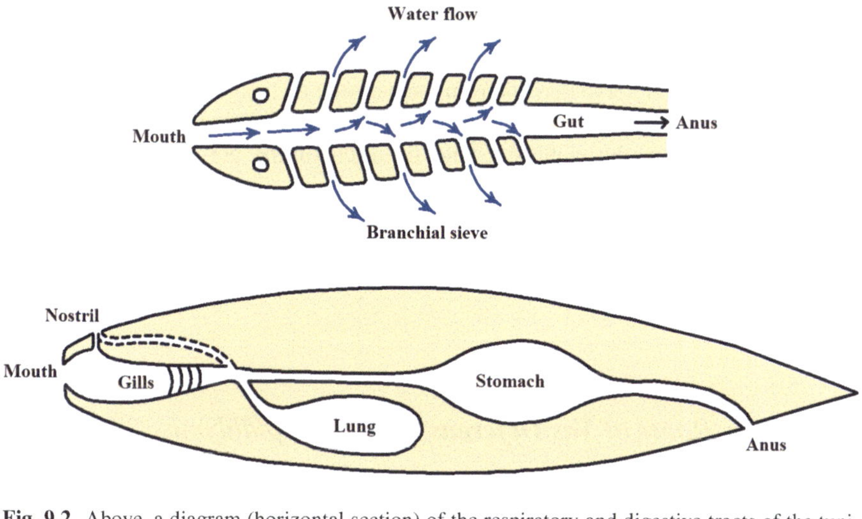

Fig. 9.2 Above, a diagram (horizontal section) of the respiratory and digestive tracts of the tunicate larvae and also of the probable ancestor of all vertebrates, showing the practical coincidence of the two types of ways in these organisms. Below a diagram (vertical section) of the condition of lungfishes, ancestors of all vertebrates, in which the lungs are originated as a diverticulum of the digestive tract, with the subsequent appearance of a new opening (the nostrils) which carries air into the respiratory tract. The images are a reworking of figures reported in Nesse and Williams (1994, p. 126)

9.1.2 The Route of the Inferior or Recurrent Laryngeal Nerve

The larynx is innervated by the left and right superior laryngeal nerves and the left and right inferior or recurrent laryngeal nerves, which are branches of the two vagus nerves (cranial nerve X).

From its origin, the superior laryngeal nerves reach the larynx directly, innervating the cricothyroid muscles.

Otherwise, the left inferior or recurrent laryngeal nerve reaches the larynx, innervating all the other intrinsic muscles of the larynx on the left, after following a longer route in which it first descends towards the heart and then returns back reaching the larynx (hence the name of recurrent, which in Latin means running back).

The right recurrent laryngeal nerve passes under the right subclavian artery, while the left recurrent laryngeal nerve passes under the arch of the aorta, in both cases following a path that does not appear logical (Netter, 2022) (Fig. 9.3).

The strange paths of the recurrent laryngeal nerves are also present in other vertebrates. In the giraffe, due to the long neck of this animal, it reaches a length of about 4.6 m (Harris et al., 2010).

The paths of the recurrent laryngeal nerves have a precise explanation traceable in the phylogeny and ontogeny of vertebrates. The laryngeal cartilages originate from the pharyngeal (or branchial) arches, 4 for the upper part and 6 for the lower part. The superior laryngeal nerve derives from the nerve of the fourth arch, while the inferior laryngeal nerve derives from the nerve of the sixth arch. The arteries of the fourth arch, which come between the two laryngeal nerves, give rise to the right subclavian artery on the right side and the arch of the aorta on the left side (Larsen, 1993).

However, this particular conformation, which could imprecisely be considered a defect of evolution, does not constitute a pathology, though it can create some problems due to the accidental cut of the inferior laryngeal nerves in case of surgical interventions.

9.1.3 The Route of Vas Deferens from the Epididymis to the Prostate

Another curious anatomical peculiarity is the path the two ductus deferentes follow from the epididymis to the prostate (Netter, 2022) (Fig. 9.4, right side).

Each duct, which serves for the passage of the seminal fluid, arises from the epididymis, which is above the testicle, collects the semen produced by the testicle, runs in the subcutaneous tissue passing over the pubic bone, then penetrates into the pelvic cavity via the inguinal canal, passes over the ureter on the same side and descends to reach the prostate. Here it joins the duct of the seminal vesicle, forming the ejaculatory duct, which ends in the urethra in the penis.

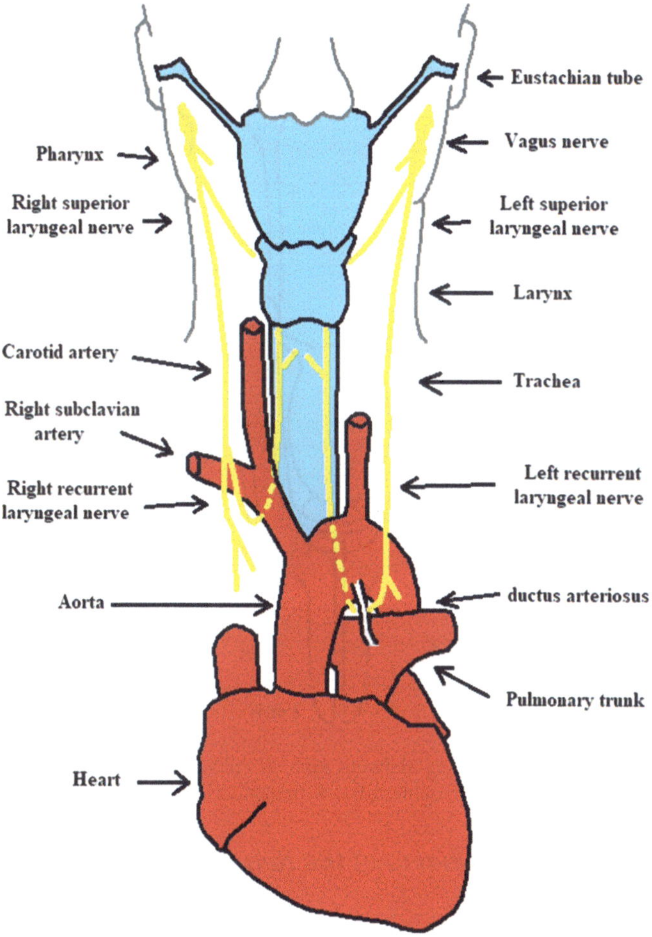

Fig. 9.3 Diagram of the path of the two recurrent laryngeal nerves. Phylogenetically and ontogenetically, the right subclavian artery and the aortic arch originate from the fourth branchial arch, while the recurrent laryngeal nerves originate from the sixth branchial arch. Therefore, the two nerves must be after the arteries above

The route of the ductus deferens appears unnecessarily stretched, and hypothetically, it could go directly from the epididymis to the prostate.

However, this lengthened path arises from the fact that the testicles are originally in the peritoneal cavity and subsequently descend into the scrotal sac. This lengthened route, like the path of the recurrent laryngeal nerve, arises from the phylogeny of the organ, and a more logical path would mean a sudden modification, i.e., an improbable *saltus*.

However, the lengthened route does not create dysfunctions and, therefore, cannot be considered a maladaptation and a cause of disease.

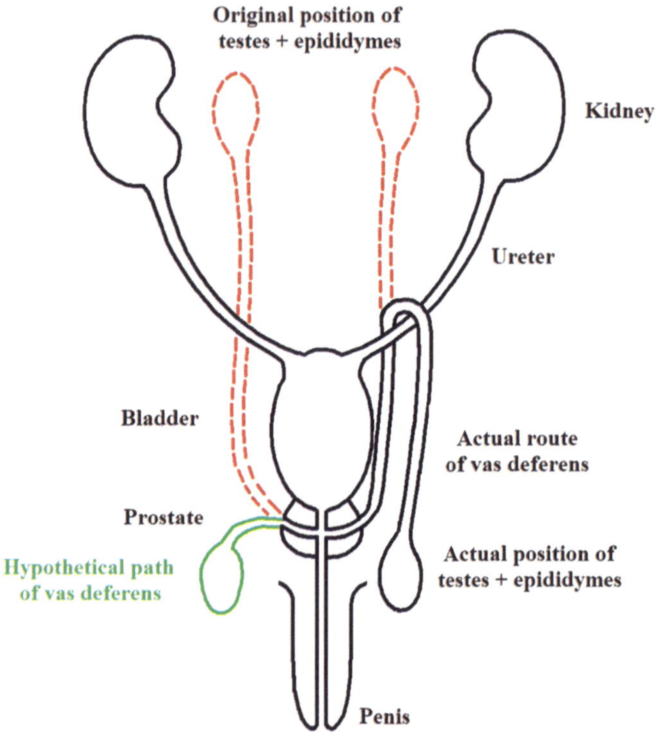

Fig. 9.4 Simplified route of the ductus deferens from the epididymis to the prostate. On the right, the real path of the ductus deferens; on the left, a hypothetical, more rational, direct route

9.1.4 Inverted Arrangement of the Retina

In vertebrates, the arrangement of the retina shows some features that may be considered illogical. In fact, the light, before reaching the photoreceptors located on the outermost part of the retina (meaning as "external" the points furthest from the geometric center of the eye), must pass through various layers of neurons and nerve fibers. Furthermore, in front of the retina, i.e., internally with respect to the photoreceptors, lie the blood vessels serving the retina, which further interfere with the perfect passage of light rays. Another drawback of this arrangement is that, in the area in which the optic nerve, central retinal artery and vein pass through the retina, there are no photoreceptors, therefore constituting the so-called blind spot (Netter, 2022) (s. Figs. 9.5, 9.6 and 9.7).

This arrangement of the retina is termed inverted retina and is practically opposite to that present in cephalopods (s. Fig. 9.8). This last retinal arrangement appears more rational, while that of vertebrates does not present any obvious evolutionary advantage.

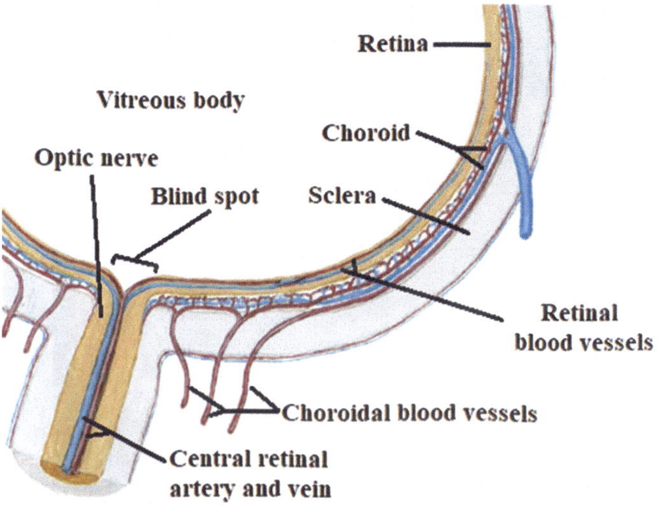

Fig. 9.5 Disposition of the retina showing: (i) the blind spot, i.e., the area of passage through the retina of the optic nerve and retinal blood vessels, and (ii) the interposition of the branches of these vessels and the ramifications of the optic nerve between the retina and the origin of the light rays

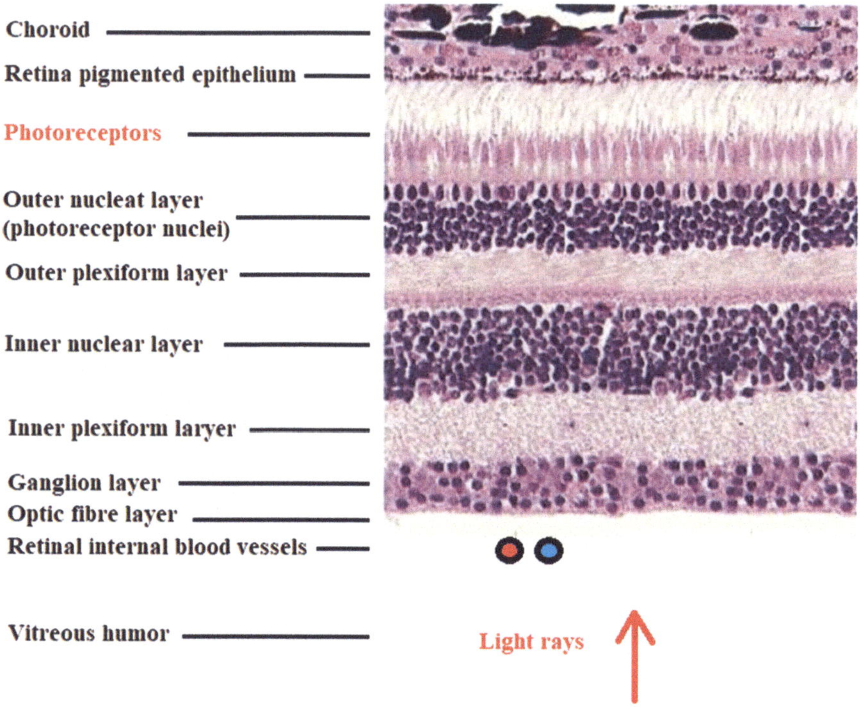

Fig. 9.6 Histology of the retina showing that the photoreceptors are located in the part furthest from the origin of the light rays

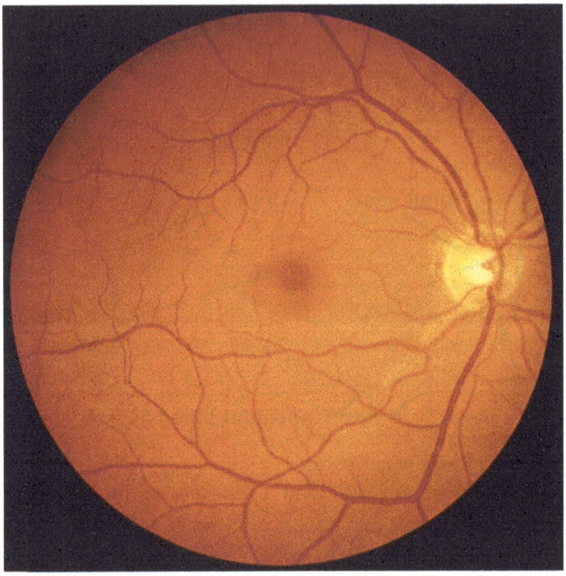

Fig. 9.7 Ophthalmoscopy of a human eye. The yellow area on the right is the optic disc, the area in which the optic nerve, the artery, and the central vein of the retina pass through the retina and then branch out over the entire retina, interposing themselves between the photoreceptors and the light rays coming from the outside. It should also be noted that the name optic nerve is improper as the whole retina is an outpouring of the CNS, and therefore, the optic nerve is, in fact, an internal nervous fascicle of the brain. The reddest area in the center is the fovea centralis, where there is the greatest density of photoreceptors and no blood vessel passes over this area in order not to reduce sensitivity, which is the maximum of the whole retina

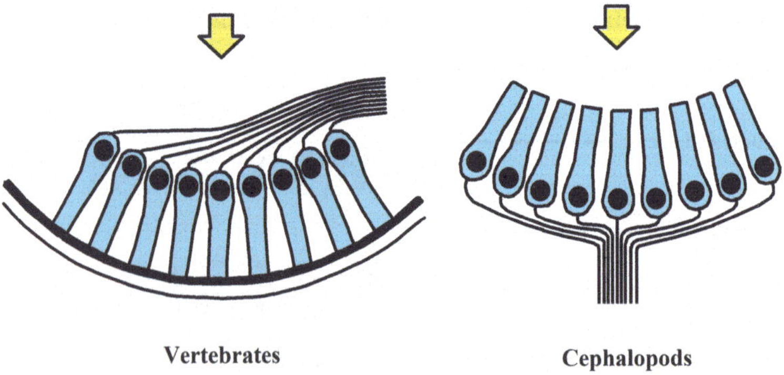

Fig. 9.8 Schematic comparison of the inverted arrangement of the retina in vertebrates and the direct arrangement in cephalopods

These two different arrangements of the retina are the consequence of the different phylogenetic development of the eye in vertebrates and cephalopods. Let us now see if, in evolutionarily normal conditions, this constitutes a problem at the origin of possible malfunctions and troubles for our species.

The problems caused by an inverted retina have been brilliantly solved by natural evolution.

In the point of retinal maximum sensitivity, the *fovea centralis* (the central part of the macula), the photoreceptors are very dense, and the layers of fibers and nerve cells are thinner and interfere little with the light rays. Furthermore, this area is avascular, i.e., it is not covered even by thin branches of blood vessels (Bringmann et al., 2018).

As regards the problem of the optic disc, which constitutes a blind spot, and the small strips where the blood vessels pass, certainly altering or blocking the passage of light rays, it is necessary to consider how the retina works.

The retina is part of the CNS and performs the initial processing of information carried by light rays. These reach the photoreceptors, and then the information is processed by various types of specialized neurons, and the result is entrusted to the neurons of the ganglion layer. An axon departs from each neuron of this layer, which, together with the other axons of the same type, constitutes the optic nerve, actually an internal fascicle of the CNS, which then reaches further brain centers (lateral geniculate nuclei and the areas of visual cortex). Here, the information is further processed, including the integration of the information coming from the two eyes (Hall & Hall, 2020).

Each optic nerve contains between 770,000 and 1.7 million nerve fibers (Jonas et al., 1992), while the rods are about 100 million and the cones three million (Hall & Hall, 2020). Consequently, the ratio between the number of photoreceptors and the number of fibers in the optic nerve is approximately 100:1. This gives an idea of the high grade of information processing in the retina.

The eye shows frequent and very rapid (saccadic) movements during which the light signal is not perceived. The signal is perceived and processed only in the phases in which the eye is fixed on a specific point (Hall & Hall, 2020). The rays coming from a specific external point after eye movements are focused on different areas of the retina, and, in some cases, the light beam can be intercepted by a blood vessel or fall on the blind spot. The visual system processes the subsequent images and eliminates the non-visible parts by integrating them with the available information. The processing is so accurate that we have no evidence of blind spots and areas where light rays are blocked by the presence of blood vessels.

In short, the visual system beautifully solves the problems arising from inverted retina arrangement.

Consequently, this arrangement cannot be considered a maladaptation (i.e., a defective adaptation), but must be evaluated as a particular conformation deriving from the phylogeny of vertebrates.

However, the presence of blood vessels in front of the retina and the weak connection between the retina and external structure (choroid) are sources of problems in particular pathological conditions. Diabetes mellitus and hypertension are often the cause of diabetic retinopathy (Lin et al., 2021) and hypertensive retinopathy (Dziedziak et al., 2022), respectively. Myopia, especially in the case of high myopia (≥ -6 diopters), is often complicated by retinal detachment (Wu et al., 2016).

The conformation of the inverted retina certainly allows the aforementioned complications in the case of particular diseases, but the conclusion that the inverted retina is a maladaptation is improper. In fact, diabetes mellitus, hypertension, and myopia are practically absent in evolutionarily normal conditions. They are caused by mismatches between the ecological niche to which we are adapted and particular alterations of the same niche (s. *Chap. 3—Diseases deriving from alterations of the ecological niche—Part I: Not regarding the holobiont*).

9.1.5 Transition to Bipedalism and Birth Canal

During the course of evolution, our species has passed from a condition in which, while walking or standing still, the body rests on four limbs (quadrupedalism) to one in which the support is only on two limbs (bipedalism) (Niemitz, 2010). In 1941, various structural drawbacks of these different conditions were described (Estabrooks, 1941), then deepened in two other books (Morgan, 1990; Nesse & Williams, 1994).

The principal suggested drawbacks were: more significant load on the lower part of the lumbar spine and the intervertebral discs, greater load and stress on the joints of the lower limbs, vulnerability for inguinal hernias due to the greater downward weight, narrowness of the pelvis with childbirth problems (increased by the greater size of the brain and so of the skull of the newborn).

Some of these difficulties and how they have been addressed by evolution are discussed in an interesting article (Gruss & Schmitt, 2015) and a special issue of *The Anatomical Record* (DeSilva & Rosenberg, 2017):

(1) The greater weight bearing on the lumbar region is contrasted by a curve of the rachis, the lumbar lordosis, which allows superior resistance to the load (Gruss & Schmitt, 2015);
(2) Due to the narrowness of the pelvic canal, in a horizontal section, this canal is not elongated as in monkeys but has a rounded shape. The head of the fetus performs a particular rotation which allows the fetus to come out (Gruss & Schmitt, 2015) (s. Fig. 9.9);
(3) Particular anatomical adaptations of the pelvis and differences between men and women contrast the difficulties associated with bipedalism and the needs for childbirth (Lewis et al., 2017; Wall-Scheffler & Myers, 2017).

However, it should be noted that:

(i) These adaptations cannot solve all problems related to childbirth. If a fetus—due to its own defect—has a body or head size (or form) incompatible with the dimensions of the pelvic canal, or if the fetus in childbirth is presented in an

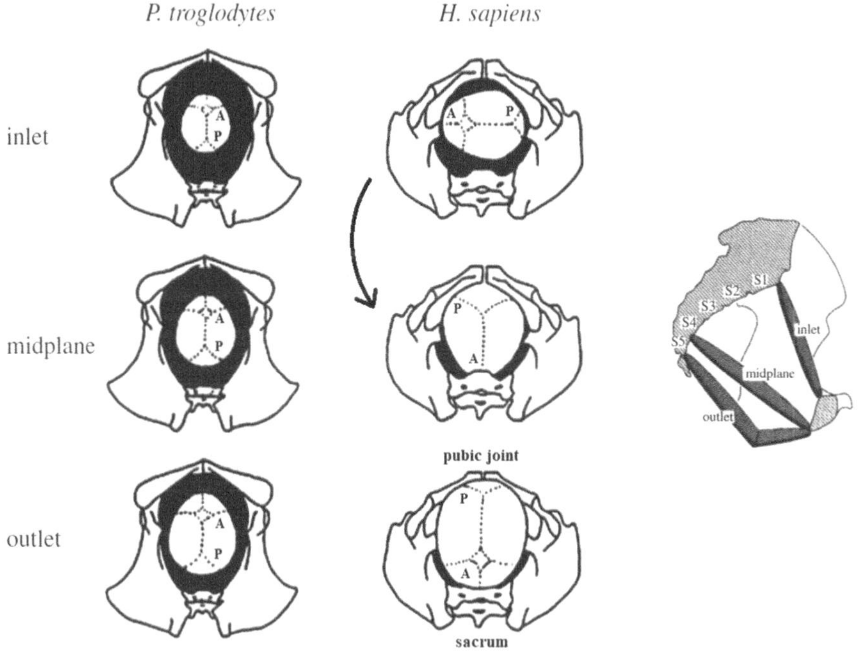

Fig. 9.9 Rotation of the fetus in the childbirth. Image modified and adapted from Gruss and Schmitt (2015)

unsuitable position for passage through this canal, the birth under evolutionarily normal conditions (therefore in the absence of surgical treatment) will be the cause of death for both fetus and mother. Indeed, in the study of a human population in the wild, three cases of death due to childbirth were reported (Hill & Hurtado, 1996).

(ii) With modern lifestyles, childbirth does not occur under conditions equivalent to those found in the wild. For example, reduced physical activity, which determines reduced development of the muscular walls of the abdomen, and the lying position of the parturient are factors that alter the physiology of childbirth.

Regarding the other adaptations to the upright position, their effectiveness should be evaluated considering populations living in conditions equivalent to those to which our species is adapted and comparing them with those of populations in modern conditions.

The study of hunter-gatherer populations, such as the Dobe!Kung, living in a territory between Botswana and Namibia (Howell, 1979) and the Ache of Paraguay (Hill & Hurtado, 1996), have not shown particular pathologies deriving from the upright position despite the fact that these people certainly do not spare the lower limbs and the back with reduced physical activity.

On the contrary, in individuals of modern populations with a sedentary lifestyle, prolonged postures that are often non-physiological, and diets far from the original ones, pathologies concerning the vertebral column and the joints of the lower limbs are frequent (Ebnezar & Rakesh, 2016). However, it would be arbitrary to attribute these pathologies not to altered lifestyles to which we are not adapted but to a vulnerability that would not have been erased by hundreds of thousands of years of natural selection.

9.2 Normal Physiological Characteristics that Become Problems in a Modified Ecological Niche

There are same physiological characteristics that do not cause diseases in the ecological niche to which we are adapted. On the contrary, they can cause serious problems in a different ecological niche for which there is no specific adaptation.

9.2.1 Necessity of Essential Nutrients

Our body needs many essential nutrients, including vitamins (s. Table 9.1), two essential fatty acids, various indispensable amino acids, choline, and various minerals (s. Table 9.2).

For some of these essential nutrients, i.e., for minerals, synthesis is impossible for any species. For the others, the synthesis in other species is possible, but our species is unable to carry it out. As a special case, for vitamin D, synthesis is possible in the skin if sufficiently exposed to ultraviolet B rays, but often, the amount obtained by synthesis must be integrated with food.

However, the question is why our body is vulnerable to the lack of many substances.

Other species are well able to synthesize molecules that are essential for us and that we are unable to synthesize. Furthermore, for health and survival, we depend on small quantities of particular minerals, for which perhaps alternative metabolic pathways would be possible that could allow us to avoid these vital needs.

In evolutionary terms, this vulnerability is perfectly explainable.

A species is best adapted to its specific ecological niche, which clearly also includes the diet. If an essential molecule is always present in a more than sufficient measure in the diet, there is no need for the synthesis of this molecule. Therefore, natural selection becomes insufficient to preserve the integrity and efficiency of a metabolic pathway, allowing this synthesis (if it exists), and the aforementioned metabolic pathway is gradually lost.

In parallel, if a metabolic pathway requires a particular mineral for the functionality of an enzyme and this mineral is present in the food in a constant and sufficient way, there is no natural selection in favor of an alternative metabolic pathway in which this need does not exist.

Table 9.1 Vitamins (Institute of Medicine, 2006)

Vitamin	Effects of inadequate intake and deficiency
Vitamin A (retinol, retinal, retinoic acid, retinoids, provitamin A carotenoids)	Xerophthalmia, night blindness, hyperkeratosis, keratomalacia, increased risk of infection, decreased immune function, and an increased risk of infectious morbidity and mortality
Vitamin B1 (thiamine)	Neurological effects, muscle weakness, weight loss, anorexia, cardiovascular effects (beriberi, Wernicke-Korsakoff syndrome)
Vitamin B2 (riboflavin)	Glossitis, seborrheic dermatitis, cheilosis, anemia (ariboflavinosis)
Vitamin B3 (niacin, nicotinamide, nicotinic acid)	Pigmented rash, diarrhea, neurological effects (pellagra)
Vitamin B5 (pantothenic acid)	Neurological effects
Vitamin B6 (pyridoxine)	Seborrheic dermatitis, microcytic anemia, epileptiform convulsions, depression, and confusion
Vitamin B7 (biotin)	Dermatitis, alopecia, conjunctivitis, neurological effects
Vitamin B9 (folic acid, folate)	Macrocytic anemia, neurological effects
Vitamin B12 (cobalamin)	Pernicious anemia, other hematological effects, neurological and gastrointestinal effects
Vitamin C (ascorbic acid)	Bleeding gums, tooth loss, hemorrhaging, vulnerability to infection, and general wasting (scurvy). In infants: bone abnormalities, hemorrhagic symptoms, anemia (infantile scurvy)
Vitamin D[a] (calciferols)	Rickets, osteoporosis, osteomalacia
Vitamin E (tocopherols and tocotrienols)	Neurological effects, skeletal myopathy, pigmented retinopathy
Vitamin K (phylloquinone and menaquinones)	Bleeding diathesis

[a]Synthesized in the skin if this is exposed to ultraviolet B rays

However, let us consider possible changes in the ecological niche, i.e., of the diet, which lead to a severe lack of an essential nutrient.

There are well-documented examples of this:

- In the past, sailors spent long periods at sea without consuming foods rich in vitamin C, such as fresh fruits and vegetables. These sailors, in an increasing state of vitamin C deficiency, exhibited a variety of symptoms, collectively described as scurvy (bleeding gums, tooth loss, hemorrhaging, vulnerability to infection, general wasting, and ultimately death);
- Today, due to the widespread and constant use of clothing, reduced exposure to sunlight is very frequent, and this causes reduced synthesis and a lack of vitamin D if this is not compensated for with an appropriate dietary supplement.

Consequently, these situations must not be considered as a consequence of a faulty evolution but as the effect of mismatches between the ecological niche to which we are adapted and new life conditions.

Table 9.2 Other essential nutrients (Institute of Medicine, 2006)

Substance	Effects of inadequate intake and deficiency
Essential fatty acids	
alpha-linolenic acid (an omega-3 fatty acid) and linoleic acid (an omega-6 fatty acid)	Scaly skin, dermatitis. Moreover, "Many studies have positively correlated essential fatty acids with … infant development" (Kaur et al., 2014)
Indispensable amino acids	
Isoleucine, leucine, lysine, methionine, phenylalanine, threonine, tryptophan, valine, and histidine	Many functional alterations, edema, failure to thrive in infants and children, poor musculature, dull skin, thin and fragile hair
Choline	Liver damage
Minerals	
Calcium	Reduced bone mass, osteopenia, osteoporosis
Chromium	Weight loss, peripheral neuropathy
Copper	Hematological alterations, osteoporosis in infants and growing children
Fluoride	Increased risk of dental caries
Iodine	Hypothyroidism, goiter, mental retardation, growth and developmental abnormalities, increased childhood mortality
Iron	Anemia, reduced physical work capacity, delayed psychomotor development in infants, impaired cognitive function
Magnesium	Symptomatic hypocalcemia, muscle cramps, neuromuscular hyperexcitability, latent tetany, seizures
Manganese	Scaly dermatitis. In various animal species: impaired growth and skeletal development, impaired reproductive function, and some metabolic alterations
Molybdenum	Infants with defects in molybdenum metabolism do not survive the first days of life or show severe neurological and other abnormalities
Phosphorus	Anorexia, anemia, muscle weakness, general debility, bone pain, rickets in children, osteomalacia in adults
Selenium	Vulnerability to illness and stress
Sulfate	In animals, growth is stunted with a deficiency of sulfate and sulfur amino acids (in particular cysteine)
Zinc	Growth retardation, alopecia, diarrhea, delayed sexual maturation and impotence, eye and skin lesions, impaired appetite
Other minerals (arsenic, boron, nickel, silicon, vanadium)	Functions in humans and effects of deficiency under scrutiny. Studies on animals suggest a role for arsenic in growth and reproduction

9.2.2 End Arteries

An end (or terminal) artery shows no anastomose with other neighboring arteries, so for the served area, there is no other circulation besides that offered by the end artery (Waugh & Grant, 2018).

A functional end artery is when there is some ineffectual anastomose.

In general, in particular for the limbs and the external areas of the body exposed to possible traumas, more peripheral arterial branches supply the same area, and in case of insufficiency of an artery, there is compensation from the nearby arteries.

However, some internal and essential body parts are served by end arteries or functional end arteries. In this regard, important examples are the two renal arteries, several cerebral arteries, the splenic artery, the penile artery, the central retinal artery, and the coronary arteries (Kao & Lee, 2009) (e.g., see Figs. 9.10 and 9.11).

Concerning the heart, the coronary arteries have been regarded as end arteries for decades. However, there are anastomoses between them, known as collateral arteries, which may provide an alternative source of blood supply to the myocardium in cases of necessity (Meier et al., 2013). It has been estimated that only one in four subjects without coronary artery disease (CAD) and only one in three subjects in the case of CAD have sufficient collateral arteries (Pohl et al., 2001; Wustmann et al., 2003). This means that the coronaries are functional end arteries in most subjects without CAD.

End arteries serving organs of vital importance is a risky condition in the event of disease leading to occlusion of the arteries because ischemia of the organs served can occur with serious functional consequences or even death. This condition of risk, which is currently at the origin of many cases of serious impairments or death,

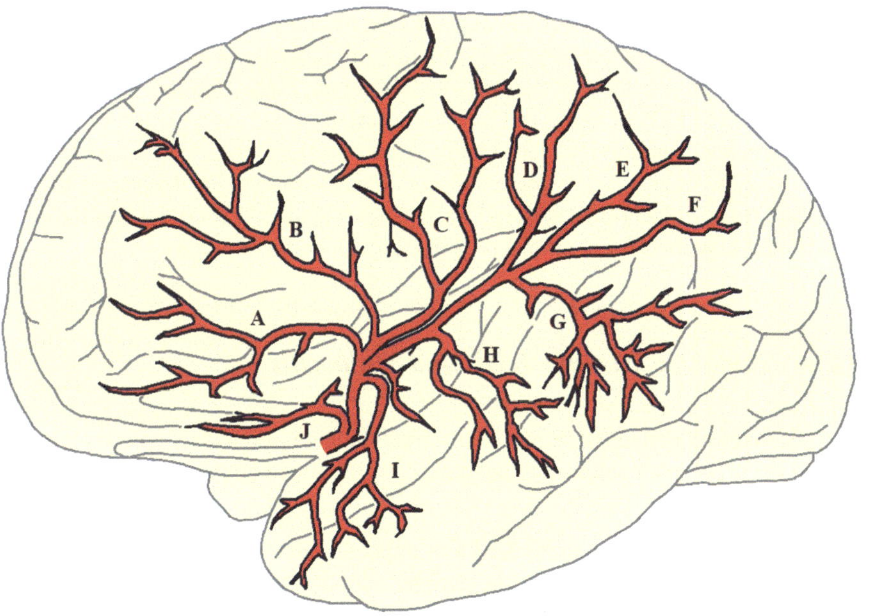

Fig. 9.10 Diagram of some end arteries of the cerebral circulation (A: orbitofrontal artery; B: precentral (pre-Rolandic) a.; C: central (Rolandic) a.; D: anterior parietal a.; E: posterior parietal a.; F: angular a.; G: posterior temporal a.; H: middle temporal a.; I: anterior temporal a.; J: middle cerebral a.)

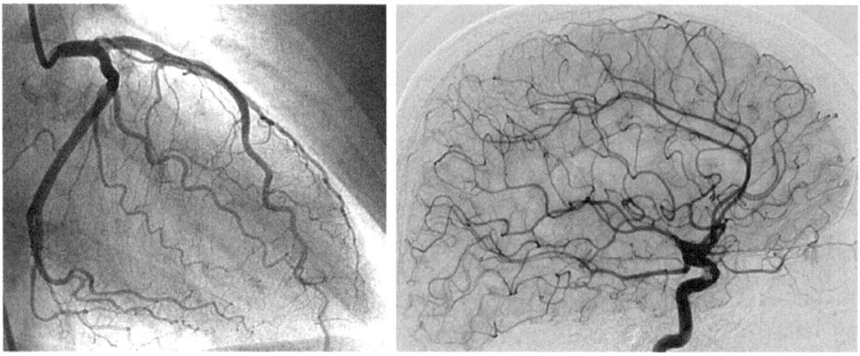

Fig. 9.11 A coronary and a cerebral angiography (on the left and on the right, respectively)

would seem to indicate that natural selection allows for an illogical serious vulnerability of the organism.

This apparent contradiction is easily explained if we remember that the reference to evolutionary normality is indispensable. In fact, in the living conditions to which we are adapted and not in modern living conditions, diseases such as diabetes mellitus, hypertension, dyslipidemia, and atherosclerotic disease, including CAD, practically do not exist, as is documented from the study of populations in the wild (Hill & Hurtado, 1996).

This does not exclude that in the rare subjects reaching advanced age in the wild, they may have fatal diseases due to obstruction of vital arteries as a consequence of the decline of progenitor endothelial cells (Hill et al., 2003), but this is part of the physiological process of aging and its manifestations.

In short, the condition of terminality (anatomical or even just functional) of many arteries serving vital organs is not a manifestation of faulty evolution. However, this can determine serious consequences in altered conditions of the ecological niche, such as those represented in many modern conditions.

9.2.3 Impacted Wisdom Teeth and Malocclusion

The problems related to the third molars (wisdom teeth) are widely spread and correlated with the phenomena of malocclusion, of which impacted wisdom teeth are part:

(i) "According to Elsey and Rock (2000) impaction of the third molar is occurring in up to 73% of young adults in Europe." (Juodzbalys & Daugela, 2013)
(ii) "The incidence of impacted wisdom teeth (third molars) is high, with some 72% of Swedish people aged 20 to 30 years having at least one impacted wisdom tooth. Impacted wisdom teeth occur because of a lack of space, obstruction, or abnormal position." (Dodson & Susarla, 2014)

(iii) "Malocclusion, defined as a handicapping dento-facial anomaly ... is one of the most common dental problems, with high prevalence ranging from 20% to 100% reported by different researchers." (Zou et al., 2018)

These pathologies are explained as a consequence of the reduced space in the jaws, but the reason for the reduced development of the jaws in a large percentage of the population is not suggested. Some report that the reduction in jaw development is due to softer diets in modern populations requiring reduced chewing efforts (Juodzbalys & Daugela, 2013). However, in evolutionary terms, this interpretation would require many generations and would not justify why natural selection fails to contrast a condition that is a source of serious inconvenience, potentially fatal in the absence of medical treatment.

The question was clearly asked in terms of evolutionary medicine in a commentary by Nesse and Williams: "If a large proportion of modern children need orthodontia and then later some require expensive and painful surgery on wisdom teeth, it implies that there is something wrong with their environment. ... Crowded and misplaced incisors and imperfectly erupted wisdom teeth may be diseases of civilization." (Nesse & Williams, 1994, p. 153).

Long before this observation, the subject was addressed by a dentist, Dr. Weston Price, in a famous work, already widely cited in the first chapter of this book, based on research conducted on many populations living in conditions close to those to which our species is adapted, i.e., evolutionarily normal conditions (Price, 1939).

In short, not wanting to repeat what was said in the chapter above, Price documented that in populations living under evolutionarily normal conditions:

(1) The development of the skeleton, and also other functions and systems of the organism, were much better than that of modern populations;
(2) The incidence of problems such as impacted wisdom teeth and malocclusion was minimal;
(3) This was also confirmed by the study of skulls preserved in museums and relating to pre-Columbian American populations and Australian and New Zealand aborigines before contact with Europeans (s. Fig. 9.12);
(4) In contemporary indigenous populations who passed within 1–2 generations from an ancestral to a modern lifestyle, a drastic modification of the state of health was observed with the diffusion in the indigenous populations of the diseases suffered by the Europeans, including dentary diseases, reduced body development and increased vulnerability to infection.

Price interpreted this physical decay, in the natives who had switched to a modern diet but also in the European peoples, as a consequence of the modified lifestyle, in particular the use of industrially processed foods with loss of essential nutritional factors, even if they were richer in calories, fats and sugars (which caused other diseases).

Apart from vitamin D, Price was unable to specify which were the nutritional factors whose deficiency was (and is) at the origin of the aforementioned dental pathologies. Recalling what was said in the previous section *§9.2.1 Necessity of*

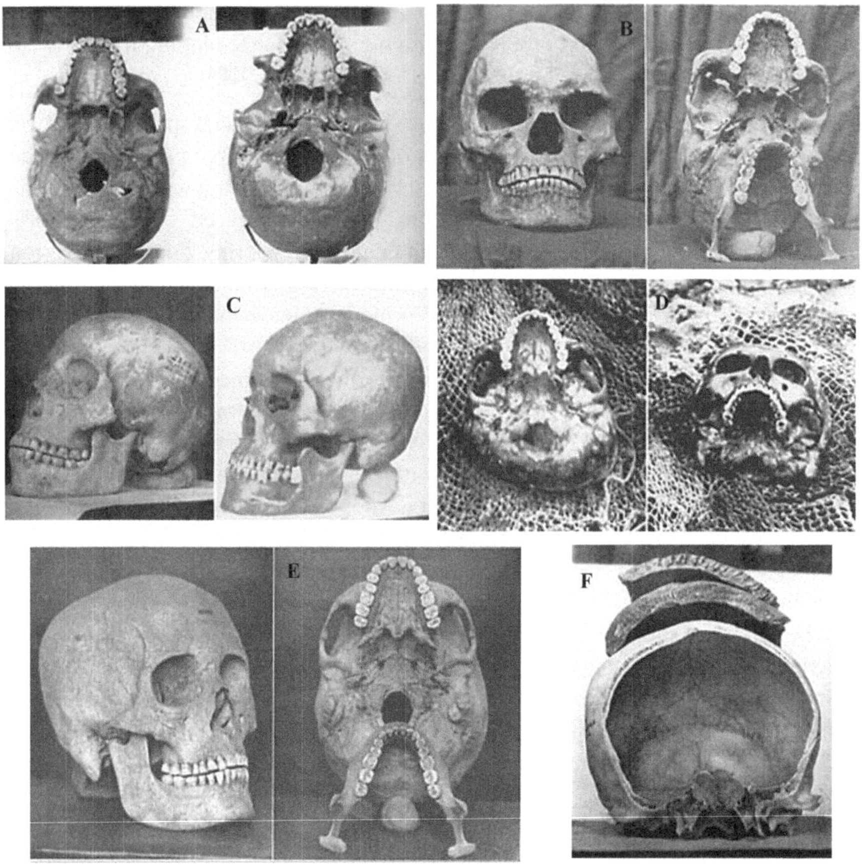

Fig. 9.12 Absence of extreme rarity of malocclusion in people following lifestyles near the ancestral one. (**a**, **c**) Skulls of primitive Indians; (**b**) Skulls of Australian Aborigines; (**d**) Skulls of ancient Chimu of Peru; (**e**) Skulls of ancient Maori of New Zealand; (**f**) Thickness of skulls of pre-Colombian Indians (top) compared with that of modern individuals (bottom). Partial images from Price (1939)

essential nutrients about various substances and minerals, it is possible to observe that for various nutrients, the deficiency causes developmental alterations. In particular, in Tables 9.1 and 9.2, there are sure or possible effects of inadequate intake and deficiency for vitamins A, B1, B12, C, D, and E, essential fatty acids and amino acids, calcium, chromium, copper, iodine, iron, manganese, molybdenum, phosphorus, selenium, sulfate, zinc, arsenic (Institute of Medicine, 2006).

However, there is no precise knowledge of the factors whose deficiencies, combined in various ways, cause the disorders reported by Price. It would be necessary to deepen Price's studies in a more systematic and scientifically rigorous way by systematically comparing populations with modern lifestyles and populations that follow lifestyles and types of nutrition as closely as possible to their ancestral ones.

In any case, as a preliminary point, when we talk about problems such as impacted wisdom teeth and malocclusion (but also about other diseases of teeth and other organs), it would be advisable to read Price's work carefully or even consult the impressive photographic documentation and the materials collected by Price and maintained by the Price-Pottenger Nutrition Foundation (www.price-pottenger.org).

A brief conclusion to this section is that wisdom tooth problems and malocclusion do not result from a defect in selection pressures or human evolution but from drastic lifestyle changes to which our species has not adapted.

9.2.4 Cecal Appendix

The cecal or vermiform appendix, or shortly appendix, is anatomically a small extroversion of the cecum. "The lack of a known function of the appendix has long been associated with the thought that the human appendix is an evolutionary remnant of a cecum that is utilized to ferment plant material. Charles Darwin, for example, noted the apparent lack of a function of the appendix in humans, and concluded that it must be an evolutionary remnant from a primate ancestor that ate leaves (Darwin, 1871). In fact, numerous researchers have argued that the appendix lacks a particular function in humans..." (Smith et al., 2009).

The fact that subjects undergoing appendectomy show no impaired function reinforced the idea that the appendix was a functionless vestigial organ, so much so that the illustrious proponents of evolutionary medicine jokingly declared: "People who recover from appendectomies seem to suffer no disadvantages from not having this part of the human body. The only functional significance of the appendix, as far as we know for sure, is to enable us to have appendicitis." (Nesse & Williams, 1994, p. 129).

However, the appendix is present in many species even phylogenetically distant from humans (Smith et al., 2009), and so the recurring presence of a vestigial organ in the absence of a function would be inexplicable: "... the identification of the structure as a 'recurrent trait' in some clades, the presence of appendix-like structures in monotremes and some non-mammalian species, and consistent features of the cecal appendix such as its narrow diameter, provide direct support for an important function of the appendix." (Laurin et al., 2011).

> Although the appendix has apparently been lost by numerous species, it has also been maintained for more than 80 million years in at least one clade, supporting the idea that the structure has a vital biological function. (Smith et al., 2009)

The modern interpretation is that the appendix has an important immunological function in relationship with the commensal intestinal bacteria: "The worm-like morphology of the appendix and its location in the gut could indicate its long-lasting immunological function by providing a 'safe house' for the commensal intestinal flora, instead of being evidence for its vestigial nature (Laurin et al., 2011; Smith et al., 2009)" (Kooij et al., 2016).

Modern lifestyles, which constitute alterations of our ecological niche, cause inflammation of the appendix (s. *§3.2.1—Diet alterations*) with serious consequences, even fatal if there is no surgical treatment, but this does not invalidate the physiological role of the appendix in evolutionarily normal conditions.

9.2.5 Fluid Retention in the Case of Heart Failure

Heart failure is a well-known syndrome, "a rapidly growing public health issue with an estimated prevalence of >37.7 million individuals globally" (Ziaeian & Fonarow, 2016), with ischemic heart disease and hypertension as significant risk factors in high-income nations (Arrigo et al., 2020).

In full-blown forms with reduced ejection fraction of the heart, there is congestive heart failure with fluid retention, particularly in the lungs and lower parts of the body, shortness of breath, and inability to sustain effort. This fluid retention is caused by a neurohormonal mechanism characterized by the activation of the sympathetic nervous system and the renin-angiotensin-aldosterone system (Hartupee & Mann, 2017).

"In the short-term, these 'neurohormonal' systems induce a number of changes in the heart, kidneys, and vasculature that are designed to maintain cardiovascular homeostasis. However, with chronic activation, these responses result in haemodynamic stress and exert deleterious effects on the heart and the circulation. Neurohormonal activation is now known to be one of the most important mechanisms underlying the progression of heart failure, and therapeutic antagonism of neurohormonal systems has become the cornerstone of contemporary pharmacotherapy for heart failure. ... neurohormonal activation is not simply a marker of disease severity, but also mediates worsening of heart failure." (Hartupee & Mann, 2017).

Nesse and Williams highlighted the apparent inconsistency between this harmful reaction of the organism and the natural selection that should contrast it, rightly observing that this reaction must be evaluated in the context of the ecological niche to which we are adapted and not in a different context such as the one constituted by modern living conditions: "In a natural environment, most instances of deficient blood pumping would result from bleeding or dehydration, in which the fluid retention mechanism would be useful indeed!" (Nesse & Williams, 1994, p. 132). In the modern context, heart failure is instead the consequence of diseases that are nonexistent in evolutionarily normal conditions or of advanced aging states, nonexistent or very rare in the same normal conditions.

Consequently, natural selection does not act against the aforementioned neurohormonal activation, which is adaptive in evolutionarily normal conditions and constitutes an example of mismatch in particular modern pathological conditions.

9.3 Diseases Caused by Conflicting Evolutionary Exigencies

In some cases, natural selection is faced with conflicting evolutionary needs for which either there is no solution, or there has not been sufficient time, in evolutionary terms, to favor effective solutions.

Some of these cases will be discussed in this section.

9.3.1 Modified Hemoglobins or Erythrocytes Protecting Against Malaria

For our species, diseases already existing in earlier times became much more frequent as a result of expansion into new habitats. It is likely that malaria, a parasitosis caused by protozoa of the genus *Plasmodium* (mainly *P. falciparum*—which causes most deaths—*P. vivax*, *P. ovale*, and *P. malariae*), has plagued humans for over 50,000 years (Joy et al., 2003). The disease was known in ancient historical times, as demonstrated by the description of the typical malaria intermittent fever of Hippocrates in *Epidemics* and *Aphorisms* (Pappas et al., 2008).

The disease spread widely as our species expanded into areas where the disease-causing protozoa were present. In fact, malaria is widely present in tropical and subtropical areas, where rains, high temperatures and humidity are frequent, together with the existence of stagnant waters which allow the proliferation of mosquitoes that are vectors of *Plasmodia* (Jamieson et al., 2006). It exerted tremendous selective pressure on our genes, in particular from the origin of agriculture, i.e., from about 10,000 years ago (Hedrick, 2011; Kwiatkowski, 2005).

In the world, malaria is the most important parasitosis and, after tuberculosis, the second infectious disease for both morbidity and mortality. For 2021, the WHO estimated 247 million new cases of malaria, with 619,000 deaths (WHO, 2022).

The evolutionary connection between malaria and particular hemoglobin variants that countered the disease was proposed in the 1940s (Haldane, 1949) and later was defined as the 'malaria hypothesis' (Lederberg, 1999).

Natural selection has favored various mutations, or more precisely, groups of mutations, which by various mechanisms are able to curb the spread and the most serious manifestations of the disease, but often show serious, even fatal, drawbacks. Among these mutations, we have:

- **Sickle-cell or Hemoglobin S (HbS) anemia**

The first genetic condition demonstrated to have such characteristics was sickle-cell anemia, which, in the heterozygous form, allowed survival of the disease while, in the homozygous form, it caused fatal anemia. Consequently, with regard to the frequency of mutations causing sickle-cell anemia, there was a balance between diffusion caused by the greater survival of heterozygous subjects and elimination of homozygous subjects (Allison, 1954a, 1954b; Livingstone, 1958). In 2015, sickle-cell anemia in the homozygous form caused about 114,800 deaths (Wang et al., 2016).

- **Hemoglobin C (HbC) anemia**

HbC is an abnormal hemoglobin that provides protection from malaria caused by *P. falciparum* (Travassos et al., 2015). The heterozygous form does not have the high morbidity and mortality associated with homozygotic HbC (Nagel & Steinberg, 2001).

- **Hemoglobin E (HbE) anemia**

HbE is an abnormal hemoglobin that determines a weak union between α- and β-globin and so an instability when there is a high amount of oxidant. This hemoglobin, widespread in Southeast Asia, would have a protective role against malaria infection, but the evidence is not yet considered sufficient (Ha et al., 2019).

- **Thalassemia (α- and β-thalassemia)**

Over 200 mutations determine various types of thalassemia, generally distinguished between α- and β-thalassemia, which show different degrees of clinical manifestations and protection from malaria (Cao & Galanello, 2010; Origa, 2017). In 2015, it was estimated that there were 280 million people with thalassemia mutations and about 439,000 with severe forms of the disease (Vos et al., 2015).

- **Glucose-6-phosphate dehydrogenase deficiency (G6PDD)**

G6PDD predisposes to hemolysis and is present in more than 400 million people worldwide (Cappellini & Fiorelli, 2008). Individuals with G6PDD are vulnerable, to varying degrees and severity, to hemolytic crises in various circumstances, particularly after the consumption of fava beans, a condition termed favism (Beretta et al., 2023). G6PDD confers protection against malaria (Mehta et al., 2000), which explains why it is not eliminated by natural selection.

- **Pyruvate kinase deficiency (PKD)**

PKD is a genetic alteration of the enzyme pyruvate kinase and predisposes to hemolysis. After G6PDD, the most common cause of enzyme-deficient hemolytic anemia is PKD (Gallagher & Glader, 2016). About the effects of *P. falciparum*, this enzyme deficiency protects against infection and replication in human erythrocytes (Ayi et al., 2008).

- **Hereditary elliptocytosis (ovalocytosis) (HE)**

HE is a group of genetic blood disorders characterized by red cells with altered cytoskeleton and membrane protein integrity and an elliptical shape, sometimes referred to as elliptocytes or ovalocytes (Soderquist & Bagg, 2013). HE severe forms predispose to hemolytic anemia, but the hypothesized association between this disorder and protection against malaria is not proven (Kotepui et al., 2023).

– **Southeast Asian ovalocytosis (SAO)**

SAO, although similar to hereditary elliptocytosis, is distinct from it. SAO confers some resistance to cerebral falciparum malaria and is present in some areas of Malaysia and Papua New Guinea (Allen et al., 1999).

– **Duffy antigen receptor negativity**

The antigens of the Duffy blood group, in particular the Duffy antigen receptor for chemokines (DARC), serve as receptors for *Plasmodium vivax* penetration into erythrocytes. The absence of these antigens, which shows some protection against *P. vivax* infection, is frequent in East Africa (Popovici et al., 2020) and rare among non-African areas such as India (Gai et al., 2019).

The worldwide geographic spread of these genetic modifications is shown in Fig. 9.13. Their exposure and discussion in this section are to be considered partial and summary. They only intend to indicate how selective pressures represented by pathogenic agents can be of extreme importance and impact in determining genetic modifications and pathological conditions.

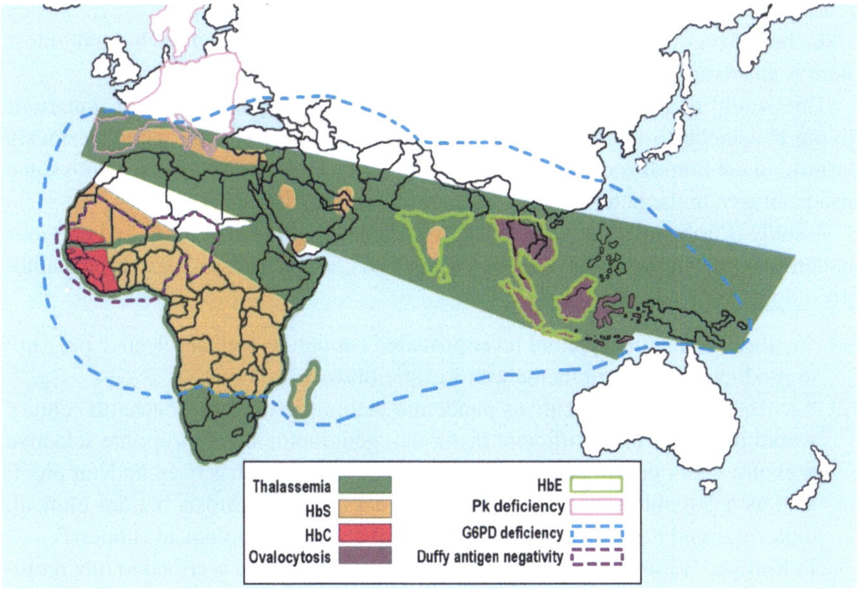

Fig. 9.13 Geographic spread of the main erythrocyte abnormalities with resistance to malaria (elaboration of two images released into the public domain by Armando Moreno Vranich—Files: https://commons.wikimedia.org/wiki/File:Red_Blood_Cell_abnormalities.png; and https://commons.wikimedia.org/wiki/File:Red_Blood_Cell_abnormalities_2.png)

9.3.2 Cystic Fibrosis

Mutations of the gene cystic fibrosis transmembrane conductance regulator (CFTR) in the homozygous form cause cystic fibrosis (CF), a fatal disease without medical treatment, while in the heterozygous form determine the condition of an apparently healthy carrier of the CF gene (Shteinberg et al., 2021).

The homozygous condition is considered rare as it occurs in 1 in 3000 people of European descent (O'Sullivan & Freedman, 2009). However, in Ireland, the heterozygous condition (CF carrier) reaches an estimated rate of 1 in 20 people (Farrell et al., 2007), so CF mutations cannot be considered rare.

About the ΔF508 mutation, which causes most cases of CFTR impairment, it is estimated a lower bound of about 580 past generations for the age of the first mutation and its multi-millennial persistence suggests a selective advantage for the heterozygous condition (Wiuf, 2001).

Moreover, although CF carriers are considered healthy, i.e., without an increased risk for CF-related conditions or other diseases, a study showed an increased risk for many CF-related conditions (Miller et al., 2020), so there is a likely negative selection for CF genes even in the heterozygous condition.

Indeed, it is not explainable how the death of homozygous individuals for the CF genes and the likely negative selection against CF carriers could allow the frequency of the heterozygous condition up to 1 out of 20 people estimated for Ireland unless there is an advantage for CF carriers.

This would mean for CF mutations a condition analogous to that of mutations giving resistance to malaria in the heterozygous condition while being seriously harmful in the homozygous state (Polgreen & Comellas, 2022), and even with some disadvantages in the heterozygous state (Miller et al., 2020).

A study (Poolman & Galvani, 2007) based on CF gene frequencies and previous epidemiological histories of various populations, as well as on aspects of pathophysiology, showed that:

(1) "neither cholera nor typhoid fever provided enough historical selective pressure to produce the modern incidence of cystic fibrosis.";
(2) "... the European tuberculosis pandemic beginning in the seventeenth century would have provided sufficient historical, geographically appropriate selective pressure under conservative assumptions. Tuberculosis has been underappreciated as a possible selective agent in producing cystic fibrosis but has clinical, molecular and now historical, geographical and epidemiological support.";
(3) In Europe, "From 1600 until 1900, 20% of all deaths are ascribed to tuberculosis", and this is compatible with "CF incidence (greater than 1 in 3000)";
(4) "If tuberculosis mortality in India had been approximately 70% of that in Europe, it would produce the observed Indian CF incidence. This lower level of tuberculosis in India is historically plausible (Bates & Stead, 1993)":
(5) In CF carriers, resistance to tuberculosis would arise from the diminished activity of the enzyme arylsulphatase B, depriving *M. tuberculosis* of a necessary nutrient, which counters its spreading (Tobacman, 2003);

(6) "There is also clinical evidence of an association between CF and resistance to tuberculosis. A lower tuberculosis mortality rate has been observed among heterozygotes versus controls … Additionally, the low incidence of tuberculosis in CF patients, particularly in relation to non-tubercular mycobacteria, has been noted by a number of investigators …"

9.3.3 *Morning Sickness*

Morning sickness has already been discussed in *§6.2 Troubles related to pregnancy – 1) Morning sickness*, and it is useless to repeat what has already been written in that section.

It is only worth highlighting that this physiological manifestation is shaped by conflicting evolutionary exigencies. On the one hand, pregnant women must consume as much food as possible that is nutritious and rich in essential substances in order to satisfy the particular needs of pregnancy. As an opposite need, avoiding foods that can be dangerous for the pregnancy and the fetus, especially in the critical weeks of the first organ formation, is important. Morning sickness leads to avoiding dangerous foods. Therefore, it contrasts with the first requirement but satisfies the second one as much as possible.

It should also be noted that the mechanisms of nausea and vomiting in reaction to various types of substances or possible foods are important mechanisms present in all ages and conditions that allow one to avoid dangerous or potentially dangerous foods, even at the cost of a reduced diet. These mechanisms in pregnancy are amplified and appropriately modulated for the different needs of this condition.

9.3.4 *Restrictions in the Biological Availability of Iron*

Problems related to iron metabolism have already been discussed in *§6.2 Troubles related to pregnancy—2) Iron deficiency or withholding*.

Without needlessly repeating what has been said in that section, it is worth emphasizing that the physiology of iron metabolism is also shaped by conflicting evolutionary needs.

On the one hand, iron is indispensable for human physiological needs at every stage of life and even more so during pregnancy, in particular. On the other hand, iron is also indispensable for microbial species, which can be harmful if their multiplication is not slowed down. These conflicting demands shape the metabolic mechanisms that hold back the free availability of iron. The need to limit iron availability to block the possible proliferation of pathogens is even stronger in pregnancy, which explains the more significant restrictions on iron metabolism in pregnancy.

9.4 Enzymatic Deficiencies That Are Not Defects

Some enzymatic deficiencies are not defects but problems deriving from new living conditions. Two widespread examples of these "defects" are discussed in the following sections.

9.4.1 Lactose Intolerance

Most mammals, including our species before the domestication of milk-producing animals, usually cease to produce lactase, an enzyme responsible for the digestion of lactose in milk, and become lactose intolerant after weaning (Swallow, 2003).

In some populations, with the availability of milk produced from farmed animals around 10,000 years ago, the ability to digest milk also in adults became important. So, in several populations independently, the persistence of lactase activity even after weaning was favored by natural selection (Ingram et al., 2009; Ségurel & Bon, 2017).

Lactase persistence is common in people of European descent (in particular from Northern and Central Europe) but also in some parts of Africa and the Middle East (s. Fig. 9.14). However, about 65% of individuals worldwide do not have lactose persistence, i.e., are lactose intolerant (Bayless et al., 2017; Storhaug et al., 2017).

In evolutionary terms, lactose intolerance cannot be regarded as a genetic defect. In reality, in adults, it is the normal ancestral condition for our species, while the

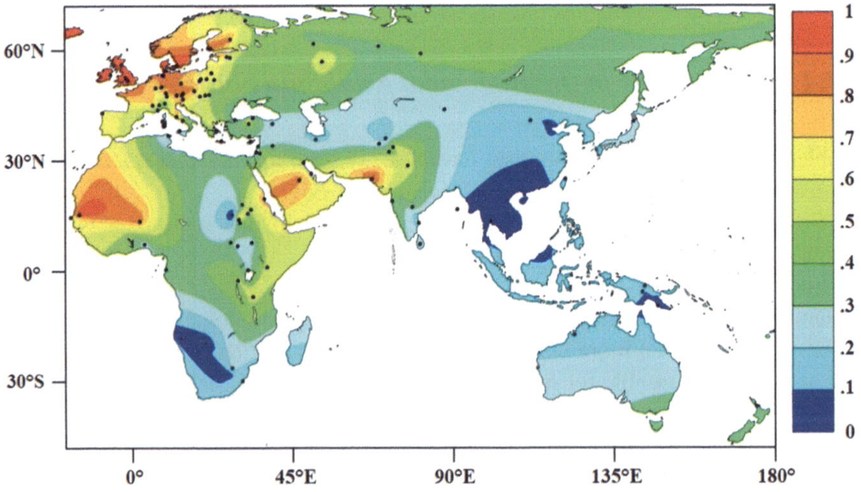

Fig. 9.14 Interpolated map of Old World lactase persistence phenotype frequencies. Dots indicate collection locations. Image from Itan et al. (2010), reproduced under the conditions of Creative Commons Attribution License (http://creativecommons.org/licenses/by/2.0)

ability to digest milk is normal only before weaning. Lactase persistence is a valuable adaptation—in some populations—to the availability of milk from domesticated animals.

9.4.2 Altered Metabolization of Alcohol

An important mitochondrial metabolic pathway transforms alcohol into acetaldehyde by the enzyme alcohol dehydrogenase (ADH) and then transforms acetaldehyde into acetic acid by the enzyme aldehyde dehydrogenases 2 (ALDH2) (Fig. 9.15). "The second step, catalyzed by ALDH2, is the rate-limiting step in ethanol metabolism, and it takes ~1 h to metabolize the amount of ethanol that is found in a single alcoholic beverage." (Chen et al., 2014).

The ALDH2*2 mutation is the predominant dysfunctional variant of ALDH2 and is present in an estimated 540 million East Asians (about 8% of the world population) (Brooks et al., 2009). There are other dysfunctional variants, among which five common ALDH2 variants present in groups of Latino, Finnish, African, and South Asian ancestries, with allele frequencies ranging from 0.6% to 2.5% and an estimated 120 million people (about 1.5% of the world population) (Chen et al., 2020).

Subjects with ALDH2 deficiency, in case of alcohol intake, show elevated blood acetaldehyde levels and toxic manifestations due to this substance (facial flushing, palpitation or tachycardia, nausea, headache, and muscle weakness), as already demonstrated since 1981 (Brooks et al., 2009; Harada et al., 1981).

The defective metabolization of aldehyde and its toxic effects in general lead to avoiding alcohol intake (Edenberg et al., 2019), but this can be overcome by the development of tolerance for aldehyde toxicity due to cultural and social factors (Brooks et al., 2009; Yokoyama et al., 2002), which can also increase their influence with changing conditions. E.g., a study reported that among Japanese alcoholics, those with inactive heterozygous ALDH2 were 2.5% in 1979, 8.0% in 1986, and 13.0% in 1992 (Higuchi et al., 1994).

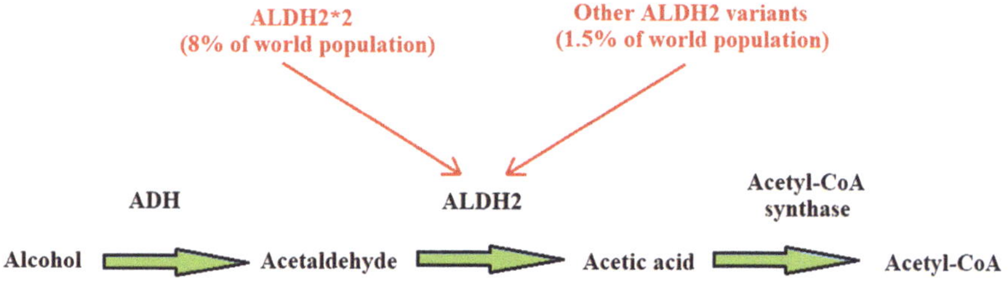

Fig. 9.15 The metabolic pathway of alcohol. Acetaldehyde shows toxic effects if, by effect of dysfunctional variants of ALDH2, it is not transformed into acetic acid

The "subgroup of ALDH2*2 heterozygous alcohol drinkers is at higher risk of many alcohol-related diseases compared to ALDH2*2 heterozygous individuals who avoid alcohol consumption." (Chen et al., 2022).

"The spectrum of different diseases associated with ALDH2 deficiency and aldehyde toxicity is well covered by many recent reviews, and includes a higher risk of cancer and cardiovascular disease … Indeed, East Asians carrying the ALDH2*2 variant are at higher risk of esophageal, head and neck, and lung cancer if they regularly consume alcohol compared to people with the wild-type ALDH2 (Baan et al., 2007; IARC Working Group on the Evaluation of Carcinogenic Risks to Humans, 2010; Im et al., 2022)." (Chen et al., 2022).

While moderate alcohol consumption appears to have a protective effect against coronary heart disease and stroke (Larsson et al., 2016; Roerecke & Rehm, 2012), in ALDH2*2 subjects, this protection is not shown and even with a low alcohol intake, there is a strong positive relation with ischemic or hemorrhagic stroke, and with higher blood pressure (Millwood et al., 2019).

A genetic study has shown that, in the Japanese population, the *ALDH2*2* variant strongly changed its allele frequency within the past 2000–3000 years (Okada et al., 2018).

The high frequency of ALDH2*2 in East Asia and ALDH2 mutations with analogous effects in various other populations and the positive millennial selection for ALDH2*2 demonstrated for the Japanese population require one or more advantages to justify this selection. There are various hypotheses in this regard:

(1) ALDH2 deficiency appears to give protection against active (i.e., clinical) tuberculosis (Park et al., 2014). This protective effect could be explained by the adverse reactions caused by alcohol consumption in ALDH2*2 subjects as these reactions "likely lead to reduced alcohol abuse, which in turn helps to maintain better general health and reduces the risk of TB …" (Park et al., 2014);
(2) For the observed strong relation between ALDH2*2 and hepatitis B viral infection in East Asia, it was suggested that the aversion to alcohol caused by ALDH2 deficiency exerted a protective effect against the negative evolution of hepatitis B in active forms of the disease and then liver cirrhosis and liver cancer (Lin & Cheng, 2002);
(3) The protective effect of *ALDH2*2* on atrial flutter, ischemic stroke, and cirrhosis (Sakaue et al., 2021) could be explained in the same way;
(4) As previously reported, in a similar way, the aversion to alcohol caused by ALDH2 deficiency allows people to avoid the cases of cancer associated with alcohol abuse. On the contrary, in subjects with ALDH2 deficiency who have developed a tolerance for aldehyde toxicity, the risks related to aldehyde toxicity alcohol abuse are significantly increased.

It has been observed that the historical frequency increase of ALDH2*2 in East Asia corresponds to the period of agricultural domestication of rice and so to an increased possibility of producing alcoholic beverages: "Grain was a common ingredient used to make alcohol. Some researchers have theorized that rice farmers may have had earlier access to the excess grain used to make alcohols like rice wine as early as 9000 years ago (Henrich, 2015; Peng et al., 2010)" (Zhu et al., 2021).

These data certainly need further investigation. Among other things, it is necessary to explain or document better:

(i) The causes of the spread of ALDH2 variants different from ALDH2*2 in areas other than the Far East;
(ii) Why the diffusion of ALDH2 variants, despite a large number of individuals presenting them, is limited to some world areas or populations.

However, a conclusion is already possible. The dysfunctional ALDH2 mutations are not to be considered enzymatic defects but constitute an evolutionary adaptation to new life conditions, which have been going on for millennia.

9.5 Childbirth and Newborn's First Months of Life

Modifications of the ecological niche, which may determine possible mismatches and diseases, can concern any aspect of life. In this regard, two considerations are fundamental:

(1) Any modification must be regarded as a possible cause of mismatch, which implies that it must be considered an alteration of the ecological niche until the opposite is eventually demonstrated. Deeming a modification as positive on the basis of unproven assumptions must be absolutely avoided, especially when expressing opinions or advice presented as scientific;
(2) As already discussed in previous chapters of this work, each of us is part of a holobiont of which countless other species are also part. Furthermore, in the relationships between mother and newborn, it is necessary to remember the complications of the existence of both the mother's holobiont and that of newborn's holobiont in formation.

Where one ignores or underestimates these considerations, fundamental for all evolutionary medicine, errors of evaluation can be made, and wrong conclusions can be reached, which certainly cannot be considered of scientific value.

These same considerations take on particular importance in the study of the periods of life of mother and newborn that extend from before to after birth.

9.5.1 Before Birth

There are alterations in the physiology of the mother and her holobiont that are often overlooked, underestimated, or even mistakenly considered normal:

– **Alterations of the holobiont**

• *Modifications of gut microbiota*

As antibiotics act against bacteria, it could be assumed that—barring specific intolerances, allergies, and side effects—they are not harmful to the user. However, antibiotics severely alter the intestinal microbiota (Blaser, 2016) to an extent and in ways that are only partially known and vary depending on the antibiotic used (Maier et al., 2021). Comparing the gut microbiota of individuals of a hunter-gatherer population living in conditions close to the original ones (Hadza people, Tanzania), and in any case not using antibiotics, with those of a modern population (urban Italians) which uses antibiotics, there is an extraordinary difference between the two groups, especially with a huge depletion of the variety of the microbiota in individuals of the second population (Schnorr, 2015) (Fig. 9.16).

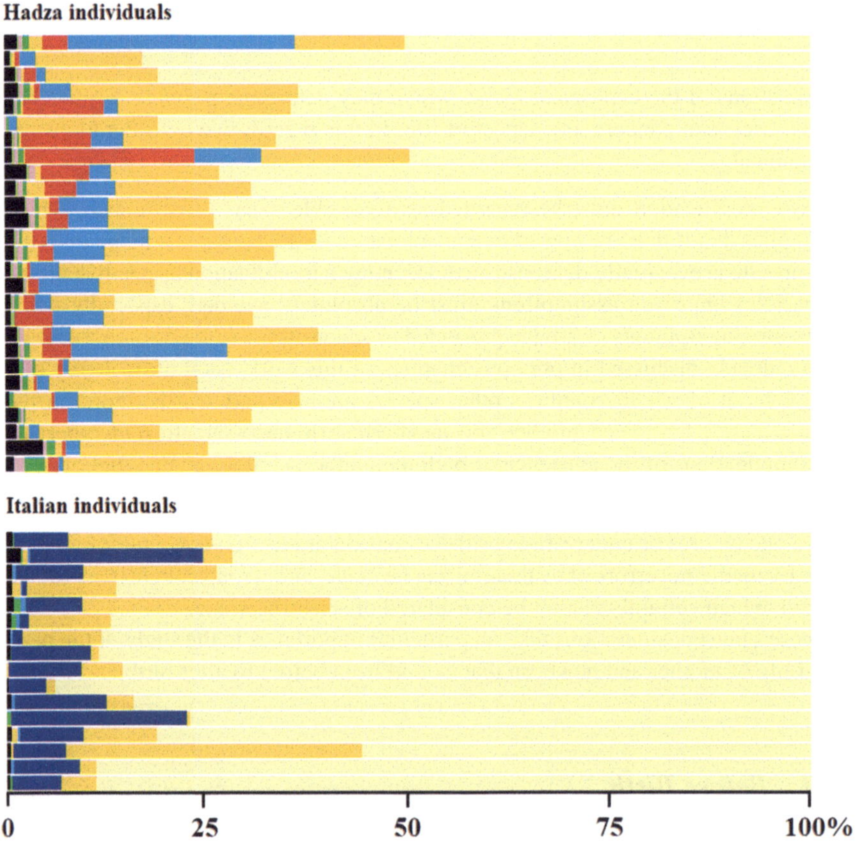

Fig. 9.16 Comparisons between the microbiota of individuals of a hunter-gatherer people (Hadza, Tanzania) and a modern population (urban Italians). Each horizontal bar shows the gut microbiota of an individual. The different colors represent different phyla of microbes. The extreme impoverishment of the variety of species is evident in the modern population. Figure from Schnorr (2015), modified and redrawn

The use of antibiotics before, during, and after childbirth is considered useful for preventing and fighting infections, assuming this does not cause particular damage, but it is not considered that the effects of antibiotics modify the microbiota of the parts of the body in which they act.

- *Modifications of skin microbiotas*

The skin microbiotas are continuously modified by the use of soaps and detergents, especially for the private parts. The use of these substances is considered a factor that is surely non-pathogenic and protects the state of health.

- *Absence of intestinal parasites*

Macroparasites are usually rare or absent in modern populations. This is regarded as a factor that improves health and is certainly not causing disease.

With regard to the modifications of the holobiont, it was discussed in *Chap. 4—Diseases deriving from alterations of the ecological niche—Part II: Regarding the holobiont*, that they are not at all indifferent to the state of health and indeed they are an evident or probable cause of widespread illnesses (e.g., allergies, autoimmune diseases, obesity, autism, skin diseases).

– **Reduced physical activity with poor tone of the abdominal walls**

Under natural conditions, there is certainly continuous physical activity, and the poor tone of the abdominal muscles is unlikely. Strong abdominal muscles are likely to contribute to efficient delivery. Nonetheless, a common medical recommendation for pregnant women was to avoid physical exertion for fear that this could trigger a premature birth. However:

– "Exercise contributes significantly to maternal and fetal wellbeing during pregnancy. Traditionally women were advised to refrain from exercise during pregnancy, but newer evidence has shown this to be false. Theoretically, there were concerns about premature labor and the risks of delivering smaller infants to women who exercised during pregnancy. A recent meta-analysis that included over 2000 women showed that aerobic exercise and moderate-intensity strength/toning exercise performed 3 or 4 days per week throughout the pregnancy was not associated with an increased risk for preterm births or low birth weight infants for normal-weight women with singleton, uncomplicated pregnancies." (Cooper & Yang, 2022);
– "Pilates exercise during pregnancy improved the labor process and increased maternal satisfaction of childbirth process, without causing complications for the mother and baby." (Ghandali et al., 2021);
– "… higher physical activity and lower sedentary time were associated with improved health outcomes. Encouraging pregnant women to increase their physical activity and decrease their sedentary time, may be important factors to improve maternal and fetal/child health outcomes." (Meander et al., 2021);
– "Exercising women generally have similar or stronger PFM [pelvic floor muscles] strength and larger levator ani muscles than non-exercising women, but this

does not seem to have a greater risk of obstructed labor or childbirth. Additionally, women that specifically train their PFM while pregnant are not more likely to have outcomes associated with obstructed labor." (Bø & Nygaard, 2020)

Furthermore, a sedentary lifestyle is associated with some maternal diseases and harmful consequences for the fetus:

– "A sedentary lifestyle during pregnancy is associated with the significant health risk of deep venous thrombosis, which threatens the lives of the infant and mother. A sedentary lifestyle can lead to obesity or at least significantly contribute to persistent obesity. Obesity is of epidemic proportions in the U.S., and obesity during pregnancy carries elevated risks for major complications. Patients with obesity are also more likely to suffer spontaneous abortions. They also have a higher risk of neural tube defects, including cleft palate, spina bifida, and hydrocephalus. Obesity also puts the pregnant patient at risk for gestational diabetes, preeclampsia, sleep apnea, macrosomia, preterm birth, and even stillbirth. The higher the women's BMI, the greater the risk of stillbirth. Exercise is safe in these women and is encouraged." (Cooper & Yang, 2022)

– **Excessive iron availability**

In pregnancy, slight anemia is physiological but is often treated with iron therapy because it is interpreted as a consequence of an iron deficiency. It was shown in section *§6.2 Troubles related to pregnancy—2) Iron deficiency or withholding*, that reduced availability of iron is a defense against pathogenic infections, which is particularly important in pregnancy, and that the administration of iron can be dangerous in certain cases.

9.5.2 During Childbirth

Even the childbirth period shows various alterations of the ecological niche, often induced or determined by medical practice.

– **Childbirth under aseptic conditions**

For vaginal delivery, it appears proper and necessary to recommend aseptic conditions of motherly intimate parts, as this is considered a sure improvement compared to primitive conditions. However, the first colonization of the newborn, which is the first phase of the formation of his holobiont, occurs during childbirth through contact with maternal mucous and skin surfaces. Consequently, the asepsis of these surfaces hinders the correct formation of the newborn's holobiont, particularly if this is associated with the physical separation of the newborn from the mother and the use of artificial milk. All this cannot be considered irrelevant to the future health of the newborn, as already discussed in *§4.1.3 Alterations of the holobiont—C: Alterations of the microbiome at birth.*

– **Childbirth in a horizontal and not upright position**

The natural position for birth is the upright one, with the mother crouched as if she had to evacuate in a squat toilet (Fig. 9.17). This position is the one that best exploits the force of gravity and appears to be more physiological and advantageous than the position preferred by the gynecologist (lying on the back with the legs raised with respect to the pelvis) (Berta et al., 2019; Desseauve et al., 2017; Gupta et al., 2017).

– **Episiotomy**

Episiotomy is the surgical enlargement of the vaginal orifice during childbirth to avoid harmful vaginal lacerations. This practice is widespread, but there is no evidence of its usefulness except in particular cases (Carroli & Belizan, 2000):

(1) "It is one of the most commonly performed surgical procedures in the world. … Maternal benefits of episiotomy include reduced risk of perineal trauma, subsequent pelvic floor dysfunction and prolapse, urinary incontinence, faecal incontinence, and sexual dysfunction. Potential benefits for the foetus are thought to include a shortened second stage of labour. However, an episiotomy can also

Fig. 9.17 The Dumbarton Oaks birthing figure (left side) shows a woman giving childbirth in a squatted position. Although considered for a long time an Aztec pre-Columbian artwork, it is likely fake (MacLaren Walsh, 2008), i.e., a modern artifact inspired by Aztec manuscripts (right side; image from the *Codex Borbonicus*, folio 13, original in the Bibliotheque de l'Assembée Nationale, Paris)

lead to potential adverse consequences, including the extension to a third- or fourth-degree tear, anal sphincter dysfunction, and dyspareunia." (Choudhari et al., 2022);

(2) "Out of 235 patients, immediate complications [of episiotomy] were reported in 10 (4.3%) patients. These included perineal tears, postpartum haemorrhage, extended episiotomy, perineal pain, inability to pass urine or stool and vaginal haematoma. Early complications including wound infection, gaping wound and resuturing of wound were reported in 21 (8.9%) patients." (Khan et al., 2020);

(3) Episiotomy gives rise to many legal claims on charges that it has been practiced improperly: "... most episiotomy-related lawsuits stemmed from a routine use of that procedure, which is almost universally advised against ..." (Zaami et al., 2021).

A rigorous study comparing vaginal deliveries in populations living in primitive conditions and modern populations in which the episiotomy practice is used would be useful. From general considerations, an apparent difference between primitive and modern populations includes the mother's age at the first childbirth, which is associated with higher tissue elasticity at younger ages and, oppositely, with higher traumatization risk at older ages (Rahmanou et al., 2016). Other concerns include a relatively frequent delivery of the subsequent children, the period between childbirths being shorter than the one required to lose muscular extension and memory, and thus, the adaptations acquired during the previous childbirths. In this regard, the contemporary means of birth control, although nowadays indispensable, have some predictable adverse effects, leading to the conditions in which the episiotomy practice might seem justified. However, in the past decades, several pelvic floor training exercises, massage complexes, and devices aimed at preventing perineal trauma have been developed (Cao et al., 2023; Leon-Larios et al., 2017). Such atraumatic approaches show promising results and can be an adequate alternative to episiotomy.

– **Cesarean section**

Childbirth by cesarean section (CS) as an alternative to the vaginal route is, in many cases, essential to save the life of the child or the mother, or both, or to avoid dangerous or disabling outcomes. In other cases, CS is appropriate to avoid risk conditions.

However, CS rates have been increasing for decades. Often, this type of childbirth is practiced on the basis of improper motivations, and it causes complications:

(1) "A caesarean section (CS) can be a life-saving intervention when medically indicated, but this procedure can also lead to short-term and long-term health effects for women and children. Given the increasing use of CS, particularly without medical indication, an increased understanding of its health effects on women and children has become crucial ... The prevalence of maternal mortality and maternal morbidity is higher after CS than after vaginal birth. CS is associated with an increased risk of uterine rupture, abnormal placentation, ectopic pregnancy, stillbirth, and preterm birth, and these risks increase in a

dose-response manner. There is emerging evidence that babies born by CS have different hormonal, physical, bacterial, and medical exposures, and that these exposures can subtly alter neonatal physiology. Short-term risks of CS include altered immune development, an increased likelihood of allergy, atopy, and asthma, and reduced intestinal gut microbiome diversity." (Sandall et al., 2018)

(2) "In present-day obstetrics, cesarean delivery occurs in one in three women in the United States, and in up to four of five women in some regions of the world. … In the United States, as the cesarean delivery rate has increased, maternal mortality and morbidity have also risen steadily over the last three decades …" (Clarel & Bruce, 2021)

(3) "The caesarean section (CS) rate continues to increase across high-income, middle-income and low-income countries. … Latest available data (2010-2018) from 154 countries covering 94.5% of world live births shows that 21.1% of women gave birth by caesarean worldwide, averages ranging from 5% in sub-Saharan Africa to 42.8% in Latin America and the Caribbean. … Projections showed that by 2030, 28.5% of women worldwide will give birth by CS (38 million caesareans of which 33.5 million in low-middle income countries annually) ranging from 7.1% in sub-Saharan Africa to 63.4% in Eastern Asia." (Betran et al., 2021)

The increasing rate of CSs is due to various factors, often overlooked:

(i) Precise definition of the date and time of delivery, according to the wishes and convenience of doctor and patient;
(ii) Greater remuneration of CSs;
(iii) Assumed greater safety of CS than natural delivery in any case;
(iv) Underestimation of possible damages for the mother;
(v) Underestimation of possible damage to the development of the newborn (Sandall et al., 2018).

– **Use of oxytocin in childbirth**

Various hormones, including progesterone, estrogens, prostaglandins, corticotropin-releasing hormone, relaxin, and oxytocin, cooperate for the onset of labor (Vannuccini et al., 2016).

Oxytocin, in particular, is important, not just for labor: "Especially the onset of labor, the progress of labor and initial breastfeeding are mediated by oxytocin. In the maternal brain it controls the initiation of the mother-infant bond and the mother's emotional responses towards her child." (Walter et al., 2021).

Oxytocin is secreted by the neurohypophysis with pulses increasing in frequency and amplitude during childbirth, with a maximum of three pulses every 10 min just before delivery (Uvnäs-Moberg et al., 2019).

Synthetic oxytocin is widely used to stimulate delivery, to make it faster (in theory, to reduce the suffering of such an event), and also to reduce postpartum bleeding (Charles et al., 2019): "Synthetic oxytocin is widely used throughout labor and postpartum care in modern obstetrics. Yet research on the implications beyond labor of maternal exposure to perinatal synthetic oxytocin is rare." (Bell et al., 2014).

Regarding the use of oxytocin to stimulate and accelerate childbirth:

(1) "Two aspects should be considered when administering synthetic oxytocin during childbirth. First, the dosage of synthetic oxytocin should not exceed the physiological blood plasma concentration of 9 mU/min (Fuchs et al., 1983; Uvnäs-Moberg et al., 2019). A recent study by Daly and colleagues (Daly et al., 2020) has shown that the dosage of synthetic oxytocin during the onset of labor and the subsequent birth varies substantially between clinics and countries. In Germany, dosages up to 27 IU within eight hours [i.e., >56 mU/min] have been documented. Secondly, a constant infusion of synthetic oxytocin results in a flattening of the natural oxytocin pulses (Uvnäs-Moberg et al., 2019)." (Walter et al., 2021)
(2) "Use of early oxytocin resulted in a statistically significant reduction in the mean duration in labour of approximately two hours but did not increase the normal delivery rate. … For women making slow progress in spontaneous labour, treatment with oxytocin as compared with no treatment or delayed oxytocin treatment did not result in any discernable difference in the number of caesarean sections performed." (Bugg et al., 2013)
(3) "Fifty percent of the results (17 of 34 measures) demonstrated an association between exogenous oxytocin and less optimal breastfeeding outcomes, while 8 of 34 measures (23%) reported no association." (Erickson & Emeis, 2017)
(4) "While the notion of drug abuse tends to be applied more to substances used outside of medical settings, there is increased concern that synthetic oxytocin is being overused in maternity care settings." (Wickham, 2015)
(5) "Although it is not recommended by the World Health Organization (WHO), synthetic oxytocin is routinely used in many Brazilian maternity services to boost and speed contractions, which has been reported in several studies (Diniz, 2001; Hirsch, 2015; Tornquist, 2004; Leal & Gama, 2014). In a study performed at a public maternity hospital in Salvador [Brazil], McCallum and dos Reis (2008) observe that oxytocin … was dreaded by women for increasing their pain and discomfort. Diniz (2005) similarly shows how the pain of childbirth becomes largely iatrogenic; in other words, it is amplified when oxytocin or other routine interventions are implemented. Today in Brazil, activists in the 'humanized childbirth' movement and researchers in collective health in general are warning of excessive interventions during childbirth." (Nucci et al., 2018)

The use of oxytocin is part of a group of practices that have been strongly criticized because they increasingly medicalize the physiological event of childbirth with advantages for healthcare professionals and hospital facilities while often damaging those who suffer them:

"Meanwhile, the minority condemning birth interventions (especially the use of oxytocics to accelerate or induce labor) most commonly argued that they embodied a dangerous attitude that only addressed the convenience of the obstetric physician, as illustrated in the excerpts below from two different articles:

> The acceleration of labor that progresses normally, without complications, only for the convenience of the attending staff, is not justified. We refer to the use of oxytocic drugs of any kind for this purpose (Novis, 1957, p. 1015). In the United States, certain obstetricians gather all their patients who are about to give birth and induce labor early so they are not deprived of excursions, fishing trips, travel, and their own weekend pleasures. And so the pregnant women are meekly hospitalized in larger or smaller batches to be subjected to the unprecedented violence of an extemporaneous delivery, only so that their obstetrician can enjoy the delights of the weekend! … (Correa, 1955, p. 26). …

The amniotic sac was seen as a 'hindrance to the smooth progress of childbirth' (Ribeiro, 1962), to be broken by the obstetrician (in a procedure known as amniotomy) to accelerate delivery. As can be seen below, this was yet another way to intervene in the birth process and increase the active role of the obstetrician:

> Early artificial rupture of the membranes is routine in any labor planning. In doing so, 'we take the initiative' in the labor process, precipitating the appearance of physiological moments of myometrial activity, which otherwise would take place with 'undesirable natural slowness' (Greele, 1961), p. 97; highlights ours.

As for positioning the woman to give birth, 'lithotomy' (lying belly-up, with legs open and flexed) was preferable because:

> it widens and better exposes the perineum and gives 'the doctor freedom of movement.' It facilitates local infiltration of anesthesia, as well as episiotomy and its suture (Adeodato Filho, 1963), p. 11; emphasis added.

In this case, it should be noted that greater freedom of movement for the doctor necessarily opposes the laboring mother's freedom of movement. Therefore, like Emily Martin (Martin, 2006), we see that physicians increasingly interfere in labor with the use of different techniques and technologies. At the same time, the woman was relegated to the background and seen as possessing a 'machine' that was to be 'repaired' or 'improved' by the obstetrician. One drawback of accelerating labor with oxytocin is an increase in pain, which in turn could be reduced with the use of anesthetics. …

Episiotomy was widely defended among the authors analyzed …

> Fortunately, its [episiotomy] use is increasingly growing. The classic and obsolete maneuvers to 'protect the perineum' now give way to the wider practice of episiotomy. We will not expound considerations on the subject, which currently seems to be a **pacifying stitch**. … (Ribeiro, 1962), p. 110–111; emphasis in the original" (Nucci et al., 2018)

9.5.3 First Days and Months of Life of the Newborn

The early stages of the newborn's life suffer from the consequences of harmful changes to evolutionarily normal living conditions, often induced on the basis of false assumptions and presented as useful improvements.

– **Use of infant formula**

For many years, infant formula (or, simply, formula) has been presented as a valid alternative to breast milk, and its use has become widespread for questionable advantages related to social status, convenience, and modernity (Fildes, 1985). The use of infant formula has declined in recent decades, but considering its long-known disadvantages and harmful consequences (Cunningham, 1995), it is still too widespread:

(1) "In low-income and middle-income countries, only 37% of children younger than 6 months of age are exclusively breastfed. With few exceptions, breastfeeding duration is shorter in high-income countries than in those that are resource-poor." (Victora et al., 2016)
(2) "In a systematic review published in 2003, Jones et al. (2003) predicted that exclusive breastfeeding through the first six months could prevent 13% of childhood deaths in children younger than 5 years of age." (Kim, 2021);
(3) "About one in three neonates in low-income and middle-income countries receive prelacteal feeds, and only one in two neonates are put to the breast within the first hour of life. Prelacteal feeds are strongly associated with delayed initiation of breastfeeding. Self-reported insufficient milk continues to be one of the most common reasons for introducing commercial milk formula (CMF) and stopping breastfeeding." (Pérez-Escamilla et al., 2023)
(4) "Key structural barriers that undermine the breastfeeding environment include gender inequities; harmful sociocultural infant-feeding norms; income growth and urbanisation; corporate marketing practices and political activities that weaken breastfeeding protection policies; labour markets that poorly accommodate women's reproductive rights and care work, reflecting major gender inequities; and poor health care that continues to undermine breastfeeding, including the medicalisation of birthing and infant care." (Pérez-Escamilla et al., 2023), references omitted.

The advantages of breastfeeding are solid and evident and, moreover, mother's milk has a sophisticated composition that modern and elaborated formulas have only partially managed to imitate:

(A) "The importance of breast milk for the growing infant is undisputed; breastfeeding decreases infantile mortality by tenfold and decreases the incidence of infectious diseases. Despite its recognized benefits, the structural richness of breast milk has also impeded the characterization of the multiple effects of milk components on infant physiology." (Hennet & Borsig, 2016);
(B) "The World Health Organization recommends exclusive breastfeeding for the first six months of life, followed by continued breastfeeding with appropriate complementary foods for up to two years or beyond" (Salminen et al., 2020);
(C) "We can adjust and further improve infant formula, but we will not be able to mimic breast milk or the art of breastfeeding because of its complexity and its effect on infant physiology, including reducing the risk of obesity, among other benefits (Hennet & Borsig, 2016). Breast milk is the gold standard of infant

nutrition and provides a mixture of unique components required for the optimal development of the infant. Breast milk is beneficial for short- and long-term health of both infants and their mothers. Therefore, the support and promotion of breastfeeding by healthcare professionals are of paramount importance." (Kouwenhoven et al., 2022);

(D) The benefit of breastfeeding is not limited to newborns only: "For nursing women, breastfeeding gave protection against breast cancer and it improved birth spacing, and it might also protect against ovarian cancer and type 2 diabetes. The scaling up of breastfeeding to a near universal level could prevent 823,000 annual deaths in children younger than 5 years and 20,000 annual deaths from breast cancer. Recent epidemiological and biological findings from during the past decade expand on the known benefits of breastfeeding for women and children, whether they are rich or poor." (Victora et al., 2016);

(E) Formula feeding is associated with gut microbiota alterations in the newborn and a higher risk of infections (Pärnänen et al., 2022);

(F) In breast milk, over 70 different types of special oligosaccharides (HMOs) promoting the growth of beneficial intestinal bacteria were identified some years ago (Wu et al., 2010; Wu et al., 2011). The number of identified HMOs is now about 200, of which more than 160 types are characterized in detail (Urashima et al., 2018), and they have shown multiple complex functions (Salminen et al., 2020). HMOs quantitatively constitute the third most abundant component of human milk (Wegh et al., 2019). The addition of suitable oligosaccharides to the infant formula tries to compensate at least in part for their deficiency of HMOs (Salminen et al., 2020; Estorninos et al., 2022);

What has been said so far overlooks the great importance of colostrum, or first milk, i.e., the mother's milk from the first days of lactation. Colostrum, markedly different in composition from ordinary milk, is rich in a long series of substances vital for the infant's defense against infection and the development of the immune system (Kim, 2021).

– **Insufficient physical contact or closeness of the newborn with the mother**

Physical contact or close proximity between newborn and mother from the first moments after birth and for many months is the rule in natural conditions. It is of fundamental importance for the correct development of the newborn.

There are some modern habits, often presented as improvements for the infant and mother, which undermine this contact with regrettable negative consequences for the infant:

(1) The newborn is often removed from the mother immediately after delivery for presumed better health care for both mother and child, and taken to a postnatal ward where its health conditions would be better guaranteed. Subsequently, it is brought back to the mother for short periods to allow natural breastfeeding, except for an insufficiency of the mother's milk, which forces it to be integrated or replaced with infant formula. This practice compromises the correct breastfeeding initiation (Moore et al., 2016).

Furthermore, suppose the baby in the postnatal ward cries. In that case, it is appeased with the administration of infant formula, which hinders the initiation and continuation of normal breastfeeding (Royal College of Midwives, 2002). Finally, it should be remembered that mother-infant physical contact is important for the baby's sensorineural development and the formation of his microbiomes (Ball & Klingaman, 2008.)

(2) On the basis of preconceived ideas on the best way to stop crying in a newborn and foster self-reliance, it was recommended to let the newborn sleep not next to the mother (or rather in another room), following certain temporal patterns for breastfeeding. Also, if he cried before the scheduled time had elapsed, it was necessary to let him cry: "… babies should not be dependent upon caregivers to calm them … John B. Watson … in the 1930s instructed parents on the fostering of self-reliance in their offspring … The method of infant 'sleep training' known as 'controlled crying', advocated in the United States by Richard Ferber (Ferber, 1985) and the United Kingdom by Gina Ford (Ford, 1999) – whereby parents progressively delay responding to their infants' cries, and then respond only in a limited and prescribed manner - are popular modern descendants of Watson's authoritarian approach to infant care. … [However,] … formula feeding, infant scheduling, and controlled crying have all been associated with adverse infant outcomes … and all conflict with the evolutionary predisposition of human infants to be cared for by responsive, emotionally and physically available mothers who provide breast milk according to infant need." (Ball & Klingaman, 2008, p. 228).

- **Presumed danger of suffocation of the newborn if the bed is shared with the parents**

The idea that the newborn sleeping next to the mother runs the risk of being crushed and suffocated by the mother, or even by the father, has been raised in the past (e.g., Byard, 1994). Hence, it has been repeatedly advised that the infant should not sleep in the same bed as the mother (or parents) (Drago & Dannenberg, 1999; Kemp et al., 2000; Scheers et al., 2003).

However:

(1) risk factors for sudden infant death syndrome (SIDS) resulted in being the prone position of the newborn, parental smoking, poverty, and young maternal age (Fleming, 1994);

(2) the hypothesis that close proximity in the bed between mother and newborn could be a risk factor for the newborn and a possible cause of SIDS has been challenged several times (O'Hara et al., 2000; McKenna & Gartner, 2000; McKenna & McDade, 2005). Already in 1986, McKenna contested it on the basis of evolutionary arguments, in particular, because the close contact between mother and newborn is important for the correct initiation and continuation of breastfeeding (McKenna, 1986);

(3) The importance of close mother-infant interaction during sleep and awakening for lactation has been the subject of multiple studies (McKenna et al., 1990; McKenna et al., 1993; McKenna et al., 1997; McKenna & Mosko, 2001)

The position of mother and newborn can be summarized as follows: "Mothers sleep in a lateral position, facing their baby, and curled up around it. Infants, positioned level with their mother's breasts, sleep in the space created between the mother's arm (positioned above her infant's head) and her knees (drawn up under infant's feet) (Baddock et al., 2006; Ball, 2006; Richard et al., 1996)" (Ball & Klingaman, 2008).

The position of the mother folded around the newborn prevents her from crushing the newborn with a rotation of the body during sleep. Furthermore, if the father sleeps on the side of the newborn's shoulders, he provides protection against possible external aggressors (insects, small animals). Moreover, it is unlikely that he would pose a danger to the baby from any movement in sleep as he would meet the mother's arms or knees before squeezing the baby.

In a study (Ball, 2006) on sleeping positions: the breast-fed infant (i) is in the supine (40%) or lateral (54%) position but never (0%) in the prone position; (ii) is almost always (75%) in physical contact with the mother; (iii) always faces at mother's chest level (100%); the mother is curled around the newborn (49%) but never the father (0%). The positions changed in the case of formula-fed infants. In particular, the newborn was in a supine position in 83% of the cases, with the face at the height of the mother's chest in 29% of the cases, while at the height of the mother's face or chin in 71%. These data showed that the position of mother and infant was not random and that there were also substantial differences in natural or formula feeding.

The proximity between the newborn's face and the mother's breasts is an important factor in facilitating natural lactation. The frequency with which the newborn stimulates the mothers nipples by sucking or touching them is important for the secretion of prolactin and, therefore, for the production of the mother's milk and the continuation of breastfeeding (Ball & Klingaman, 2008).

References

Adeodato Filho, J. (1963). Condução do parto. *Revista de Ginecologia e d'Obstetrícia, 7*, 1–16.

Allen, S. J., Clegg, J. B., Alpers, M. P., Mgone, C. S., Peto, T. E., O'Donnell, A., et al. (1999). Prevention of cerebral malaria in children in Papua New Guinea by southeast Asian ovalocytosis band 3. *The American Journal of Tropical Medicine and Hygiene, 60*(6), 1056–1060. https://doi.org/10.4269/ajtmh.1999.60.1056

Allison, A. C. (1954a). The distribution of the sickle-cell trait in East Africa and elsewhere, and its apparent relationship to the incidence of subtertian malaria. *Transactions of the Royal Society of Tropical Medicine and Hygiene, 48*(4), 312–318. https://doi.org/10.1016/0035-9203(54)90101-7

Allison, A. C. (1954b). Protection afforded by sickle-cell trait against subtertian malareal infection. *British Medical Journal, 1*(4857), 290–294. https://doi.org/10.1136/bmj.1.4857.290

Arrigo, M., Jessup, M., Mullens, W., Reza, N., Shah, A. M., Sliwa, K., & Mebazaa, A. (2020). Acute heart failure. *Nature Reviews. Disease Primers, 6*(1), 16. https://doi.org/10.1038/s41572-020-0151-7

Ayi, K., Min-Oo, G., Serghides, L., Crockett, M., Kirby-Allen, M., Quirt, I., et al. (2008). Pyruvate kinase deficiency and malaria. *The New England Journal of Medicine, 358*(17), 1805–1810. https://doi.org/10.1056/NEJMoa072464

Baan, R., Straif, K., Grosse, Y., Secretan, B., El Ghissassi, F., Bouvard, V., et al. (2007). Carcinogenicity of alcoholic beverages. *The Lancet. Oncology, 8*(4), 292–293. https://doi.org/10.1016/s1470-2045(07)70099-2

Baddock, S. A., Galland, B. C., Bolton, D. P., Williams, S. M., & Taylor, B. J. (2006). Differences in infant and parent behaviors during routine bed sharing compared with cot sleeping in the home setting. *Pediatrics, 117*(5), 1599–1607. https://doi.org/10.1542/peds.2005-1636

Ball, H. L. (2006). Parent-infant bed-sharing behavior: Effects of feeding type and presence of father. *Human Nature, 17*(3), 301–318. https://doi.org/10.1007/s12110-006-1011-1

Ball, H. L., & Klingaman, K. (2008). Breastfeeding and mother-infant sleep proximity. In W. R. Trevathan, E. O. Smith, & J. J. McKenna (Eds.), *Evolutionary medicine and health* (pp. 226–241). Oxford University Press. ISBN 978-019530706.

Bates, J. H., & Stead, W. W. (1993). The history of tuberculosis as a global epidemic. *The Medical Clinics of North America, 77*(6), 1205–1217. https://doi.org/10.1016/s0025-7125(16)30188-2

Bayless, T. M., Brown, E., & Paige, D. M. (2017). Lactase Non-persistence and Lactose Intolerance. *Current Gastroenterology Reports, 19*(5), 23. https://doi.org/10.1007/s11894-017-0558-9

Bell, A. F., Erickson, E. N., & Carter, C. S. (2014). Beyond labor: The role of natural and synthetic oxytocin in the transition to motherhood. *Journal of Midwifery & Women's Health, 59*(1), 35–42.: quiz 108. https://doi.org/10.1111/jmwh.12101

Beretta, A., Manuelli, M., & Cena, H. (2023). Favism: Clinical features at different ages. *Nutrients, 15*(2), 343. https://doi.org/10.3390/nu15020343

Berta, M., Lindgren, H., Christensson, K., Mekonnen, S., & Adefris, M. (2019). Effect of maternal birth positions on duration of second stage of labor: Systematic review and meta-analysis. *BMC Pregnancy and Childbirth, 19*(1), 466. https://doi.org/10.1186/s12884-019-2620-0

Betran, A. P., Ye, J., Moller, A. B., Souza, J. P., & Zhang, J. (2021). Trends and projections of caesarean section rates: Global and regional estimates. *BMJ Global Health, 6*(6), e005671. https://doi.org/10.1136/bmjgh-2021-005671

Blaser, M. J. (2016). Antibiotic use and its consequences for the normal microbiome. *Science, 352*(6285), 544–545. https://doi.org/10.1126/science.aad9358

Bø, K., & Nygaard, I. E. (2020). Is physical activity good or bad for the female pelvic floor? A narrative review. *Sports Medicine, 50*(3), 471–484. https://doi.org/10.1007/s40279-019-01243-1

Bringmann, A., Syrbe, S., Görner, K., Kacza, J., Francke, M., Wiedemann, P., & Reichenbach, A. (2018). The primate fovea: Structure, function and development. *Progress in Retinal and Eye Research, 66*, 49–84. https://doi.org/10.1016/j.preteyeres.2018.03.006

Brooks, P. J., Enoch, M. A., Goldman, D., Li, T. K., & Yokoyama, A. (2009). The alcohol flushing response: An unrecognized risk factor for esophageal cancer from alcohol consumption. *PLoS Medicine, 6*, e1000050. https://doi.org/10.1371/journal.pmed.1000050

Bugg, G. J., Siddiqui, F., & Thornton, J. G. (2013). Oxytocin versus no treatment or delayed treatment for slow progress in the first stage of spontaneous labour. *The Cochrane Database of Systematic Reviews, 6*, CD007123. https://doi.org/10.1002/14651858.CD007123.pub3

Byard, R. W. (1994). Is co-sleeping in infancy a desirable or dangerous practice? *Journal of Paediatrics and Child Health, 30*(3), 198–199. https://doi.org/10.1111/j.1440-1754.1994.tb00618.x

Cao, A., & Galanello, R. (2010). Beta-thalassemia. *Genetics in Medicine, 12*(2), 61–76. https://doi.org/10.1097/GIM.0b013e3181cd68ed

Cao, X., Yang, Q., Wang, Q., Hu, S., Hou, L., Sun, M., et al. (2023). PFMT relevant strategies to prevent perineal trauma: A systematic review and network meta-analysis. *Archives of Gynecology and Obstetrics, 308*(2), 387–401. https://doi.org/10.1007/s00404-022-06769-w

Cappellini, M. D., & Fiorelli, G. (2008). Glucose-6-phosphate dehydrogenase deficiency. *Lancet, 371*(9606), 64–74. https://doi.org/10.1016/S0140-6736(08)60073-2

Carroli, G., & Belizan, J. (2000). Episiotomy for vaginal birth. *The Cochrane Database of Systematic Reviews, 2*, CD000081. https://doi.org/10.1002/14651858.CD000081

Charles, D., Anger, H., Dabash, R., Darwish, E., Ramadan, M. C., Mansy, A., et al. (2019). Intramuscular injection, intravenous infusion, and intravenous bolus of oxytocin in the third stage of labor for prevention of postpartum hemorrhage: A three-arm randomized control trial. *BMC Pregnancy and Childbirth, 19*(1), 38. https://doi.org/10.1186/s12884-019-2181-2

Chen, C. H., Ferreira, J. C., Gross, E. R., & Mochly-Rosen, D. (2014). Targeting aldehyde dehydrogenase 2: New therapeutic opportunities. *Physiological Reviews, 94*(1), 1–34. https://doi.org/10.1152/physrev.00017.2013

Chen, C.-H., Ferreira, J. C. B., Joshi, A. U., Stevens, M. C., Li, S.-J., Hsu, J. H.-M., et al. (2020). Novel and prevalent non-East Asian ALDH2 variants; Implications for global susceptibility to aldehydes' toxicity. *eBioMedicine, 55*, 102753. https://doi.org/10.1016/j.ebiom.2020.102753

Chen, C.-H., Kraemer, B. R., & Mochly-Rosen, D. (2022). ALDH2 variance in disease and populations. *Disease Models & Mechanisms, 5*(6), dmm049601. https://doi.org/10.1242/dmm.049601

Choudhari, R. G., Tayade, S. A., Venurkar, S. V., & Deshpande, V. P. (2022). A review of episiotomy and modalities for relief of episiotomy pain. *Cureus, 14*(11), e31620. https://doi.org/10.7759/cureus.31620

Clarel, A., & Bruce, K. Y. (2021). Cesarean section one hundred years 1920-2020: The good, the bad and the ugly. *Journal of Perinatal Medicine, 49*(1), 5–16. https://doi.org/10.1515/jpm-2020-0305

Cooper, D. B., & Yang, L. (2022). Pregnancy and exercise. In *StatPearls [Internet]*. StatPearls Publishing.

Correa, C. (1955). Editorial: Assistência ao parto. *Anais Brasileiros de Ginecologia, 7*, 25–32.

Cunningham, A. S. (1995). Breastfeeding: Adaptive behaviour for child health and longevity. In P. Stuart-Macadam & K. Dettwyler (Eds.), *Breastfeeding: Biocultural perspectives* (pp. 243–263). Aldine de Gruyter. ISBN 978-3110151671.

Daly, D., Minnie, K. C. S., Blignaut, A., Blix, E., Vika Nilsen, A. B., Dencker, A., et al. (2020). How much synthetic oxytocin is infused during labour? A review and analysis of regimens used in 12 countries. *PLoS One, 15*(7), e0227941. https://doi.org/10.1371/journal.pone.0227941

Darwin, C. (1871). *The descent of man and selection in relation to sex*. John Murray.

DeSilva, J. M., & Rosenberg, K. R. (2017). Anatomy, development, and function of the human pelvis. *Anatomical Record (Hoboken, N.J.), 300*(4), 628–632. https://doi.org/10.1002/ar.23561

Desseauve, D., Fradet, L., Lacouture, P., & Pierre, F. (2017). Position for labor and birth: State of knowledge and biomechanical perspectives. *European Journal of Obstetrics, Gynecology, and Reproductive Biology, 208*, 46–54. https://doi.org/10.1016/j.ejogrb.2016.11.006

Diniz, C. S. G. (2001). Entre a técnica e os direitos humanos: possibilidades e limites da humanização da assistência ao parto. In *Tese (Doutorado em Medicina) – Universidade de São Paulo*. Paulo.

Diniz, C. S. G. (2005). Humanização da assistência ao parto no Brasil: os muitos sentidos de um movimento. *Ciência e Saúde Coletiva, 10*(3), 627–637.

Dodson, T. B., & Susarla, S. M. (2014). Impacted wisdom teeth. *BMJ Clinical Evidence, 2014*, 1302.

Doody, D. P. (2004). Foreign body aspiration. In H. C. Grillo (Ed.), *Surgery of the trachea and bronchi* (1st ed., pp. 707–718). BC Decker Inc.

Drago, D. A., & Dannenberg, A. L. (1999). Infant mechanical suffocation deaths in the United States, 1980-1997. *Pediatrics, 103*(5), e59. https://doi.org/10.1542/peds.103.5.e59

Dziedziak, J., Zaleska-Żmijewska, A., Szaflik, J. P., & Cudnoch-Jędrzejewska, A. (2022). Impact of arterial hypertension on the eye: A review of the pathogenesis, diagnostic methods, and treatment of hypertensive retinopathy. *Medical Science Monitor, 28*, e935135. https://doi.org/10.12659/MSM.935135

Ebnezar, J., & Rakesh, J. (2016). *Textbook of orthopedics* (5th ed.). Jaypee Brothers Medical Pub. ISBN 978-9386056689.

Edenberg, H. J., Gelernter, J., & Agrawal, A. (2019). Genetics of alcoholism. *Current Psychiatry Reports, 21*, 26. https://doi.org/10.1007/s11920-019-1008-1

Elsey, M. J., & Rock, W. P. (2000). Influence of orthodontic treatment on development of third molars. *The British Journal of Oral & Maxillofacial Surgery, 38*(4), 350–353. https://doi.org/10.1054/bjom.2000.0307

Erickson, E. N., & Emeis, C. L. (2017). Breastfeeding outcomes after oxytocin use during childbirth: An integrative review. *Journal of Midwifery & Women's Health, 62*(4), 397–417. https://doi.org/10.1111/jmwh.12601

Estabrooks, G. H. (1941). *Man, the mechanical misfit*. Macmillan Co.

Estorninos, E., Lawenko, R. B., Palestroque, E., Sprenger, N., Benyacoub, J., Kortman, G. A. M., et al. (2022). Term infant formula supplemented with milk-derived oligosaccharides shifts the gut microbiota closer to that of human milk-fed infants and improves intestinal immune defense: A randomized controlled trial. *The American Journal of Clinical Nutrition, 115*(1), 142–153. https://doi.org/10.1093/ajcn/nqab336

Farrell, P., Joffe, S., Foley, L., Canny, G. J., Mayne, P., & Rosenberg, M. (2007). Diagnosis of cystic fibrosis in the Republic of Ireland: Epidemiology and costs. *Irish Medical Journal, 100*, 557–560.

Ferber, R. (1985). *Solve your child's sleep problems*. Simon, & Schuster Books. ISBN 978-0671620998.

Fildes, V. (1985). *Breasts, bottles, and babies: A history of infant feeding*. Edinburgh University Press.

Fleming, P. J. (1994). Understanding and preventing sudden infant death syndrome. *Current Opinion in Pediatrics, 6*(2), 158–162. https://doi.org/10.1097/00008480-199404000-00006

Ford, G. (1999). *The contented little baby book*. Vermilion. ISBN 978-0091869144.

Fuchs, A. R., Goeschen, K., Rasmussen, A. B., Rehnström, J. V., Saling, E., & Fuchs, F. (1983). Cervical ripening with intracervical prostaglandin-E2 gel. I. Clinical results and effect on plasma levels of oxytocin and 13,14-dihydro,15-ketoprostaglandin-F2 alpha. *American Journal of Perinatology, 1*(1), 64–69. https://doi.org/10.1055/s-2007-1000055

Gai, P. P., van Loon, W., Siegert, K., Wedam, J., Kulkarni, S. S., Rasalkar, R., et al. (2019). Duffy antigen receptor for chemokines gene polymorphisms and malaria in Mangaluru, India. *Malar Journal, 18*(1), 328. https://doi.org/10.1186/s12936-019-2966-9

Gallagher, P. G., & Glader, B. (2016). Diagnosis of pyruvate kinase deficiency. *Pediatric Blood & Cancer, 63*(5), 771–772. https://doi.org/10.1002/pbc.25922

Ghandali, N. Y., Iravani, M., Habibi, A., & Cheraghian, B. (2021). The effectiveness of a Pilates exercise program during pregnancy on childbirth outcomes: A randomised controlled clinical trial. *BMC Pregnancy and Childbirth, 21*(1), 480. https://doi.org/10.1186/s12884-021-03922-2

Greele, F. (1961). Condução do parto. *Revista de Ginecologia e d'Obstetrícia, 2*, 96–102.

Gruss, L. T., & Schmitt, D. (2015). The evolution of the human pelvis: Changing adaptations to bipedalism, obstetrics and thermoregulation. *Philosophical Transactions of the Royal Society of London. Series B, Biological Sciences, 370*, 20140063. https://doi.org/10.1098/rstb.2014.0063

Gupta, J. K., Sood, A., Hofmeyr, G. J., & Vogel, J. P. (2017). Position in the second stage of labour for women without epidural anaesthesia. *The Cochrane Database of Systematic Reviews, 5*(5), CD002006. https://doi.org/10.1002/14651858.CD002006.pub4

Ha, J., Martinson, R., Iwamoto, S. K., & Nishi, A. (2019). Hemoglobin E, malaria and natural selection. *Evolution, Medicine, and Public Health, 2019*(1), 232–241. https://doi.org/10.1093/emph/eoz034

Haldane, J. B. S. (1949). The rate of mutation of human genes. *Hereditas, 35*(S1), 267–273.

Hall, J. E., & Hall, M. E. (2020). *Guyton and hall textbook of medical physiology* (14th ed.). Elsevier Science Health Science. ISBN 978-0323597128.

Hanada, H. (2016). Specific problems with accidental deaths in Japan – Cardiac arrest from hot baths and foreign body airway obstruction in elders. *Circulation Journal, 80*(7), 1523–1524. https://doi.org/10.1253/circj.CJ-16-0500

Harada, S., Agarwal, D. P., & Goedde, H. W. (1981). Aldehyde dehydrogenase deficiency as cause of facial flushing reaction to alcohol in Japanese. *The Lancet, 2*(8253), 982. https://doi.org/10.1016/s0140-6736(81)91172-7

Harris, T., Goddard, J., & Thompson, P. (Eds.). (2010). *Mammal anatomy: An illustrated guide* (pp. 74–75). Marshall Cavendish Corporation. ISBN 978-0761478829.

Hartupee, J., & Mann, D. L. (2017). Neurohormonal activation in heart failure with reduced ejection fraction. *Nature Reviews. Cardiology, 14*(1), 30–38. https://doi.org/10.1038/nrcardio.2016.163

Hedrick, P. W. (2011). Population genetics of malaria resistance in humans. *Heredity, 107*(4), 283–304. https://doi.org/10.1038/hdy.2011.16

Hennet, T., & Borsig, L. (2016). Breastfed at Tiffany's. *Trends in Biochemical Sciences, 41*(6), 508–518. https://doi.org/10.1016/j.tibs.2016.02.008

Henrich, J. (2015). *The secret of our success: How culture is driving human evolution, domesticating our species, and making us smarter*. Princeton University Press. ISBN 978-0691178431.

Higuchi, S., Matsushita, S., Imazeki, H., Kinoshita, T., Takagi, S., & Kono, H. (1994). Aldehyde dehydrogenase genotypes in Japanese alcoholics. *The Lancet, 343*, 741–742. https://doi.org/10.1016/s0140-6736(94)91629-2

Hill, K., & Hurtado, A. M. (1996). *Ache life history*. Aldine De Gruyter. ISBN 978-0202020363.

Hill, J. M., Zalos, G., Halcox, J. P., Schenke, W. H., Waclawiw, M. A., Quyyumi, A. A., & Finkel, T. (2003). Circulating endothelial progenitor cells, vascular function, and cardiovascular risk. *The New England Journal of Medicine, 348*(7), 593–600. https://doi.org/10.1056/NEJMoa022287

Hirsch, O. (2015). O parto "natural" e "humanizado" na visão de mulheres de camadas médias e populares no Rio de Janeiro. *Civitas-Revista de Ciências Sociais, 15*(2), 229–249.

Howell, N. (1979). *Demography of the Dobe !Kung*. Aldine De Gruyter. ISBN 978-0202306496 (2nd edition, 2000).

IARC Working Group on the Evaluation of Carcinogenic Risks to Humans. (2010). Alcohol consumption and ethyl carbamate. *IARC Monographs on the Evaluation of the Carcinogenic Risks to Humans, 96*, 3–1383. https://www.ncbi.nlm.nih.gov/books/NBK326568/

Im, P. K., Yang, L., Kartsonaki, C., Chen, Y., Guo, Y., Du, H., et al. (2022). Alcohol metabolism genes and risks of site-specific cancers in Chinese adults: An 11-year prospective study. *International Journal of Cancer, 150*, 1627–1639. https://doi.org/10.1002/ijc.33917

Ingram, C. J., Mulcare, C. A., Itan, Y., Thomas, M. G., & Swallow, D. M. (2009). Lactose digestion and the evolutionary genetics of lactase persistence. *Human Genetics, 124*(6), 579–591. https://doi.org/10.1007/s00439-008-0593-6

Institute of Medicine. (2006). *Dietary reference intakes: The essential guide to nutrient requirements*. The National Academies Press. ISBN 978-0309100915. doi:10.17226/11537.

Itan, Y., Jones, B. L., Ingram, C. J., Swallow, D. M., & Thomas, M. G. (2010). A worldwide correlation of lactase persistence phenotype and genotypes. *BMC Evolutionary Biology, 10*, 36. https://doi.org/10.1186/1471-2148-10-36

Jamieson, A., Toovey, S., & Maurel, M. (2006). *Malaria: A traveller's guide* (p. 30). Penguin Random House Struik. ISBN 978-1770073531.

Jonas, J. B., Schmidt, A. M., Müller-Bergh, J. A., Schlötzer-Schrehardt, U. M., & Naumann, G. O. (1992). Human optic nerve fiber count and optic disc size. *Invest Ophthalmology & Visual Science, 33*(6), 2012–2018.

Jones, G., Steketee, R. W., Black, R. E., Bhutta, Z. A., Morris, S. S., & Bellagio Child Survival Study Group. (2003). How many child deaths can we prevent this year? *The Lancet, 362*(9377), 65–71. https://doi.org/10.1016/S0140-6736(03)13811-1

Joy, D. A., Feng, X., Mu, J., Furuya, T., Chotivanich, K., Krettli, A. U., et al. (2003). Early origin and recent expansion of Plasmodium falciparum. *Science, 300*(5617), 318–321. https://doi.org/10.1126/science.1081449

Juodzbalys, G., & Daugela, P. (2013). Mandibular third molar impaction: Review of literature and a proposal of a classification. *Journal of Oral & Maxillofacial Research, 4*(2), e1. https://doi.org/10.5037/jomr.2013.4201
Kao, L., & Lee, T. (2009). *Surgery: PreTest self-assessment and review* (12th ed., p. 61). McGraw-Hill Medical. ISBN 978-0071598637.
Kaur, N., Chugh, V., & Gupta, A. K. (2014). Essential fatty acids as functional components of foods - A review. *Journal of Food Science and Technology, 51*(10), 2289–2303. https://doi.org/10.1007/s13197-012-0677-0
Kemp, J. S., Unger, B., Wilkins, D., Psara, R. M., Ledbetter, T. L., Graham, M. A., et al. (2000). Unsafe sleep practices and an analysis of bedsharing among infants dying suddenly and unexpectedly: Results of a four-year, population-based, death-scene investigation study of sudden infant death syndrome and related deaths. *Pediatrics, 106*(3), E41. https://doi.org/10.1542/peds.106.3.e41
Khan, N. B., Anjum, N., Hoodbhoy, Z., & Khoso, R. (2020). Episiotomy and its complications: A cross sectional study in secondary care hospital. *The Journal of the Pakistan Medical Association, 70*(11), 2036–2038. https://doi.org/10.5455/JPMA.290331
Kim, Y. J. (2021). Immunomodulatory effects of human colostrum and milk. *Pediatric Gastroenterology, Hepatology & Nutrition, 24*(4), 337–345. https://doi.org/10.5223/pghn.2021.24.4.337
Kooij, I. A., Sahami, S., Meijer, S. L., Buskens, C. J., & Te Velde, A. A. (2016). The immunology of the vermiform appendix: A review of the literature. *Clinical and Experimental Immunology, 186*(1), 1–9. https://doi.org/10.1111/cei.12821
Kotepui, K. U., Mahittikorn, A., Masangkay, F. R., & Kotepui, M. (2023). Association between ovalocytosis and Plasmodium infection: A systematic review and meta-analysis. *Scientific Reports, 13*(1), 7164. https://doi.org/10.1038/s41598-023-34170-3
Kouwenhoven, S. M. P., Muts, J., Finken, M. J. J., & Goudoever, J. B. V. (2022). Low-protein infant formula and obesity risk. *Nutrients, 14*(13), 2728. https://doi.org/10.3390/nu14132728
Kwiatkowski, D. P. (2005). How malaria has affected the human genome and what human genetics can teach us about malaria. *American Journal of Human Genetics, 77*(2), 171–192. https://doi.org/10.1086/432519
Larsen, W. J. (1993). *Human embryology* (pp. 318–323). Churchill Livingstone. ISBN 978-0443087245. Retrieved February 26, 2013.
Larsson, S. C., Wallin, A., Wolk, A., & Markus, H. S. (2016). Differing association of alcohol consumption with different stroke types: A systematic review and meta-analysis. *BMC Medicine, 14*(1), 178. https://doi.org/10.1186/s12916-016-0721-4
Laurin, M., Everett, M. L., & Parker, W. (2011). The cecal appendix: One more immune component with a function disturbed by post-industrial culture. *Anatomical record (Hoboken, N.J.), 294*(4), 567–579. https://doi.org/10.1002/ar.21357
Leal, M. D. C., & Gama, S. G. (2014). *Nascer no Brasil: sumário executivo temático da pesquisa*. Escola Nacional de Saúde Pública.
Lederberg, J. (1999). J. B. S. Haldane on infectious disease and evolution. *Genetics, 153*(1), 1–3. https://doi.org/10.1093/genetics/153.1.1
Leibniz, G. (1705). *Nouveaux Essais sur l'entendement humain*, IV, 16: "la nature ne fait jamais des sauts", with the Latin translation in Carolus Linnaeus (1751). *Philosophia Botanica*, 1st ed., Chapter III, § 77, p. 27.
Leon-Larios, F., Corrales-Gutierrez, I., Casado-Mejía, R., & Suarez-Serrano, C. (2017). Influence of a pelvic floor training programme to prevent perineal trauma: A quasi-randomised controlled trial. *Midwifery, 50*, 72–77. https://doi.org/10.1016/j.midw.2017.03.015
Lewis, C. L., Laudicina, N. M., Khuu, A., & Loverro, K. L. (2017). The human pelvis: Variation in structure and function during gait. *Anatomical record (Hoboken, N.J.), 300*(4), 633–642. https://doi.org/10.1002/ar.23552

Lin, Y. P., & Cheng, T. J. (2002). Why can't Chinese Han drink alcohol? Hepatitis B virus infection and the evolution of acetaldehyde dehydrogenase deficiency. *Medical Hypotheses, 59*(2), 204–207. https://doi.org/10.1016/s0306-9877(02)00253-0

Lin, K. Y., Hsih, W. H., Lin, Y. B., Wen, C. Y., & Chang, T. J. (2021). Update in the epidemiology, risk factors, screening, and treatment of diabetic retinopathy. *Journal of Diabetes Investigation, 12*(8), 1322–1325. https://doi.org/10.1111/jdi.13480

Livingstone, F. B. (1958). Anthropological implications of sickle cell gene distribution in West Africa. *American Anthropologist, 60*(3), 533–562. https://doi.org/10.1525/aa.1958.60.3.02a00110

MacLaren Walsh, J. (2008). The Dumbarton Oaks Tlazolteotl: Looking beneath the surface. *Journal de la Société des américanistes, 94-1*, 7–43. https://doi.org/10.4000/jsa.8623

Maier, L., Goemans, C. V., Wirbel, J., Kuhn, M., Eberl, C., Pruteanu, M., et al. (2021). Unravelling the collateral damage of antibiotics on gut bacteria. *Nature, 599*(7883), 120–124. https://doi.org/10.1038/s41586-021-03986-2

Martin, E. (2006). *A mulher no corpo: uma análise cultural da reprodução*. Garamond. ISBN 978-8576170990.

McCallum, C., & dos Reis, A. P. (2008). Users' and providers' perspectives on technological procedures for "normal" childbirth in a public maternity hospital in Salvador, Brazil. *Salud Pública de México, 50*(1), 40–48. https://doi.org/10.1590/s0036-36342008000100010

McKenna, J. J. (1986). An anthropological perspective on the sudden infant death syndrome (SIDS): The role of parental breathing cues and speech breathing adaptations. *Medical Anthropology, 10*(1), 9–92. https://doi.org/10.1080/01459740.1986.9965947

McKenna, J. J., & Gartner, L. M. (2000). Sleep location and suffocation: How good is the evidence? *Pediatrics, 105*(4 Pt 1), 917–919.

McKenna, J. J., & McDade, T. (2005). Why babies should never sleep alone: A review of the co-sleeping controversy in relation to SIDS, bedsharing and breast feeding. *Paediatric Respiratory Reviews, 6*(2), 134–152. https://doi.org/10.1016/j.prrv.2005.03.006

McKenna, J. J., & Mosko, S. (2001). Mother-infant cosleeping: Toward a new scientific beginning. In R. Byard & H. Krous (Eds.), *Sudden infant death syndrome: Problems, puzzles, possibilities* (pp. 259–272). Arnold Publishing. ISBN 978-0340759172.

McKenna, J. J., Mosko, S., Dungy, C., & McAninch, J. (1990). Sleep and arousal patterns of co-sleeping human mother/infant pairs: A preliminary physiological study with implications for the study of sudden infant death syndrome (SIDS). *American Journal of Physical Anthropology, 83*(3), 331–347. https://doi.org/10.1002/ajpa.1330830307

McKenna, J. J., Mosko, S. S., & Richard, C. A. (1997). Bedsharing promotes breastfeeding. *Pediatrics, 100*(2 Pt 1), 214–219. https://doi.org/10.1542/peds.100.2.214

McKenna, J. J., Thoman, E. B., Anders, T. F., Sadeh, A., Schechtman, V. L., & Glotzbach, S. F. (1993). Infant-parent co-sleeping in an evolutionary perspective: Implications for understanding infant sleep development and the sudden infant death syndrome. *Sleep, 16*(3), 263–282. https://doi.org/10.1093/sleep/16.3.263

Meander, L., Lindqvist, M., Mogren, I., Sandlund, J., West, C. E., & Domellöf, M. (2021). Physical activity and sedentary time during pregnancy and associations with maternal and fetal health outcomes: An epidemiological study. *BMC Pregnancy and Childbirth, 21*(1), 166. https://doi.org/10.1186/s12884-021-03627-6

Mehta, A., Mason, P. J., & Vulliamy, T. J. (2000). Glucose-6-phosphate dehydrogenase deficiency. *Best Practice & Research Clinical Haematology, 13*(1), 21–38. https://doi.org/10.1053/beha.1999.0055

Meier, P., Schirmer, S. H., Lansky, A. J., Timmis, A., Pitt, B., & Seiler, C. (2013). The collateral circulation of the heart. *BioMed Central Medicine, 11*, 143. https://doi.org/10.1186/1741-7015-11-143

Miller, A. C., Comellas, A. P., Hornick, D. B., Stoltz, D. A., Cavanaugh, J. E., Gerke, A. K., et al. (2020). Cystic fibrosis carriers are at increased risk for a wide range of cystic fibrosis-related conditions. *Proceedings of the National Academy of Sciences of the United States of America, 117*(3), 1621–1627. https://doi.org/10.1073/pnas.1914912117

Millwood, I. Y., Walters, R. G., Mei, X. W., Guo, Y., Yang, L., Bian, Z., et al. (2019). Conventional and genetic evidence on alcohol and vascular disease aetiology: A prospective study of 500 000 men and women in China. *The Lancet, 393*, 1831–1842. https://doi.org/10.1016/S0140-6736(18)31772-0

Moore, E. R., Bergman, N., Anderson, G. C., & Medley, N. (2016). Early skin-to-skin contact for mothers and their healthy newborn infants. *Cochrane Database of Systematic Reviews, 11*(11), CD003519. https://doi.org/10.1002/14651858.CD003519.pub4

Morgan, E. (1990). *The scars of evolution*. Souvenir Press. ISBN 978-0285629967.

Nagel, R. L., & Steinberg, M. H. (2001). Hemoglobin SC disease and HbC disorders. In M. H. Steinberg, B. G. Forget, D. R. Higgs, & R. L. Nagel (Eds.), *Disorders of hemoglobin: Genetics, pathophysiology, and clinical management* (1st ed., pp. 756–785). Cambridge University Press. ISBN 978-0521632669.

Nesse, R. M., & Williams, G. C. (1994). *Why we get sick: The new science of Darwinian medicine*. Times Books. ISBN 978-0812922247.

Netter, F. H. (2022). *Netter atlas of human anatomy: A systems approach* (8th ed.). Elsevier Science Health Science. ISBN 978-0323760287.

Niemitz, C. (2010). The evolution of the upright posture and gait—A review and a new synthesis. *Naturwissenschaften, 97*, 241–263. https://doi.org/10.1007/s00114-009-0637-3

Novis, A. (1957). Meios aceleradores do parto. *Revista de Ginecologia e d'Obstetrícia, 7*, 1015–1020.

Nucci, M., Nakano, A. R., & Teixeira, L. A. (2018). Synthetic oxytocin and hastening labor: Reflections on the synthesis and early use of oxytocin in Brazilian obstetrics. *História, ciências, saúde--Manguinhos, 25*(4), 979–998. https://doi.org/10.1590/S0104-59702018000500006

O'Sullivan, B. P., & Freedman, S. D. (2009). Cystic fibrosis. *The Lancet, 373*(9678), 1891–1904. https://doi.org/10.1016/s0140-6736(09)60327-5

O'Hara, M., Harruff, R., Smialek, J. E., & Fowler, D. R. (2000). Sleep location and suffocation: How good is the evidence? *Pediatrics, 105*(4 Pt 1), 915–917. https://doi.org/10.1542/peds.105.4.915-a

Okada, Y., Momozawa, Y., Sakaue, S., Kanai, M., Ishigaki, K., Akiyama, M., et al. (2018). Deep whole-genome sequencing reveals recent selection signatures linked to evolution and disease risk of Japanese. *Nature Communications, 9*, 1631. https://doi.org/10.1038/s41467-018-03274-0

Origa, R. (2017). β-Thalassemia. *Genetics in Medicine, 19*(6), 609–619. https://doi.org/10.1038/gim.2016.173

Pappas, G., Kiriaze, I. J., & Falagas, M. E. (2008). Insights into infectious disease in the era of Hippocrates. *International Journal of Infectious Diseases, 12*(4), 347–350. https://doi.org/10.1016/j.ijid.2007.11.003

Park, J. Y., Elshami, A. A., Kang, D. S., & Jung, T. H. (1996). Plastic bronchitis. *The European Respiratory Journal, 9*, 612–614.

Park, S. K., Park, C. S., Lee, H.-S., Park, K. S., Park, B. L., Cheong, H. S., & Shin, H. D. (2014). Functional polymorphism in aldehyde dehydrogenase-2 gene associated with risk of tuberculosis. *BMC Medical Genetics, 15*, 40. https://doi.org/10.1186/1471-2350-15-40

Pärnänen, K. M. M., Hultman, J., Markkanen, M., Satokari, R., Rautava, S., Lamendella, R., et al. (2022). Early-life formula feeding is associated with infant gut microbiota alterations and an increased antibiotic resistance load. *The American Journal of Clinical Nutrition, 115*(2), 407–421. https://doi.org/10.1093/ajcn/nqab353

Peng, Y., Shi, H., Qi, X. B., Xiao, C. J., Zhong, H., Ma, R. L., & Su, B. (2010). The ADH1B Arg47His polymorphism in East Asian populations and expansion of rice domestication in history. *BMC Evolutionary Biology, 10*, 15. https://doi.org/10.1186/1471-2148-10-15

Pérez-Escamilla, R., Tomori, C., Hernández-Cordero, S., Baker, P., Barros, A. J. D., Bégin, F., Lancet Breastfeeding Series Group, et al. (2023). Breastfeeding: Crucially important, but increasingly challenged in a market-driven world. *The Lancet, 401*(10375), 472–485. https://doi.org/10.1016/S0140-6736(22)01932-8. Erratum in: *The Lancet*, 2023, 401(10380), 916.

Pohl, T., Seiler, C., Billinger, M., Herren, E., Wustmann, K., Mehta, H., et al. (2001). Frequency distribution of collateral flow and factors influencing collateral channel development. Functional collateral channel measurement in 450 patients with coronary artery disease. *Journal of American College of Cardiology, 38*, 1872–1878. https://doi.org/10.1016/S0735-1097(01)01675-8

Polgreen, P. M., & Comellas, A. P. (2022). Clinical phenotypes of cystic fibrosis carriers. *Annual Review of Medicine, 73*, 563–574. https://doi.org/10.1146/annurev-med-042120-020148

Poolman, E. M., & Galvani, A. P. (2007). Evaluating candidate agents of selective pressure for cystic fibrosis. *Journal of the Royal Society Interface, 4*(12), 91–98. https://doi.org/10.1098/rsif.2006.0154

Popovici, J., Roesch, C., & Rougeron, V. (2020). The enigmatic mechanisms by which Plasmodium vivax infects Duffy-negative individuals. *PLoS Pathogens, 16*(2), e1008258. https://doi.org/10.1371/journal.ppat.1008258

Price, W. A. (1939). *Nutrition and physical degeneration*. Paul B. Hoeber. [Other editions in 1945, 1970, 1989, 1997, 2004, 2006, 2008, 2011; the last edition – ISBN 978-0916764203 - has been used].

Rahmanou, P., Caudwell-Hall, J., Kamisan Atan, I., & Dietz, H. P. (2016). The association between maternal age at first delivery and risk of obstetric trauma. *American Journal of Obstetrics and Gynecology, 215*(4), 451.e1–451.e7. https://doi.org/10.1016/j.ajog.2016.04.032

Ribeiro, P. (1962). Assistência ao parto normal. *Revista de Ginecologia e d'Obstetrícia, 10*, 103–112.

Richard, C., Mosko, S., McKenna, J. J., & Drummond, S. (1996). Sleeping position, orientation, and proximity in bedsharing infants and mothers. *Sleep, 19*(9), 685–690. https://doi.org/10.1093/sleep/19.9.685

Roerecke, M., & Rehm, J. (2012). The cardioprotective association of average alcohol consumption and ischaemic heart disease: A systematic review and meta-analysis. *Addiction, 107*(7), 1246–1260. https://doi.org/10.1111/j.1360-0443.2012.03780.x

Royal College of Midwives. (2002). *Successful breastfeeding* (3rd ed.). Churchill Livingstone. ISBN 978-0443059674.

Sakaue, S., Kanai, M., Tanigawa, Y., Karjalainen, J., Kurki, M., Koshiba, S., et al. (2021). A cross-population atlas of genetic associations for 220 human phenotypes. *Nature Genetics, 53*, 1415–1424. https://doi.org/10.1038/s41588-021-00931-x

Salih, A. M., Alfaki, M., & Alam-Elhuda, D. M. (2016). Airway foreign bodies: A critical review for a common pediatric emergency. *World Journal of Emergency Medicine, 7*(1), 5–12. https://doi.org/10.5847/wjem.j.1920-8642.2016.01.001

Salminen, S., Stahl, B., Vinderola, G., & Szajewska, H. (2020). Infant formula supplemented with biotics: Current knowledge and future perspectives. *Nutrients, 12*(7), 1952. https://doi.org/10.3390/nu12071952

Sandall, J., Tribe, R. M., Avery, L., Mola, G., Visser, G. H., Homer, C. S., et al. (2018). Short-term and long-term effects of caesarean section on the health of women and children. *The Lancet, 392*(10155), 1349–1357. https://doi.org/10.1016/S0140-6736(18)31930-5

Scheers, N. J., Rutherford, G. W., & Kemp, J. S. (2003). Where should infants sleep? A comparison of risk for suffocation of infants sleeping in cribs, adult beds, and other sleeping locations. *Pediatrics, 112*(4), 883–889. https://doi.org/10.1542/peds.112.4.883

Schnorr, S. L. (2015). The diverse microbiome of the hunter-gatherer. *Nature, 518*(7540), S14–S15. https://doi.org/10.1038/518S14a

Ségurel, L., & Bon, C. (2017). On the evolution of lactase persistence in humans. *Annual Review of Genomics and Human Genetics, 18*(1), 297–319. https://doi.org/10.1146/annurev-genom-091416-035340

Shteinberg, M., Haq, I. J., Polineni, D., & Davies, J. C. (2021). Cystic fibrosis. *The Lancet, 397*(10290), 2195–2211. https://doi.org/10.1016/S0140-6736(20)32542-3

Smith, H. F., Fisher, R. E., Everett, M. L., Thomas, A. D., Randal Bollinger, R., & Parker, W. (2009). Comparative anatomy and phylogenetic distribution of the mammalian cecal appendix. *Journal of Evolutionary Biology, 22*(10), 1984–1999. https://doi.org/10.1111/j.1420-9101.2009.01809.x

Soderquist, C., & Bagg, A. (2013). Hereditary elliptocytosis. *Blood, 121*(16), 3066. https://doi.org/10.1182/blood-2012-09-457788

Storhaug, C. L., Fosse, S. K., & Fadnes, L. T. (2017). Country, regional, and global estimates for lactose malabsorption in adults: A systematic review and meta-analysis. *The Lancet. Gastroenterology & Hepatology, 2*(10), 738–746. https://doi.org/10.1016/S2468-1253(17)30154-1

Swallow, D. M. (2003). Genetics of lactase persistence and lactose intolerance. *Annual Review of Genetics, 37*, 197–219. https://doi.org/10.1146/annurev.genet.37.110801.143820

Tobacman, J. K. (2003). Does deficiency of arylsulfatase B have a role in cystic fibrosis? *Chest, 123*(6), 2130–2139. https://doi.org/10.1378/chest.123.6.2130

Tornquist, C. (2004). *Parto e poder: o movimento pela humanização do parto no Brasil.* Tese (Doutorado) – Universidade Federal de Santa Catarina, Florianópolis.

Travassos, M. A., Coulibaly, D., Laurens, M. B., Dembélé, A., Tolo, Y., Koné, A. K., et al. (2015). Hemoglobin C trait provides protection from clinical falciparum malaria in Malian children. *The Journal of Infectious Diseases, 212*(11), 1778–1786. https://doi.org/10.1093/infdis/jiv308

Urashima, T., Hirabayashi, J., Sato, S., & Kobata, A. (2018). Human milk oligosaccharides as essential tools for basic and application studies on Galectins. *Trends in Glycoscience and Glycotechnology, 30*, SJ11–SJ24. https://doi.org/10.4052/tigg.1734.1SJ

Uvnäs-Moberg, K., Ekström-Bergström, A., Berg, M., Buckley, S., Pajalic, Z., Hadjigeorgiou, E., et al. (2019). Maternal plasma levels of oxytocin during physiological childbirth - A systematic review with implications for uterine contractions and central actions of oxytocin. *BMC Pregnancy and Childbirth, 19*(1), 285. https://doi.org/10.1186/s12884-019-2365-9

Vannuccini, S., Bocchi, C., Severi, F. M., Challis, J. R., & Petraglia, F. (2016). Endocrinology of human parturition. *Annales d'Endocrinologie (Paris), 77*(2), 105–113. https://doi.org/10.1016/j.ando.2016.04.025

Victora, C. G., Bahl, R., Barros, A. J., França, G. V., Horton, S., Krasevec, J., Lancet Breastfeeding Series Group, et al. (2016). Breastfeeding in the 21st century: Epidemiology, mechanisms, and lifelong effect. *The Lancet, 387*(10017), 475–490. https://doi.org/10.1016/S0140-6736(15)01024-7

Vos, T., Barber, R. M., Bell, B., Bertozzi-Villa, A., Biryukov, S., Bolliger, I., GBD 2013 Mortality and Causes of Death Collaborators, et al. (2015). Global, regional, and national incidence, prevalence, and years lived with disability for 301 acute and chronic diseases and injuries in 188 countries, 1990-2013: a systematic analysis for the Global Burden of Disease Study 2013. *The Lancet, 386*(9995), 743–800. https://doi.org/10.1016/s0140-6736(15)60692-4

Wall-Scheffler, C. M., & Myers, M. J. (2017). The biomechanical and energetic advantages of a mediolaterally wide pelvis in women. *Anatomical record (Hoboken, N.J.), 300*(4), 764–775. https://doi.org/10.1002/ar.23553

Walter, M. H., Abele, H., & Plappert, C. F. (2021). The role of oxytocin and the effect of stress during childbirth: Neurobiological basics and implications for mother and child. *Frontiers in Endocrinology (Lausanne), 12*, 742236. https://doi.org/10.3389/fendo.2021.742236

Wang, H., Naghavi, M., Allen, C., Barber, R. M., Bhutta, Z. A., Carter, A., GBD 2015 Mortality and Causes of Death Collaborators, et al. (2016). Global, regional, and national life expectancy, all-cause mortality, and cause-specific mortality for 249 causes of death, 1980–2015: A systematic analysis for the Global Burden of Disease Study 2015. *The Lancet, 388*(10053), 1459–1544. https://doi.org/10.1016/S0140-6736(16)31012-1

Waugh, A., & Grant, A. (2018). *Ross & Wilson anatomy and physiology in health and illness* (p. 83). Elsevier Science Health Sciences. ISBN 978-0702072840.

Wegh, C. A., Geerlings, S. Y., Knol, J., Roeselers, G., & Belzer, C. (2019). Postbiotics and their potential applications in early life nutrition and beyond. *International Journal of Molecular Sciences, 20*, 4673. https://doi.org/10.3390/ijms20194673

WHO. (2022). *World malaria report 2022*. World Health Organization. ISBN 978-92-4-006489-8.

Wickham, S. (2015). Synthetic oxytocin: Looking beyond the benefits. *The Practising Midwife, 18*(1), 34–35.

Wiuf, C. (2001). Do delta F508 heterozygotes have a selective advantage? *Genetical Research, 78*(1), 41–47. https://doi.org/10.1017/S0016672301005195

Wu, S., Grimm, R., German, J. B., & Lebrilla, C. B. (2011). Annotation and structural analysis of sialylated human milk oligosaccharides. *Journal of Proteome Research, 10*(2), 856–868. https://doi.org/10.1021/pr101006u

Wu, P. C., Huang, H. M., Yu, H. J., Fang, P. C., & Chen, C. T. (2016). Epidemiology of Myopia. *Asia-Pacific Journal of Ophthalmology (Philadelphia), 5*(6), 386–393. https://doi.org/10.1097/APO.0000000000000236

Wu, S., Tao, N., German, J. B., Grimm, R., & Lebrilla, C. B. (2010). Development of an annotated library of neutral human milk oligosaccharides. *Journal of Proteome Research, 9*(8), 4138–4151. https://doi.org/10.1021/pr100362f

Wustmann, K., Zbinden, S., Windecker, S., Meier, B., & Seiler, C. (2003). Is there functional collateral flow during vascular occlusion in angiographically normal coronary arteries? *Circulation, 107*, 2213–2220. https://doi.org/10.1161/01.CIR.0000066321.03474.DA

Yokoyama, A., Kato, H., Yokoyama, T., Tsujinaka, T., Muto, M., Omori, T., et al. (2002). Genetic polymorphisms of alcohol and aldehyde dehydrogenases and glutathione S-transferase M1 and drinking, smoking, and diet in Japanese men with esophageal squamous cell carcinoma. *Carcinogenesis, 23*(11), 1851–1859. https://doi.org/10.1093/carcin/23.11.1851

Zaami, S., Zupi, E., Lazzeri, L., Centini, G., Stark, M., Malvasi, A., et al. (2021). Episiotomy: A medicolegal vicious cycle. *Panminerva Medica, 63*(2), 224–231. https://doi.org/10.23736/S0031-0808.20.03946-4

Zhu, C., Talhelm, T., Li, Y., Chen, G., Zhu, J., & Wang, J. (2021). Relationship between rice farming and polygenic scores potentially linked to agriculture in China. *Royal Society Open Science, 8*, 210382. https://doi.org/10.1098/rsos.210382

Ziaeian, B., & Fonarow, G. C. (2016). Epidemiology and aetiology of heart failure. *Nature Reviews. Cardiology, 13*(6), 368–378. https://doi.org/10.1038/nrcardio.2016.25

Zou, J., Meng, M., Law, C. S., Rao, Y., & Zhou, X. (2018). Common dental diseases in children and malocclusion. *International Journal of Oral Science, 10*(1), 7. https://doi.org/10.1038/s41368-018-0012-3

Chapter 10
Implementation of Evolutionary Medicine

Already two centuries ago, Erasmus Darwin, Charles's grandfather, observed that "to conduct the practice of medicine … as a great number of unconnected facts are difficult to be acquired, and to be reasoned" there was the need for a "theory founded upon nature, that should bind together the scattered facts of medical knowledge, and converge into one point of view the laws of organic life, …" (Darwin, 1794-1796).

This need has not disappeared and indeed has increased enormously with the growth of medical knowledge: "Many general medicine textbooks now tend to be so large and structured in such a way that no one can realistically be expected to read one from cover to cover. Yet, at the same time, there is a strong onus on the author - or frequently, teams of authors - to include, as far as is possible, everything of practical relevance. … However easy access to information becomes, a certain sense of explication is missing; while the 'whats' and 'hows' abound, the 'whys' and 'wherefores' are relatively few. If the latter constitute the framework within which the detail can be better understood and remembered, than it is questionable whether the way in which medical information is currently approached is ideal. … Certainly, having the right intellectual framework within which to understand something is one of the best aids to mastering a subject. The best teachers are not necessarily those who know the most about a given subject; rather, they are the ones who, among other things, can reveal meaning amid a mass of data." (Lewis, 2008).

A possible rational response to this need is the deep-rooted integration of current medicine with the concepts of evolutionary medicine, i.e., the necessary transformation of current medicine into evolutionary medicine.

G. Libertini et al., *The Essential Revolution of Evolutionary Medicine*, Advances in Studies of Aging and Health 4, https://doi.org/10.1007/978-3-031-84877-3_10

10.1 Evolutionary Medicine, a New Medical Discipline, or a New Medicine?

Some definitions of evolutionary medicine are as follows:

- "... evolutionary or Darwinian medicine uses an evolutionary perspective to understand why the body is not better designed and why, therefore, diseases exist at all. (Nesse & Williams, 1994; Stearns, 1999; Williams & Nesse, 1991)" (Nesse, 2001);
- "... it is Lappé (1994) who gives the most concise (even axiomatic) description of the aims of a medicine that has fully assimilated evolutionary theory when he suggests that '[w]e need to understand the evolutionary origins of disease in order to fully comprehend their prevention' and '[w]e need to incorporate evolutionary models of disease in order to treat them effectively' (p. 9)" (Lewis, 2008);
- "Evolutionary medicine is the enterprise of using evolutionary biology to address the problems of medicine" (Nesse, 2008).

These definitions point to evolutionary medicine as a rational and useful method for analyzing medical problems and suggesting appropriate treatments. From this point of view, knowledge of the concepts of evolutionary medicine is rightly recommended as a necessary and fundamental part of the knowledge set that a doctor must have (Nesse, 2008).

However, this approach is limiting.

In biology, the knowledge of evolutionary mechanisms is not simply a discipline that contributes to the knowledge of biological phenomena but constitutes the fundamental basis for understanding all biological phenomena. This assertion is condensed in Dobzhansky's famous statement, "Nothing in biology makes sense, except in the light of evolution" (Dobzhansky, 1973).

A possible reformulation of this concept in a form such as 'knowledge of the mechanisms of evolution contributes to the knowledge of biological phenomena', would limit Dobzhansky's concept: evolution is not simply an important phenomenon that contributes to the understanding of biological mechanisms; more precisely, evolution is the fundamental phenomenon underlying the origin and development of living beings and must be taken into account for any biological phenomenon.

If medicine, in a synthetic definition, is a branch of biology aimed at developing the best methods for preventing and treating diseases, the extension of Dobzhansky's concept to medicine is rational, as expressed some years ago in the already mentioned famous aphorism: "Nothing in medicine makes sense, except in the light of evolution" (Varki, 2012).

Therefore, if we consider the knowledge of evolutionary mechanisms only as a contribution to the necessary knowledge of the physician, this is an irrationally restricted conception.

As for biology in general, for medicine evolutionary mechanisms constitute the fundamental phenomena at the basis of pathological phenomena and must be taken into account for any disease.

In this conception, it is limiting to maintain that medicine must be enriched by a new discipline, evolutionary medicine, which will make important contributions to understanding many diseases.

More broadly, it is necessary that the core concepts of medicine be those of evolutionary mechanisms and that the knowledge of these concepts must inspire the understanding and, therefore, the prevention and treatment of all diseases.

So, it is not only desirable to enrich current medicine with the concepts of evolutionary medicine but also to transform current medicine into evolutionary medicine.

This point does not mean that the knowledge of evolutionary mechanisms is necessary or valuable for every detail of medicine, just as such knowledge is not necessary or useful for every detail of biology. On the contrary, this means that both biology and medicine must be seen in the context of evolutionary mechanisms, having them as an essential general point of reference because "having the right intellectual framework within which to understand something is one of the best aids to mastering a subject" (Lewis, 2008).

It is important to deepen the general meaning of this assertion. Medical problems can be tackled in three ways that are quite different in terms of approach, methods, and implications.

(1) **Current medicine**. Manifestations, physio-pathological mechanisms, epidemiology, prevention of the diseases, and the best methods to treat and prevent them are studied;
(2) **Evolutionary medicine with a limited approach**. Current medicine must be enriched with concepts and methods defined by evolutionary medicine. This approach provides important insights into the primary (i.e., ultimate) causes of disease and consequently into disease prevention and treatment. Furthermore, to understand various diseases, the study of the mechanisms of natural selection allows us to obtain valuable information regarding their pathogenetic mechanisms. According to this approach, evolutionary medicine should be studied in the context of medical studies and would constitute an important enrichment of medical knowledge.
(3) **Evolutionary Medicine without undue limitations**. All medicine should be reformulated in terms of evolutionary medicine. The concepts of evolutionary medicine do not constitute an addition to the concepts of current medicine but constitute a rational pivotal basis for understanding the primary causes of diseases and developing the best means to prevent and treat them. As explained in the first chapter of this book (*Chap. 1—Introduction*), in the reformulation of medicine, diseases do not follow the traditional classification—based on criteria such as: (i) affected organs or systems; (ii) the skills of the various specialists; and (iii) divisions originated from tradition—but follow a rational classification based on the primary causes of the illnesses. This classification allows a rational elaboration for the prevention and treatment of diseases and satisfies Erasmus Darwin's aspirations for a rational understanding of diseases, i.e., not based on analogies between symptoms or pathophysiological manifestations.

> The rational classification of diseases based on the evolutionary mechanisms that originate them, overcoming empirical and tradition-based criteria, is analogous to the classification of species based on their origin through natural evolution, overcoming previous classifications based on superficial or external analogies.

Another aspect that is different in the last approach compared to the other two is the completely different interpretation of aging phenomenon. In fact, aging is considered an adaptive phenomenon, i.e., a phenomenon determined by evolution and genetically determined and modulated, within the phenomena defined as phenoptotic, and not a degenerative phenomenon that natural selection is unable to control. This subject is widely discussed in *Chap. 8—Physiological phenomena that cause troubles and suffering or even death—Part III: Slow phenoptosis or aging* and in other works such as (Libertini et al., 2021).

It could be objected that the evolutionary medicine approach is too radical compared to that of current medicine.

A similar objection was formulated, with strong consensus on the part of many scholars, against the daring theory of evolution by natural selection when it was proposed for the first time. However, the history of biology has shown that Darwin's theory of evolution was not a simple enrichment of biology but something that involved a profound rethinking and reformulation of all biology.

Something similar is strongly needed for medicine. Or, more precisely, since medicine is an applied branch of biology, the revolution already widely implemented for the other branches of biology must extend to medicine.

10.2 Identification of the Primary Causes of a Disease

The identification of the primary cause, or causes, of a disease is not a simple enrichment of knowledge about the disease, but the fundamental information that absolutely must be known.

Knowledge of the primary causes of a disease is not synonymous with understanding the pathogenic mechanism. It is possible to have good knowledge of the pathogenetic mechanism without being aware of the primary causes. Conversely, it is also possible to have evidence of the primary causes without a precise understanding of the pathogenic mechanisms.

However, for the knowledge of the primary causes of a disease, it is essential to have some specific information, which can be summarized as follows:

(1) In conditions that are normal in evolutionary terms, i.e., in the ecological niche to which the species is adapted, or at least in conditions close to the aforementioned normal conditions, what is the disease frequency?
(2) In modern conditions, i.e., generally mostly distant from the ancestral ones, what is the disease frequency?

(3) In the transition from conditions that were normal, or close to normal ones, to the current ones, how has the frequency of the disease changed?
(4) Comparing populations living in different conditions, more or less close to those existing in the ancestral conditions, are there variations in the disease frequency?
(5) What specific alterations of normal conditions are most correlated with the onset and frequency of the disease?
(6) If groups of individuals are brought back to life conditions closer to normal in evolutionary terms, if the disease has a reversible mechanism, what happens to the frequency of the disease or its intensity?

Let us consider the case of a disease X that: (i) has a minimum frequency under normal conditions; (ii) has considerable frequency under modern conditions; (iii) showed or nowadays shows a notable frequency increase in the transition from conditions closer to the ancestral ones to those of modern life; (iv) shows a notable difference in frequency between populations in different conditions, in particular with less frequency in populations with lifestyles closer to the ancestral ones; (v) with possible or known specific alterations of the ecological niche related to the frequency increase; (vi) shows, if reversible, an improvement with the restoration of conditions closer to normal ones.

For this disease, it is rational to hypothesize, regardless of its pathogenetic mechanism, that:

(A) a small fraction of the cases is due to genetic alterations, which can be of various types. These alterations are generally not preventable, and the cases due to them are also present in normal evolutionary conditions;
(B) in other cases, disease X is a secondary manifestation of other diseases (Y_1, … Y_n), and consequently, the primary causes (and the possible preventive measures) must be sought in the origin of each disease Y;
(C) most cases are due to alterations of the ecological niche, and therefore, disease X, in general, can be prevented by correcting the alterations above of the ecological niche;
(D) in the case of the alterations above of the ecological niche, within a population there will be an individual variety in the vulnerability to the development of the disease. This vulnerability must absolutely not be labeled as a primary cause of the disease. It is a concomitant factor that, if due to genetic factors, is not preventable and, as a rule, should not be the first target of a treatment;
(E) to the extent that disease X is reversible, it is generally possible to counteract the course of the disease by restoring normal conditions.

For current medicine and for evolutionary medicine, there are two opposite general description of disease X:

- For current medicine, disease X has a multifactorial pathogenesis and there are important genetic factors contributing at varying degrees to the vulnerability for the disease, sometimes with a family history of the same disease. In addition, specific lifestyle characteristics that need to be modified are important risk factors for the onset, severity, and course of disease X.

- For evolutionary medicine, disease X, apart from a small fraction of cases due to genetic alterations, is caused by alterations of the ecological niche that determine the disease directly or through another disease Y. The disease is, therefore, largely preventable with appropriate corrections of specific alterations of the ecological niche. Genetic conditions that result in vulnerability to, or even familiarity with, the disease are not primary causes of the disease and are not preventable. However, they should not be considered a reason or a pretext that invalidates or diminishes the importance of prevention. When the disease is established, to the extent that it is reversible, correction of the alterations in the ecological niche should contribute positively to the course and outcome of the disease.

What has been said above is by no means an abstraction or a simple theoretical exercise.

Let us look at a specific example of a disease that matches the description of disease X.

Essential hypertension is in clear correlation with alterations of the ecological niche related to the modern lifestyle.

As already highlighted in *Chap. 1—Introduction*, in a 1972 work (Truswell et al., 1972), it was shown that both systolic and diastolic blood pressure increased in relation to age in the contemporary London population. In the same work, it was highlighted that in a population of hunter-gatherers, the blood pressure values did not change with age.

In recent years, it has been estimated that in the US, in 2017, 46% of people were affected by hypertension (Whelton et al., 2018). The frequency of hypertension varies strongly according to human populations, for example, going from low frequencies in rural India (3.4% for men and 6.8% for women) to high frequencies in Poland (68.9% for men and 72.5% for women) (Kearney et al., 2004), while in a bushman population, in 1972, the frequency was irrelevant (Truswell et al., 1972).

The hypertension epidemic is caused by changes in the ecological niche to which our species has not adapted. E.g., excessive salt intake (Bragulat & de la Sierra, 2002), Feng et al., 2017).

It is also true that the resistance to the development of hypertension is influenced by individual genetic characteristics. In fact, many genetic variants (Ehret et al., 2011) and genetic loci (Kato et al., 2015) related to blood pressure increase are known. However, these cases of genetic vulnerability to the development of hypertension should not make us forget that essential hypertension does not develop due to the mere existence of these genetic variants in the absence of alterations in the normal ecological conditions.

Regarding the correlation between age and hypertension, this correlation exists in modern populations with a Western diet lifestyle (Carrera-Bastos et al., 2011). However, it has been known for a long time that in life conditions similar to those to which our species is adapted, there is no increase in blood pressure in relation to age (Truswell et al., 1972).

In short, in a description of hypertension in terms of evolutionary medicine, this disease (or rather syndrome), apart from few cases deriving from genetic alterations, is due to alterations of the ecological niche, which are correctable both before the onset of the disease and after the onset of the disease.

An analogous discussion can be proposed for many diseases in which the leading primary cause is constituted by alterations of the ecological niche. Various examples have been discussed in *Chap. 3—Diseases deriving from alterations of the ecological niche—Part I: Not regarding the holobiont.*

A particular category of diseases is when the alterations of the ecological niche concern the holobiont. Examples of this condition have been discussed in *Chap. 4—Diseases deriving from alterations of the ecological niche—Part II: Regarding the holobiont.*

The examples given in these chapters by no means constitute an exhaustive discussion of each disease. Furthermore, there are undoubtedly many other diseases depending at least in part on alterations of the ecological niche that need to be defined and studied in depth.

However, these examples are part of a general knowledge of the primary causes of all diseases, which is a necessary precondition for the implementation of evolutionary medicine.

10.3 Difficulties Against the Implementation of Evolutionary Medicine

Medicine arises from the need to alleviate the suffering of those who are ill and restore their health. This demand has been actual in the past and is still true in modern times, and medical care has always been oriented towards these goals.

To the extent that a doctor (or a health facility) manages to have good results regarding these purposes, economic rewards, career progression, consensus, and social authority follow. The same is true for scientific research in the medical field. A new pharmacological, surgical, or different treatment means success, honors, and economic returns if effective.

In particular, in the case of a severe illness that leads to death, if the doctor manages to make even only a part of the patients survive or heal, this is considered a great success, which in various ways is the object of gratification both by the patients and the community.

Suppose now that a doctor manages to prevent the cases of the aforementioned severe or deadly disease through preventive measures. With rational logic, this doctor should be appreciated and gratified much more than the one who treats the cases in which the disease manifests itself, also obtaining healing in some patients.

However, it is more probable that whoever prevents the mentioned deadly disease will have much lower appreciation and rewards than whoever cures them.

In fact, the rewards for the effective prevention of a disease do not coincide with those of an effective cure of the same disease.

A few rare people will appreciate the preventive work of a doctor, but most will not be aware of it at all.

No one whose disease has been prevented will go to his medical office to be treated, so gratefully paying the due fees.

No one will thank him for a successful recovery.

No healthy person will acknowledge the suffering he avoided.

Furthermore, career advancements, the possibility of scientific publications, academic prestige, and prizes for scientific activity will be much less probable.

On the contrary, if a doctor and a healthcare facility devote their entire activity to treating illnesses, the economic return and social prestige are certain. Furthermore, if the treatments are innovative, success in terms of scientific prestige is also likely.

Let us now consider the logic of evolutionary medicine.

In cases where prevention is not possible, for example, in the majority of serious malformations and developmental alterations on a genetic basis, evolutionary medicine can and must proceed with the same methods and objectives as current medicine.

However, in many diseases and most cases, the primary causes of the diseases are well explainable as consequences of alterations of the ecological niche and are therefore preventable.

Consequently, evolutionary medicine aims at the prevention of such diseases and considers their cure, when effective prevention has not been pursued or has not been possible, as a way to reduce the consequences of a failure.

This strategy implies some significant difficulties for those who want to pursue the logic and methods of evolutionary medicine because prevention is often:

(i) not understood or underestimated;
(ii) in contrast with deeply rooted beliefs or lifestyle habits;
(iii) impossible because it requires changes in living conditions beyond the decision-making possibilities of the doctor or the health facility;
(iv) not profitable or clearly bankrupt in economic terms for the doctor or health facility pursuing it.

In short, even if evolutionary medicine arises from rational evaluations, its concrete application appears to be hampered by insurmountable difficulties.

A pragmatic person can argue with full reason that evolutionary medicine, even if founded on rational foundations, is unsustainable and inapplicable in practical terms. In other words, evolutionary medicine could be considered an interesting theoretical exercise that, in practice, is of little interest to medicine and is discouraged for a medical professional.

10.4 Necessity of Evolutionary Medicine

Modern medicine is generally considered and portrayed as a modern triumph of scientific progress with countless extraordinary achievements. According to this description, numberless diseases that were once not understood and could not be treated are now widely understood and treated more and more effectively. Surely, many facts about diseases need to be explored and explained, but this does not diminish the enormous progress achieved in medicine. It is also true that there is a need for better treatments for many diseases and for innovative treatments for diseases that are still incurable today, but the enormous advances in medicine (Goldman & Cooney, 2023; Jameson, 2018) bode well that these insufficiencies will be overcome with further advances in medicine.

It is likely that this description of progress and effectiveness would be widely accepted and shared by a large part of the population and also of the medical world.

Unfortunately, it is only partially true and omits fundamental data that lead to opposite considerations.

In fact, it is true that the knowledge and methods relating to the understanding and treatment of diseases are enormously advanced, making possible previously unthinkable therapeutic methods.

It is also true, however, that many diseases treated today did not exist before modern times or had sporadic occurrences. Furthermore, despite the effectiveness of the treatments, these are often not decisive in all cases or have only a partial effect. Moreover, while the capacities and possibilities of curing diseases grow continuously, both the number of diseases and the number of sick people appear to grow even more.

Many diseases or groups of diseases that derive from alterations of the ecological niche and are enormously widespread today have been discussed in Chaps. 3 and 4 of this work already mentioned above. Among them, there are troubles within the definitions of neoplastic diseases, allergic and autoimmune diseases, diabetes mellitus type 2, refractive defects (myopia, astigmatism, hypermetropia), urolithiasis, cholelithiasis, cholecystitis, pancreatitis, caries, malocclusion, obesity, hypertension, cardiovascular diseases, constipation, hemorrhoids, anal fissures, colon diverticulosis, appendicitis, hepatopathies by alcoholic consumption, hepatic steatosis, illnesses caused by tobacco smoking or air pollution, hearing loss by excessive exposure to noise, eating disorders, anxiety, depression, infection epidemics, autism spectrum disorders, colon diseases, dermatopathies and diseases of mucous membranes caused by alterations of local microbiota, esophagitis, gastritis and gastroduodenal ulcers.

In the aforementioned chapters, many data have been reported regarding their high frequency in the modern era, their absence or minimum frequency in ancestral (i.e., normal) conditions, the variation of their frequency according to the living conditions of a population, the increase in their frequency in the transition from normal conditions to contemporary ones. These data demonstrate that the diseases above, and many others, depend on alterations of the ecological niche and so are preventable by the correction of these alterations.

It is true that (i) for each disease further data are surely useful or necessary; (ii) in some cases, the data are sufficient to suggest disease origin from alterations of the ecological niche but not to demonstrate it in an accurate and certain way; and (iii) in other cases, there are no studies aimed at demonstrating by epidemiological data the difference between normal and modern conditions and, consequently, the origin of the disease in alterations of the ecological niche.

However, what is evident and indisputable, as resulting at least partially from the data reported in Chaps. 3 and 4, is the enormous overall impact of these illnesses on the state of health of modern populations.

It is also known and indisputable that for many of these diseases, there are treatments that allow either a decisive cure or at least a reduction or slowing down of symptoms and progression of the disease. Yet, it is also true that in many cases, the treatments are only palliative or partially reduce the consequences of the diseases. Moreover, in many cases worldwide, due to economic or organizational shortcomings, the possible treatments are applied in insufficient, discontinuous, or ineffective ways or even not applied at all.

The Global Burden of Diseases, Injuries, and Risk Factors Studies (GBD)—e.g., (GB 2019 Diseases and Injuries Collaborators, 2020)—demonstrate the enormous burden of some of these diseases and their serious overall effect on the health status of populations worldwide.

These considerations, albeit in an approximate way, show that a large part of the burden of suffering, loss of human lives, and social and economic damage caused by diseases is determined by alterations of the ecological niche.

A simple summary of what has happened and is happening with regard to the overall state of health worldwide is as follows.

We have gone from times and living conditions in which medicine was practically non-existent to the modern era in which medicine has reached levels of knowledge and effectiveness that were once inconceivable.

At the same time, there has been, and continues to be, the onset of previously non-existent diseases or the spread of previously infrequent and sporadic diseases. This has dramatically increased the number of people affected by one or more illnesses previously non-existent or rare.

These diseases are countered or only partially overcome by an increasingly effective medicine.

We, therefore, have a paradoxical and contradictory situation.

With modern medical treatments, we are winning many battles better and more and more, but the number of clashes with diseases is growing more and more. Despite the more effective treatments, the numbers of sick people, victims of diseases, suffering, and outcomes of diseases are increasing. Furthermore, the resulting social and economic burden is also increasing proportionately.

The optimist could say that medicine is always more successful. The pragmatist will say that the success of modern medicine is like Pyrrhic victories: it wins, but the number of wounded, dead, and suffering increases, as well as the associated costs.

It is necessary to radically modify the objectives and methods of medicine so that the results can be effective as an overall measure and not relative to individual cases.

10.5 Effective Implementation of Evolutionary Medicine

What has been discussed in the previous sections highlights a huge underlying contradiction.

The doctor who wants to consider and apply the principles of evolutionary medicine will have difficulty preventing diseases in subjects who follow well-rooted lifestyles and beliefs that contrast with the rational logic of prevention. Furthermore, treating a disease is much more advantageous than preventing in terms of economic and professional return. Consequently, the doctor is strongly discouraged from applying evolutionary medicine.

The same goes for healthcare facilities. For example, surgery is far more rewarding than preventing the conditions that lead to the need for surgery. A hypothetical healthcare facility preventing 100,000 cases of coronary heart disease will be an economic failure compared to another facility treating 1000 cases of coronary heart disease by coronary angioplasty, coronary artery bypass grafts, and expensive methods of diagnosis and treatment.

However, the increasing diffusion of alterations of our ecological niche and the scarce diffusion of preventive strategies, often associated with a lack of knowledge of the primary causes of a disease and so of effective preventive measures, determine the growing incidence of a large number of diseases, multiplies the need for treatments that are only partially and not in all cases effective, expands the costs and the individual and social burden of suffering and conditions of disability.

The solution to this seemingly unsolvable contradiction is actually the one already applied in the past and in modern times for other health problems.

It is necessary to distinguish between:

(i) preventive measures that must be developed and primarily applied by the Public Authority (PA), with the indications, collaboration, consent, and support of the Scientific Biomedical World (SBW), doctors, and health facilities;
(ii) curative measures that must be applied by the doctors and health facilities, with the scientific indications of the SBW and the guidelines, consent, and support of the PA.

To develop the best strategies for disease prevention, the active contributions of the biomedical academic world and of public or private institutions specifically financed or incentivized by the PA (here collectively defined as the Scientific Biomedical World, or SBW) are essential. In order to define the best strategies to fight diseases, a preliminary step is the objective definition by the SBW of the primary causes of each disease and of the best methods to prevent each of them and fight the pathogenic mechanism. These results will be transmitted to the PA, which—compatibly with other economic, political, cultural, or social needs—will seek the best and most effective and feasible application, among other things, through:

(A) appropriate laws and regulations that discourage or curb lifestyles and habits that are the primary cause of each illness;
(B) continuous information campaigns supporting lifestyles and habits that avoid or reduce the primary causes of disease.

In the measures adopted, in general, prohibitions relating to lifestyles and habits harmful to health must be avoided based on the principles that (i) everyone has the right to be informed carefully and continuously of the risks of actions harmful to health; and (ii) nevertheless, no one must be obliged to live according to established rules, even if this is aimed to protect one's health.

However, exceptions are possible and necessary, especially when it is necessary to protect the safety of a person from accidental events, e.g., in environments such as the workplace, and when it is necessary to defend the health of third parties from any harmful behavior. For example, obligations relating to:

- use of seat belts and presence of airbags and ABS in motor vehicles;
- use of protective helmets when going on two-wheeled vehicles,
- provisions for the prevention of accidents in the workplace;
- limits on the production, circulation, and sale of poisonous or dangerous substances;
- limits on the sale, possession and use of personal weapons;
- regulations and limitations on the use of drugs;
- prohibitions on the sale of alcohol to minors;
- special precautions to protect against infections in cases of sanitary procedures and in other cases where there is a risk of infections;
- regulations and characteristics of buildings to prevent structural failures, confer earthquakes resistance, and avoid possible accidents.

After the appropriate evaluations, the PA will have to fully inform the doctors and health facilities of what is suggested by the SBW to prevent diseases. Doctors and healthcare facilities will actively collaborate as far as possible in disseminating information about disease prevention and limitation. Such actions will not limit or invalidate the main task of doctors and health facilities, namely the treatment of diseases, especially in cases where prevention has failed and it is therefore necessary to oppose and limit the consequences of this failure.

Moreover, the PA will inform and guide all non-medical, public or private, organizations and structures and any other subject about non-medical actions, indications, and obligations that concern disease prevention.

An organizational scheme inspired by these principles is shown in Fig. 10.1.

As far as aging is concerned, the topic deserves the mention of some clarifications already discussed in *Chap. 8—Physiological phenomena that cause troubles and suffering or even death—Part III: Slow phenoptosis or aging*.

It is important to distinguish between:

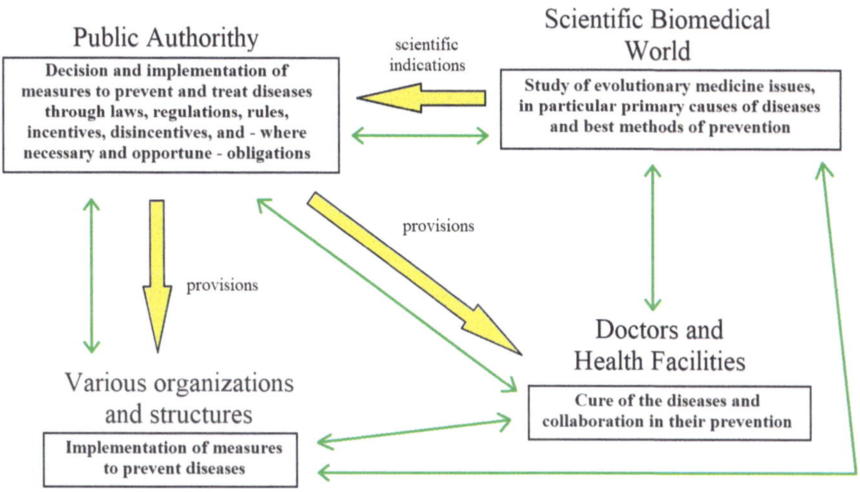

Fig. 10.1 Organizational scheme for the implementation of evolutionary medicine. Green lines indicate the exchange of information and collaboration

(A) acceleration of the aging process and age-related progressive effects of diseases, in particular, those caused by alterations of the ecological niche;
(B) physiological aging.

Indeed:

- as regards (A), prevention is possible and necessary.
- as regards (B), prevention is not possible. Even an individual with an ideal lifestyle will age and eventually die as a result of events related to age-related ever-increasing vulnerability. As physiological aging is not a disease, actions that curb or even block aging should not be considered as a cure for a disease. In a neutral way, they can be defined as anti-aging actions. In theory, if aging is a physiological process, it is possible to modify or slow down this process. A recently published work is suggested for this topic, which requires specific discussions (Libertini et al., 2021).

As for the phenomena discussed in:

- *Chapter 6—Physiological phenomena that cause troubles and suffering or even death—Part I: Non-phenoptotic phenomena*;
- *Chapter 7—Physiological phenomena that cause troubles and suffering or even death—Part II: Phenoptotic phenomena different from slow phenoptosis*;

it is essential to remember that these phenomena are not diseases and that, as discussed in the chapters above, it is useful or necessary to counteract them only in particular cases.

These phenomena should be deepened by the SBW for appropriate suggestions to the PA.

As regards the diseases deriving from relations with other living beings (s. Chap. 5), particular attentions are required, e.g.:

(1) The use of antibiotics should be limited as much as possible not only to avoid the onset and spread of resistant microbial strains but mainly to avoid alterations related to damages to our holobiont;
(2) In particular, the routine use of antibiotics for preventive purposes before and after any surgical intervention must be evaluated from the point of view of both the actual need and that of the alterations of the holobiont;
(3) It is essential to re-discuss the practice of eradication of helminths as they are part of our holobiont from ancestral times;
(4) It is necessary to evaluate the effects of antivirals on the balances of our holobiont;
(5) In general, every action that alters or can alter our holobiont must be critically evaluated in view of the ancestral balances within the holobiont and avoiding the prejudice that every action against components of the holobiont is useful or at least harmless for the host;
(6) Vaccines must be used to the maximum as they represent a targeted method to curb specific pathogens favored by the current high population densities and the frequency and speed of people's movements. At the same time, since vaccines target specific pathogens, they do not harm the usual components of our holobiont.

An example of how it is indispensable to consider the primary causes of diseases and not limit oneself to the treatment of diseases, disregarding, among other things, the consequences of such treatments in evolutionary terms, is given by the examination of mortality from infectious diseases and the effects of largely non-curative actions that mattered in this regard.

For the USA, data on infectious disease mortality from 1900 to 1995 are available (Armstrong et al., 1999), as shown in Fig. 10.2.

These data show that mortality from infectious diseases was very high at the beginning of the twentieth century, reaching about 800 deaths/100,000/year (d/y). In 1980, this mortality had decreased to 36 d/y and then recovered slightly in the following years. For those who want to believe that this great success was the result of ever better medical treatments, in particular the use of ever more powerful antimicrobial substances, it is opportune to highlight some facts. The first substances with antibacterial effects, sulfonamides, were used for the first time in 1937; the use of penicillin began in 1941–1943, starting from the theaters of war; modern antibiotics were discovered or synthesized in the following decades.

Most of the decline in mortality from infectious diseases (from about 800 d/y in 1900 to less than 200 d/y in 1943, i.e., 75% reduction in mortality) occurred before penicillin use began. With the beginning of penicillin use, the previous rate of reduction in mortality continued at an unchanged rate before declining in 1953–1980. After 1980, mortality rose again despite the discovery or synthesis of new potent antibiotics.

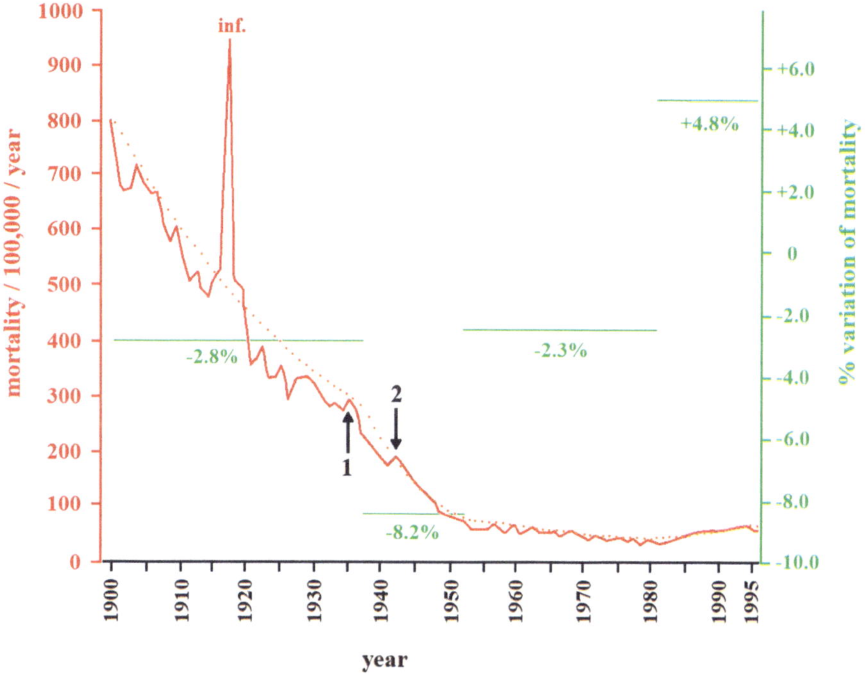

Fig. 10.2 Overall trends in infectious disease mortality rate and percent variation of mortality rate in the USA from 1900 to 1996. Data from (Armstrong et al., 1999). Red line: mortality rates / 100,000 / year; orange dots: smoothed mortality rates, disregarding the substantial mortality increase around 1918 due to the influenza pandemic (inf.); green lines: % variation of mortality in the periods 1900–1937, 1938–1952, 1953–1980, and 1981–1995; arrow 1: the beginning of sulphonamide use; arrow 2: the beginning of penicillin use

These data indicate that the use of antibiotics could not have had an influence on the reduction of mortality before the 1940s, and that in the following years, this influence was small: "… most of this change cannot be attributed to the deployment of antibiotics! By 1940, infectious disease mortality had already dropped to about 210 deaths per 100,000 people per year. Antibiotics did even not become available for clinical use until 1941! …" (Generaux & Bergstrom, 2005).

The high mortality for infectious diseases at the beginning of the twentieth century was likely caused by very poor economic, social, and hygienic conditions. The sharp decline in mortality until 1980 was determined by improved living conditions and not by medical curative actions, which, moreover, were practically very limited before the use of antibiotics: "… Innovations in food handling and preparation - most notably refrigeration and pasteurization, both implemented widely in the United States during the first half of the 20th century - reduced foodborne disease transmission. Huge improvements in nutrition followed the discovery of the key role of certain minerals in the human diet, including iodine and vitamin D … Better nutrition produced a tremendous decline in the proportion of the American population that was chronically malnourished and therefore highly susceptible to infectious disease." (Generaux & Bergstrom, 2005).

Furthermore, the massive and increasingly extensive use of antibiotics has led, as a consequence of known and predictable evolutionary mechanisms, to the significant problem of infections with germs resistant to more and more antibiotics: "In the United States alone, at least 200,000 people and probably far more suffer from a hospital-acquired infection every year. The associated mortality is considerable; the Center for Disease Control has estimated that 90,000 U.S. residents die each year from nosocomial infections. To place this number in context, AIDS/HIV kills approximately 17,000 per year in the United States, influenza 37,000 per year, and breast cancer roughly 40,000 per year. As large as these numbers are, some estimates suggest that the actual magnitude of the problem could be up to tenfold higher." (Bergstrom & Feldgarden, 2008).

This example suggests that in the fight against infectious diseases:

- preventive measures are much more important than direct medical care;
- in the use of substances with effects on microorganisms, it is necessary to evaluate the evolutionary effects on species sensitive to the substances used for the foreseeable selection of resistant strains;
- an integrated approach is indispensable in which (i) prevention is of the utmost importance, and (ii) medical treatment is necessary, with appropriate insights and precautions, in cases where prevention has failed.

10.6 Conclusion

For those who have followed the academic courses of current medicine and then practiced a medical profession for years, observing the methods learned, often with outstanding professionalism and dedication, the discussion of evolutionary medicine might seem like a useless new complication. Why learn and even follow concepts that are certainly useful in biological disciplines but appear very distant from daily medical practice? These concepts can certainly be helpful in the field of research and for some specific medical problems. However, their general application in the medical field could appear to be a useless dispersion of attention and energy for those already in constant struggle with difficulties and responsibilities of all kinds.

Nevertheless, the enormous worldwide diffusion of many diseases, the fact that the cure for a large part of the cases is exceeded by the emergence of new cases, and the birth and spread of entire categories of new diseases oblige the most attentive doctor to careful reflection.

While medicine consistently achieves new therapeutic successes, it appears to be increasingly succumbing to the spread of new disease cases. In modern times, the statistical "normality" for an average individual is to be afflicted by multiple diseases simultaneously and be considered healthy if none of these are severe or fatal.

In fact, according to common feeling, if an individual, for example, is affected 'only' by caries, malocclusion, myopia, overweight, latent type 2 diabetes, hemorrhoids, mild scoliosis, and hearing impairment, all common and partly treatable diseases, but not by serious or life-threatening illness, why should we consider him ill?

Modern medicine has at its disposal innumerable means, which were non-existent before, to cure, more or less well, a large number of diseases. While there are certainly diseases for which effective treatments have yet to be defined, on the whole medicine has achieved capabilities that were once unthinkable.

A doctor who has successfully treated a large number of diseases for years implements effectively the great capabilities of modern medicine. Why should he care about the large number of new sick people and diseases? After all, these are new opportunities to treat other patients, to apply effective treatment methods, and, last but not least, to achieve ever greater professional satisfaction.

However, a more attentive observer, whether medical or non-medical, with an objective and more general evaluation, will note that despite the greater effectiveness and remarkable progress of medicine and the significant and growing economic commitment to this, there is no corresponding improvement in the general state of health.

Therefore, the need for a different approach to studying, preventing, and treating diseases is evident.

Let us consider the case of autoimmune diseases, already discussed in detail in *Chap. 4—Diseases deriving from alterations of the ecological niche—Part II: Regarding the holobiont*, in *§4.1.2 Alterations of the holobiont—B: Elimination of intestinal parasites*. These diseases, which were rare or non-existent up to a few decades ago, are now more and more frequent.

The current therapeutic approach is to use anti-inflammatories, steroids, immunosuppressants, and, lately and more effectively, specific monoclonal antibodies. In particular, the use of these antibodies allows the remission of serious and disabling autoimmune diseases and represents one of the most active fields of pharmacological research: "The global therapeutic monoclonal antibody market was valued at approximately US$115.2 billion in 2018 and is expected to generate revenue of $150 billion by the end of 2019 and $300 billion by 2025. Thus, the market for therapeutic antibody drugs has experienced explosive growth as new drugs have been approved for treating various human diseases, including many cancers, autoimmune, metabolic and infectious diseases." (Lu et al., 2020).

Monoclonal antibodies are mainly used for the treatment of many types of cancer and autoimmune diseases (Lu et al., 2020), two groups of diseases of enormous impact that are probably caused in most cases by alterations of the ecological niche (s. Ch. 4 and 5 of this book).

The top-selling monoclonal antibody, adalimumab ($19.9 billion in revenue in 2018), has long been used for the treatment of rheumatoid arthritis (FDA approval in 2002), psoriatic arthritis (2005), ankylosing spondylitis (2006), juvenile idiopathic arthritis (2008), psoriasis (2008), Crohn's disease (2010), ulcerative colitis (2012), hidradenitis suppurativa (2015), uveitis (2018) (Lu et al., 2020).

As examples of autoimmune diseases treated with monoclonal antibodies, we have:

ankylosing spondylitis (Ebrahimiadib et al., 2021); antiphospholipid syndrome (Yun et al., 2023); autoimmune diseases of the skin ("pemphigus vulgaris, cutaneous lupus erythematous, dermatomyositis, systemic sclerosis, thyroid dermopathy, autoimmune pemphigoid diseases, and cutaneous vasculitis diseases.") (Ly et al., 2023) or inflammatory skin diseases ("bullous autoimmune diseases, eczema, prurigo, alopecia areata, chronic spontaneous urticaria, Netherton syndrome and a variety of other chronic inflammatory skin diseases.") (Olbrich et al., 2023); autoimmune encephalitis (Yang & Liu, 2021); autoimmune hemolytic anemia (Berentsen et al., 2023; Hill & Hill, 2018); chronic autoimmune demyelinating neuropathies (Briani & Visentin, 2022); immune thrombocytopenia (Lucchini et al., 2019); multiple sclerosis (Katsavos & Coles, 2018; Krajnc et al., 2022); myasthenia gravis (Alabbad et al., 2020; Song et al., 2022); neuromyelitis optica spectrum disorder (Nie & Blair, 2022); pemphigus vulgaris (Didona et al., 2022); primary Sjögren's syndrome (Fasano & Isenberg, 2019); rheumatoid arthritis (Kesharwani et al., 2019); systemic lupus erythematosus (Deng, 2021; Joy et al., 2022); type 1 diabetes (Ke et al., 2021); thyroid-associated ophthalmopathy (Slentz et al., 2020); and uveitis (Burek-Michalska & Turno-Kręcicka, 2020).

An optimist might say we are on the right track to control and defeat autoimmune diseases.

Unfortunately, an objective observer might point out that:

(1) there is an increasing spread of these diseases;
(2) often, even the most effective treatments are not always such as to curb or block the diseases;
(3) certain types of autoimmune diseases cause death or serious impairment before effective action can be taken;
(4) the use of monoclonal antibodies involves heavy and increasing costs;
(5) the social burden of these diseases is also increasing.

This demonstrates that the therapeutic approach alone is limitedly effective for these diseases and with growing and increasingly less sustainable economic and social costs.

The same autoimmune diseases analyzed with the criteria of evolutionary medicine have as a presumable cause, serious alterations of our holobiont, particularly the eradication of macroparasites.

Attempts to cure some autoimmune diseases by temporarily reintroducing macroparasites are underway (Arai & Lopes, 2022; Lothstein & Gause, 2021; Shi et al., 2022; Shields & Cooper, 2022).

Unfortunately, these attempts encounter two basic difficulties:

(i) often, the reintroduction of macroparasites is late and fails to obtain a complete regression of the disease;
(ii) based on the preconception that the presence of macroparasites is usually harmful and, therefore, they must always be eliminated, the therapeutic reintroduc-

tion of macroparasites uses species that do not take root permanently in humans and thus avoids the reintroduction of species that are present under normal evolutionary conditions.

These data show that for autoimmune diseases:

(A) the solely therapeutic approach is insufficient, with limited efficacy and, moreover, very expensive;
(B) greater clarity is needed in defining the primary causes of these diseases, overcoming preconceived ideas;
(C) therapeutic approaches based on the reintroduction of macroparasites appear promising but should be more precocious and based on species of macroparasites to which our species is adapted;
(D) the spread of autoimmune diseases requires more effective solutions with bearable costs.

This example, limited to a class of diseases that is, however, rampant and of great importance, shows how a typical evolutionary medicine approach is needed to solve this type of problem.

For these observations regarding autoimmune diseases, a possible correct criticism is that further studies are needed to demonstrate that their primary cause is in alterations of the ecological niche (in the specific case, alterations of the holobiont constituted by the eradication of macroparasites). This objection can be extended to all cases in which the spread of a type of disease appears to be consequent to alterations of the ecological niche.

The answer is simple and direct:

- Further studies are certainly needed. Any scientific development has its basis in the doubts that the researcher must have and in the pursuit of evident proofs;
- These studies are indispensable and urgent, considering the growing number of cases and the seriousness of the consequences for the people affected and for society in general;
- Pending definitive studies, it is easy to predict that the majority of research funding will continue to go to treatments that can provide enormous gains and benefits for those pursuing such studies. This implies that studies to define better the primary causes of such diseases and aimed at prevention methods will be considered less attractive.

These considerations about autoimmune diseases can certainly extend to other diseases. The evaluations and decisions in this regard fully invest the meaning and importance one will want to attribute to evolutionary medicine and its implementation as the fulcrum of future medicine.

In fact, there are two possible alternative ways:

(1) Pretend that evolutionary medicine, its indications, and solutions are only of theoretical value with little significance compared to the big results and enormous economic weight of current medicine. Consequently, continue with the

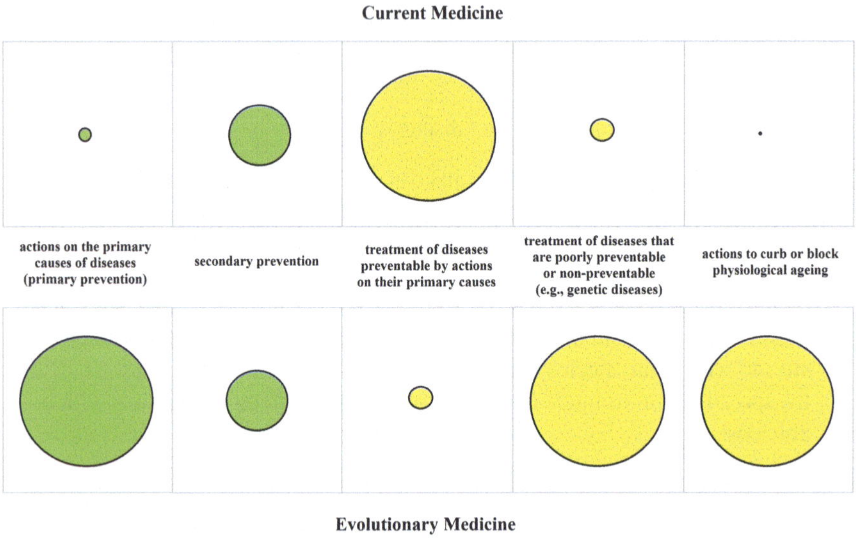

Fig. 10.3 In the transformation from current medicine to evolutionary medicine, the main objective becomes the primary prevention of diseases, which allows us to limit the commitment to treating preventable diseases. This does not necessarily mean a downsizing of the field of medicine, as considerable efforts will be focused on the treatment of non-preventable diseases and actions—now under development—to limit or stop aging. The size of the circles summarily indicates the relative effort for each goal. Green indicates preventive actions, while yellow indicates other types of action (treatment of diseases and anti-aging actions)

methods and results of current medicine despite the increase in cases of illness and in the associated costs and suffering;

(2) Reorganize the whole of medicine and health systems according to the indications of evolutionary medicine, as suggested in the previous section. The guiding principle of medicine must be the priority prevention of diseases by intervening on their primary causes. Clearly, when action on the primary causes has not been possible, and in cases where this is impossible, medicine must aim at curing diseases.

Figure 10.3 summarizes the comparison between how resources are used today with current medicine and how they could be distributed in the case of the full and coherent implementation of evolutionary medicine.

It should be noted that this implementation does not imply a downsizing of the total resources destined for medicine but their redistribution based on rational logic.

The main differences from the current situation are as follows:

(A) Shifting commitments from treating preventable diseases to preventing them;
(B) Greater commitment to the treatment of diseases that are little or not at all preventable (e.g., genetic diseases);

(C) Commitment to curb or block aging within the framework of the paradigm that explains aging as an adaptive and programmed physiological phenomenon. This approach must not be confused with the primary prevention of the factors that determine an acceleration of the physiological aging process.

At this point, an attentive reader might ask whether evolutionary medicine is the reality of future medicine or an ideal goal that is interesting and challenging but unattainable.

The answer is that evolutionary medicine represents a rational reformulation and reorganization of medicine and health system based on objective scientific foundations. However, in the implementation of everything, rationality is subordinated to the will, which is conditioned by a thousand impulses and is not always rational.

References

Alabbad, S., AlGaeed, M., Sikorski, P., & Kaminski, H. J. (2020). Monoclonal antibody-based therapies for myasthenia gravis. *BioDrugs: Clinical Immunotherapeutics, Biopharmaceuticals and Gene Therapy, 34*(5), 557–566. https://doi.org/10.1007/s40259-020-00443-w

Arai, T., & Lopes, F. (2022). Potential of human helminth therapy for resolution of inflammatory bowel disease: The future ahead. *Experimental Parasitology, 232*, 108189. https://doi.org/10.1016/j.exppara.2021.108189. Erratum in: *Experimental Parasitology*, 2021, 234, 108218.

Armstrong, G. L., Conn, L. A., & Pinner, R. W. (1999). Trends in infectious disease mortality in the United States during the 20th century. *JAMA, 281*, 61–66. https://doi.org/10.1001/jama.281.1.61

Berentsen, S., Fattizzo, B., & Barcellini, W. (2023). The choice of new treatments in autoimmune hemolytic anemia: How to pick from the basket? *Frontiers in Immunology, 14*, 1180509. https://doi.org/10.3389/fimmu.2023.1180509

Bergstrom, C. T., & Feldgarden, M. (2008). The ecology and evolution of antibiotic-resistant bacteria. In S. C. Stearns & J. C. Koella (Eds.), *Evolution in health and disease* (2nd ed.). Oxford University Press. ISBN 978-0199207459.

Bragulat, E., & de la Sierra, A. (2002). Salt intake, endothelial dysfunction, and salt-sensitive hypertension. *Journal of Clinical Hypertension (Greenwich), 4*(1), 41–46. https://doi.org/10.1111/j.1524-6175.2002.00503.x

Briani, C., & Visentin, A. (2022). Therapeutic monoclonal antibody therapies in chronic autoimmune demyelinating neuropathies. *Neurotherapeutics, 19*(3), 874–884. https://doi.org/10.1007/s13311-022-01222-x

Burek-Michalska, A., & Turno-Kręcicka, A. (2020). Adalimumab in the treatment of non-infectious uveitis. *Advances in Clinical and Experimental Medicine, 29*(10), 1231–1236. https://doi.org/10.17219/acem/125431

Carrera-Bastos, P., Fontes-Villalba, M., O'Keefe, J. H., Lindeberg, S., & Cordain, L. (2011). The western diet and lifestyle and diseases of civilization. *Research Reports in Clinical Cardiology, 2*, 15–35. https://doi.org/10.2147/RRCC.S16919

Darwin, E. C. (1794-1796). *Zoonomia, or The Laws of Organic Life.* Printed for J. Johnson, 2nd corrected ed., London [Available from Project Gutenberg, https://www.gutenberg.org/files/15707/15707-h/15707-h.htm and https://www.gutenberg.org/files/27600/27600-h/27600-h.htm].

Deng, W. P. (2021). Recent advances in the treatment of systemic lupus erythematosus with belimumab in children. *Zhongguo Dang Dai Er Ke Za Zhi/Chinese journal of contemporary pediatrics [Chinese, English], 23*(10), 1069–1074. https://doi.org/10.7499/j.issn.1008-8830.2107153

Didona, D., Paolino, G., Di Zenzo, G., Didona, B., Pampena, R., Di Nicola, M. R., & Mercuri, S. R. (2022). Pemphigus vulgaris: Present and future therapeutic strategies. *Dermatology Practical & Conceptual, 12*(1), e2022037. https://doi.org/10.5826/dpc.1201a37

Dobzhansky, T. (1973). Nothing in biology makes sense except in the light of evolution. *The American Biology Teacher, 35*(3), 125–129.

Ebrahimiadib, N., Berijani, S., Ghahari, M., & Pahlaviani, F. G. (2021). Ankylosing spondylitis. *Journal of Ophthalmic & Vision Research, 16*(3), 462–469. https://doi.org/10.18502/jovr.v16i3.9440

Ehret, G. B., Munroe, P. B., Rice, K. M., Bochud, M., Johnson, A. D., Chasman, D. I., The International Consortium for Blood Pressure Genome-Wide Association Studies, et al. (2011). Genetic variants in novel pathways influence blood pressure and cardiovascular disease risk. *Nature, 478*(7367), 103–109. https://doi.org/10.1038/nature10405

Fasano, S., & Isenberg, D. A. (2019). Present and novel biologic drugs in primary Sjögren's syndrome. *Clinical and Experimental Rheumatology, 37 Suppl 118*(3), 167–174.

Feng, W., Dell'Italia, L. J., & Sanders, P. W. (2017). Novel paradigms of salt and hypertension. *Journal of the American Society of Nephrology, 28*(5), 1362–1369. https://doi.org/10.1681/ASN.2016080927

GBD 2019 Diseases and Injuries Collaborators. (2020). Global burden of 369 diseases and injuries in 204 countries and territories, 1990-2019: A systematic analysis for the Global Burden of Disease Study 2019. *Lancet, 396*(10258), 1204–1222. https://doi.org/10.1016/S0140-6736(20)30925-9. Erratum in: *Lancet*, 2020 Nov 14, 396(10262), 1562.

Generaux, D. P., & Bergstrom, C. T. (2005). Evolution in action: Understanding antibiotic resistance. In J. Cracraft & R. W. Bybee (Eds.), *Evolutionary science and society: Educating a new generation* (BCSC/AIBS Symposium, Washington, DC) ISBN: 1-929614-23-3.

Goldman, L., & Cooney, K. A. (Eds.). (2023). *Goldman-Cecil Medicine* (27th ed.). Elsevier Science Health Science. ISBN 978-0323930383.

Hill, A., & Hill, Q. A. (2018). Autoimmune hemolytic anemia. *Hematology. American Society of Hematology. Education Program, 2018*(1), 382–389. https://doi.org/10.1182/asheducation-2018.1.382

Jameson, J. L. (Ed.). (2018). *Harrison's principles of internal medicine* (20th ed.). McGraw-Hill Education. ISBN 978-1259644030.

Joy, A., Muralidharan, A., Alfaraj, M., Shantharam, D., Cherukuri, A. S. S., & Muthukumar, A. (2022). The role of belimumab in systemic lupus erythematosis: A systematic review. *Cureus, 14*(6), e25887. https://doi.org/10.7759/cureus.25887

Kato, N., Loh, M., Takeuchi, F., Verweij, N., Wang, X., Zhang, W., et al. (2015). Trans-ancestry genome-wide association study identifies 12 genetic loci influencing blood pressure and implicates a role for DNA methylation. *Nature Genetics, 47*(11), 1282–1293. https://doi.org/10.1038/ng.3405

Katsavos, S., & Coles, A. (2018). Alemtuzumab as treatment for multiple sclerosis. *Cold Spring Harbor Perspectives in Medicine, 8*(10), a032029. https://doi.org/10.1101/cshperspect.a032029

Ke, Q., Kroger, C. J., Clark, M., & Tisch, R. M. (2021). Evolving antibody therapies for the treatment of type 1 diabetes. *Frontiers in Immunology, 11*, 624568. https://doi.org/10.3389/fimmu.2020.624568

Kearney, P. M., Whelton, M., Reynolds, K., Whelton, P. K., & He, J. (2004). Worldwide prevalence of hypertension: A systematic review. *Journal of Hypertension, 22*(1), 11–19. https://doi.org/10.1097/00004872-200401000-00003

Kesharwani, D., Paliwal, R., Satapathy, T., & Das Paul, S. (2019). Rheumatoid arthritis: An updated overview of latest therapy and drug delivery. *Journal of Pharmacopuncture, 22*(4), 210–224. https://doi.org/10.3831/KPI.2019.22.029

Krajnc, N., Bsteh, G., Berger, T., Mares, J., & Hartung, H. P. (2022). Monoclonal antibodies in the treatment of relapsing multiple sclerosis: An overview with emphasis on pregnancy, vaccination, and risk management. *Neurotherapeutics, 19*(3), 753–773. https://doi.org/10.1007/s13311-022-01224-9

Lappé, M. (1994). *Evolutionary medicine-rethinking the origins of diseases*. Sierra Club Books. ISBN 9780871565198.

Lewis, S. (2008). Evolution at the intersection of biology and medicine. In W. R. Trevathan, E. O. Smith, & J. J. McKenna (Eds.), *Evolutionary medicine and health: New perspectives* (pp. 399–415). Oxford University Press. ISBN 978-0195307061.

Libertini, G., Corbi, G., Conti, V., Shubernetskaya, O., & Ferrara, N. (2021). *Evolutionary gerontology and geriatrics: Why and how we age* (Advances in studies of aging and health) (Vol. 2). Cham. ISBN 978-3030737740.

Lothstein, K. E., & Gause, W. C. (2021). Mining helminths for novel therapeutics. *Trends in Molecular Medicine, 27*(4), 345–364. https://doi.org/10.1016/j.molmed.2020.12.010

Lu, R. M., Hwang, Y. C., Liu, I. J., Lee, C. C., Tsai, H. Z., Li, H. J., & Wu, H. C. (2020). Development of therapeutic antibodies for the treatment of diseases. *Journal of Biomedical Science, 27*(1), 1. https://doi.org/10.1186/s12929-019-0592-z

Lucchini, E., Zaja, F., & Bussel, J. (2019). Rituximab in the treatment of immune thrombocytopenia: What is the role of this agent in 2019? *Haematologica, 104*(6), 1124–1135. https://doi.org/10.3324/haematol.2019.218883

Ly, S., Nedosekin, D., & Wong, H. K. (2023). Review of an anti-CD20 monoclonal antibody for the treatment of autoimmune diseases of the skin. *American Journal of Clinical Dermatology, 24*(2), 247–273. https://doi.org/10.1007/s40257-022-00751-7

Nesse, R. M. (2001). How is Darwinian medicine useful? *The Western Journal of Medicine, 174*(5), 358–360. https://doi.org/10.1136/ewjm.174.5.358

Nesse, R. M. (2008). The importance of evolution for medicine. In W. R. Trevathan, E. O. Smith, & J. J. McKenna (Eds.), *Evolutionary medicine and health: New perspectives* (pp. 416–431). Oxford University Press. ISBN 978-0195307061.

Nesse, R. M., & Williams, G. C. (1994). *Why we get sick: The new science of Darwinian medicine*. Times Books. ISBN 978-0812922247.

Nie, T., & Blair, H. A. (2022). Inebilizumab: A review in neuromyelitis optica spectrum disorder. *CNS Drugs, 36*(10), 1133–1141. https://doi.org/10.1007/s40263-022-00949-7

Olbrich, H., Sadik, C. D., Ludwig, R. J., Thaçi, D., & Boch, K. (2023). Dupilumab in inflammatory skin diseases: A systematic review. *Biomolecules, 13*(4), 634. https://doi.org/10.3390/biom13040634

Shi, W., Xu, N., Wang, X., Vallée, I., Liu, M., & Liu, X. (2022). Helminth therapy for immune-mediated inflammatory diseases: Current and future perspectives. *Journal of Inflammation Research, 15*, 475–491. https://doi.org/10.2147/JIR.S348079

Shields, V. E., & Cooper, J. (2022). Use of helminth therapy for management of ulcerative colitis and Crohn's disease: A systematic review. *Parasitology, 149*(2), 145–154. https://doi.org/10.1017/S0031182021001670

Slentz, D. H., Nelson, C. C., & Smith, T. J. (2020). Teprotumumab: A novel therapeutic monoclonal antibody for thyroid-associated ophthalmopathy. *Expert Opinion on Investigational Drugs, 29*(7), 645–649. https://doi.org/10.1080/13543784.2020.1772752

Song, Z., Zhang, J., Meng, J., Jiang, G., Yan, Z., Yang, Y., et al. (2022). Different monoclonal antibodies in myasthenia gravis: A Bayesian network meta-analysis. *Frontiers in Pharmacology, 12*, 790834. https://doi.org/10.3389/fphar.2021.790834

Stearns, S. C. (Ed.). (1999). *Evolution in health and disease* (1st ed.). Oxford University Press. ISBN 978-0198501102.

Truswell, A. S., Kennelly, B. M., Hansen, J. D. L., & Lee, R. B. (1972). Blood pressure of !Kung bushmen in northern Botswana. *American Heart Journal, 84*(1), 5–12. https://doi.org/10.1016/0002-8703(72)90299-2

Varki, A. (2012). Nothing in medicine makes sense, except in the light of evolution. *Journal of Molecular Medicine (Berlin, Germany), 90*(5), 481–494. https://doi.org/10.1007/s00109-012-0900-5

Whelton, P. K., Carey, R. M., Aronow, W. S., Casey, D. E., Collins, K. J., Dennison Himmelfarb, C., et al. (2018). 2017 ACC/AHA/AAPA/ABC/ACPM/AGS/APhA/ASH/ASPC/NMA/PCNA guideline for the prevention, detection, evaluation, and management of high blood pressure in adults: A report of the American College of Cardiology/American Heart Association Task Force on Clinical Practice Guidelines. *Hypertension, 71*(6), e13–e115. https://doi.org/10.1161/HYP.0000000000000065

Williams, G. C., & Nesse, R. M. (1991). The dawn of Darwinian medicine. *The Quarterly Review of Biology, 66*(1), 1–22. https://doi.org/10.1086/417048

Yang, J., & Liu, X. (2021). Immunotherapy for refractory autoimmune encephalitis. *Frontiers in Immunology, 12*, 790962. https://doi.org/10.3389/fimmu.2021.790962

Yun, Z., Duan, L., Liu, X., Cai, Q., & Li, C. (2023). An update on the biologics for the treatment of antiphospholipid syndrome. *Frontiers in Immunology, 14*, 1145145. https://doi.org/10.3389/fimmu.2023.1145145

The manufacturer's authorised representative in the EU is Springer Nature Customer Service Centre GmbH, Europaplatz 3, 69115 Heidelberg, Germany. If you have any concerns regarding our products, please contact ProductSafety@springernature.com

Printed and bound by CPI Group (UK) Ltd, Croydon, CR0 4YY
01/10/2026
02225949-0014